Archaeology
Basic Field Methods

R. Michael Stewart
Temple University

KENDALL/HUNT PUBLISHING COMPANY
4050 Westmark Drive Dubuque, Iowa 52002

Book Team

President and Chief Executive Officer Mark C. Falb
Vice President, Director of National Book Program Alfred C. Grisanti
Editorial Development Supervisor Georgia Botsford
Developmental Editor Tina Bower
Vice President, Production Editorial Ruth A. Burlage
Production Manager Jo Wiegand
Production Editor Carrie Maro
Permissions Editor Colleen Zelinsky
Design Manager Jodi Splinter
Designer Deb Howes

Figures 1.2, 1.3, 1.8, 1.9, 5.9, 5.11, 5.12, 8.33, 9.37, 9.72, 9.73, 9.74, 9.76: Rob Tucher and the Cultural Resource Group, Louis Berger and Associates, Inc.
Figures 2.1 and 8.36: Charles Bello
Figures 2.2, 4.14 and 9.7: Edward Morin
Figure 2.9: Colonel Howard MacCord, Sr.
Figures 4.1, 4.2, 4.3, 4.5, 4.7, 6.1, 6.2, 6.3, 6.4, 6.5, 6.6, 6.9, 6.10, 6.12, 6.13, 6.16, 6.19, 6.21, 6.41, and 6.42: United States Geological Survey
Figures 4.3, 4.6: Pennsylvania Department of Conservation and Natural Resources, Bureau of Topographic and Geologic Survey
Figure 4.8: United States Department of Agriculture, Natural Resources Conservation Service
Figures 5.13, 8.34, 9.4, 9.6, 9.7, 9.17, 9.26, 9.28 and 9.77: URS Greiner, Inc.
Figure 5.10: Cyndi Mosch
Figure 6.2: State of Maryland, Department of Natural Resources, Maryland Geological Survey
Figure 6.25: Darrin Lowery
Figure 6.29: Greg Katz
Figures 6.40 and 8.28: Fennelle Miller
Figures 8.2, 8.9 and 8.20: Pete Condon
Figure 8.19: Paul Nevin
Figure 8.13: Barbara Hildebrandt
Figure 8.18: David Mudge
Figure 9.5: Joan Gardner
Figure 9.12: Chris Hummer
Figure 9.16: Jay Custer
Figure 9.71: Thomas Crist
All other Figures: R. Michael and Ali Stewart

ISBN 978-0-7872-8129-8

Printed in the United States of America
10 9 8 7 6 5

To my best friend, Ali.

About the Author

Michael Stewart was awarded a Ph.D. in Anthropology from the Catholic University of America in 1981, and is currently an Associate Professor of Anthropology at Temple University in Philadelphia, Pennsylvania. He is trained in both prehistoric and historic American archaeology and has over 20 years of experience in cultural resource management. His long-standing interest is in the archaeology of the native peoples of the Eastern Woodlands, especially the Northeast and Mid-Atlantic sections of this area. His particular interests include the ways in which hunting and gathering groups adjusted to dynamic, shifting environments following the last ice age. Dr. Stewart also studies the waxing and waning of the socially complex behavior of hunting and gathering societies between 2500 BC and AD 900, including burial ceremonialism, the reorganization of traditional forms of work, trade and exchange, and the adoption of a farming way of life and all of its social implications and ramifications. He also feels it is important to use the geosciences in understanding how archaeological deposits are formed, and in recreating paleoenvironments, a necessary step in evaluating cultures in context.

Contents

What's the Point?

Nobody ever learned how to do proper archaeological fieldwork from a book (including this one).

David Hurst Thomas, Archaeology

I fundamentally, absolutely, wholeheartedly agree with this statement! It is impossible to take a "cookbook" approach to archaeological field investigations. You can't walk out into the woods or the middle of an open field, get out some encyclopedic tome that you studied in school and look up the answer to the question, What do I do now? Any archaeological deposit, regardless of whether it is 100 or 100,000 years old, is the end result of a unique combination of cultural and natural events and processes. It follows then, that field strategies and methods must be flexible in order to cope with the variety inherent in archaeological deposits. Approaches to fieldwork must also be flexible to accommodate the variety of questions we seek to answer about the past and its peoples through the study of archaeological deposits. We never are able to excavate 100 percent of anything, nor should we want to in most circumstances. Short of total study, there is no single way to investigate a deposit that would enable us to answer all of the questions that we might eventually want to ask of it.

So why bother with a book on field methods? There is a logic that connects the goals of a field investigation with field techniques, decisions made in the field, and an understanding of what the archaeological record of the past is like. I believe that once you get a grasp of "big picture" motivations, then what gets done day-to-day in the field will make more sense. Many of the techniques typically employed in fieldwork are fairly straightforward and you will learn a lot about them from this book. What is more difficult to learn is **when** to use a technique, or **how** to adapt it or a battery of techniques to a particular field situation. In other words, how do we make decisions about what to do in the field? While a book cannot always tell you what to do in a specific situation, it can construct a framework for solving problems and making decisions. Dealing with a specific field situation then becomes a matter of being aware of the kinds of information that you need to collect and plug into the framework.

This book introduces field methods and the decision-making process with simple language. It provides a basis for understanding the links between the nature of archaeological evidence and the recognition of that evidence in the field, and the techniques involved in the search for, and recovery of, archaeological evidence in context. Even the most well-planned field school or course in field methods cannot provide encyclopedic coverage of archaeological methods and related issues. This book is designed to engage readers with different backgrounds and interests, and lead them into more detailed studies. Anyone beginning their study of archaeology, participants in field schools, novice field technicians, and those simply curious about the workings of archaeology will find much of interest here.

Because of my background and experiences, this field manual will be most relevant to archaeologists working in North America on deposits related to Native Americans. However, many of the basic approaches described here are of equal relevance to those more interested in colonial and historic sites archaeology. Shipwreck and underwater archaeology, although significant endeavors, are not included here.

Chapters 1 and 2 discuss the nature of archaeology and archaeological evidence, building a foundation for understanding fieldwork. These are especially important for persons who have no prior experience or background in archaeology. Being in the field can be frustrating if you are uncomfortable with recognizing artifacts, or what constitutes archaeological evidence. Equally important is the discussion of the concept of context, the relationship between an object, other objects, and the landscape, since at its most basic level archaeological fieldwork is the recognition of evidence, and the recovery and interpretation of that evidence in context.

What should you know before you get to the field, and why are you going in the first place? What broader skills are needed for success in archaeological fieldwork? These issues are addressed in chapters 3–7. There is no substitute for a clear statement of a field project's goals or its research design, since all other actions flow from these.

An introductory field manual can't anticipate all of the specific research questions that might guide a field project, so I take a general approach to the subject (chapter 3). There is a standard battery of information that any field technician should have, or be provided with, prior to going out into the field on any project (chapter 4). This includes details about local geology, soils, topography, vegetation, and previous archaeological investigations of relevance. These data assist the field technician in the recognition of archaeological evidence and understanding the logic behind decisions regarding how an area will be investigated, i.e., where a surface survey is appropriate, where excavations should be placed, how big excavations should be and how deeply they should be dug. I'll review where to look for this background information and how to begin to interpret it. Guidelines on appropriate behavior or field etiquette, gearing up for your personal needs in the field, and an awareness of the potential hazards of fieldwork are discussed in chapter 5. Maps, mapmaking, and surveying are indispensable for planning and executing fieldwork and are the focus of chapter 6. When you consider the importance of the context of evidence to archaeological interpretation, the high level of effort devoted to surveying and mapping is obvious. Every field archaeologist must have a grounding (no pun intended) in sediments, soils, and the processes that shape landscapes (geomorphology). We retrieve the material products of human behavior from these things. It is only logical that we learn as much as possible about them. Chapter 7 is devoted towards this end.

My presentation of approaches to fieldwork (chapters 8 and 9) is organized using two principles. The first is whether a field operation relates to working on the surface of a landscape or working into and beneath it (subsurface). The second organizing principal has to do with the level of the field investigation, that is, are we looking for archaeological deposits, are we testing deposits that have already been located, or are we intensively studying a particular locality? Combining the two, I believe, reflects most closely the reality of what happens in the field. Field techniques are described in detail. For example, I won't just tell you that archaeologists excavate by arbitrary levels within natural strata and define these terms. I'll go through the step-by-step procedures of how to do it and why it is important to do so in the first place. Although many of the techniques that we employ in the field sound simple, applying them in the field and deciding when to use one technique rather than another is a more complicated matter. This is where teachers have always relied on the "learn by doing" approach. Students who watch us thinking out loud and acting upon our thoughts when we're in the field see the decision-making process at work and hopefully discover the logic behind it. Decision making can vary from relatively mundane issues such as how big do we make an excavation unit and how deep do we dig it, to more complex matters such as where to place an excavation unit, and what is the best way to take it apart. Discussions of the decision-making process are embedded in each chapter of the text. Fieldwork is only as good as the notes, records, maps, and photographs generated to document it. Without documentation even the most exciting finds are useless for helping us to understand the past. Guidelines for note taking, the use of standardized record forms, photography, data collection and preservation are discussed for each phase of field investigation.

The text concludes with a chapter (10) giving readers an idea of where and how to look for fieldwork opportunities or paid employment as a field technician. I also describe what it can be like to work as a field technician with an organization doing cultural resource management (CRM) studies. CRM is where the vast majority of archaeologists in this country currently find employment, regardless of what type of academic degree they hold. Examples of job descriptions for field technicians used by employers are provided so that you can see the qualifications and range of skills valued by those in the marketplace.

Each chapter includes numerous photos and line drawings illustrating concepts and methods. Exercises are provided that hone your grasp of concepts and skills. These exercises are useful regardless of whether or not you are learning about field methods in the classroom or as a participant in a field project. Each chapter concludes with suggestions for further reading and study. Many readers of this book are, or will be, participants in an archaeological field school. You should be aware that the Register of Professional Archaeologists has recommended guidelines and standards for field schools. The range of concepts and skills presented in this field manual encompasses those found in the guidelines.

MEASUREMENTS

You will notice that many of my references to measurements in the text are in the English rather than metric system. The metric system is an international standard of measurement and employed by many archaeologists in this country and abroad. However, much of the non-archaeological data that archaeologists use in the course of their investigations is in the English system, i.e., feet divided into inches or feet divided into tenths. Engineers in this country frequently employ this latter type of scale and so construction plans, maps, and other environmental data of importance to an archaeological study are reported in "engineer-scaled" feet. Breaking feet up into tenths brings many of the advantages of the metric system to the English system of measurement. Soil surveys and maps make use of the English system although some may report data in both the English and metric systems. Geologic and topographic maps published by the state and federal government also provide scales appropriate to both systems of measurement. For archaeologists working in cultural resource management, many of the maps and environmental data that will be supplied to them by the engineers working on projects with them will use the modified English system of measurement. For these reasons I use engineer-scaled feet most frequently as a convention for describing measurements. In reality, the tapes that I use in the field are multipurpose, scaled on one side in engineer-scaled feet and on the other in meters and centimeters.

EXPERIENCE

My first field experience was as a volunteer on a weekend excavation of a shell midden in coastal Maryland. Since then I've worked in the eastern United States from southern Virginia north to Massachusetts, from the Atlantic Ocean west to the Appalachians. For a brief but enlightening time I worked in the Northwest. I've experienced fieldwork as a student, field technician, crew chief, project director, and teacher, both in the context of academically motivated projects and cultural resource management studies. My field experiences include work on Native American sites of all periods, and historic/colonial sites of the late 17th, 18th, and 19th centuries. (If you would like the excruciating details visit my homepage at www.temple.edu/anthro/stewart> and see my curriculum vitae.)

Then as now, archaeological fieldwork was to me this glorious combination of mind, body, and soul, the total workout for someone curious about the people who came before him and the lessons that we might learn from their history. I bring to this book over 26 years of field experience, passion, curiosity, and the realization that there will always be something new to learn. We never really stop being students. While nobody may have ever learned to do proper fieldwork from a book, I believe that what you are about to read will make your training experiences in the field more understandable and enjoyable.

ACKNOWLEDGMENTS

In making my "points" about archaeological fieldwork I was thankfully aided by many people and organizations. Throughout the process, Ali Stewart assisted with background research and the preparation of graphics, and was the ever-present taskmaster with the gentle hand. My special thanks go to all of the people who have participated in Temple University's annual field school in archaeology for making me be explicit about things that I have always taken for granted.

A variety of government agencies are gratefully acknowledged for their permission to use published materials: the National Park Service for job descriptions that appear in Appendix 5; the Pennsylvania Department of Conservation and Natural Resources, Bureau of Topographic and Geologic Survey for permission to use copies of geologic maps; State of Maryland, Department of Natural Resources, Maryland Geological Survey for permission to use topographic maps; the United States Geological Survey for permission to print portions of USGS topographic maps and related publications; and the United States Department of Agriculture, Natural Resources Conservation Service for permission to print portions of soil survey maps. These maps were produced by the Natural Resources Conservation Service through the National Cooperative Soil Survey Program. Information specialists at the USGS provided answers to questions about topographic maps. The Society for American Archaeology (SAA), the Register of Professional Archaeologists (ROPA), and the American Cultural Resources Association (ACRA) are thanked for permission to print the text of their ethical statements and codes of conduct. Charles Niquette coordinated the effort with ROPA and Tom Wheaton for ACRA. The National Conference of State Historic Preservation Officers allowed the information contained in Appendix 3 to be used.

The kindness of the many individuals and organizations who provided photos used in the text is greatly appreciated and includes: Fennelle Miller and the Washington State Department of Natural Resources, Ellensburg, Washington; John Hotopp and Rob Tucher, of the Cultural Resource Group, Louis Berger, Inc., East Orange, New Jersey; Charles Bello, Cultural Resource Consulting Group, Highland Park, New Jersey; Barbara Hildebrandt, Stewartsville, New Jersey; Pete Condon, Portales, New Mexico; Patty Jo Watson and Cyndi Mosch, Department of Anthropology, Washington University, St. Louis; Paul Nevin, York, Pennsylvania; David Mudge, New Jersey Department of Transportation, Trenton; Stephen Tull, Ed Morin, and the URS Corporation, Florence, New Jersey; Darrin Lowery, Tilghman Island, Maryland; Greg Katz, Philadelphia, Pennsylvania; staff of the former Cultural Resource

Management Program at the University of Pittsburgh, Pennsylvania; Joan Walker and Bill Gardner of Thunderbird Research Associates, Winchester, Virginia; Jay Custer and the Center for Archaeological Research at the University of Delaware, Newark; Howard MacCord Sr., Richmond, Virginia; and Chris Hummer, Berwyn, Pennsylvania.

I am grateful to John Foss, University of Tennessee, Institute of Agriculture, who allowed me to borrow freely from his *Field Guide to Soil Profile Description*. Job descriptions that appear in Appendix 5 are courtesy of Richard Hunter of Hunter Research, Inc., Trenton, New Jersey, and Steve Tull of the URS Corporation, Florence, New Jersey. Burial data forms were kindly provided by Tom Crist courtesy of The Public Archaeology Laboratory, Inc., and Kise, Straw & Kolodner, Inc., Philadelphia. A number of individuals provided advice regrading dress codes for field archaeologists when I solicited them over the Internet including: Charles Niquette, Michele Wilson, Joseph Schulderein, Erwin Roemer, Slim Zyniecki, Dana Vaillancourt, Steve Cressman, Robert Jesse, Loretta Lautzenheiser, Susan Baldry, Jim Rudolph, Sean Hess, John Dendy, and John Doershuk. Brad Koldehoff, Ed Morin, and George Miller led me to useful reference materials on a variety of subjects.

DIG DEEPER

- Barker, Philip. 1996. *Techniques of Archaeological Excavation* (3rd ed). London: B.T. Batsford Ltd. A very comprehensive text written from the perspective of European sites with structural ruins.
- Hester, Thomas R., Harry J. Shafer, and Kenneth L. Feder. 1997. *Field Methods in Archaeology* (7th ed). Mountain View, CA: Mayfield Publishing Company. This is an updated version of the Heizer and Graham (1967) and Hester, Heizer and Graham (1975) texts that were used by older generations of archaeologists, myself included.
- Orser, Charles E., Jr. and Brian M. Fagan. 1995. *Historical Archaeology.* New York: Harper Collins. A basic introduction to the discipline with sections related to carrying out fieldwork.
- Palmer, Marilyn and Peter Neaverson. 1998. *Industrial Archaeology: Principles and Practice.* London: Routledge. A basic introduction to the discipline with sections related to carrying out fieldwork. One of the few synthetic references on the subject.
- *Journal of Field Archaeology*
  Published quarterly by Boston University for The Association for Field Archaeology, it contains articles on widely applicable techniques and procedures, as well as site- and area-specific field studies. It can be difficult reading for someone with no background in archaeology.

❑ A number of texts are out of print but worth looking at if you can find them in a used bookstore. These include:

- Dancey, William S. 1981. *Archaeological Field Methods: An Introduction.* Minneapolis, MN: Burgess Publishing Company. A brief but relatively comprehensive text. The chapters on the formation of the archaeological record and research design are especially worth reading.
- Joukowsky, Martha. 1980. *A Complete Manual of Field Archaeology.* Englewood Cliffs, NJ: Prentice-Hall. A lengthy and fairly comprehensive publication. The coverage of mapmaking and surveying in archaeology is very useful.
- Meighan, Clement W. 1961. *The Archaeologist's Note Book.* San Francisco: Chandler Publishing Company. An interesting compendium of types of standardized record forms of use to archaeologists.
- Noel-Hume, Ivor. 1983. *Historical Archaeology: A Comprehensive Guide for Both Amateurs and Professionals to the Techniques and Methods of Excavating Historical Sites.* New York: Alfred A. Knopf. The classic reference dealing with one man's pursuit of colonial archaeology. Provides an interesting contrast to the ways in which many prehistoric sites are excavated.
- **http://www.tamu.edu/anthro/news.html** A website compiling recent press articles and releases dealing with archaeology and anthropology. It is a great place to visit on a weekly basis if you want to keep up with breaking news. *Anthropology in the News.*

Basic Definitions and Assumptions

A work that cost days, weeks, or years of toil has a right to existence. To murder a man a week before his time we call a crime; what are we to call the murder of years of his labor? Every tablet, every little scarab is a portion of life solidified; so much will, so much labor, so much living reality. . . The work of the archaeologist is to save lives. . .

Sir Flinders Petrie, Methods and Aims in Archaeology

ARCHAEOLOGY DEFINED

Very simply, archaeology is one means of studying the human past. The precise goals or purpose of an archaeological study of humans can vary depending upon the interests or theoretical perspective of the researcher, as the sample of quotations below make clear.

Archaeology . . . "furnishes a sort of history of human activity, provided always that the actions have produced concrete results and left recognizable material traces. It turns into history whenever it remembers that the objects it studies embody the thoughts of human beings and societies" (V. Gordon Childe, Progress and Archaeology, *1944:1–2).*

Archaeology is . . . "the systematic study of antiquities as a means of reconstructing the past" (Grahame Clark, Archaeology and Society, *1957:17).*

". . . the three aims of archaeology - reconstruction of culture history, reconstruction of extinct lifeways, and the delineation of culture process" (Lewis Binford, New Perspectives in Archaeology, *1968:8)*

"Evolutionary archaeology should be understood as an explanatory framework that accounts for the structure and change evident in the archaeological record in terms of evolutionary processes (natural selection, flow, mutation, drift) either identical to or analogous with those processes as specified in neo-Darwinian evolutionary theory" (Robert Dunnell, American Antiquity, *1978:197).*

"Reconstruction of past forms of society and explanation as to how they evolve and transform themselves are the goals that almost universally guide contemporary archaeological research" (Philip Kohl, Dictionary of Marxist Thought, *1983:28).*

". . . the history of archaeology, the different uses of archaeology around the world and in our societies, and the role of archaeology in relation to the destruction and conservation of the heritage show that archaeology is and should be part of a social process. The aim of archaeology is not secure knowledge of the past for its own sake but secure knowledge of the past that is socially responsive in the present" (Ian Hodder, Processual and Postprocessual Archaeologies, *1991:30).*

In these views, archaeology concerns itself with learning the details of everyday life as well as significant or unique events, arranging these reconstructions in chronological sequences to create histories, attempting to understand or explain why things happened the way that they did, and perhaps underneath it all, discovering processes that relate to all human societies and cultures. Some maintain that, in the words of Bruce Trigger (1990), "changing social conditions influence not only the questions archaeologists ask but also the answers that they are predisposed to find acceptable" (p. 13). Others note that archaeology can be a means of promoting particular social or political agendas. In Israel, for example, archaeological studies of the past figure prominently in ongoing claims to land, and the validation of religious traditions (Elon 1996).

Any interpretation requires making some type of assumptions. Those dealing with technology, subsistence, economy, settlement, or the basic descriptive details of a people's way of life are more closely tied to the evidence of the activity or behavior. It is at higher levels of interpretation and explanation, where assumptions about the basic nature and workings of society are brought to bear, that archaeologists most frequently disagree. Explanations dealing with the origins of agriculture, the development of class-based society, the rise of cities and the state can be very different depending on whether you see society and culture through the eyes of a cultural ecologist, evolutionist, or Marxist (see Trigger [1990] for summaries of these and other theoretical viewpoints).

But few disagree about the nature of the things that we deal with in our pursuit of the past—material evidence and its context. We can compose, then, a definition of archaeology that links research endeavors at their most basic level. ***Archaeology*** is the study of the human past, the basis of which is material evidence (artifacts, ecofacts, human remains) and its context. In turn, ***fieldwork*** can be characterized as the recognition, observation, and collection of archaeological evidence in context.

DOCUMENTS AND ARCHAEOLOGY

Archaeology can contribute to an understanding of people, places, and events of both prehistory and history. This can encompass the time from over three million years ago when the first ancestors of humans appeared in Africa, to the 20th century. The importance of archaeology for the study of prehistory should be obvious, but why would it be useful for the historic period, the time for which we have documents, maps, written records, and oral traditions to consider? You might be thinking, 'because the written or documentary record of the past is incomplete,' and you would be correct. Not only is this record incomplete in that specific people, cultures, places, things, and times may not be represented, but the records that do exist don't provide the range of perspectives that we might find valuable in an analysis or reconstruction of the past. Thomas Jefferson's account of the life of the slaves on his Virginia plantation, Monticello, is informative in many respects but provides us the view of a white male, Virginian slave owner who was also a politician struggling with notions of civil rights. How different might slave life at Monticello seem if we had the perspective of a male slave, a female slave, or a white farm hand working on the property to consider (cf., Gruber 1990; Kelso 1997)? Each of these points of view adds to our greater understanding, but each reflects the perspective of the individual resulting from things like their age, sex, position in society, and how they wish to be perceived by others. Even documents that seem to be purely descriptive can be biased. Archaeology provides additional perspectives and missing details, and attempts to do so in an objective fashion. Together, archaeology and the documentary record can be powerful tools for learning about the past.

A very simple example of how archaeology can fill in the gaps not covered by documents has to do with maps. From the perspective of a 20th century citizen of the United States one would assume that a map represents a fairly accurate depiction of its subject matter. During the second half of the 19th century a large number of county and municipal maps, published as atlases, were created by itinerant surveyors and companies in the eastern United States. These maps provide an incredible amount of detail. Streams, roads, buildings, and even the names of landowners or who is living in a specific building are often provided! Figure 1.1 is an example of this type of map showing the Indian Spring area of western Maryland, circa 1877. If you were asked to locate the archaeological remains of all late 19th century buildings in Indian Spring, wouldn't this map be all that you would need to guide your fieldwork? The answer would be no.

Some of the county or municipal atlases produced during the late 19th century were done by "subscription." That is, the surveyor or mapmaker would include a general set of details for a given area, but additional detail would depend on who was subscribing to the map, or paying a fee to have their property, business, or house located and identified. If you weren't subscribed you wouldn't appear on the map. For an archaeologist, these 19th century maps depict only a sample of what may be out there to be found during a field investigation. Were you to design a field strategy based only on the available maps, relevant archaeological deposits could easily be missed. When deciding how to use a document, archaeologists, like historians, must consider the background of the person creating the document, why and how it was created, and for what audience.

FIGURE 1.1.

Detail from "Indian Spring District No. 15" in *An Illustrated Atlas of Washington County, Maryland.* Compiled, drawn and published from actual surveys by Lake, Griffing and Stevenson, 1877, page 11.

MATERIAL EVIDENCE

ARTIFACTS

In the most comprehensive sense, an ***artifact*** is anything made, used, or altered in some way by humans. A spearpoint chipped from stone, a pottery vessel, fragments of a glass wine bottle, a leather shoe, all of these are artifacts, as is the debris created in the process of manufacturing them and the tools employed in the manufacturing process. Artifacts need not have a formal shape. The sharp chip of rock struck from a flint core and used as a cutting implement is as much an artifact as a leaf-shaped knife blade made from the same material and set in a handle of bone. The fist-sized cobble plucked from a streambed and used "as-is" to pound and grind grain is an artifact. The array of artifacts that a people produce can be referred to as ***material culture***.

Artifacts can include objects that have not been physically modified. A nodule of naturally occurring flint picked off the surface and transported to a nearby camp and cached by a prehistoric artisan is an artifact, regardless of whether or not it is eventually used to make something. The removal of the nodule from its natural setting or context, and its transportation by a human constitutes "alteration" of the object, and brings it within our definition of an artifact. Near Trenton, New Jersey amateur archaeologists have excavated portions of the ruins of the home of Charles Conrad Abbott (Stanzeski 1974). Abbott was a naturalist and student of Indian prehistory who lived during the 19th and early 20th centuries. He amassed a collection of geological specimens and Indian artifacts, and figured prominently in early debates about the antiquity of Indian peoples in America (cf., Joyce et al. 1989; Kraft 1993; Meltzer 1983). When Abbott's home, Three Beeches, burned to the ground in 1914, his collections were in the house and became incorporated in the ruins. A geode (a hollow nodule of rock with crystal-lined walls) recovered during the excavation was recognized as an artifact, even though it was initially a product of nature. Archaeologists sometimes refer to artifacts that have been transported, but not worked, as ***manuports.***

Artifacts that are not portable, or whose nature is changed by their excavation or removal from the field, are called ***features.*** The term can refer to single things as well as clusters and associations of things (Figures 1.2 to 1.6). A dug pit filled with trash, the ruins of a building, a temple mound, and an artificially created landscape are all examples of features. A pile of fire-cracked rock representing a former hearth is a feature defined by the close association of individual artifacts, as is the cluster of chips and hammerstones resulting from the production of stone spearheads.

FIGURE 1.2.

A cluster of fire-broken rock representing a hearth feature in a prehistoric Native American site, Abbott Farm National Landmark, New Jersey.

FIGURE 1.3.

The darker colored matrix fills a shallow, basin-shaped pit. The excavated cross-section seen in the foreground reveals the depth and profile of this pit feature.

ECOFACTS

Ecofacts are any type of evidence related to the physical environment in which archaeological deposits are situated. When you consider the many components of an environmental system, like geology, sediments, soils, landscape, plants, animals, and climate, there are many types of data that could and do get collected as part of field investigations. For example, vegetation can be represented by the visible remains of plant parts, as artifacts like matting or baskets, as residues adhering to other artifacts, as charcoal preserved in hearths, or as microscopic pollen (the organic male reproductive cells produced by a plant) and phytoliths (silica bodies produced within the cell walls of plants). Obviously, there are some types of material evidence related to the environment that straddle the definitions of ecofacts and artifacts.

FIGURE 1.4.

The dark-colored conical shape shown in this cross-section represents the burned and disintegrated remains of a wooden post once part of a Native American dwelling in a farming village situated along the Susquehanna River in Pennsylvania. Prior to sectioning, this feature appeared as a dark-colored oval stain on the floor of the excavation unit.

FIGURE 1.5.

Monumental types of features are represented by the two burial mounds seen in this photo, one on the right (with stairs leading up its side), and one in the left background. The picture was shot from the top of a larger mound. The mounds were constructed by people of the early Mississippian culture (AD 900–1200) and are part of the Ocmulgee National Monument, located in Georgia.

FIGURE 1.6.

The ruins of a historic pueblo in Colorado, another example of more monumental types of features.

The evaluation of published information, and the collection and analysis of additional environmental data is essential to the success of any archaeological project. One reason for this concern relates to the mechanics of planning and carrying out fieldwork. Imagine that your job is to locate all archaeological sites created during the past 11,000 years that are located within a 3000-acre property including hilltops, slopes, and floodplain areas adjacent to streams. Where might archaeological deposits be buried? Could you find them at or near existing surfaces? How deeply will you have to dig in each of these environments in order to search for artifacts and features that could be up to 11,000 years old? Where might examinations of exposed surfaces suffice? During your excavations you discover artifacts buried three feet below the existing surface. The artifacts are found throughout a distinctive soil that is one foot thick. How did these artifacts get buried? Do the artifacts represent a single occupation or multiple occupations? Is the deposit horribly disturbed, or are artifacts laying where they were originally lost or discarded? Does the layering of the soil and the depth of the deposit provide evidence about the possible age of the artifacts? To answer these and other related questions, you must have an elementary understanding of the nature of the environments in which you are working and how they may have evolved over time, and be actively collecting and evaluating environmental data while in the field (Figures 1.7 and 1.8).

FIGURE 1.7.

The darker soil at the top of this profile from southern Virginia is about one foot thick and has been plowed repeatedly by farmers over the past 200 years. It occurs above soils estimated to have been weathering for 10,000 or more years. Any previously layered sedimentary or archaeological deposits have been churned by the action of plowing. Artifacts found in this plowzone may represent single or multiple occupations occurring over the past 10,000 years.

FIGURE 1.8.

The darker layers in this photo represent a series of A horizons, or "surface" soils, that were buried as a result of flooding. The lack of visible weathering in the sediments upon which each rests implies that the time between floods was relatively short, perhaps less than 500 years in each case, an assessment confirmed by radiocarbon dates. This site is less disturbed than the one shown in the previous photo and has artifacts situated in deposits that allow archaeologists to examine discrete occupations representing short periods of time. Abbott Farm National Landmark, New Jersey.

Human beings, like any other animal, are part of an ecosystem. No matter how sophisticated our technologies and systems of production become, we cannot escape our ties to the physical world. The environment presents us with opportunities and limitations. While it does not strictly determine what we do or how we do it, the environment does affect our lives. How people use and interact with the environment is influenced by their needs, desires, level of technology and the social relations bound up in the maintenance, recreation and use of technology (the mode of production). How culture conditions the way in which people perceive their environment underlies these influences. Therefore, a thorough study or interpretation of any society or culture of the past, whether we are describing a way of life or trying to explain why things happened, requires environmental reconstruction. Because we are part of an ecosystem, we always have an effect on the environments in which we live. So an understanding of cultural systems will, in turn, aid studies of the environment and environmental change.

HUMAN REMAINS

The human body and the archaeological contexts in which remains are found can provide a remarkable record of both physical and social life. Information about diet, disease, types of work performed, medical and healing practices, a variety of social behaviors (e.g., tattooing, alteration of the body for aesthetic purposes, warfare), social status, perceptions about life and death, and demography can be derived from appropriate types of analyses (cf., Beck 1995; Larsen 1997; Renfrew and Bahn 1996:403–440; Schwartz 1998). Some of these topics, like disease and demography, can only be addressed through the study of human remains. In our zeal to recover such information we must not forget that we are dealing with fellow humans whose remains deserve to be treated with respect, if excavated at all. Concern for the treatment of the dead is embodied in a variety of state and federal regulations and laws. These are summarized below in the discussion of ethical concerns for archaeologists.

CONTEXT AND ITS IMPORTANCE

> *The essential value of antiquities, apart from their purely artistic interest, lies in the circumstances in which they are found. The inexperienced traveler is apt to pick up a number of objects haphazard, without accurately noting their find-spots, and even, getting tired of them, as a child of flowers that he has picked, to discard them a mile or two away. If the first act is a blunder, the second is a crime; it is better to leave them lying in place.*
>
> *—G.F. Hill, How to Observe in Archaeology*

The concept of context is central to the practice of any type of archaeology. In fact, archaeology has been called the science of context (e.g., Butzer 1980; Schoenwetter 1981). Concern for the context of archaeological evidence is one of the things that separates the professional practitioner from the collector looking for "goodies" to arrange on shelves or put into frames for display. ***Context*** can be characterized in a variety of useful ways:

- the situation in which something occurs or exists;
- the three-dimensional location of material evidence (provenience); the spatial relationships that exist among and between artifacts or any kind of material evidence, and the matrix or sediments in which that evidence occurs;
- the chronological placement, or location in time of material evidence;
- the geographic or environmental situation in which material evidence exists; and
- the "location" of the behaviors represented by material evidence within a social or cultural system.

As a group, these characterizations describe the three fundamental components of context, the spatial, the chronological, and the behavioral. Michael Schiffer (1972, 1987:3–4, 1996) uses the term ***archaeological context*** to refer to the situation of archaeological evidence in space and time, and ***systemic context*** to aspects of an artifact's manufacture, use, and meaning during its actual participation in a behavioral or cultural system. Context brings additional meaning to artifacts that is not inherent in the artifacts themselves. The exact same artifact can take on different meanings depending upon the context in which it is found. How often have you heard politicians complaining that their opponents have misused quotes? "That's not what I meant when I said that. My remarks were taken out of context." It is the same with artifacts or any type of material evidence; context literally can change the meaning assigned to objects or any type of material evidence.

Consider this typical scenario. An artifact collector walks into an archaeologist's office with his/her latest find seeking answers to some pretty basic questions—How old is this? What is it? What is it made of? Without fail, the first words out of the archaeologist's mouth will be, "Where did you find it?" Why? The answer relates to context and its interpretative potential.

Physical objects by themselves are sources of all sorts of information. The raw material(s) from which an artifact is fashioned is often readily identified by an archaeologist or other specialist. The type of the material, and the shape and condition of the object can imply things about how it was made or used. The style or form of the object may be one that has been seen repeatedly in contexts that have been dated elsewhere, and thus its age can be estimated by analogy with these other finds. But knowing where something was found and with what it was associated increases interpretive possibilities, and enhances the reliability of statements made strictly on the basis of examining the physical characteristics of the object.

Knowing where on the landscape an artifact was found, and whether it was picked off the surface or pulled from a bank along a stream, may tell us how old the object might be, especially if we know something about the landscape's stratigraphy and general age (published soil surveys are helpful here, see chapters 4 and 7). Locational information might tell us if the object is part of an intact archaeological deposit, or has been displaced from its original context.

The raw material from which an artifact is made becomes an even more significant attribute for analysis when we consider context. Where are natural sources of the material from which an artifact is made? How far distant are these sources from where the artifact was found, and what can we infer from this spatial relationship? Is the material procured directly by the same people making, using, and discarding the artifact, or indirectly through trade with one or more intermediaries involved in the procurement and production process?

A few years ago, a stone spearhead typically used by the earliest inhabitants of the Delaware River Valley was found on the surface at Island Beach State Park, New Jersey (Bello and Cresson 1995). The "fluted" style of the projectile is known to be over 10,000 years old. The collectors who found the piece were excited.

The archaeologists to whom they showed the artifact were more circumspect. They knew that the find spot is a barrier island fronting the Atlantic Ocean, and is frequently subjected to the ravages of storms and high water. In fact, the size and shape of the island has changed over time as a result of the natural processes that act upon it, and it is unlikely that any portion of the island now above water is more than a few thousand years old. The archaeologists therefore doubted that the spearhead was found in its original context, an inference supported by its highly weathered condition and the lack of any associated artifacts. They conjectured that it had been washed onto the beach from some distant point and did not represent a portion of an intact archaeological deposit.

Do other, equally suitable materials occur in the same areas as the raw materials that are selected for use? Why aren't these materials exploited? Where do other objects of similar form, age and material occur? Looking at the nature and distribution of natural resources and comparing this with archaeological data about **what** is being used, **how** it is being used, and **where** it is being used is standard procedure. These comparisons can ultimately provide us with information on technology, the organization of production, cultural preferences regarding the desirability of a material, the extent of group territories and settlement movements, contact between groups, trade and exchange, and the human ecology of resources held in common by people.

Developing answers to these questions requires a broad range of archaeological data, even more contextual information, and a thorough knowledge of the natural world. What begins as a consideration of spatial and chronological context leads to statements and hypotheses regarding behavior and behavioral contexts.

Context is a flexible or "telescoping" concept in that you can consider the context of a thing or piece of evidence on a variety of scales ranging from the very small to the very large: within an archaeological deposit or intrasite, between deposits or intersite, regional, interregional, at single points in time **(*synchronic*)** or though time **(*diachronic*)**.

The stone artifacts seen in Figure 1.9 are identifiable as the result of a prehistoric artisan working cobbles of chert (a flint-like, cryptocrystalline type of rock) derived from a stream deposit. Similar unworked cobbles can be found in stream deposits near the excavations where the artifacts were discovered. Thus we can say that the raw material used in production is locally available. The chipping debris in each pile was spatially associated (when it was encountered in excavations everything was in two small clusters), and a number of pieces can be refitted to one another. A number of the flakes retain the weathered cortex of the outer surface of the cobble. These observations indicate that the flakes in each cluster are derived from a single cobble. The stream-rolled cobbles of quartzite in the photo show signs of battering on one end. This damage, and each cobble's spatial association with the cluster of flakes suggests that they are "hammerstones" used in the working of the chert cobble. A closer examination of the flakes from each cluster, and the proportion of the original chert cobble represented by this debris, implies that the stoneworker's purpose was simply to produce relatively large, sharp-edged flakes, rather than attempting to shape the cobble into a tool with a formal shape like a spearhead or knife.

Similar clusters of flakes and hammerstones are distributed across the archaeological deposit in the same excavation level and are arguably contemporaneous. From this evaluation of intrasite contexts it might be inferred that stoneworking was practiced by a number of individuals in the group that created the site's deposits, rather than by a specialist working on behalf of the larger group.

Do similar artifacts and contextual associations occur at other sites of comparable age in the local area or the broader region? If the answer is yes, our evaluation of intersite contexts for these data suggests that some of these localities may have been occupied by our original group under study, and/or that the technology and the organization of the work involved may be shared by groups across the larger region.

In turn, we can take the same artifacts, find them in different contexts, and come to the conclusion that certain work or technology is not shared and that craft specialization may be evident. Evidence for tool

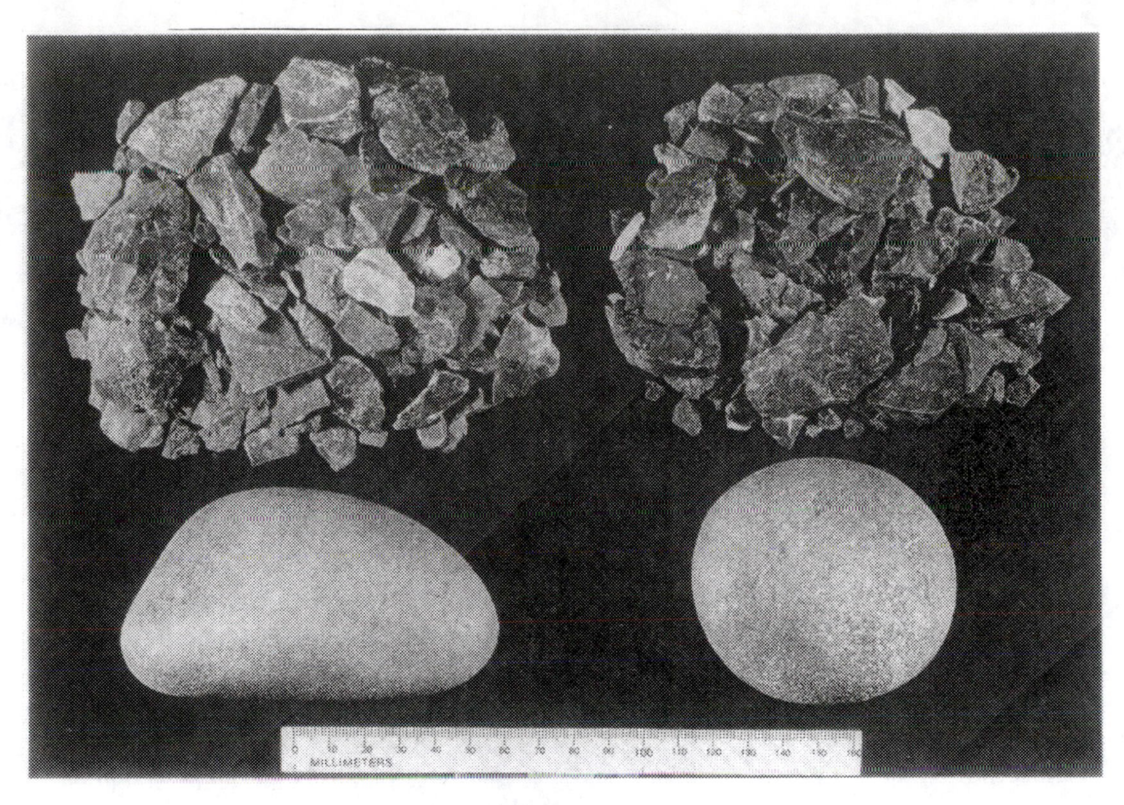

FIGURE 1.9.

Chert flakes and hammerstones from two artifact clusters in an Archaic Indian site, Abbott Farm National Landmark, New Jersey.

manufacturing and the end product of the manufacturing process are associated with all of the dwellings excavated at Village B shown in Figure 1.10. The intrasite contexts of this evidence imply that some member of every household was making stone tools, and that this technology was shared throughout the community. At Village A, although everyone in the community appears to have used triangular flint knives, evidence of the manufacturing of these tools is confined to a single dwelling. This suggests that there is a craft specialist serving the village and raises all sorts of questions like what does this specialist get in return for her knives, and does she have a special social status in the village?

INTERPRETIVE IMPORTANCE OF CONTEXT

In the above examples, the specific nature of artifacts and their spatial and chronological associations provide a means of interpreting a technological activity and showing how it may be integrated into the larger society. An example drawn from modern life makes the same point about the interpretive importance of context, and may be more understandable to the student new to archaeology.

Figure 1.11 depicts an object that should be very familiar. From the perspective of an archaeologist of the far-future there are a number of things that can be inferred from the penny itself. It is made of an alloy, a synthetic combination of metals. The alloy does not occur in nature so we may infer that the culture manufacturing the penny had knowledge of metallurgy. The culture obviously has a recording system. We can observe numbers, a date (1998), and expressions of two different languages (English and Latin). The words "In God We Trust" reflect a belief system, and "United States" implies the existence of integrated, complex governments. Some males in the culture wear beards and a particular type of clothing. There is a representation of a building with a distinctive architectural style.

Interpretations of the penny expand when considered in the contexts in which they might be found at a single point in time (synchronic perspective), for example, sites across the United States dating to 1998. Archaeologists would find them in contexts, like private and public banks, that would indicate their

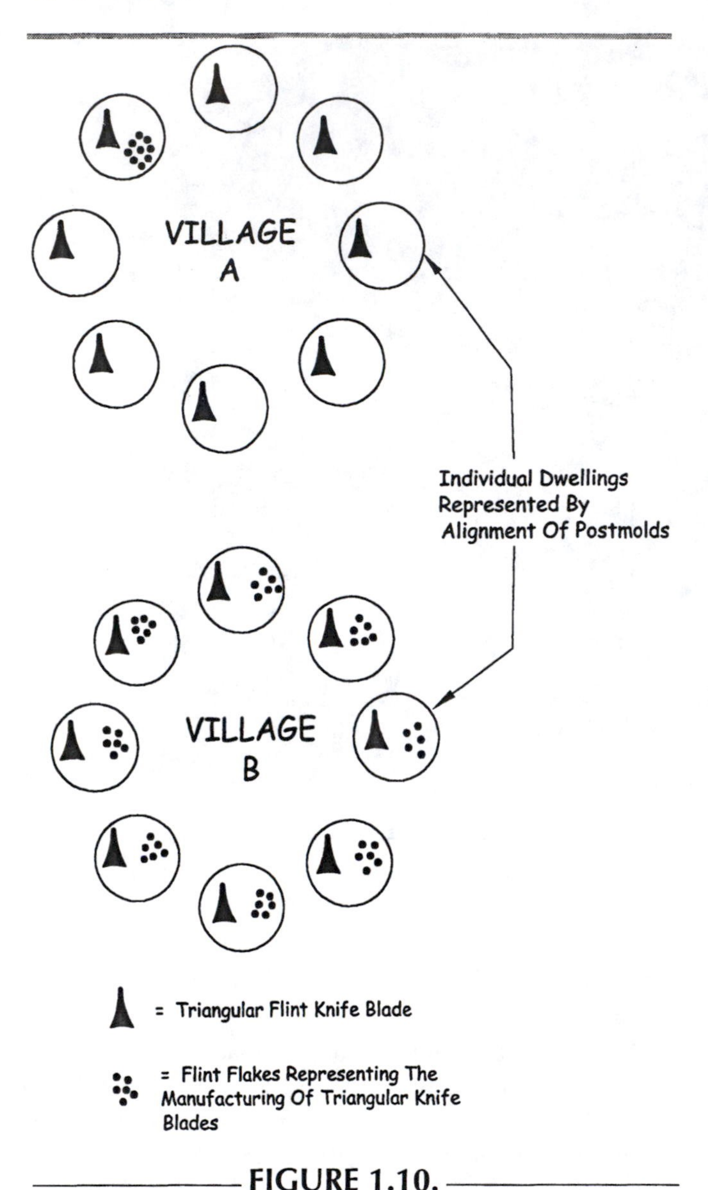

FIGURE 1.10.

Plan of hypothetical site excavations that show the context of finished tools and manufacturing debris.

importance as currency. They would be found associated with other types of coins, and would far outnumber them. But pennies would also be found in the gutters of streets along with trash, with other coins in basins surrounding fountains in central portions of "malls," and occasionally on the eyes of a corpse in a tomb. While the initial inferences made about the penny stay with the artifact wherever it is found, these contexts add to, and alter its overall meaning.

As more individual finds from different sites are considered, different dates are noted on the pennies, although the size and shape remains the same. This suggests that this type of artifact has been in use over

FIGURE 1.11.

A familiar artifact—obverse and reverse sides of a 1998 Lincoln cent from the United States.

an extended period of time and that its size, shape, and decoration have some meaning that transcends the passage of time. It also suggests that the artifacts are mass produced, given the incredibly similar physical attributes that each possesses, and hints at a particular level of technological achievement for the culture making them.

It could also be noted that pennies with the earliest dates are made of copper, those with a date of 1943 are made of a zinc-coated steel, those with 1944 to 1961 dates are composed of an alloy of copper and tin, those dated 1962 to 1981 are a copper and zinc alloy, and those postdating 1981 are fashioned from copper-plated zinc (Office of Public Correspondence 1998). This suggests that changes occurred in the perception of the value or utility of copper, or the technology used in the production of pennies, or even the economic system within which pennies function. The value of these metals and alloys could be compared with those used in fashioning the other types of coins with which pennies are associated at some sites. Archaeologists would increase their knowledge of the relative economic worth of the penny. It would be learned that the sources of the raw materials from which pennies are made come from various areas within the United States and abroad, hinting at widespread and complex trade networks. Archaeological research would locate only a handful of sites (mints) where the pennies were produced, and yet the artifacts are found throughout the country. An intricate distribution system would be inferred.

As future archaeologists studied the contexts in which pennies occur through time (diachronic perspective) there would be more to say, and some of the questions raised as a result of synchronic analysis would be partially answered. The architectural style of the structure depicted on the 1998 object would be found to be an ancient design that is reused through time, especially in the construction of special-purpose buildings. It would be learned that the man depicted on the coin (Abraham Lincoln) lived and died long before the coin bearing his image was first produced in 1909. He obviously has symbolic importance and his image appears on many things, some monumental. Changes in penny production would be seen to correspond in some way with political and economic changes. For example, archaeologists might learn that zinc-coated steel pennies were created so that copper resources could be devoted to manufacturing things needed by the armed forces engaged in World War II. And what does the longevity of the penny in the currency of the 20th century imply about the prevailing economic system and government?

The examples above only begin to illustrate what a powerful interpretative tool context can be in archaeological research. Making use of it begins with attention to where and how evidence is found in the field.

ARCHAEOLOGICAL SITES

The traditional focus of archaeological fieldwork has been the ***site,*** a spatial concentration of material evidence or artifacts. It is a term that archaeologists use to organize, group, and separate the material evidence that they encounter in the field. However simple this definition may seem, the term gets used in many different ways (cf., Dunnell and Dancey 1983; Ebert 1988, 1992; Hayden 1993:64–66; Thomas 1998:95; Webster et al. 1993:18). How big of a spatial concentration of evidence do you need to designate something as an archaeological site? How many artifacts? How closely spaced must artifacts be in order for them to be part of the same site? Of different sites? Can the chemical signature of buried sediments, which has been interpreted as the residue of human activity, qualify an area as a site? Should a consideration of a deposit's research potential or significance figure into the decision to call it a site or not?

In theory we are interested in any archaeological evidence that has the potential to inform us about the past, no matter the size of the deposit or the number of artifacts that it contains. Understanding the nature and range of human activities and how they are distributed across the landscape is at the root of reconstructing lifeways and attempting to explain the developments of the past (Dunnell and Dancey 1983; Thomas 1975). Giving attention to any place where material evidence is found has been characterized as siteless archaeology.

All states maintain a ***State Historic Preservation Office*** (SHPO) that is a repository for information related to archaeological sites and other types of cultural resources. The locations of archaeological deposits are recorded on United States Geological Survey topographic maps (see chapter 6). Descriptive information is also maintained on site forms (see chapter 4). Usually anyone can register an archaeological site with their SHPO. It generally requires filling out a site form and providing a map location. Each new site is then given an official, three-part designation. 45KI23 is the designation of the Duwamish Site located in the Puget Sound area of Washington State (Campbell 1981). The number 45 refers to Washington's position in an alphabetical listing of the 48 states prior to the inclusion of Alaska and Hawaii (Alaska and Hawaii received numbers 49 and 50, respectively). The letters KI stand for King County, the county in which the site is located, and the number 23 means that Duwamish is the twenty-third archaeological site officially recorded for King County. For comparison, the twenty-first recorded site in Litchfield County, Connecticut is 6LF21 (Moeller 1980).

In practice, lots of small, scattered or isolated archaeological finds get relegated to a nonsite category in the bureaucratic systems that manage cultural resources in the public interest. Individual finds of projectile points, or a handful of pressure flakes representing an ancient hunter resharpening a spearhead, may not be accorded site status. The same is true of the broadly scattered pieces of historic pottery and glass that are often found in fields surrounding a farmstead. However, this information can be very useful when seen in a larger spatial and behavioral context (Figure 1.12; Custer 1988). Unfortunately, how we talk about archaeological evidence and the terms that we use can effect the ways in which others perceive its importance. This can become critical in cultural resource management studies (see chapter 3) where legal issues and the allocation of funds hinge on the perceived significance of an archaeological deposit.

The ability of any archaeological evidence to increase our knowledge of the past is not self- evident. It requires different scales of contextual analysis. Go into the field willing to deal with any and all archaeological evidence that you encounter.

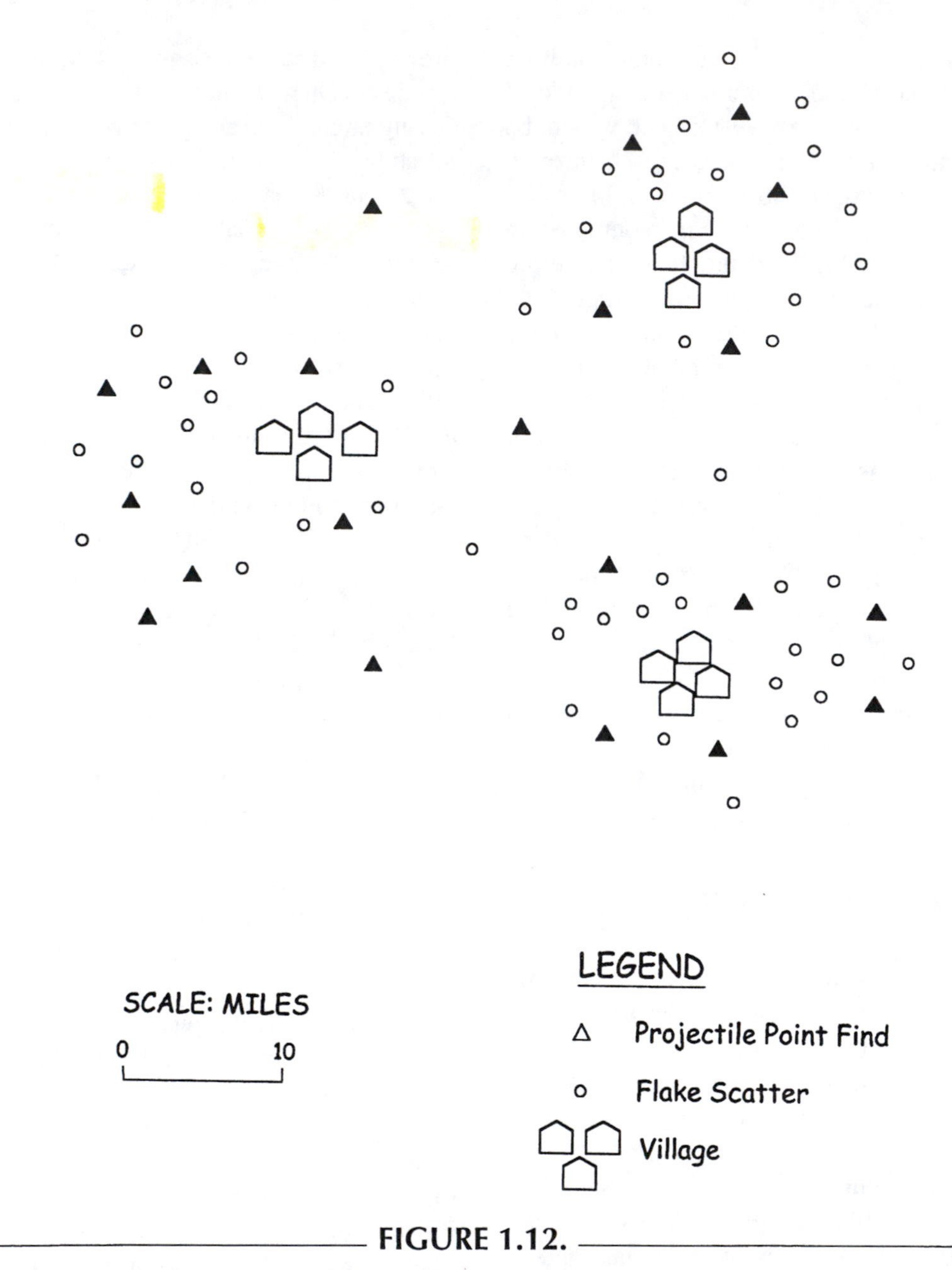

FIGURE 1.12.

The drawing shows the distribution of all archaeological deposits found within a hypothetical project area. The spatial relationship between village locations and isolated and low-density finds of artifacts can help to define their exploitative territories. The lithic scatters and isolated finds of projectile points in this example may be the material fingerprint of short-term hunting and gathering forays out from the respective villages.

ETHICAL CONCERNS

What constitutes acceptable behavior for an archaeologist working in the field or elsewhere? What standards exist that allow us to decide what are appropriate activities versus those that are not appropriate? These are ethical concerns.

Archaeologists are faced with complex responsibilities because of the many arenas in which the discipline is practiced today. There are many "publics" who have a stake in the management of archaeological deposits, their study, interpretation, publication, and presentation to both professional and nonprofessional audiences (cf., King 1998; Lynott 1997; McManamon 1991; McManamon and Hatton 1999). There is the obvious interest of the community of archaeologists, museum staffs, and other scholars interested in aspects of the past, teachers, and their students. Much archaeological research is funded by federal, state and local governments in the public interest (see chapter 3). The tax-paying public therefore has (or should have) an interest in how this aspect of cultural heritage is managed on their behalf. Cultural and ethnic groups, especially Native Americans, may be concerned about all aspects of archaeological practice. Archaeological sites and their interpretation may be part of the economic development strategies of some communities. Certainly the bureaucrats, managers, lawyers, and other parties involved in complying with the laws and procedures established by government have a vested interest in the practice of archaeology. Artifact collectors and commercial antiquities dealers are intrinsically concerned about goings-on in the world of archaeology.

The often unique situations in which archaeologists work require that statements of ethical principles be framed at a general level. Professional standards and controls are needed to hold archaeologists accountable if for no other reason than that research is being performed in the public trust. Ethical standards have evolved over the history of the discipline as sources of funding, legislation regarding antiquities and cultural resources, and the people and clients served have become more complex. The issue of international looting of artifacts, the related destruction of archaeological sites, and scholarly research based on looted materials stimulated a review and revision of previous ethical standards (McGimsey 1995:13; Wylie 1995).

PROFESSIONAL ORGANIZATIONS

Text of the ethical principles and code of conduct promoted by three important organizations are included in Appendix 1. The Society for American Archaeology (SAA) is an international organization dedicated to the research, interpretation, and protection of the archaeological heritage of the Americas. The Register of Professional Archaeologists (ROPA) is involved in a variety of activities whose general purpose is to advance professionalism and high standards in archaeology. Its creation was approved by the membership of SAA. Together SAA and ROPA support the establishment and maintenance of a profession-wide code of conduct, standards of research performance, and grievance procedure. The American Cultural Resources Association (ACRA) promotes the professional, ethical and business practices of the cultural resources industry, including all of its affiliated disciplines, for the benefit of the resources, the public, and the members of the association. In addition, the Secretary of the United States Department of the Interior has proposed standards for archaeological and historic preservation activities, and the people involved in them. These can be found in the *Federal Register, Volume 48, No.190, Part IV,* and the *Code of Federal Regulations, 36 CFR, Part 61.*

PRESERVATION ETHIC

At the core of all of these principals, standards, and codes is a critical assumption about the archaeological record of the past. Archaeological deposits are a finite resource. There are only so many of them out there. The archaeological investigation of a deposit literally "consumes" it. Granted, we do our best to collect all of the information that we believe to be relevant, and use the best methods to collect it. But anyone familiar with the history of archaeology as a discipline will point out that the methods and technologies useful in fieldwork have changed, in turn altering views about what constitutes useful data to collect.

Prior to the advent of radiocarbon dating in the late 1940s (Willey and Sabloff 1993:183–185), few archaeologists collected the bits and pieces of charcoal that they encountered in their excavations. Why bother? Shouldn't a brief note in the field book regarding the presence of burned wood be sufficient? Today, wood charcoal is important as a medium for deriving radiocarbon determinations of age and dendrochronology, as well as providing information about local and regional environments and climate change. In turn, the kinds of research questions that archaeologists have considered to be important have changed through time and have had an impact on the type of data collected in the field. The growing importance of understanding the environment and human ecology led archaeologists to collect and analyze not only the most visible remains of plants and animals, but pollen, phytoliths, and residues adhering to artifacts. The flotation of samples of excavated matrix, a standard field technique today (see chapter 9), was one outcome of this interest that dates from the 1960s (Struever 1968). There is a dynamic relationship between theory, what we would like to learn from the past, what constitutes archaeological data, and the methods appropriate for data collection and analysis.

The finite and irreplaceable nature of the archaeological record is embodied in the central Stewardship Principle promoted by the SAA, and all other principles are in some way derived from it (see Appendix 1). The assumptions embodied in the Stewardship Principle have also been referred to as the "preservation ethic" by archaeologists. The principle further requires that archaeologists use the record of the past for the benefit of all people, and work for its long-term preservation.

This and other principles and standards have implications for fieldwork and obligations for those who engage in it. Many of these may seem most relevant for people who design and direct archaeological research, and not the student, field technician, or person just beginning their experience with fieldwork. Do not ignore them. No matter what position you occupy, understanding the structure of an archaeological project and the responsibilities of others can only benefit your personal performance and growth.

PERFORMANCE STANDARDS

Of course, those who participate in fieldwork should be aware of ethical standards and codes of performance, as well as federal, state, and local laws regarding fieldwork, the collection and ownership of artifacts, and the treatment of human remains. Fieldwork should be organized in accordance with some scientific plan or research design (see chapter 3). Those who work in the field should have adequate training, supervision, and access to necessary resources. This requires keeping up-to-date with advances in archaeological methods and techniques, and recognizing personal limitations. We need to know which specialities can contribute to archaeological fieldwork and why, learn their basic vocabulary so that we can ask specialists questions in their own "language" and understand the answers that are given, and know when it's time to get them involved with a project. The United States Secretary of the Interior Standards require that those responsible for archaeological projects minimally have a graduate degree in archaeology, anthropology, or a closely related field, one year of full-time professional experience or specialized training, and one year of supervisory experience. Individual states often have their own formal standards for who can carry out archaeological research within state boundaries, and guidelines for how different types of fieldwork should be organized.

Adequate records must be kept during fieldwork, and provisions must be made for the long-term preservation and storage of artifacts and records following the conclusion of fieldwork. Archaeologists must be accountable for sharing and reporting the results of their work to as wide an audience as possible. Don't forget that we are consuming a limited resource and often doing so with public funds. Interaction with the public and the production of popular versions of technical reports are often written into the contracts governing cultural resource management studies. Offering public tours of excavations and providing volunteer opportunities are some ways that archaeological projects are shared and communicated with the public. Consultations with Native Americans or any community that has a special interest in a site or investigation should be initiated.

Cultural resource management studies have created vast mountains of technical reports that are on file with various government agencies and the SHPO of every state. While any professional can gain access to these reports, it is often not as simple as going to the local library, and the nonprofessional community rarely gets the chance to see them. There are a variety of venues in which archaeologists can publish the results of their work including local, state, and national journals of archaeology and anthropology, as well as popular magazines. Both small and large presses publish book-length treatments of archaeological topics. Taking that extra step beyond the production of a technical report requires the concerted effort of individuals, and often provides little reward beyond knowing that you are doing your best to realize the full value of the archaeological deposits that you have consumed.

ETHICS AND BUSINESS OBLIGATIONS

What are our duties towards the clients or sponsors who pay for our work? Can we be both advocates for cultural resources and advocates for clients? Do professional, legal, and business obligations and ethics clash? Archaeology carried out in the context of cultural resource management studies adds to the ethical concerns and performance standards of which an archaeologists must be aware (see the ACRA code in Appendix 1). Behaving responsibly requires:

- ☑ understanding and acknowledging the needs of your clients,
- ☑ keeping certain types of information confidential,
- ☑ knowing employment and labor laws and regulations,
- ☑ and being concerned about employee safety, education and advancement.

In short, ethical standards must deal with the realities of providing a service in the world of commerce.

Ethics and Descendant Communities

Widespread interaction and constructive dialog between archaeologists, Native Americans, and other descendant communities is a fairly recent development in the United States. All sides have much to gain from cooperation and can be a powerful force for education and historic preservation. What is our relationship to the descendant communities whose ancestors we study? What are the interests of descendant communities in archaeological practice and interpretation? How might perspectives of descendant communities enhance archaeological interpretations? These are some of the big questions being considered (Ferguson 1996). Certainly the ethical principles of stewardship, accountability, public education and outreach demand that constructive relationships be established.

The issue of how to handle human remains and sacred objects has been one stimulus for rethinking the relationship between archaeology, archaeologists, native peoples, and the social and political contexts in which we all live and work. Human remains, associated grave goods, sacred objects, and sacred landscapes must be seen in a different light than things that are typically thought of as archaeological evidence or cultural resources. This is made clear in existing federal policy and legislation. Consultation is a procedural and legal mandate, as well as an ethical concern.

The *American Indian Religious Freedom Act of 1978* is a government policy designed to protect and preserve for American Indians, Eskimos, Aleuts, and Native Hawaiians their inherent right of freedom to believe, express, and exercise their traditional religions. It allows access to sites, use and possession of sacred objects, and freedom to worship through ceremonial and traditional practices. The federal government, its agencies and departments need to consult with traditional religious leaders to see how government practices affect Native American cultural and religious practices. A presidential executive order (13007) issued in 1996 deals more specifically with Indian sacred sites. Government sponsored archaeology is a concern here.

Of special importance is the *Native American Graves Protection and Repatriation Act, 1990* (also see the Department of the Interior, 1993; McManamon 1992), referred to as NAGPRA. It requires that federal agencies inventory human remains and associated grave goods in their holdings. Anyone who has been involved in a federal undertaking, used federal money, required a federal permit, etc. that resulted in the collection of human remains or grave goods must do the same. The same must be done with any human remains or funerary objects discovered or originating on federal property. Culturally affiliated tribes must be presented with inventories of sacred objects, human remains and grave goods. The return (repatriation) of sacred objects, human remains, and grave goods occurs at the request of the culturally affiliated tribe, which must be federally recognized. The discovery of human remains, grave goods, or sacred objects during an investigation (on federal or tribal lands, or involving a federally supported action) requires consultation between affiliated or potentially affiliated Native Americans and the other parties involved about the treatment and disposition of the remains or objects involved. The sale or purchase of Native American human remains is illegal, regardless of whether or not they were found on federal or Indian lands. Work is currently underway to develop rules for dealing with culturally unaffiliated remains of Native Americans.

Religious freedom is an inherent right of citizens of the United States. It is unfortunate that laws were needed to extend this right to the native peoples of this country. Regardless of the scientific value, or value to the public of human remains, grave goods, etc., native rights are privileged. In other words, there is no "public trust" when culturally affiliated human remains, burial goods, or sacred objects are involved.

Archaeology is a discipline of the broad field of anthropology and most archaeologists consider themselves to be anthropologists. Excerpts from the ethical statements of the broader anthropological community are of interest here.

> *"Anthropological researchers have primary ethical obligations to the people, species, and materials they study and to the people with whom they work. These obligations can supersede the goal of seeking new knowledge, and can lead to decisions not to undertake or to discontinue a research project when the primary obligation conflicts with other responsibilities, such as those owed to sponsors or clients . . . Anthropological researchers must do everything in their power to ensure that their research does not harm the safety, dignity, or privacy of the people with whom they work, conduct research, or perform other professional activities."* (Code of Ethics of the American Anthropological Association, 1998)

> *"Our primary responsibility is to respect and consider the welfare and human rights of all categories of people affected by decisions, programs of research in which we take part . . . It is our ethical responsibility,*

to the extent feasible, to bring to bear on decision making, our own or others, information concerning the actual or potential impacts of such activities on all whom they might affect" (National Association of Practicing Anthropologists, Ethical Guidelines for Practitioners, 1988).

These statements reinforce the intent of the SAA's ethical principles and existing policy and law. However, you should be aware that not every archaeologist agrees with this perspective. For some, archaeologists are the only ones who can speak adequately for their "informants", the ancient peoples who created the deposits that we study (e.g., Meighan 1996). It is therefore a disservice to these informants, and the public who supports archaeological research and recognizes the value of cultural heritage, to repatriate human remains and artifacts prior to their thorough study by scientists. Friends of America's Past (Portland, Oregon) and the American Committee for the Preservation of Archaeological Collections (Whittier, California) are nonprofit organizations that actively promote similar views, maintaining that archaeological evidence and knowledge of the past is owned by no one and must be preserved for the future. The membership of these organizations includes professionals and nonprofessionals.

What you think and feel about ethical principles and codes of conduct are important! If you don't have an opinion about the subject, you should get one. Sooner or later you will have to act on your personal standards. The topic is so crucial that developing more college courses on archaeological and professional ethics is a major recommendation of the Society for American Archaeology for teaching in the 21st century (Bender and Smith 2000). Think hard before you consume an archaeological resource. Design and participate in research that makes the most of the archaeological deposits being investigated for all concerned peoples. Work as gently and as efficiently as you can.

APPLYING YOUR KNOWLEDGE

1 2 3 4 5 6 7 8 9 10 11 12

1. Pick an everyday object and describe it as an archaeologist of the future might. What can you say about the object and the culture that created it on the basis of the object alone? Look for that object in different contexts; how does context change your ability to say things about the object and the cultural system within which it exists?

2. Background research for many archaeological investigations, especially survey projects, involve interviewing artifact collectors and examining their collections (see chapter 4). Is the collection of artifacts from surface sites by nonprofessionals in this country looting? There are well-intentioned collectors who have just as big an impact on site destruction as commercial relic hunters. Are archaeologists to ignore this extensive and often detailed source of information on the nature and distribution of archaeological sites? Noncommercial collectors will often will their collections to local or state museums. Should these donations be accepted? There is already a rift between archaeologists and many artifact collectors. Do we widen this rift or close it? How?

3. The accidental discovery in 1996 of human remains over 9000 years old along the Columbia River in Washington State has focused international attention on the relationship between archaeology, science, and the concerns of Native Americans. Dubbed Kennewick Man, the controversy over these remains continues today. It is a good case study for initiating discussions of the interplay of ethics, archaeological practice, and the interests of Native Americans and nonprofessionals. What is the position of the Native Americans regarding how Kennewick Man should be treated? The position of the scientists? What is the legally defined relationship between Native Americans and scientists in determining how human remains and artifacts should be handled? What type of relationship should exist based on the tenets of anthropological and archaeological codes of ethics? What rights and concerns, if any, should the general public have in the matter?

Continued

❑ Note: Background information on Kennewick Man can be obtained from a number of sources. The U.S. Department of the Interior maintains a website with pertinent information: **www.cr.nps.gov/aad/kennewick/index.htm** Visit the website maintained by the Tri-City Herald, one of the major newspapers that has been covering the story from its beginning: **http://www.tri-cityherald.com/bones** The website maintained by Friends of America's Past also includes links to relevant and up-to-date material: **http://www.friendsofpast.org** Three articles are also pertinent:

- Chatters, James C. 1997. Encounter with an Ancestor. *Anthropology Newsletter,* January, 9–10.
- Chatters, James C. 2000. The Recovery and First Analysis of an Early Holocene Human Skeleton from Kennewick, Washington. *American Antiquity* 65(2):291–316.
- Slayman, Andrew. 1997. A Battle Over Bones. *Archaeology* 50(1):16–23.
- Read more about NAGPRA and ethical concerns in general (see Appendix 2) for your discussion of Kennewick Man.

❑ Note: Remember that when using Internet resources, addresses can change. If any of the addresses listed below fail to work, use a search engine and the key words in the website of interest.

DIG DEEPER

Introduction to Archaeology

❑ There are a number of great texts that provide introductions to archaeology, its linkage with anthropology and other disciplines, how archaeologists think and look at the past, and identify and solve problems related to ancient culture and society. My three favorites are listed below. David Hurst Thomas' book is the most fun of the three to read.

- Renfrew, Colin and Paul Bahn. 1996. *Archaeology: Theory, Methods, and Practice.* New York: Thames and Hudson.
- Sharer, Robert and Wendy Ashmore. 1993. *Archaeology: Discovering Our Past.* Mountain View, California: Mayfield Publishing.
- Thomas, David Hurst. 1998. *Archaeology.* Fort Worth, Texas: Harcourt Brace and Company.

❑ There are numerous archaeological sites with public access that you can visit to begin to get an idea of their type, variety and context. Guides to federal and state parks are helpful in locating sites to visit. Also consult tourist bureaus or chambers of commerce. In addition, there are a number of published guides that you might consult. Examples include:

- Folsom, Franklin and Mary Elting Folsom. 1993. *America's Ancient Treasures: A Guide to Archaeological Sites and Museums in the United States and Canada.* Albuquerque: University of New Mexico Press.
- McDonald, Jerry N. and Susan L. Woodward. 1987. *Indian Mounds of the Atlantic Coast: A Guide to Sites from Maine to Florida.* Newark, OH: McDonald and Woodward Publishing Company.
- Thomas, David Hurst. 1999. *Exploring Ancient Native America: An Archaeological Guide.* New York and London: Routledge.

Continued

- Wilson, Josleen. 1980. *The Passionate Amateur's Guide to Archaeology in the United States.* New York: Collier Books.
- Woodward, Susan L. and Jerry N. McDonald. 1986. *Indian Mounds of the Middle Ohio Valley: A Guide to Adena and Ohio Hopewell Sites.* Newark, OH: The McDonald and Woodward Publishing Company.

Ethical Concerns

❑ There are a variety of books and articles that present different perspectives on ethics in archaeology, the collecting and displaying of material culture, and interacting with Native peoples. The suggestions below will get you started and lead you to other literature of interest.

- Lynott, Mark. 1997. Ethical principles and archaeological practice: Development of an ethics policy. *American Antiquity* 62(4):589–599.
- Lynott, Mark J. and Alison Wylie (eds.). 1995 Ethics in American Archaeology: Challenges for the 1990s. Special Report. Washington, DC: Society for American Archaeology.

  These two references present the background and results of the Society for American Archaeology's recent review and updating of ethical principles for the profession.
- Vitelli, Karen D. 1996. *Archaeological Ethics.* Walnut Creek, CA: AltiMira Press.

❑ One of the more recent collections of position papers and case studies representing different points of view.

❑ There are numerous Internet sources on various ethical issues: the archaeological politics of private collecting, commercial treasure hunting, and professional archaeology: **http://wings.buffalo.edu/anthropology/Documents/ lootbib.html** Ethics and professional standards in anthropology: **http://www.aaanet.org** Click on "Ethics." Handbook on Ethical Issues in Anthropology: **http://www.ameranthassn.org/sp23.htm** Education and Government Affairs Section of the Society for American Archaeology (SAA) Website: **http://www.saa.org/ Education/PubEd/pubed.html** Links to Native American tribes that maintain websites can be found at **http://www 2.cr.nps.gov/tribal/other.htm**

- Rose, Jerome C., Thomas J. Green and Victoria Green. 1996. NAGPRA: Osteology and the repatriation of skeletons. *Annual Review of Anthropology,* 25:81–103.
- A well-balanced presentation considering the sensitivity of the subject for archaeologists and Native Americans. A collection of short articles representing diverse views can also be found in a special report of *Federal Archaeology, Volume 7, No. 3.* Washington, DC: National Park Service, Departmental Consulting Archaeologist and Archaeological Assistance Program.
- Internet sources include the National Archaeological Data Base, which includes the latest information and rules related to NAGPRA. **http://www.cast.uark.edu/other/nps/nadb/nadb_al.html**
- Repatriation of Skeletal Material and Cultural Patrimony: **http://www.uoknor.edu/aiq/aiq202.html**
- Swidler, Nina, Kurt E. Dongoske, and Roger Anyon (eds.). 1997. *Native Americans and Archaeologists: Stepping Stones to Common Ground.* Walnut Creek, CA: AltaMira Press.

❑ An important collection of position papers and case studies representing different points of view regarding the interaction of native peoples with archaeologists and the scientific community. An equally enlightening collection of short articles appears in a special, double issue of *Common Ground: Archaeology and Ethnography in the Public Interest, Volume 2, Number 3/4.* Washington, DC: National Park Service.

CHAPTER 2

The Archaeological Record and the Recognition of Evidence

What seest thou else In the dark backward and abysm of time?

—William Shakespeare, The Tempest—

HOW THE ARCHAEOLOGICAL RECORD IS FORMED

The ***archaeological record*** is shorthand for referring to material evidence, its context, and all of the processes responsible for their creation and present condition. These characteristics are embodied in archaeological deposits. The archaeological record is both formed and transformed by a variety of cultural and natural processes. Any human behavior that creates some type of material evidence helps to form the archaeological record. In turn, human behavior (cultural processes) can alter or transform the condition of pre-existing material evidence and its context. This is also true of natural processes. The same processes that create and transform the physical world in which we and our ancestors have lived can, and have, acted to form and preserve archaeological deposits, or are responsible for varying degrees of their alteration and destruction.

There are innumerable cultural and natural processes that could be listed as having a potential effect on the formation and transformation of the archaeological record. Our hypothetical example on the following page deals with only a few but illustrates the realities with which all archaeologists must cope. The archaeological record is subject to change. It rarely, if ever, retains the character it had when human behavior generated a material fingerprint.

The archaeological record is a contemporary phenomena. When we encounter an archaeological deposit in the field, it represents everything that has happened to it from the time of the original behavior that resulted in the creation of material evidence to the present. The nature of the archaeological record can be very different depending upon when we encounter it. If we discovered and investigated our hypothetical site prior to 1990, it would have been unplowed and revealed layering or stratigraphy that would have aided our determinations of the age of the deposit, and the reconstruction of the natural processes affecting the deposit and shaping the landscape through time. It would have contained recognizable deer bone, an unbroken hammerstone, a flake tool, and microflakes. Our ability to accurately interpret the site would be pretty good. By the end of 1999, there are fewer types of artifacts, a smaller number of artifacts, and the context of the finds has been dramatically altered. Interpretation of this evidence would lack the detail, clarity, and confidence of the pre-1990 effort. You may sometimes hear archaeologists referring to a well-preserved deposit representing a single occupation as a "snapshot" of time. This doesn't mean that time has somehow been frozen and that everything is just as it was in the past. Rather, it denotes a high ability to recognize and control for all of the processes responsible for the deposit's formation and transformation.

Imagine a scene that was undoubtedly enacted many times in ancient North America. A lone hunter stalks a deer grazing in an open field along a stream. The hunter hurls his spear and brings down the deer. Approaching the kill, he removes a chunk (core) of flint from his leather pouch. Searching the ground, he finds a fist-sized cobble deposited by the stream during a previous flood. He uses the cobble (hammerstone) to strike the core. Small, nearly unnoticeable chips of flint (microflakes) are detached from the core along with a large, sharp-edged flake. Discarding the hammerstone, he uses the large flake (flake tool) to dress out the deer, removing its entrails and isolating the meatier portions of the carcass. The flint core goes back into the pouch. During the butchering, the flake tool occasionally strikes bone, resulting in the detachment of a few microflakes from along the sharp edge being employed. Before the butchering is complete, the edge of the flake tool becomes dull. The hunter removes an antler tine from his pouch and uses the pointed end to pressure flakes off the stone tool, rejuvenating its dulled edge. Replacing the antler in his pouch, he completes the butchering and discards the flake tool. Securing the hindquarters of the deer together with a leather thong, he hefts them over his shoulder and heads back to the camp that he left earlier in the day.

At this moment in time, let's say 1000 years ago, material evidence of the hunting and butchering activity left "on-site" consists of some deer remains (soft tissue and a partial skeleton), a hammerstone, a flake tool, and microflakes resulting from the production, use, and maintenance of the flake tool. Human behavior (or cultural processes) produced a deposit of artifacts or material evidence. That night, a small pack of wolves find the remains of the deer and consume most of the remaining soft tissue and some of the bone (deposit transformed by a natural process). The following night, a foraging skunk passes what is left of the deer and stops to gnaw a portion of foreleg. The skunk consumes some skin and tendon and leaves gnaw marks on the bone (deposit transformed by natural process). Over the next several weeks of hot and humid weather, the remaining soft tissue of the deer decays and is consumed by insects, leaving only whatever bone is left (deposit transformed by natural process). The bone dries out and begins to crack over the ensuing months (deposit transformed by natural process).

Six months after the original kill, the nearby stream floods. It rises over its bank and creates a pool of sediment-laden water on the landscape where the deer bone, hammerstone, flake tool, and microflakes are located. The floodwater recedes, leaving behind 10 centimeters of silt that cover the artifacts (deposit transformed by natural process). Burial below the silt slows down the weathering of the deer bone. Time passes, the sediments weather, vegetation grows, dies, and decays on the existing surface. In the next 1000 years, two additional floods effect the landscape and add another 15 centimeters of sediment to the surface (deposit transformed by natural processes). The already slowed weathering of the deer bone is further retarded. By the year 1990, the deer bone, hammerstone, flake tool, and microflakes are buried about 30 centimeters below the existing surface of the land.

A modern farmer repeatedly plows, to a depth of one foot (about 30–35 centimeters), the field in which the archaeological deposit is buried. Plowing moves the artifacts up and down and from side to side in the ground. It brings the remaining deer bone, the hammerstone, and some of the microflakes to the surface. The hammerstone is struck by the plow and fragmented. Only one of the fragments bears the damage resulting from the cobble's original use as a hammer. By bringing the bone to the surface, plowing promotes its rapid mechanical, chemical, and biological decay. By 1995, the bone no longer exists and the only artifacts left in the deposit are the fragmented hammerstone, flake tool, and microflakes. Plowing also homogenizes the color, texture, and structure of the soils or sediments in which the artifacts originally occurred (i.e., their matrix). It blends and mixes the once visible artifact-bearing layer as well as those that formed on top of it (deposit transformed by cultural process).

The hammerstone fragments and some microflakes are churned back underground by repeated plowing, and the flake tool is brought to the surface. An artifact collector walking the field during the winter of 1999 sees the flake tool, recognizes it as an artifact, picks it up and takes it home (deposit transformed by cultural process). At the turn of the millennium, the archaeological deposit consists of fragments of the hammerstone and microflakes.

The archaeological record is patchy. Not all sites are subject to the same processes that act to preserve or destroy material evidence, so not all types of artifacts or material evidence are going to be consistently preserved. Artifacts based on organic materials are typically underrepresented. Our ability to reconstruct the physical environment in which people lived may be possible in some areas where plant and animal remains are preserved in some form, but not in other areas where preservation is poor. Not all cultural behavior may be encoded in some type of material evidence amenable to study by archaeologists. But it is premature to list behaviors that we will never be able to reconstruct or examine. Because of the variability in formation processes of the archaeological record, some types of behavior may be represented in one region and not another, or at one site and not another. The archaeological record is not a direct reflection of human behavior, as a variety of processes act on material evidence and its context following its creation. It is up to archaeologists to translate material evidence into behavior through objective interpretation of evidence.

As I noted in chapter 1, archaeology and the archaeological record have the ability to provide perspectives on the past that are not found in documents and that get around the bias of the persons who created documents. William Dancey (1981:17) says it best when he notes that:

> *It is possible to "falsify history" through the written record, but it would be a rare thing indeed if people intentionally set out to create an archaeological record that reflected behavior as they wanted it to be viewed by later generations. This fact is what makes archaeology such a potentially powerful tool for understanding the past, in both prehistoric and historic periods.*

The potential for artifacts and ecofacts to be preserved once they are a part of an archaeological deposit is conditioned by many variables that relate to mechanical, chemical, and biological weathering or breakdown. Variables include:

- ☑ what an object is made of, its hardness, density, permeability, moisture/water content, and chemistry.
- ☑ whether the object has been modified, and if so, how has modification affected the object's hardness, density, internal structure or chemistry, and permeability.
- ☑ the nature of the depositional environment, its type, chemistry, oxygen content, climate, moisture content, related plants, animals and organisms, and how all of these attributes fluctuate over the short-term.
- ☑ the nature of the long-term, postdepositional environment, its type, chemistry, oxygen content, climate, moisture content, related plants, animals and organisms, and how all of these attributes fluctuate.

Things fashioned from rock, many metals, and synthetic materials (e.g., alloys, ceramics, and glass) are obviously more durable than those involving organic materials. Remember though that with time even rock weathers and metals corrode and rust. Porous materials that can absorb water or already have a high water content will be very susceptible to mechanical destruction brought on by repeated cycles of drying and wetting, or freezing and thawing. These same attributes can also contribute to chemical and biological processes that result in the breakdown and decay of materials. Any modification to an object that inhibits its ability to absorb and lose moisture fosters preservation. For example, dense types of hardwoods will outlive more porous species. Wood that has been charred will outlast uncharred wood since the charring process drives off moisture and hardens the remaining wood. The soft tissue of an animal can disappear relatively quickly, while the bone will last longer. Of the bone, denser elements like teeth have a better chance of being preserved than other skeletal elements. Low-fired, porous types of pottery (e.g., terra cotta, earthenware) will fragment at a higher rate than more highly fired ceramics (e.g., stoneware, porcelain).

FIGURE 2.1:

Exposing whale bone at the Summer Bay Site in the Aleutian Islands, Alaska. The density of the bone and long-term trends in climate worked to preserve these organic remains.

The depositional environment of material evidence is equally critical. Where, how, and with what was something discarded, lost, or cached? Materials left at or near the surface are most susceptible to the effects of climate, the interaction of water, sunlight, soils, vegetation, and organisms, and the general behavior of animals. Water and other fluids are mediums that can promote mechanical breakdown as well as chemical and biological processes. The degree of acidity that a soil or sediment possesses (pH values of 1 to 6) can affect the rate of decay; acid environments foster decay while environments with a neutral or basic pH (values of 7 to 14) are not as detrimental and may actually aid in the preservation of artifacts. Depositional environments containing high percentages of salts or oils can act to preserve some types of organic artifacts.

Organic artifacts discarded in association with quantities of shell, a situation that characterizes large numbers of sites in coastal areas, have a much increased chance of being preserved. The weathering of the calcium carbonate-rich shell creates a chemical environment with high/basic pH values. Organic artifacts discarded in association with some metals can be preserved by the rust and oxidizing materials derived from them as they corrode.

Warm weather tends to promote chemical reactions and the destructive activities of organisms, insects and vegetation, cold weather less so (Figure 2.1). In conjunction with the acid soils that often occur, tropical environments are some of the most destructive of organic artifacts. The degree to which environmental conditions fluctuate over the short-term also has an effect. Daily and seasonal extremes in temperature and precipitation promote the mechanical breakdown of materials to a greater degree than conditions that remain relatively stable over the long-term. The consistently dry environments of many rockshelters and caves are well know to archaeologists for their potential to contain preserved organic artifacts. The same can be said for depositional environments that are consistently underwater, waterlogged, wet, or cold (Figure 2.2). The public is probably most aware of the dramatically preserved organic materials derived from these latter environments. Shipwrecks, the remarkably preserved bodies of the Bog People of Iron Age Europe, the Iceman from the European Alps, Peruvian mummies from the

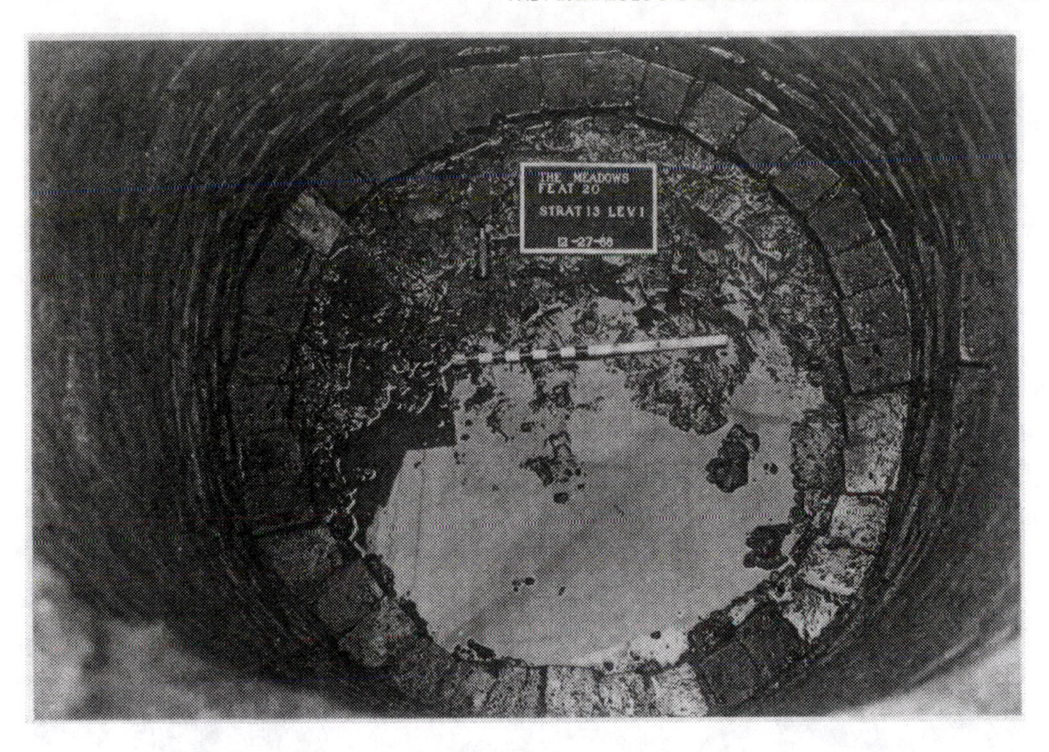

FIGURE 2.2:

Abandoned wells often were convenient places to discard trash during historic times. The below-ground context, and the moist and occasionally waterlogged depositional environment of the well, aids in the preservation of organic artifacts. This example dates from the 1860s–1870s and was associated with working class households in Philadelphia (Lee Decker et al 1993).

heights of the Andes mountains, and the innumerable examples of animals found frozen in glacial ice have captured the popular imagination in recent years and provided important insights into the past.

The potential for artifact preservation, especially of organic artifacts, is enhanced by rapid or intentional burial, or the removal of materials from exposed surface contexts where weathering can proceed fairly quickly. Butchered animal bones that get discarded in a pit that is backfilled will long outlive similar bone left at the surface. Artifacts deposited upon a surface that is soon buried will be better preserved than those on surfaces that remain relatively stable for long periods of time. Rockshelters, caves, and the interiors of structures provide environments in which artifacts are sheltered from many of the weathering processes that take place in the open. The implications for organic artifacts are clear, but differences in depositional environments also affect more durable materials. The refitted stone tool shown in Figure 2.3 is made from metarhyolite, a material that weathers more rapidly than many other types of rock used by prehistoric artisans. The upper, more darkly colored half of the tool was discarded (and subsequently excavated) from the interior of a rockshelter. The more weathered, lighter-colored half of the artifact was discarded and found in the open area in front of the rockshelter.

The nature of a depositional environment can change over time and influence the preservation of archaeological evidence. Natural and cultural processes can bury the original deposit more deeply and remove it further from the weathering processes active near surfaces, or they can expose once buried deposits. The weight and compaction of the deposits above the matrix in which artifacts occur may foster mechanical stress and alter drainage characteristics. Changes in climate and concomitant changes in drainage, plant and animal communities may occur.

How does an understanding of formation processes influence the ways that we search for and investigate archaeological deposits in the field? Since formation processes can be so variable from place to place, we must not expect to be able to use the same strategies to investigate every archaeological site. Some sites will require more attention and care than others because of the nature of artifact preservation or the integrity of their contexts. Developing a basic understanding of natural processes and attention to environmental reconstruction is a first step that can be taken towards recognizing what has happened to an archaeological deposit,

The Windover Site, 8BR246 (Doran and Dickel 1988; Doran et al. 1990), located in Florida, is a great example of how a variety of the factors discussed above contribute to the preservation of human remains and organic artifacts. At the 7000 ±year-old site are burials that include over 168 individuals and plant fiber textiles representing garments, bags, and matting, bone and antler implements, floral and faunal remains. The burials were emplaced in a wooded marsh, below water, sometimes being staked down to hold them in place. Most individuals appeared to have been buried within 48 hours of death, so rapid burial in a consistently wet environment would have aided preservation. When discovered and excavated, the burials were in a peat deposit more than two meters below the bottom of a shallow pond. The anaerobic (oxygen-poor) environment and the neutral pH of the water and peat worked together to foster long-term preservation.

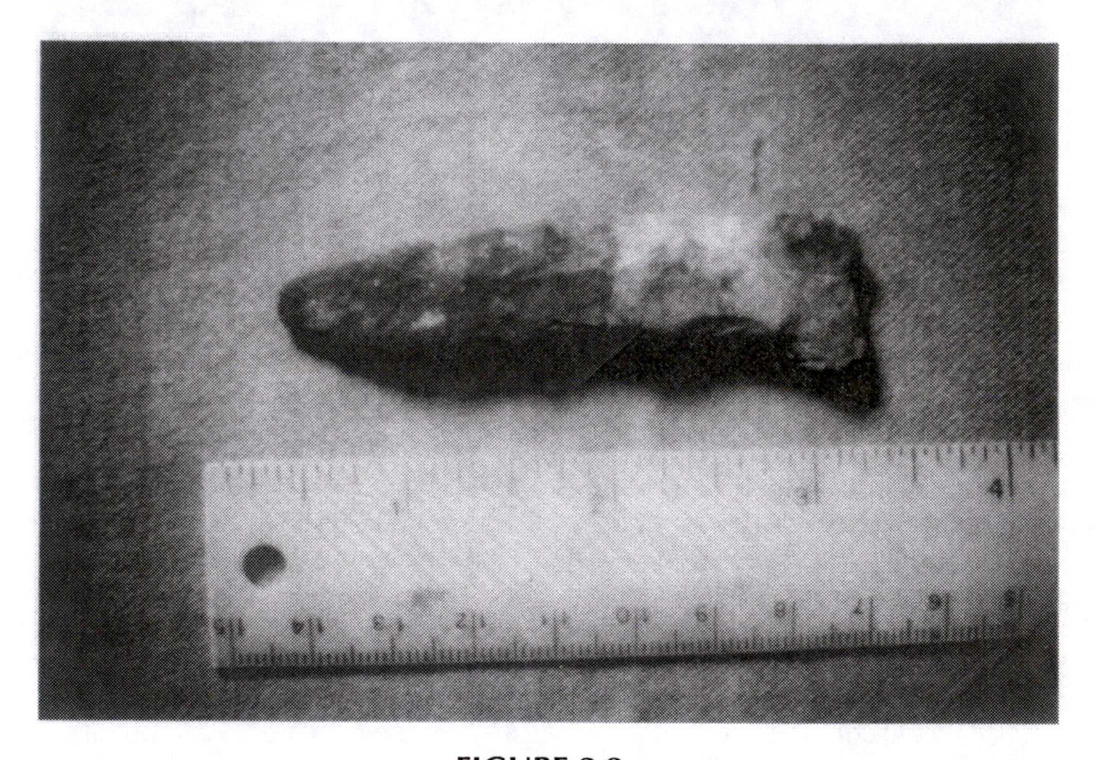

FIGURE 2.3:

Refitted pieces of a broken stone tool from a rockshelter in the Blue Ridge Mountains of Maryland. The upper half of the artifact was found in excavations inside of the rockshelter while the lower half occurred in excavations in the open area in front of the rockshelter.

evaluating the degree to which it has been disturbed, and devising a field strategy appropriate for its investigation. Processes of interest include:

- ☑ biological activity (the life cycle of vegetation, behavior of animals and organisms),
- ☑ chemical reactions,
- ☑ climate,
- ☑ fluvial processes (those involving surface water),
- ☑ colluvial processes (those involving slopes and the action of gravity),
- ☑ aeolian processes (those involving the wind),
- ☑ glaciation,
- ☑ volcanism,
- ☑ earthquakes and faulting,
- ☑ and fluctuations in sea level.

All around you, every day, are examples of how modern life impacts the landscape. As an archaeologist you should become attuned to these activities and add them to your catalog of cultural processes that have the potential to affect archaeological deposits. As you read history, anthropological accounts of living cultures (***ethnography***), archaeological perspectives on living cultures (***ethnoarchaeology***), and archaeological reconstructions of cultures, you will get an appreciation of the range of behaviors that humans are capable of and the type of material evidence that they might generate. The ongoing analysis of archaeological evidence feeds back into an understanding of deposit formation and transformation.

Things fall apart, decay, and even vanish from archaeological view. Most of the artifacts that you will find in the field will be broken or damaged in some way. Only vague physical traces of organic artifacts may be encountered. It will help if you are already familiar with what whole objects look like, the technology involved in their manufacture, how they are used, how they might be expected to break, and the debris these processes create. ***Taphonomy***, the study of the processes that affect organic remains after their death, can be expanded to include considerations of the processes that affect all manner of material things following their discard. So we can talk about the taphonomy of bone as well as the taphonomy of an abandoned house or discarded knife. In this more global sense, taphonomy is synonymous with transformation processes as the term has been used in our discussion of the archaeological record.

There are procedures that can be used in the systematic search for hard-to-see artifacts and ecofacts. ***Flotation*** is a method of using water or other fluids in combination with fine-meshed screen to separate small fragments of small (micro) artifacts and plant and animal remains from excavated matrix. It was developed specifically to cope with these degraded, easy-to-miss types of evidence. In some cases it may be necessary to systematically evaluate the chemistry of excavated matrix in order to recognize chemical signatures of organic evidence that has completely decayed (see chapter 9 for a discussion of these and other techniques used for the recovery of plant remains).

Field technicians need an elementary understanding of conservation techniques in order to be prepared to stabilize artifacts found in a delicate state of preservation (see chapter 9). Excavating and removing artifacts from their matrix changes their depositional environment. Exposure can cause them to lose or absorb moisture rapidly, initiating shrinking or swelling and mechanical stress. Any type of artifact made from porous material is susceptible, although organic materials tend to suffer the most. I know excavators who have literally watched newly exposed bone disintegrate before their eyes as it rapidly lost the moisture that it had retained for so long. Underwater archaeologists probably have the best appreciation of these matters since so much of what they recover requires conservation because of the effects of overlong immersion in water and salts.

RECOGNIZING ARCHAEOLOGICAL EVIDENCE

The definition of what constitutes an artifact or material evidence is fairly straightforward. But how do you recognize this evidence in the field? How do you answer the question, Is this an artifact? Don't confuse this with the closely related question of **What** is it? These are different issues. Recognizing something as the byproduct of human action is the first step. You will encounter things that you know are artifacts or features, but not be able to readily explain what they are, what they were used for, or what they meant to the people making and using them. Many of us study the archaeology of cultures alien or different from our own. There is no reason why we should assume that the material evidence created by these cultures will automatically be recognizable to us.

Archaeologists have a framework for recognizing evidence. It is constructed from:

- ☑ analogs with things in our own culture,
- ☑ documentary and pictorial records,
- ☑ knowledge of ethnography,
- ☑ ethnohistory,
- ☑ ethnoarchaeology,
- ☑ the cumulative achievements of archaeology,
- ☑ experimental archaeology,
- ☑ the natural and material sciences, and
- ☑ evaluation of context and associations.

There are things that you will instantly recognize as artifacts because they have analogs in objects still found today (Figure 2.4). But the farther back in time that we go, the fewer the opportunities to employ such analogies. Care must be taken not to extend the analogy of form/shape to function, or other inferences that go beyond that of recognizing something as an artifact. Shared form does not have to equate with the use of similar raw materials, manufacturing processes, use, symbolic or ideological meaning, or any social relations related to these things. And since artifacts are but a first step in developing objective reconstructions of the past and learning about social relations, we don't want to blur our archaeological vision with strict, all-encompassing analogies.

FIGURE 2.4:

The objects in this photo span over 300 years of time. The youngest in age is at the top of the picture and is probably the most familiar. The one below it on the left dates from the 1930s and is ceramic, as are the next three, dating from the 19th and 18th centuries respectively. The oldest, at the bottom, is Native American and fashioned from catlinite or pipestone. All are recognizable as artifacts called pipes. Analogies in the form of objects is one way to recognize artifacts, and in some cases provide information about their function.

The ethnographies of living peoples that anthropologists and others have produced since the 19th century provide a broad look at the tremendous diversity of lifestyles, technologies, material culture, and belief systems created by humans worldwide. They encompass studies of hunter-gatherers, herders, fishing peoples, farmers, and societies organized in bands, tribes, chiefdoms, and states. Many ethnographies of Native American peoples were written as the United States government and non-native settlers expanded across the country. Earlier accounts were generated by explorers, traders, missionaries, and early colonists. Scholars have used these accounts, other documentary evidence such as maps and deeds, and oral histories to create ***ethnohistory***. An ethnohistory attempts to recreate the life of a people, but unlike the anthropologist doing ethnography, the ethnohistorian cannot directly observe or question the people being studied. Ethnography, ethnohistory, history in general, and documentary and pictorial records will convey to the student of archaeology a sense of the things that people do and create. Studies relevant to the region where you are working will of course be the most helpful.

Archaeologists attempting to use ethnographies as a source of information for modeling how human behavior translates into some form of material evidence were often frustrated. Descriptions of material culture were not always included in an ethnography, or were not detailed enough, or were not clearly linked with a specific type of behavior or activity. To remedy this, some archaeologists began doing ethnographic fieldwork with an archaeological bent. They would live with and observe a people, but with an eye focused on how behavior results in the creation and transformation of material evidence. Called ***ethnoarchaeology***, the intent is not simply to see how people make and use things with which we may not be familiar, but to better understand how different cultures perceive, organize, and transform space; how, where, and why objects are discarded; and how subsequent activities of people may affect pre-existing material evidence.

There is a fairly vast archaeological literature that describes, illustrates and interprets native and non-native material culture. Of course, evaluations of context and associations have played a role in deriving these interpretations. Becoming familiar with the work that has been completed in your area of interest will aid in your ability to recognize artifacts and other material evidence in the field. There are a variety of general works that illustrate the technologies involved in the production of different types of artifacts (sample references are listed at the end of the chapter). Archaeological deposits are littered with debris generated by the procurement of raw materials, the manufacturing, use, and maintenance of artifacts and features. Acquaint yourself with these byproducts (and the technologies that create them) because they far outnumber the finished items that you will encounter in the field. Displaying and interpreting material culture is the bread and butter of museums and is an easy way for you to begin to reap the benefits of existing archaeological research.

One of the techniques used to better understand the archaeological record is the design and completion of replicative experiments. The scope of such experiments is wide ranging and includes the manufacturing and use of artifacts, the observation of the processes affecting artifacts and features once they are abandoned or discarded, and the simulation of archaeological deposits and contexts for the purpose of examining their transformation over time. One of the results of experimentation is that it expands existing ideas about the types and condition of material evidence and contexts that we might expect to find in the field. As I've already noted, the debris created as a result of various manufacturing processes especially is worthy of our attention, and is less likely to have a form or shape recognizable to the uninformed.

From the natural and material sciences archaeologists gain background about the makeup of the natural world, its operative processes, and the things that these processes are capable of producing. How do natural products differ from what humans create using natural resources? How can we tell if a rock has been shaped by humans or simply modified by some natural process? Is the landscape we are examining a natural one or has it been shaped in some way by humans? As simple as it may sound, a lot of archaeological evidence can be recognized in the field as a result of answering 'no' to the question, Is it natural? The question can be applied to landscapes, features of the landscape, rocks and minerals, plants and animals. It should be apparent that having background in the subjects that I've summarized above would be helpful, but detailed knowledge of the natural world is essential. The examples below illustrate my point.

All of the rocks in Figure 2.5 have had their exterior surfaces worn smooth by being tumbled for a long period of time in a stream. All have been subsequently used as hammerstones in the fashioning of chunks of rock into tools. This left distinctive damage on the ends of the cobbles that wore through their stream-smoothed exteriors. The isolated location of this damage distinguishes them from cobbles that might have experienced accidental battering in nature. The accidental battering would have damaged a greater number of areas on the cobbles.

Is the stream-rolled cobble shown in Figure 2.6 an artifact? It has no visible damage, use wear, or signs of intentional modification. There are no residues on its surfaces. Is it an artifact? The correct answer would be,

FIGURE 2.5:

Stream-rolled cobbles that have been used as hammerstones in the manufacturing of stone tools.

FIGURE 2.6:

A stream-rolled cobble exhibiting no damage or intentional modification. The context in which an object is found can lead to its identification as an artifact.

'it depends on where it was found.' Found in a streambed or along the banks of a river with hundreds of similar objects, the cobble is easily interpreted as a product of nature that has never been made, used, or altered by a human. It makes sense in the context in which it occurs. What do we say about this selfsame cobble when found on a mountaintop where geologists and environmental scientists tell us no stream, river, or glacier has ever flowed? In this context it is identifiable as an artifact. Its presence on the mountaintop is the product of human activity by default. We may never be able to say what the cobble was to be used for, why it was transported to the mountaintop, or what it may have meant to the person who transported it. But our understanding of the natural world and context makes it recognizable as an artifact.

Without a prior understanding of regional geology you might not recognize all of the rocks depicted in Figure 2.7 as artifacts. The rock is a Miocene sandstone that weathers at a dramatic rate once freshly broken or worked and left exposed at the surface. Area archaeologists often ignored the material in their work until they began finding artifacts with easily recognizable shapes made from it, and realized what they had been missing.

Vegetation can be considered as an artifact in some cases, and provides evidence suggesting the presence of archaeological features. Old-growth cedar trees along the Northwest Coast still show scars from where Native Americans systematically removed wood from them for planking generations ago (e.g., Stewart 1996). Historic archaeologists are wary of ornamental plants, shrubs, or trees growing in unnatural settings or contexts that may be near former house sites, linear or patterned arrangements of old trees hinting at road alignments, property or field boundaries (e.g., Schlereth 1980: 147–159). The spruce trees seen in the foreground and background of Figure 2.8 are out of place in this upland environment that typically supports a temperate deciduous forest. In this same cluster are apple and pear trees. All are **out** of context in an environment **not** altered by humans. The trees flank the location where a structure belonging to a 19th-century farmstead once stood.

Patterned growth in vegetation or crops (***crop marks***) can also provide clues about what types of archaeological features may exist below ground. Below-ground features influence vegetation at the surface because they have an impact on drainage characteristics and soil chemistry. Things like backfilled pits, cellar holes, or

FIGURE 2.7:

Highly weathered prehistoric artifacts (top row) of Miocene sandstone from the Eastern Shore of Maryland, and freshly broken pieces of the same material (bottom row).

FIGURE 2.8:

The spruce and fruit trees in this photograph are not species typically found in the forests that develop in this type of upland environment. They are intentional plantings flanking the former location of a 19th-century structure.

buried foundations might hold water longer than the surrounding undisturbed soils, and thus foster thicker or lusher vegetation at the surface (e.g., Figure 2.9). Or these features might retain less moisture than the surrounding matrix and be nutrient-poor, so the vegetation growing over them would be thin or stunted in comparison with surrounding vegetation. In either case, the result is patterned growth that appears unnatural to the eye. In some cases, patterned vegetation indicative of subsurface features can be seen from the ground. A lush stand of ferns growing over a backfilled well is easily spotted. Examining landscapes from the air or evaluating aerial photographs provides the best results when large features are involved, like structures or entire settlements.

Archaeological features may be recognizable as anomalies in soil color or texture at the surface, as well as below ground in excavations. The dark stain in which the person is standing in Figure 2.10 is the former location of a dump or midden associated with a late prehistoric Indian farming village, 18WA23. The dark color of the sediments relates to the high degree of organic material that was discarded and decayed here when the village was in existence. The stain corresponds with the highest densities of pottery and bone fragments found on-site. The tops of backfilled pits, or any archaeological feature that originally involved a subsurface disturbance, can sometimes be visible as soil anomalies at an existing surface. This is most often the case in plowed or cultivated fields where subsurface deposits are periodically churned to the surface. Features recognizable as soil anomalies are usually encountered in excavations.

Natural processes layer sediments and soils in specific ways (Figure 2.11a). Sediments and the soils that develop from them take on a number of characteristics as time passes. They become more compacted and can develop structure (aggregates of soil particles, see chapter 7). Humans do all sorts of things to disturb the natural status quo, like digging pits for the storage of goods, the disposal of trash, the burial of the dead, the construction of foundations, wells, and drainage facilities. This digging interrupts or breaks the natural layering or stratigraphy in the ground (Figure 2.11b). It can also alter the degree of compaction or structure of the soil removed from the pit, and mix sediments that were once part of discrete layers with different colors and textures.

The backfilling of a pit, whether intentional or the result of natural processes, contributes to the mixing of sediments and layering that is different from the undisturbed stratigraphy surrounding the pit (Figure 2.11c). Even if someone tried to put the excavated sediments

FIGURE 2.9:

The dark-colored rings (20–25 feet in diameter) visible in the grass in front of the cornfield probably mark the location of wigwam-like structures at the Miley Site in the Shenandoah Valley of Virginia. Excavations in the nearby cornfield revealed rings of postmolds describing structures of the same size (MacCord and Rodgers 1966). The location of the former walls of these houses have an effect on the ability of the soil to retain moisture. The dark-colored rings are where grass was more well-watered, and thus greener/darker than the surrounding growth.

FIGURE 2.10:

The dark-colored stain in which the person is standing is the former location of a dump or midden associated with a late prehistoric Indian farming village along Antietam Creek in western Maryland. Sediments, too, can be artifacts if they are modified or altered by humans.

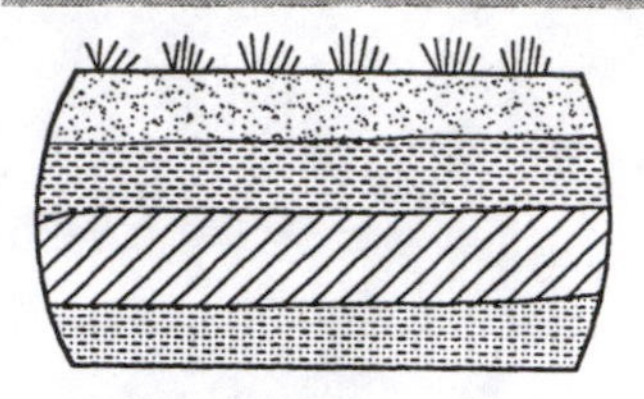

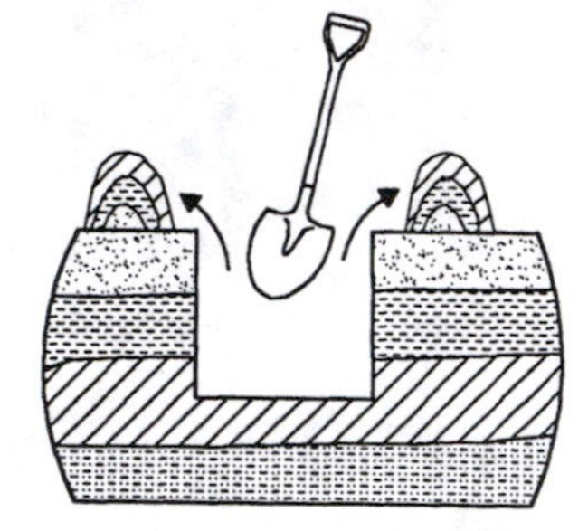

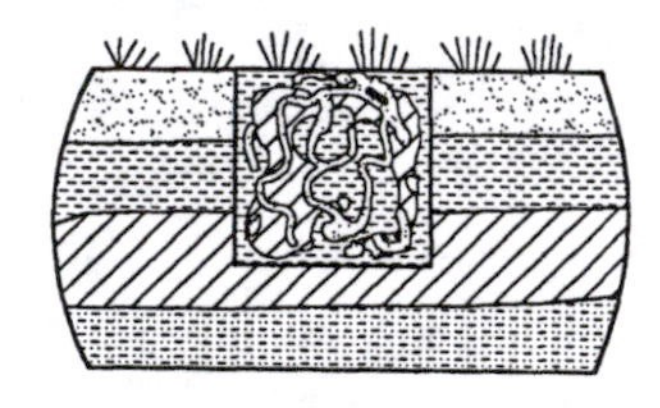

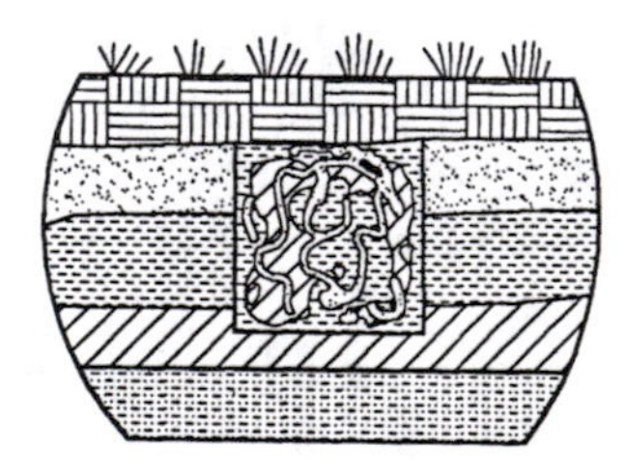

FIGURE 2.11:

The creation of a pit feature recognizable as a soil anomaly.

back into the ground in the same order in which they were naturally arranged, it would be difficult to replicate the degree of compaction or structure of the original, undisturbed strata. Backfilling can also incorporate new sediments and artifacts, differentiating pit fill even further from the adjacent, undisturbed stratigraphy. As time passes, the top of the backfilled pit can be obscured by other deposits and soil development (Figure 2.11d).

An archaeological excavation adjacent to our hypothetical pit would begin by removing the first visible layer or stratum (Figure 2.12a). At the base of this first

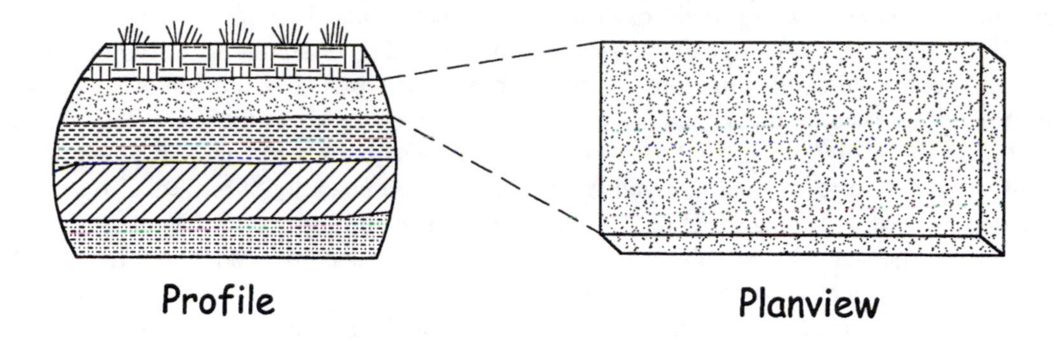

a. Undisturbed natural sediments seen in excavation after The removal of the surface layer.

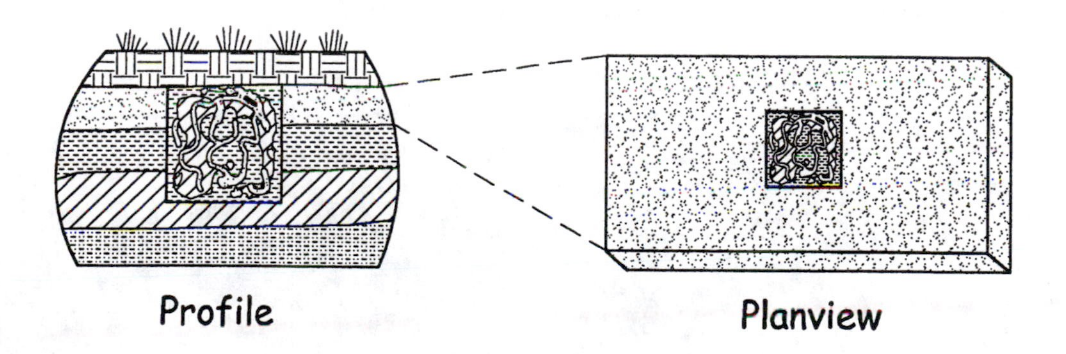

b. Top of pit feature revealed as a soil anomaly after the removal of the surface layer

FIGURE 2.12:

Recognizing a pit feature in an excavation unit.

excavation level we would see the consistent color and texture of the undisturbed subsoil. An excavation unit opened over the location of the pit feature would encounter a different situation (Figure 2.12b). With the first layer removed, the floor of the unit would show a square pattern of sediments with a color and texture distinctive from that of the surrounding matrix. The top of the backfilled pit is revealed.

The recognition of pit features or other excavations can be complicated by the character of the natural stratigraphy. Pits dug into single sedimentary layers, or through layers that are similar in color, texture, and structure may not result in a colored anomaly, visible because of its contrast with the surrounding matrix. These features are discernible, however, as textural or structural anomalies. Also, be aware that a variety of noncultural processes can create and later fill holes in the ground, or create subsurface disturbances. Again, knowledge of the natural world and your archaeological skills will help you distinguish between natural and cultural phenomena.

The stepped "mound" in Figure 2.13 will be recognized as an "artifact" or "feature" by many of you. You may even have contributed to its construction, or one like it. It is a landfill of succesive layers of garbage and sediment fill, and represents a constructed type of landscape. It is identifiable as an artifact of human activity because it is unnatural in the context of the natural landscape or topography. Although of modern vintage, this example underscores the importance of knowing natural trends in topography and what causes them.

When I first saw the "mound" pictured in Figure 2.14 I was so excited that I almost drove my car into a ditch. I was doing field work and gathering data to write an Indian prehistory of a large valley system in western Maryland. The mound stood out as unnatural in the context of the local topography. For months I had pored over topographic and geologic maps and spent countless hours driving and hiking around the valley to learn about the natural environment. I had a great sense of what should be natural. My documentary research had turned up references to burial mounds that had once been in the area but had been destroyed by farmers and curiosity seekers over the years. The couple in the nearby farmhouse fed my excitement when they told me that, yes, everyone called it an Indian burial mound, but no one had ever dug into it. What a find! Even the project geologist was convinced it was the real thing when he first saw it.

Soon though, our knowledge of the natural world changed our opinion about the mound. Upon closer inspection we noticed portions of exposed bedrock on one side of the mound and auger borings into its top revealed a natural layering of sediments, not the unnatural stratification that one would expect from humans mounding earth a basketload at a time. Further research showed the mound to be an unusual geological feature, an erosional remnant of the surrounding sandstone and shale bedrock. I later discovered other examples of the same phenomenon during my field work. This doesn't denigrate the usefulness of understanding the natural world in archaeological field work, it dramatically underscores it.

FIGURE 2.13:

The stepped mound seen here is a part of a sanitary landfill. It is a modern example of a human-built landscape/artifact/feature.

FIGURE 2.14:

The feature shown in this photo was originally thought to be an Indian mound because of its unnatural appearance in the context of the surrounding landscape. Soil and geologic analysis indicated that it is actually an unusual natural, not cultural feature.

IMPROVING YOUR SKILLS

Developing the skills useful in the recognition of archaeological evidence is a career-long endeavor. Being a voracious reader and a consumer of ethnographies and archaeological texts is a tremendous help. There are innumerable television programs, videos, and films dealing with ethnographic and archaeological subjects. Although the interpretive quality of these productions varies greatly, they are nonetheless full of images of material culture, past and present. Visit museums.

In many areas of the country there is always some type of cultural resource management study in progress. Contact your State Historic Preservation Office to learn about ones taking place in your area. These projects often make provisions for public tours and displays, and occasionally volunteer opportunities. Even spending a single day as a volunteer on an archaeological site (see chapter 10) will enhance your understanding of artifacts and features.

Those majoring in anthropology and archaeology in college will have access to a variety of relevant courses. Many schools now offer course work in sediments, soils, and geomorphology (landscape evolution) designed for the archaeologist. If you haven't already taken an introductory course in physical geology, physical geography, or earth science, do so.

If you don't already, spend more time outside. Buy field guides for the trees, plants, and animals for the region in which you live and spend time learning their names and how to recognize them. Which plants prefer wet habitats, which prefer dry? Are all of the species in your area native ones, or have some been introduced? Think about why the character of a forest or field changes over space.

Buy the 7.5 minute series topographic map and a geologic map for an area that you frequent. Take them with you on walks and refer to them constantly. This will give you a better idea of how mapped data translate into three-dimensional landscapes and other features. Areas of steep topography, the banks and channels of streams, may provide you with a view of rock deposits. Chapters 4 and 6 provide background on these types of maps and their uses.

APPLYING YOUR KNOWLEDGE

1. If you are taking an archaeology course or participating in a field school, ask your instructor to arrange a trip to your state museum or State Historic Preservation Office to examine the artifact collections stored there. Your instructor may also know artifact collectors who would be willing to have their collections examined.

2. Try your hand at experimentally replicating one or more of the technologies that you will likely encounter in the field (e.g., stone boiling, flint knapping, pottery construction). Don't expect instant success. The greatest value of this exercise is to acquaint you with production processes, the debris it can create and how it differs from things produced in nature. Remember that the waste byproducts/artifacts involved in production are more numerous and better represented in archaeological deposits than the finished end products.

3. Spend a day cataloging activities that you observe around you that have the potential to transform the landscape, and therefore, an archaeological deposit. What kind of activities are the most widespread? Which are the most destructive?

4. Visit a park or other open space available to the public. Tour the area on foot and try to identify landscapes, plants, and animals there that don't seem to be a part of the natural environment. Explain your reasoning for why you think these things are out of context.

Continued

5. Select and read an ethnography dealing with a North or South American culture. The purpose of the exercise is to describe what an archaeological site representing a settlement of your culture might look like. What are the natural processes that might have resulted in the burial or nonburial of artifacts and features? What types of artifacts and features are on-site? How are they distributed? How do they reflect or not reflect the activities and behaviors that once occurred on-site? To what degree do the deposits at your site reflect other portions of the culture's territory and the activities performed at other types of settlements? Where did the materials from which the artifacts are manufactured originate in your culture's world?

6. There are a variety of archaeological sites on state and federal lands that are open to the public that you might visit to get a feel for their setting and variety. The National Park Service and state departments of parks and recreation are sources of information on such locations. There are also publications that provide information on many of the more well-known places open to the public. I've listed a few of these below. Each is mindful of the fact that sites are fragile resources and need to be treated with care.

DIG DEEPER

Touring Archaeological Sites

- Wilson, Josleen. 1980. *The Passionate Amateur's Guide to Archaeology in the United States.* New York: Collier Books.
- McDonald, Jerry N. and Susan L. Woodward. 1987. *Indian Mounds of the Atlantic Coast: A Guide to Sites from Maine to Florida.* Newark, OH: McDonald and Woodward Publishing Company.
- Woodward, Susan L. and Jerry N. McDonald. 1986. *Indian Mounds of the Middle Ohio Valley: A Guide to Adena and Ohio Hopewell Sites.* Newark, OH: McDonald and Woodward Publishing Company.

Formation Processes

- Schiffer, Michael B. 1996. *Formation Processes of the Archaeological Record.* Salt Lake City: University of Utah Press. This is the single best place to start your study of formation processes. Read Schiffer then move on to the other works noted below and those listed in their bibliographies. An earlier version of Schiffer's text (1987) was published by the University of New Mexico Press.
- Goldberg, Paul, David T. Nash, and Michael D. Petraglia (eds.). 1993. *Formation Processes in Archaeological Context.* Monographs in World Prehistory No. 17. Madison, WI: Prehistory Press.
- Micozzi, Marc S. 1991. *Postmortem Change in Human and Animal Remains: A Systematic Approach.* Charles C. Thomas, Springfield, IL.

❑ For more background on the role of chemistry, archaeology, and artifact preservation check out the following references.

- Lambert, Joseph B. 1997. *Traces of the Past: Unraveling the Secrets of Archaeology Through Chemistry.* Reading, MA: Addison-Wesley.

Continued

❑ There are a variety of texts that deal with natural processes and their implications for archaeological research. For a good introduction I recommend the following reference.

- Waters, Michael R. 1992. *Principles of Geoarchaeology: A North American Perspective.* Tucson: University of Arizona Press.

❑ If you are looking for schools that offer specialized course work relating archaeology to geology, geomorphology, or soil science (collectively referred to as geoarchaeology), check out the *Guide to Anthropology Departments* published by the American Anthropological Society, Washington D.C. The guide lists faculty and their interests, which will give you a hint about the focus of the department. You can then look for websites for appropriate schools where course listings and descriptions may be online. The Archaeometry Laboratory at the University of Minnesota (Duluth) publishes a directory of graduate programs in archaeological geology and geoarchaeology. Remember that in many cases you can take college courses without being enrolled in a degree program. Contact the Continuing Education department of the school in which you are interested.

Artifacts and Related Technologies

❑ The following references provide textual and pictorial introductions to different types of material culture and the processes underlying them.

- Andrefsky, William Jr. 1998. *Lithics: Macroscopic Approaches to Analysis.* Cambridge, UK: Cambridge University Press.
- Hodges, Henry. 1971. *Artifacts: An Introduction to Early Materials and Technology* (4th ed). London: John Baker.
- *Miller, George L., Olive Jones, Lester A. Ross, and Terecita Majewski (compilers). 1991. Approaches to Material Culture Research for Historical Archaeologists.* California, Pennsylvania. Society for Historical Archaeology.
- Noel-Hume, Ivor. 1974. *A Guide to Artifacts of Colonial America.* New York: Alfred A. Knopf.
- Rice, Prudence M. 1987. *Pottery Analysis: A Sourcebook.* Chicago: University of Chicago Press.
- Rye, Owen S. 1981. *Pottery Technology.* Taraxacum Manuals in Archaeology 4. Washington, D.C.: Taraxacum.
- Sutton, Mark Q. and Brooke S. Arkush. 1998. *Archaeological Laboratory Methods: An Introduction* (2nd ed). Dubuque, IA: Kendall/Hunt Publishing Company.
- Turnbaugh, Sarah Peabody and William A. Turnbaugh. 1997. *Indian Baskets.* Atglen, PA: Schiffer Publishing Ltd.
- Wendrich, Willemina. 1991. *Who Is Afraid of Basketry: A Guide to Recording Basketry and Cordage for Archaeologists and Ethnographers.* City the Netherlands: Centre for Non-Western Studies, Leiden University.
- Whittaker, John C. 1994. *Flintknapping: Making and Understanding Stone Tools.* Austin: University of Texas Press.
- Wisseman, Sarah U. and Wendell S. Williams. 1993. *Ancient Technologies and Archaeological Materials.* Langhorne, PA: Gordon and Breach Science Publishers.

CHAPTER 3

Fieldwork: Motivations and Design

Conceptions without experience are void;
experience without conceptions is blind.

Albert Einstein

WHY DO FIELDWORK?

The practical reality of archaeological research enables fieldwork to be broken down into three general types of endeavors. The logical place to start is the discovery or location of archaeological sites or deposits, a process referred to as an ***archaeological survey, site survey,*** or ***reconnaissance.*** Once discovered, the focus of fieldwork shifts to ***testing*** the site, learning more about its physical attributes and assessing its potential to contribute to ongoing research. These first two steps may be combined in a single field effort, or may be sequentially staged. ***Intensive investigations*** of a locality represent the most focused research endeavor, employing information derived from previous work to design a strategy that will realize the value of a site to enhance our understanding of the past. Not all site investigations are taken to this level.

All fieldwork should contribute in some way to our knowledge of the past because of the finite nature of the archaeological record and the ethical principle of stewardship. This is made possible by the use of research designs and the collection of, minimally, data related to the horizontal and vertical dimensions of an archaeological deposit, the range of material culture included in the deposit, any internal patterning, deposit age, formation processes, and deposit integrity.

RESEARCH

Fieldwork can be initiated solely to collect information related to a question, problem, or issue of importance in contemporary archaeology, ranging from the methodological to the theoretical. Knowing what's hot and what's not entails attending professional meetings, interacting with colleagues, and doing a lot of reading. Agencies, foundations, or any organization that provides funding for archaeology have an impact on the types of problems or research questions that receive attention. There are, however, lots of unfunded, small-scale research projects in progress that cover a broad range of topics with potential local, state, and national significance.

TRAINING

Training students should never be the only motivation for carrying out fieldwork. Numerous field schools in archaeology are offered through colleges and universities on an annual basis (see chapter 10) and are (or should be) integrated into research programs. This integration is the ethical responsibility of any professional providing training experiences, and it's good teaching because it shows students in the most direct way possible the linkage between theory, ideas, and practice.

ACCIDENTAL DISCOVERIES

A lot of fieldwork follows on the heels of accidental discoveries or finds made by the public, artifact collectors, and amateur archaeologists. In these cases you are not heading out into the field with a formal research design. But the perceived significance of a find and the level of effort that you are willing to expend in a follow-up investigation reflect your understanding of what has been found, and how it might enhance existing frameworks of history and prehistory. Of course, some unplanned finds are the seeds from which large projects grow, and any concerted field effort needs to have a research design.

Cultural Resource Management Studies

The federal and state governments, as well as many local municipalities, have recognized that the past and cultural heritage have value, and have taken the responsibility to manage cultural resources in the public interest. Without question, investigations carried out in compliance with local, state and federal laws—***cultural resource management studies***—account for the lion's share of archaeological field projects in this country, as well as jobs, committed personnel and monies devoted to archaeology (Green and Doershuk 1998:122–123; King 1987). The development of a professional and national consciousness regarding historic preservation dates from the 19th century in this country. Federal laws and policy enacted since the end of World War II are most important for an understanding of current practices (cf. Fowler 1982, 1986; King 1998). Appendix 2 provides a brief summary of the more important of these. They condition the practice of archaeology in the field and beyond.

It's not my intention to provide an in-depth look at the intricacies of cultural resource management (CRM); there are numerous articles and books that do so. But to be unaware of the basic framework of cultural resource management places the person interested in archaeological fieldwork at a disadvantage. Too much research is driven by compliance with the law for this aspect of the discipline to be ignored by any student or practitioner. The summary here conveys an outline of the way that the system works. The impact of cultural resource management on the design and implementation of fieldwork is woven throughout the remaining chapters of this book.

Archaeological resources comprise only a part of what falls under the umbrella of cultural resources, as King (1998:9) elegantly notes:

> *. . . the corpus of "cultural resources", is a big, complex, intricate mosaic of things and institutions and values, beliefs and perceptions, customs and traditions, symbols and social structures. And it's integral to what makes people and communities communities, so it's charged with a great deal of emotion. . . By "cultural resources", then, I mean those parts of the physical environment—natural and built—that have cultural value of some kind to some sociocultural group.*

Archaeologists working in cultural resource management may interact professionally with, or be part of teams that include historians, architectural historians, engineers, folklorists, cultural anthropologists, physical anthropologists, a variety of natural scientists, managers, planners, government officials, and of course, members of the public and interested communities. The archaeology practiced by professionals in the context of cultural resource management is meant to be the same archaeology practiced by those in colleges, universities, and museums. In fact, it can be much more. Existing ethical standards and codes of conduct are meant to apply to all archaeologists, no matter where they practice.

From an archaeological perspective, there are a number of themes that run through much of the relevant legislation guiding today's work. Knowledge of the past is valuable, and not just to professionals. An archaeological resource can have significance from an academic or scholarly point of view, but importance can also stem from its value to living cultures. In turn, not all resources will have significance or value. Observations embedded in the ethical principle of stewardship are taken to heart. Archaeological resources are finite, fragile, and nonrenewable, and even professional investigation results in their "destruction." Preservation of sites should be just as important as realizing their scientific value through investigations guided by well-thought-out research designs. Human remains and grave goods deserve to be treated with respect.

Federal agencies, those responsible for the management of public and Indian lands, anyone requiring or using a federal permit, or anyone using federal monies have to consider the impact that their activities might have on archaeological resources. Inventories of existing archaeological resources have to be created before other decisions and procedural steps can be taken. Archaeology performed in the context of cultural resource management is not meant to be salvage, investigating sites one step ahead of the bulldozers. It is designed to be proactive and part of a negotiated procedure that gives a voice to all people who are concerned with a project in some capacity. Archaeology and cultural resource management are meant to be an integral part of overall planning.

The National Historic Preservation Act (1966, amended 1980, 1986, 1999) has been called the cornerstone law of historic preservation in the United States.

It builds upon and clarifies previous laws defining the government's responsibilities regarding cultural resources related to sponsored projects. It established the National Register of Historic Places, which is a listing of objects, structures, sites, districts (collections of individual properties), archaeological resources, Indian and Hawaiian sacred sites, and landscapes that have cultural or design values. In order to be listed, a resource must be determined to be significant at the local, state, or national level.

The Act created State Historic Preservation Officers (SHPO). Each SHPO is appointed by the governor of a state and oversees historic preservation or cultural resource management activities in that state (Appendix 3). This is usually done out of a State Historic Preservation Office, also referred to as the SHPO. The initial review and processing of National Register nominations is handled by the SHPO and a State Historic Preservation Board that it establishes. Today, there are also Tribal Historic Preservation Officers (THPO) with responsibilities for Indian lands. The powers of the Advisory Council on Historic Preservation (ACHP) were expanded regarding the formulation of regulations for Section 106 reviews and consulting throughout this process. The ACHP is an independent federal agency that advises the President and Congress about historic preservation matters.

The Act includes statements about the funding of activities. In particular it notes that federal agencies may waive the 1 percent cap of total project costs to support historic preservation activities that was established by previous legislation.

Section 106 of the Act provides the best crash course in what drives the archaeological component of cultural resource management. In a nutshell, the federal government and its agencies must take into account how their undertakings could affect historic properties. An "undertaking" can be any project involving construction, rehabilitation and repair, demolition, or the transfer of federal property. Any activity that uses federal monies or loans, or requires a federal license or permit is also considered to be a federal undertaking. A "historic property" is defined as one listed in, or eligible for inclusion in the National Register. If a project will have an adverse effect on a historic property, it must be determined how the adverse effect might be avoided, minimized, or mitigated. Throughout this process, the federal agency involved must consult with the SHPO. The ACHP must also be given an opportunity to comment on any undertaking, although recent revisions to the Act state that the council will no longer routinely review decisions agreed to by a federal agency, SHPO, and/or THPO.

At the outset, the agency responsible for a project identifies and contacts the appropriate SHPO or THPO for consultation. Plans are made to involve the public and other consulting parties. Determining whether there are historic properties that might be affected by a project is a phased operation. Recall that in the language of the Act, a historic property is one that is listed in, or eligible for inclusion in the National Register. For the purpose of this discussion, we're concerned about the existence of archaeological resources of National Register quality.

You can think of the early stages of the Section 106 process as asking and answering a series of smaller questions that are embedded in the larger charge to consider the effects of a project on National Register or "significant" sites. First, are there any archaeological sites/deposits/resources located within the project area? Part of this question may be answered by checking the files and data banks maintained by the SHPO or THPO or other institutions that serve as archaeological data repositories. This background research might reveal sites already listed on the National Register, or in the process of being evaluated for possible inclusion in the Register. It might indicate the presence of sites that have never been evaluated using National Register criteria. Background research may also reveal that there are no known sites recorded for the project area. This latter alternative could lead to a speedy conclusion of the Section 106 review process if the lack of sites is supported by a previous archaeological investigation of the area, or other information clearly showing that there is no potential for sites to occur.

Fieldwork becomes necessary to locate and identify archaeological resources within a project area (site survey) when existing information is lacking or incomplete, and subsequently to gather enough information to determine whether each site has the potential to be nominated to the National Register (site testing; determination of eligibility). Site survey and testing may be staged as multiple operations or combined into single efforts. It depends on the nature of the federal undertaking and what is already known about the resources of the project area. These stages of fieldwork are accompanied by additional background research that might aid in the location of additional sites, and provides information for the construction of interpretative contexts in which the significance, or National Register quality, of an archaeological site can be evaluated. A variety of archaeological resources may exist within a project area and not all will require the same level of investigation. Rarely will all be shown to be significant or eligible for inclusion in the National Register of Historic Places.

Prior to initiating fieldwork, the consulting parties will develop a scope-of-work that they feel addresses the needs of the project. Individuals and companies that perform archaeological research or cultural resource management studies (contractors) will prepare research designs and budgets in response to this scope-of-work, and submit them as proposals. Minimal requirements for the conduct of archaeological research are included in federal regulations. In some cases, the SHPO/THPO have their own guidelines for the conduct of archaeological research that are more detailed than federal regulations, and which must be reflected in a contractor's research design. Proposals will be evaluated and an individual or firm will then be selected to perform the work.

Reports are prepared detailing the purpose, methods, and results of the investigations by the contractor. The type and number of reports or documents that are created often mirror the way in which the fieldwork was staged. Sometimes archaeological studies are couched in larger documents that deal with other environmental and planning issues, like environmental impact statements. Popular versions of reports using nontechnical language may be prepared for public distribution. It is difficult to circulate technical reports in public because of the cost and the sensitive information that they contain. Detailed maps showing the location of sites in conjunction with detailed descriptions of stratigraphy and the site's significance could promote unauthorized artifact collecting and digging.

The technical reports must contain the recommendations of the investigating archaeologist about the need for further action or research. At the conclusion of a site survey, no further work might be recommended because sufficient information was collected to determine that none of the sites in the project area are worthy of nomination to the National Register. Or it might be the opinion of the investigator that a small number of sites **might** have the potential to be nominated, but others don't. More work is recommended on the select few in order to make a determination of their eligibility.

Reports are reviewed by staff of the SHPO and the agency sponsoring the project to insure the quality of the work, and to guarantee that appropriate procedures have been followed. Of special importance is whether or not the reviewers agree with the evaluations and recommendations about the archaeological resources under study, and how the overall project should proceed.

In order for any historic property to be judged significant, it must be evaluated in terms of the current status of American history, archaeology, and architecture. It must have integrity of location, design, setting, materials, workmanship, feeling, or associations. For an archaeological site, this usually relates to the ability to accurately date deposits or the individual components of a deposit, and control for all of the processes responsible for the deposit's formation and transformation. An archaeological resource must then meet one or more of four criteria which are spelled out in the National Historic Preservation Act. It must:

a. be associated with events significant in the broad pattern of history;

b. be associated with persons/lives important in history;

c. embody distinctive characteristics of a type, period, method of construction, or represent the work of a master artisan or possess high artistic value; or

d. have yielded or have the potential to yield information important in history or prehistory.

Archaeological sites typically get nominated under the last criterion "d." Remember, too, that standing structures and other types of historic properties that might fall under criteria a, b, or c can have archaeological components.

It is up to the person preparing the nomination to show that a site is significant. This can be done by reference to pre-existing syntheses of history and prehistory, and demonstrating how a site refines or adds to these frameworks. Some states have developed comprehensive plans which identify important research questions and data. Likewise, the National Park Service has supported the development of "thematic contexts" that summarize all of the relevant information on a particular time period, area, or theme, and identify important research questions and data (e.g., Grumet 1992, 1995). Showing how an archaeological resource furthers the goals of a comprehensive plan or thematic context bolsters arguments for its significance in the nomination process. In many cases, however, the person preparing the nomination must gather sufficient evidence and background to fashion a context in which an archaeological resource's significance is made clear.

The issue of significance remains a ticklish one in the management of cultural resources (cf, Butler 1987; Darvill 1995; Leone and Potter 1992; Lipe 1984; Schaafsma 1989). Determining the significance of a resource requires the existence of some type of standard against which it can be judged. Where do such standards come from? Value or significance is not inherent in an archaeological deposit or cultural resource. It is ascribed. It comes from the "outside" and is subject to the concerns of the moment, be they academic, scholarly, social, political, or economic.

Although the survey or site identification stage of a cultural resource management project is not designed specifically to collect data sufficient to determine site significance, it is often the case that many sites are easily recognized as being insignificant in terms of National Register criteria. The information derived from them during the site survey is still of use, and becomes part of larger data banks. But further work would not add to an increase in useful data or a better understanding of the site's potential significance. Many cultural resource management projects do not go beyond the site identification stage. Although there are sites found within a project area, none may appear to be significant, that is, no "historic properties" exist in the terminology of National Historic Preservation Act. Therefore, the federal undertaking will have no effect on historic properties and the undertaking can proceed as proposed. The same logic applies whenever fieldwork gathers sufficient information from a site to determine its National Register eligibility, and the site does not measure up to standards.

When a site(s) or an archaeological resource(s) is considered to have potential for nomination to the National Register, and the consulting parties agree, then the next step is to determine what effect, if any, the federal undertaking will have on it. There are two possible outcomes here. A project may have no adverse effect, meaning that the undertaking will affect one or more historic properties but the effect will not be harmful. Or an undertaking may have an adverse effect, meaning that it will harm one or more historic properties. If it is determined that there will be no adverse effect, the federal undertaking can proceed as planned. A finding of an adverse effect requires additional consultation in order to find ways to make the undertaking less harmful. Mitigating an adverse effect can involve avoiding significant properties, preserving them in-place, realizing their value through intensive field investigations and subsequent analyses, or some combination of these.

Choosing to resolve adverse effects through the recovery of significant information from an archaeological site leads to a final stage of fieldwork. Like all other stages it requires the formulation of a scope-of-work for the project, and the creation of research designs and budgets by contractors who wish to vie for the job. The research design is of crucial importance here because it must clearly show how fieldwork and analysis will capture the value of the things for which the site was deemed to be significant. Once this final phase of fieldwork (mitigation) is completed, the federal undertaking may proceed. The analysis of data and the preparation of technical reports following fieldwork is a time-consuming procedure, and it is assumed that the approved research design will be followed. Construction, or whatever is behind the "adverse effect", usually begins before final reports for a project are issued. Popular reports, films, or videos may be prepared for public consumption.

Organizing cultural resource management projects into stages is of benefit to everyone concerned. The continued interaction of archaeologists, planners, engineers, federal agency and SHPO staffs, and the public provides for the efficient and flexible use of information and funds. Knowing the location of archaeological resources early in the planning process may make it possible to redesign or relocate projects to minimize their impact on sites. In other cases, knowing the impact a project might have on significant resources, and the potential cost of studying these resources, may cause some undertakings to be abandoned altogether. Don't forget that this whole process involves **all** cultural resources within a project area, not just archaeological sites.

The law notwithstanding, most everything in cultural resource management is subject to negotiation, from what happens in the local municipality through county, state, and federal levels. Law and procedures have enough built-in "wiggle-room" to allow project-specific negotiations to occur, and good and bad things can result from this. Any discipline that is lived out to any degree in the public arena (like law/lawyers; civil engineering; public education) is subject to politically and economically motivated negotiations that may bend the idealism inherent in the original vision of the pursuit. The person who denies this process has a limited view of life in the "real" world.

In my opinion, becoming and remaining successful (intellectually and economically) in the world of cultural resource management means living up to the intent of the law, not merely the letter of the law. It requires not only being a good archaeologist, but understanding how the system works (the "anthropology" of cultural resource management), and thereby understanding how to manipulate it. Manipulation is beneficial if practitioners are committed to ethical standards, resource advocacy, and learning something about the past—what I see as the intent of the law.

Making a distinction between adhering to the intent of the law versus the letter of the law may seem ridiculous to some. Doesn't the fact that we have such laws in the first place demonstrate the existence of good and proper intentions? Sure it does. But in practice, individuals with different perceptions of the law and different motivations negotiate the track of a project from start to finish, and the results can be as variable as the individuals involved. There isn't an archaeologist working in cultural resource management today who doesn't have horror stories about shoddy work and what they perceive as the bending of the law. However, none can forget that today every major advance in regional archaeology and most significant publications are tied in some way to a cultural resource management project, its data, or some initiative it spurred (e.g., Adovasio and Carlisle 1988; Green and Doershuk 1998).

RESEARCH DESIGN

What do you want to do, and how are you going to do it? This is the essence of a research design. The focus on constructing explicit research designs was an attempt by archaeologists during the 1960s to realize more of the scientific and anthropological value of the sites that were being consumed through surface collections and excavations (remember the preservation ethic and stewardship principle?). In being explicit about research goals and how they were to be achieved, archaeologists had to confront assumptions about the way they practice archaeology. Are all aspects of past societies and cultures accessible through archaeology? Are all of the questions we would like answered about the past encoded in the archaeological record? If so, in what ways? How well do interpretations of the past jibe with archaeological data? How well do field methods correspond with the questions being asked or the goals being set?

We can't say that we want to study ancient religions without being clear about how we believe religious behavior might be reflected in different types of archaeological evidence, where and how this evidence might be found. We can't say that we want to learn about Native American cultural systems and only dig major village sites in the floodplains of rivers. What about hunting camps, fishing camps, quarries, and ritual localities and the perspective that they provide? Formulating a research design requires framing a question, problem, or idea in such a way that it has implications that can be evaluated with archaeological data. It fosters clear thinking, or at least clearer thinking.

Today, archaeologists as a group are much more cognizant of the finite and fragile nature of the archaeological record and our ethical responsibilities to make the most of it. Devising and using research designs are standard practices in archaeology, and mandated by the laws and procedures governing cultural resource management.

COMMON ELEMENTS OF A RESEARCH DESIGN

1. Establish or identify goals, problems, or issues of concern. Develop hypotheses (statements that can be falsified) to be tested related to these goals, problems or issues. These need to be specific and not overgeneralized, and they must be stated in a way that suggests solutions, or testable implications.

2. Determine the information or types of data needed to address the items listed in #1.

3. Determine how you will search for and collect this information. Where will you search for this information? What are appropriate field strategies and techniques for data collection (includes collection of background data as well as fieldwork)?

4. Collect data.

5. Analyze and evaluate collected data with methods that are appropriate to the data and the project goals, problems, or hypotheses.

6. Address the project's goals with evaluated data, reformulate questions, and ask new ones.

7. Disseminate information and plan for the storage and curation of artifacts and other data.

In its strictest sense this process is an expression of hypothetic-deductive reasoning. It can be contrasted with strict inductive reasoning, where facts are gathered and as they are amassed, suggest ideas or explanations of phenomena to the researcher. Both the type of reasoning and the use of research designs in archaeological research are actually much more flexible. Archaeological reasoning (and scientific reasoning in general) represents a combination of deductive and inductive approaches. Research might begin with specific ideas that are testable with archaeological data. The very process of collecting and analyzing data may suggest elaborations of the original idea or hypothesis to the investigator, or entirely new ideas that seem to better fit the facts. In turn, this might dictate changes in field strategy and laboratory analysis. Such flexibility doesn't denigrate the process. But it does mean that the whole research design process may be reiterated during the course of a project, and it shows the value of considering multiple hypotheses when dealing with any research question or issue.

A scientific or hypothesis-testing approach to research design does not preclude exploring more abstract topics like the symbolism or belief systems with archaeological data. But whatever is done has to be operationalized in some way, showing a linkage between ideas, questions, data, and interpretation. If there are competing explanations or interpretations of the same data, how do we choose between them? We look for the one that has the clearest and most direct connections between ideas, questions, and data.

The goals, issues, or hypotheses that drive a research design can range from the simple to the complex, be site-specific or regional in scope, be accomplished within a single season or require years for completion. Research designs can be hierarchical, that is, there may be an overarching goal that can only be reached by achieving a series of more fundamental ones. Suppose you want to test the idea/hypothesis that contact between the native peoples of Florida and Spanish explorers, traders, and colonists resulted in the rapid abandonment of traditional technologies as native peoples adopted European-made goods. In order to test this hypothesis, sites and data corresponding to the time bracketing initial contact between the Indians and outsiders need to be identified. Sites dating to the time just before contact between the Indians and outsiders will need to be identified to provide baseline data for comparative purposes. Extensive archaeological survey might be necessary with its first goal being to locate sites, and **then** to isolate ones that date to the appropriate time period. Eventually, more intensive studies of individual localities may be needed to provide the quality of data necessary to address the original hypothesis.

Judging whether a federal undertaking will have an effect on significant archaeological sites necessitates first finding all sites, then studying each to determine whether they are significant, and then assessing potential impacts. At the site survey level of investigations, the types of data collected are most appropriate for constructing and testing/refining models that predict where sites will occur based on their association with environmental features. A patterned distribution of sites across the landscape may also imply things about other aspects of cultures. The types of raw materials represented by artifacts found on sites, and the styles of the artifacts themselves, can be used to begin an examination of the relationship that peoples living within the project area may have had with groups in surrounding areas, and the size and shape of their respective territories. Beyond the site survey stage of this work, research will be guided by questions specific to the archaeology of the region in which the work is taking place.

The geographic focus of cultural resource management projects is predetermined, and so places some constraints on the research issues that might be addressed. Work is confined to a bounded right-of-way, and may involve investigating areas that have not been the traditional focus of archaeological fieldwork. Although restrictive in some respects, this aspect of cultural resource management studies has certainly led to new discoveries and broadened our outlook on the past. Cultural resource management can provide funding for extensive and intensive studies, and supportive technology that is not typically available through other means.

There are numerous logistical or practical concerns that are a part of every research design, especially where fieldwork is concerned. What will be the responsibilities of the members of a project team? What permits, if any, need to be secured before fieldwork can begin? Which landowners and artifact collectors need to be interviewed? Where are appropriate places to do background research? How will people get to the field site? Where will they live? What equipment will be needed? How will health and safety concerns be managed in the field? These topics are taken up in the chapters that follow.

BUDGETS

I've also not mentioned the budget side of any research design. Putting dollar figures on specific tasks in a research design can be an art in itself, and realistically requires a lot of hands-on experience on the part of the person formulating it. Every technical proposal submitted in conjunction with a cultural resource management study, or in quest of a grant, has a matching budget proposal. The public rarely gets to see the intricate details of these since they can contain proprietary information. Budgets can be incredibly detailed, and it is very instructive to compare them with their companion technical proposals and the reports of the investigations that are eventually produced. If you are enrolled in a course in archaeological methods, see if your instructor can arrange to get copies of these types of documents. Short of creating a research design and budget of your own, and then having to live up to it, this is the best instruction in the matter that you can get. In an effort to demystify the process, the National Park Service sponsored a publication illustrating the connections between research designs, budgets, and practical reality (Carnes et al. 1986). This work is over 20 years old, and you know what inflation has been like! But it's a place to start your education about research designs and budgets. It is naive to deny that the amount of funding available for a project can put constraints on the scope of a research design, whether it's part of a cultural resource management study or an investigation funded by a more traditional granting agency or foundation. This doesn't change the ethical obligations of an archaeologist. It means thinking harder about ranking research issues and making sure they fit as precisely as possible with the archaeological resource under study. Sometimes it calls for saying 'no thanks' to a proffered contract, or postponing a project until adequate funds can be found.

For those doing archaeology with firms specializing in cultural resource management (CRM), conducting research on tight schedules and budgets can leave little time for reflection, generalization, and synthesis of data between projects. This is problematic, since determining site or resource significance in the context of any compliance project requires an understanding and synthesis of data in local, regional, and ever-broadening frameworks. The time constraints under which most projects proceed can make it difficult to experiment with, or integrate new approaches into everyday practice. Communication and interaction between CRM archaeologists and those working out of colleges, universities, and museums is vital for the health of the profession. Those on the outside of cultural resource management need to spend more time operationalizing new approaches and ideas so that they can be tested and used by CRM professionals who have many more opportunities for fieldwork and analysis. CRM professionals are in an excellent position to help academics revitalize and make more relevant their courses and training programs.

Regardless of the specific focus of a research design, all projects need to collect data related to the horizontal and vertical dimensions of an archaeological deposit, the range of material culture included in the deposit and any internal patterning, deposit age, formation processes, and deposit integrity. You also can't ignore any apparent material evidence of human action that is encountered, regardless of the research goals that got you into the field in the first place. All of this helps to ensure that the information collected can be used for a variety of purposes, now and in the future. Archaeologists are still making great use of data collected 70 to 100 years ago because of the original investigators' attention to basic description and context (e.g., Lyon 1996; Morse and Morse 1998; Stewart 1995).

SAMPLING

It is rare that an archaeologist is able to examine every portion of a project area or individual site. At some level all research designs involve sampling, whether the concern is with site survey, testing, or intensive excavations. Sampling means looking at a segment, portion, or subset of a "whole." The intention is that the sample will in some way represent the whole, and that interpretations of the sample can be applied to the whole.

Only things that are readily observable or measurable can be sampled. Suppose your job is to learn about the opinions members of your archaeology class have about artifact collecting by non-professionals. There are 40 people in the class and you don't have the time to interview all of them. You can't sample their opinions because you don't know what they are ahead of time, prior to your interviews. But you can sample the people themselves, wherein the opinions you seek will be found.

Sampling schemes can be qualitative/purposive or quantitative/probabilistic. A qualitative or purposive sample is one chosen on a subjective, nonstatistical basis. In your quest for opinions about artifact collecting held by your classmates, you interview only those people with whom you are friendly. Qualitative sampling is convenient, makes use of special circumstances in individual cases, and allows you to use your intuition. A quantitative or probabilistic sample is one derived using statistical procedures and probabilities. In your classroom survey you assign each student a number. You determine that you have the time to interview 20 people, a 50 percent sample fraction that you believe should adequately reflect the opinions of the larger group. Using a table of random numbers, you select 20 numbers. You match the selected numbers with those assigned to your classmates and begin your interviews. Quantitative samples have the advantage of being more objective and replicable than qualitative ones, and are amenable to statistical manipulation in ways that qualitative schemes are not. Your statistically random sample of classmates will enable you to generate mathematical statements of confidence in your results, i.e., what's the probability that the opinions of the people that you interviewed will reflect the range of opinions held by the entire class.

Example One

Archaeological fieldwork involves both types of approaches to sampling, applied singly or in combination, as the following examples will illustrate. In the first example, a survey to locate archaeological sites of prehistoric Indians is being planned for a 12-square-mile section of a river basin (Figure 3.1). The project area consists of two mountain ranges, and valley areas along the river and its tributaries. The time and funds available for the project are insufficient to underwrite an examination of the entire area, so it will have to be sampled. It is estimated that 25 percent of the project area could be examined. The number and location of the archaeological sites are unknown, so it is impossible to directly sample sites. The landscape and environments of the river basin, however, are readily observable and measurable in a variety of different ways. Since archaeological sites are situated on the landscape, sampling the land/environment is appropriate. In order to show the effectiveness of different approaches to sampling, we'll assume that we now

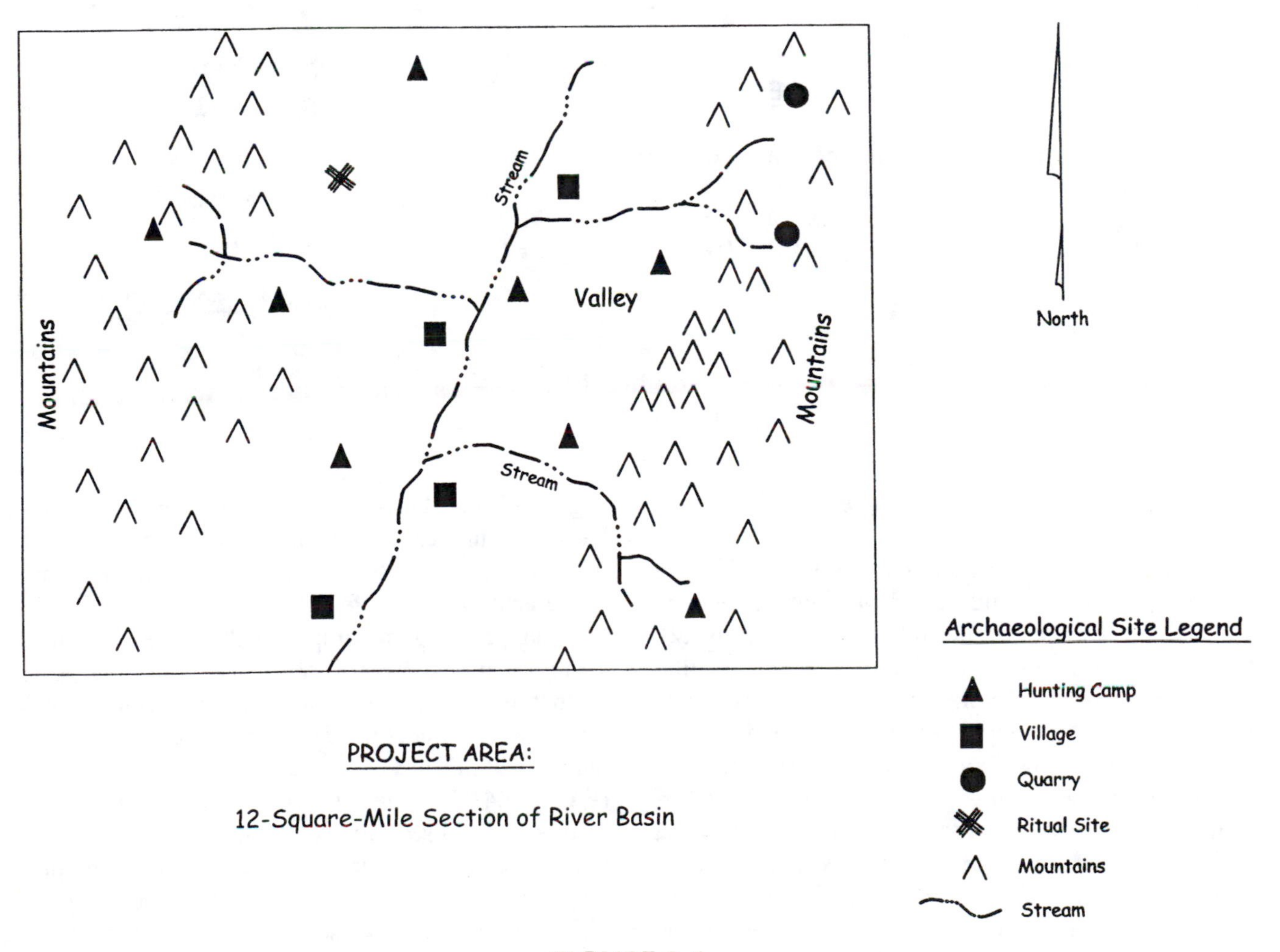

FIGURE 3.1.

Map of project area consisting of 12 square miles of a river basin.

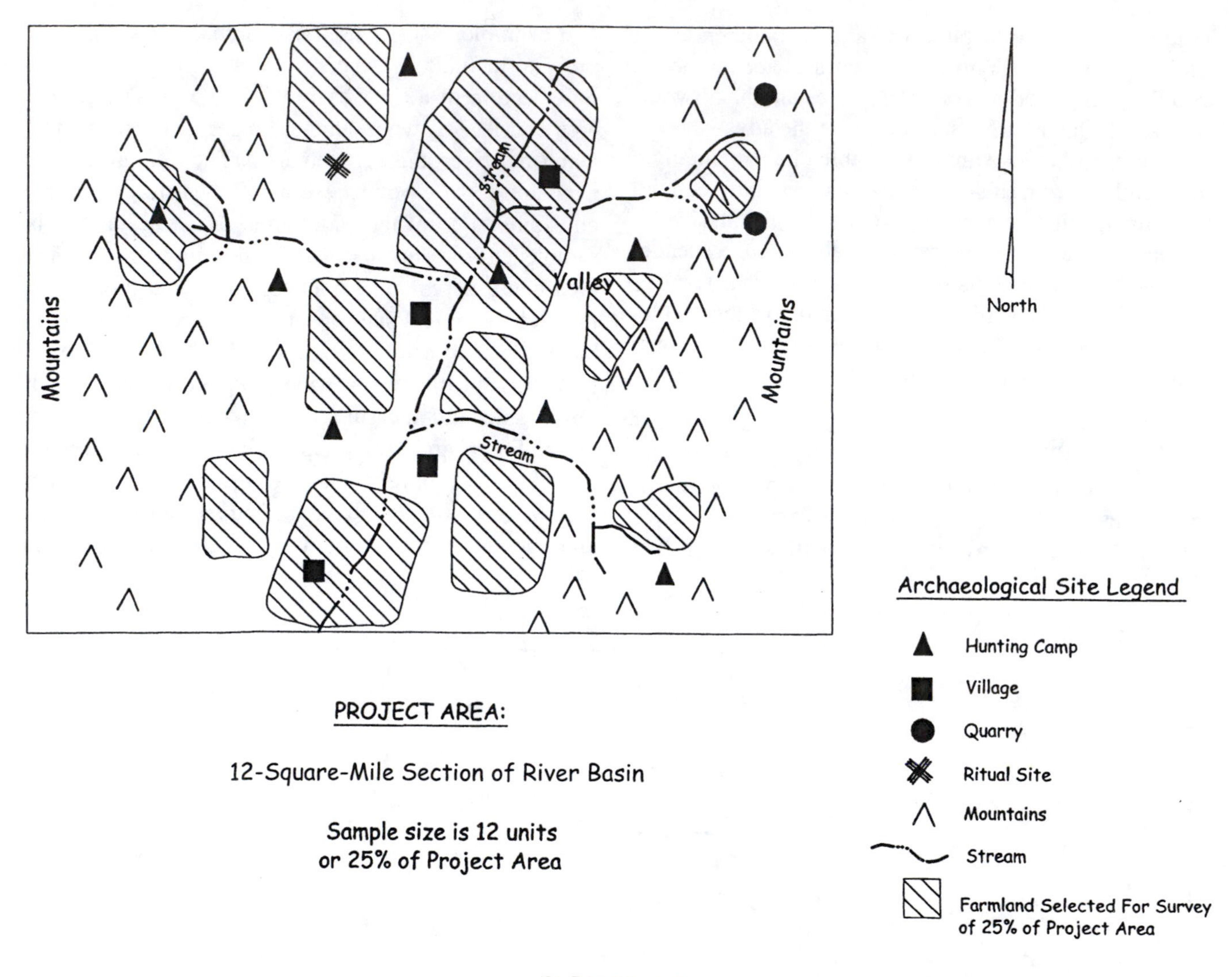

FIGURE 3.2.

Qualitative survey of project area.

know about the location and character of every site in the river basin.

One way to qualitatively sample the valley for archaeological sites would be to investigate 25 percent of all landscapes that are currently farmed (Figure 3.2). The archaeological survey of the areas in this qualitative sample results in the discovery of four sites (if any portion of a site falls within an examined area we will consider the site to be found). Two of the discovered sites are villages and two are hunting camps. A proportionately large amount of the valley segment of the project area is examined, simply because that is where most of the farms are located.

Deriving a quantitative sample of 25 percent of the project area first requires dividing the area into units of some sort to facilitate the selection of the sample. In Figure 3.3 a grid of square units has been used and each has been numbered for identification. The project area consists of 48 of these units, 12 of which comprise 25 percent of the whole.

A simple random sample of 25 percent means that 12 units must be randomly selected from the possible 48 that define the project area. Using a table of random numbers, and matching random selections with the ID numbers of the grid units, 12 sampling units are selected (Figure 3.4). The field survey of these sampling units results in the discovery of two sites, both of which are hunting camps. A proportionately large amount of mountainous environments are examined by random chance. If the random sample selection process were run again, the representation of environmental zones could change.

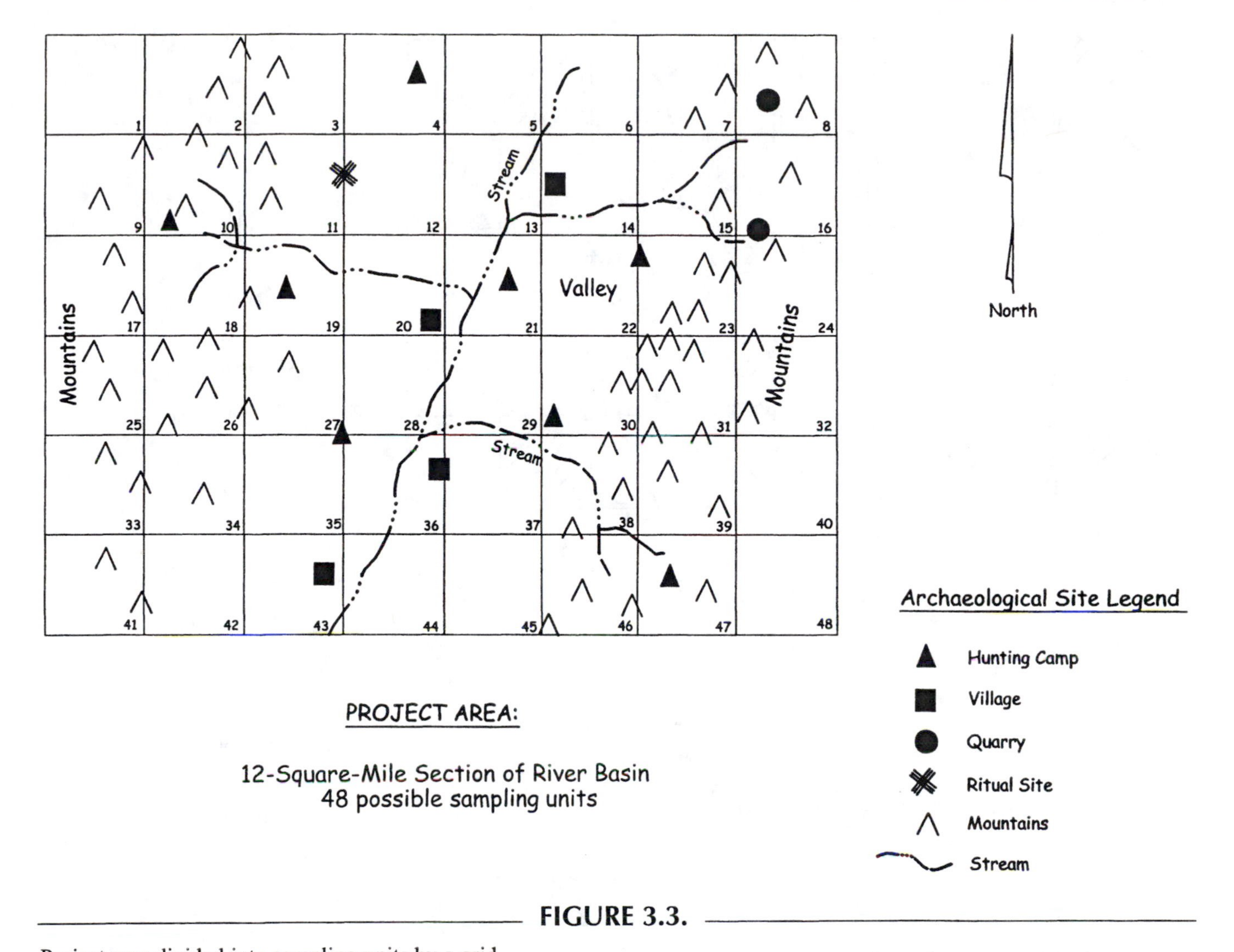

FIGURE 3.3.

Project area divided into sampling units by a grid.

Figure 3.5 shows the derivation of another type of probabilistic scheme, a stratified random sample of the project area. Stratifying a sample population means breaking it down into identifiable subpopulations from which samples can be selected. In effect, it reduces the variability between sampling units. In the scheme depicted in Figure 3.5, types of environments are used to organize or stratify the section of river basin into three pieces. The strata consist of the eastern and western mountain ranges, and the intervening valley In the simple random sample of the project area (see Figure 3.4), mountainous areas were more heavily represented than other environmental zones in the project area. If environmental variability is somehow linked to variability in the distribution and nature of archaeological sites, then the random sample will give a skewed impression of what the project area really contains. In a sense, stratifying a sample allows an investigator to build explicit assumptions about the phenomena under study into the sample selection process.

In the stratified sample we are still limited to examining only 12 of 48 possible sampling units for a 25 percent sample. Since the project area contains three strata, the 12 units must be divided equally between the three strata. Again using a table of random numbers and the ID numbers of individual sampling units, four units are selected from the eastern mountain range, four from the western range, and four from the valley. This strategy results in the discovery of six sites: three hunting camps, two quarries, and a village.

The stratified random sample resulted in the discovery of the greatest number and diversity of sites, and provided a wide-ranging view of the environmental zones in the project area and how they are used by the

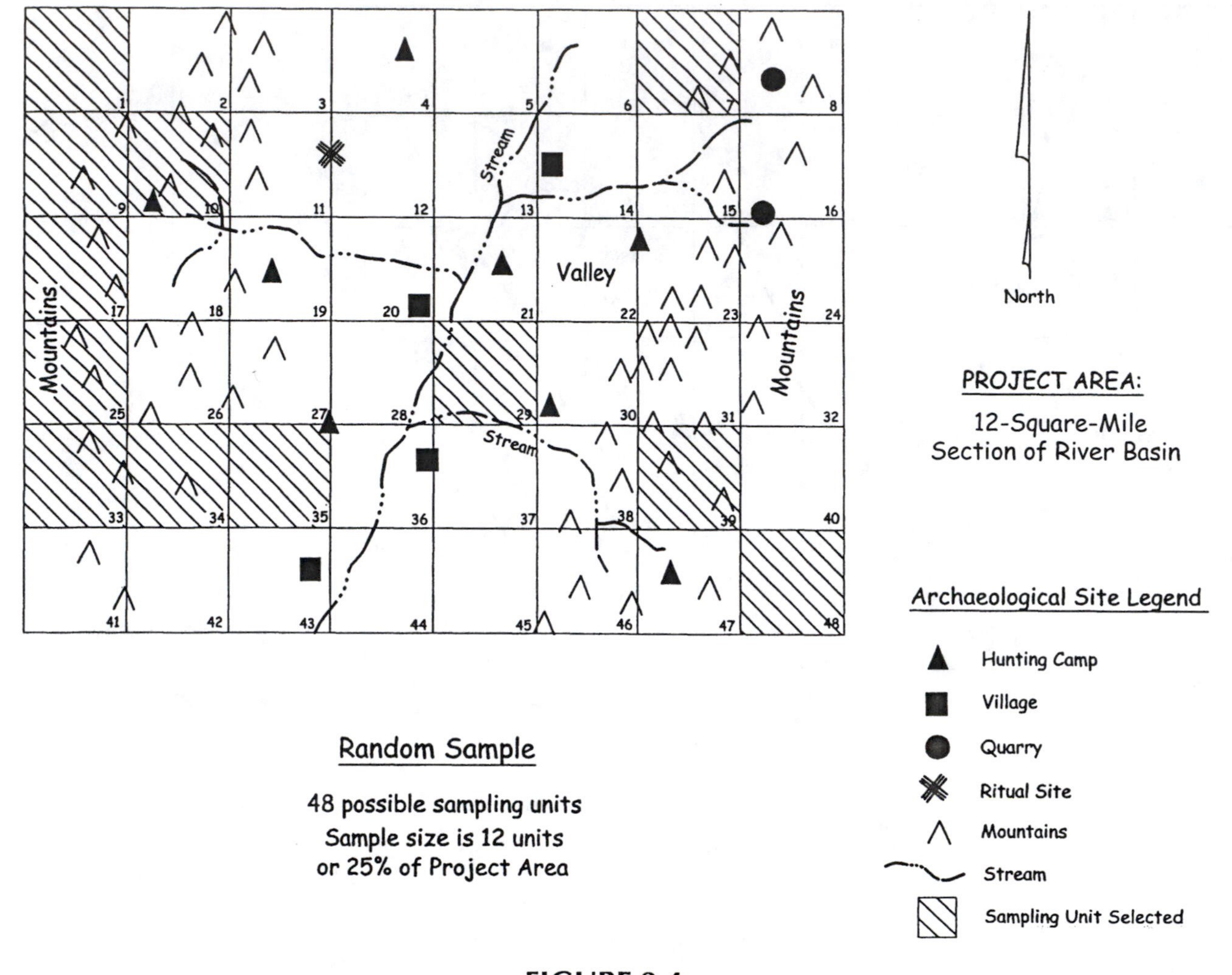

FIGURE 3.4.

Simple random sample of project area.

native inhabitants. Using the same criteria, the qualitative strategy was a close runner-up, and the simple random sample ranked third. None of the surveys resulted in the single ritual site being found.

In the qualitative scheme, surface survey would be easy because of the existence of plowed fields and extensive acreage could be examined relatively quickly. Many of the farms are situated in environments adjacent to streams, which are settings where some types of archaeological sites have been found in other regions. Fewer landowners would have to be contacted for permission to do the work since farms typically consist of large tracts of land, and the farms are easily accessed from existing roads. Qualitative schemes are also useful for first looks at areas that are large blanks on the archaeological landscape. They can provide a lot of useful information in a relatively short period of time, and make use of an archaeologist's intuitive field sense. Both of the quantitative schemes could prove to be more time consuming because of the need to locate the sampling units in "real space."

For the ultimate purpose of best understanding all aspects of native life in the project area the stratified sampling scheme gave the best results. Because of its statistical derivation, this scheme could reliably be used to develop a model for predicting where other sites might occur in areas that have not been surveyed. Simple random samples can produce widely varying results and are not used that frequently. One reason for this is that no matter where archaeologists are working today, there is enough known (or assumed) about how people distributed their activities across space that sampling schemes can be stratified on some basis. For archaeology, quantitative sampling strategies seem to work best when there is some type of patterning inherent in the phenomena being investigated. They are not

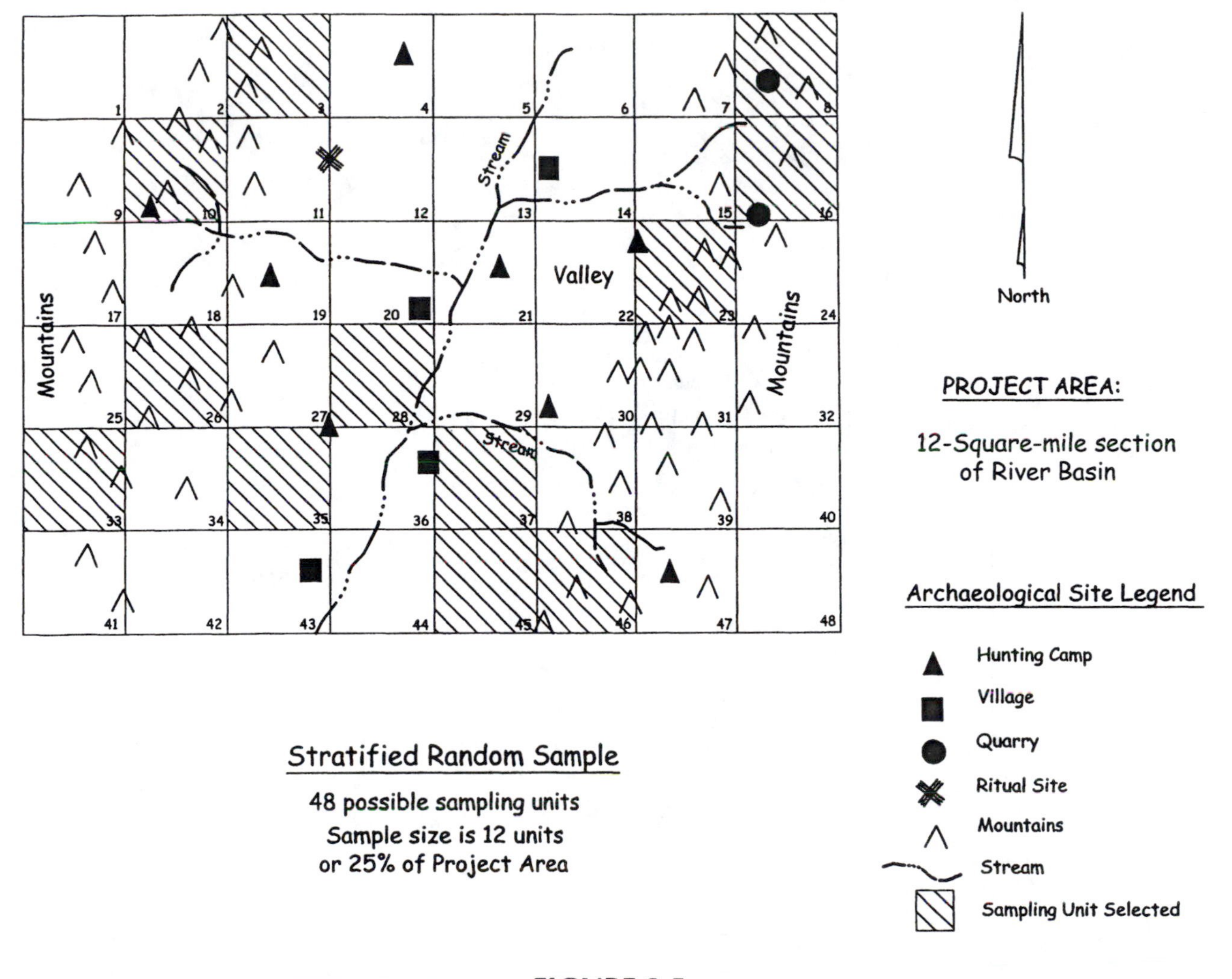

FIGURE 3.5.

Stratified random sample of project area.

well suited to dealing with unique things, like the ritual site, or sites/deposits that are highly clustered in space. In practice, survey projects employ a combination of qualitative and quantitative techniques, literally combining the best of both sampling worlds.

Example Two

Sampling is just as pertinent in the testing and intensive study of individual archaeological sites. Figure 3.6 depicts a typical situation. Surface finds of artifacts and the results of a limited number of test excavations have been used to define the horizontal extent of a Native American archaeological site. A number of basic questions about the physical attributes of the site need to be answered. What is the nature of the deposit within these boundaries? Are there buried and stratified deposits? Is the stratigraphy the same across all areas of the site? Is there internal patterning of artifacts and features at the site representing activity areas? Deciding where to place test excavations to gain the information necessary to answer these questions would involve both qualitative and quantitative approaches to sampling.

The environmental setting suggests that the landscape on which the site is located has been shaped by different types of natural processes. A large portion of the site occurs within the 500-year floodplain of the nearby stream, while an equally large section falls outside of this area. It is therefore likely that the site's stratigraphy will vary between these areas, because of the variable rate with which floods may have buried or eroded sediments, artifacts, and features. The project geomorphologist, or anyone familiar with the behavior of streams, will also know that the landform's stratigraphy

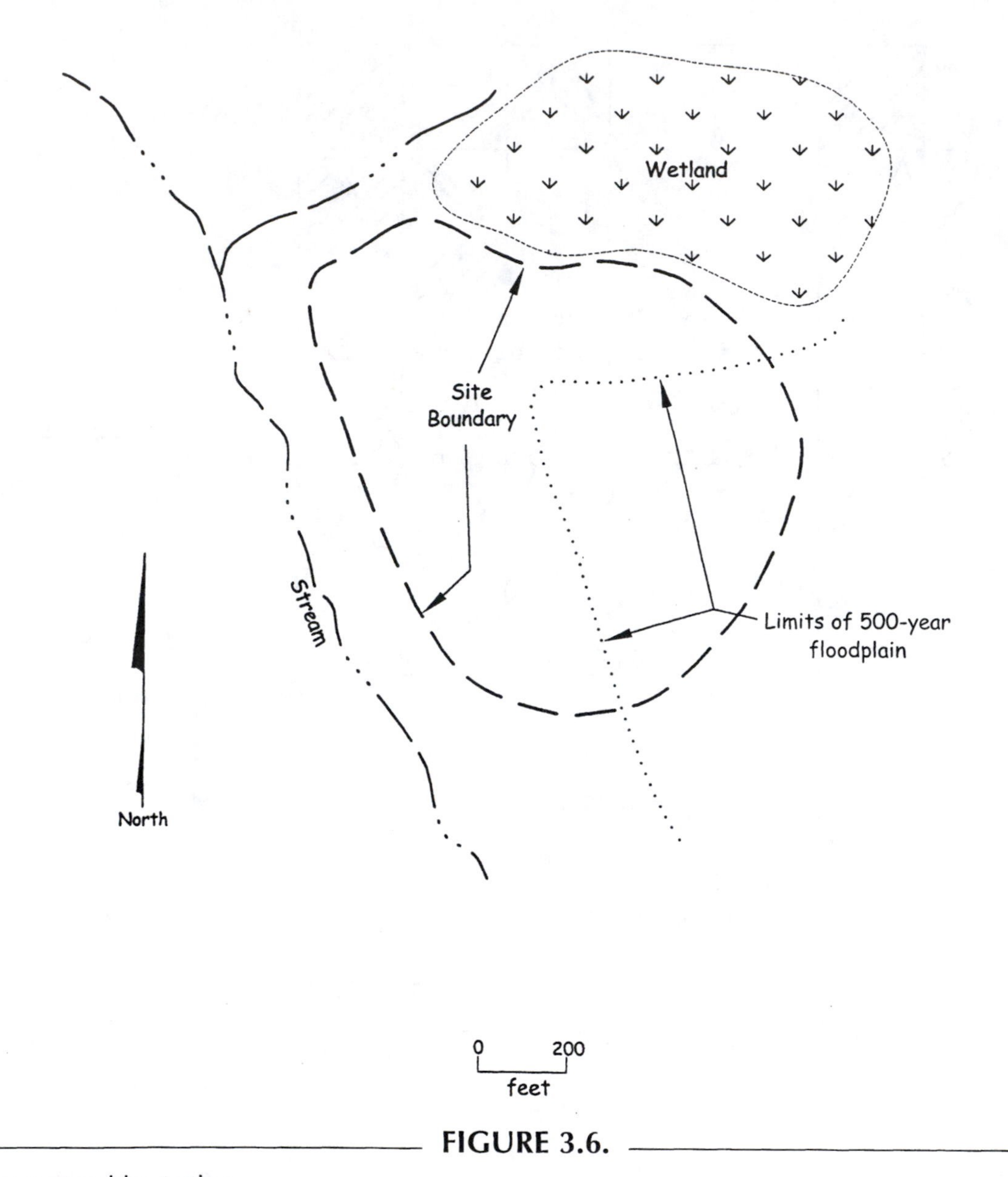

FIGURE 3.6.

Map of site area requiring testing.

will likely vary within the floodplain with increasing distance from the stream's edge, and increasing distance from the stream junction. Under these circumstances, there are specific locations where excavations need to be placed to gather data on stratigraphy and the natural processes affecting site and landscape formation. A quantitative sample of the site landscape is not a critical issue here.

A quantitative sampling scheme for the placement of excavations whose purpose is to determine whether artifacts represent patterned activities within site boundaries would be appropriate. Information derived from previous surface finds of artifacts and the results of initial excavations might suggest ways that the site area could be stratified for sampling. Previous studies of sites located in similar environmental settings might also provide insights about how to stratify the area for sampling purposes.

Example Three

In Figure 3.7 we are confronted with an 18th-century town, or complex of sites, that have been tested to a degree sufficient to locate all structures along the main road. Documentary research turned up an old map indicating the name of the property owners or residents, and the nature of some of the structures. An assessment of old probate wills provides some indication of the economic status of each of the town's residents. You want an archaeological perspective on the economic and social life of the town, but can only intensively excavate a portion of it. There are few businesses in town, so you would want to include excavations at each of them in a qualitative sample of the town, along with the church. In turn, you might stratify the residential areas by the economic status of the residents and draw a quantitative sample from them.

There are many permutations of quantitative approaches to sampling and only a few have been explored here. Deciding which are useful must be determined on a project-by-project basis (cf., Mueller 1974, 1975; Plog et al. 1978; Schiffer et al. 1978).The purpose, goals or research questions to be addressed by a project are important considerations in devising a sampling strategy, as are the time and money available for the work. What is to be done with the data resulting from the work, and the importance of it being amenable to statistical manipulation must be weighed in the decision-making process.

Sampling can be a problematic issue at times in cultural resource management archaeology. Project areas can be so large that not all landscapes within it can be examined in the same way, or examined at all during an archaeological survey. How can this be reconciled with the need to find all significant sites within a project area? Models detailing the relationship between Native American site locations and features of the environment are used to stratify project areas into high-, medium-, and low-probability zones for site occurrence. While all zones may be sampled, fieldwork often focuses on the high- and medium-probability areas.

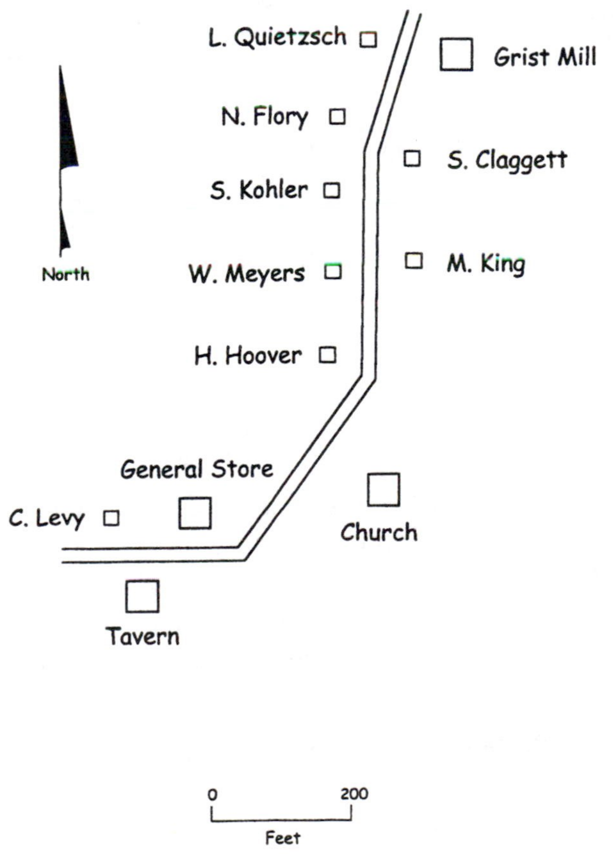

In a predictive or locational model, it is the landform or environmental zone that is the focus of sampling and analysis since archaeological sites represent an unknown quantity. The close link between the physical environment and how people organize their activities across the landscape makes this a viable strategy. Although a number of environmental and social factors are undoubtedly responsible for the distribution of sites across the landscape, it is the environmental aspects that are stressed in most models since the social behaviors and relationships of aboriginal cultures are imperfectly known. What is actually being predicted by most locational models are environmental zones within which archaeological sites are likely to occur, and not the exact location of individual sites, since they are generally much smaller than the landforms or habitats in which they are found. For example, archaeological sites may consistently occur on low-relief, well-drained ground in close proximity to stream junctions. This does not mean that if a stream junction is completely surrounded by low-relief, well-drained ground there will be continuous archaeological deposits. One or more sites may eventually be found, but their exact location within the broader environmental zone cannot be predicted consistently. The archaeologist is therefore obliged to test the entire zone before assessing the validity of the models that led to the selection of the environmental zone.

The various ways in which predictive models have been formulated, both qualitative and quantitative, raise questions about the confidence with which they may be used, and the results of surveys based upon them. In turn, it is precisely because of the need to deal with archaeological surveys of large areas in cultural resource management studies that sampling and predictive modeling have been refined to the degree that they have (cf. Custer 1979; Darsie and Keyser 1985; Kvamme 1989; Nance 1983).

FIGURE 3.7.

Map of town site requiring intensive study.

APPLYING YOUR KNOWLEDGE

1. Read an archaeological site report (American), technical report resulting from a cultural resource management study, or any detailed publication dealing with some type of fieldwork. Answer as best you can the following questions:

- ☑ What was the motivation for doing fieldwork?
- ☑ What were the goals that the fieldwork was designed to address?
- ☑ What types of background research were completed prior to beginning fieldwork?
- ☑ What were the sources of this information, i.e., library, state agency, museum, interviews with knowledgeable individuals?
- ☑ Summarize the various field methods employed.
- ☑ Which techniques were tailored specifically to the condition of the site or the types of research questions or hypotheses being addressed?
- ☑ What is the natural and cultural stratigraphy of the site?
- ☑ How were they defined?
- ☑ What methods were used to date the archaeological deposits?
- ☑ What aspect of the investigation interested you the most?
- ☑ In your opinion, did the investigation adequately address the proposed goals? Why or why not?

DIG DEEPER

Cultural Resource Management

❑ The most up-to-date and comprehensive text that you can read on the subject at the level of the federal government is Thomas King's book. For an interesting comparison you might want to look up a copy of the Schiffer and Gumerman book, now out-of-print.

- Carnett, Carol L. 1995. *A Survey of State Statutes Protecting Archaeological Resources.* Preservation Law Reporter Special Report, Archaeological Assistance Study Number 3. Washington, D.C.: National Trust for Historic Preservation.
- King, Thomas F. 1998. *Cultural Resource Law and Practice: An Introductory Guide.* Walnut Creek, CA: AltaMira Press.

Continued

- McManamon, Francis P. and Alf Hatton. 1999. *Cultural Resource Management in Contemporary Society.* New York: Routledge Press.
- Schiffer, Michael B. and George J. Gumerman. 1977. *Conservation Archaeology: A Guide for Cultural Resource Management Studies.* New York: Academic Press.

❑ Publications describing government law and policy are often available in limited numbers for free. Try contacting the following and ask for a catalog of publications:

- Publication Specialist, U.S. Department of the Interior, National Park Service, Archaeological Assistance Division, P.O. Box 37127, Washington, D.C. 20013-7127.
- National Register of Historic Places, National Park Service, P.O. Box 37127, Washington, D.C. 20013-7127.

❑ There are many Internet sources for material regarding cultural resource management, applicable law, policy, and procedures. Remember that when using Internet resources, addresses can change. If any of the addresses listed below fail to work, use a search engine and the key words in the website of interest.

- American Cultural Resources Association (ACRA): among other things, contains updates on relevant legislation. You can also subscribe to the listserv and follow a variety of on-line discussions regarding cultural resource management. **http://www.mindspring.com/~wheaton/index.html**
- Archive of Federal Legislation Regarding Cultural Resource Management: **http://archnet.uconn.edu/topical/crm/crmusdoc.html**
- Cultural Resource Management Magazine (CRM): electronic version of the publication. **http://www.cr.nps.gov/crm**
- Internet Resources for Heritage Conservation, Historic Preservation, and Archaeology: **http://www.cr.nps.gov/ncptt/irg/**
- National Archaeological Data Base: technical reports of cultural resource management projects; latest information and rules related to NAGPAR. **http://www.cast.uark.edu/other/nps/nadb/nadb_al.html**
- National Environmental Protection Act (NEPA) Call-In Service: deals with technical inquiries regarding the implementation of the law, including cultural resources issues. Click on 'NEPA Call-In products' to get to the Cultural Resource section, which uses case studies to illustrate how the law should work; click on 'Fact Sheet' for summaries of legislation. **www.gsa.gov/pbs/pt/call-in/nepa.htm**> Information also available through the National Preservation Institute **http://www.npi.org**> click on 'Tools for Cultural Resource Managers.'
- National Register of Historic Places: background information, property listings, sample nominations. **http://www.cr.nps.gov/nr/**
- Section 106 of the National Historic Preservation Act, Revisions, 1997: **http://www.mindspring.com/~wheaton/sect106regs.html**
- State Historic Preservation Offices (SHPO) Data Base: citations and summaries of state legislation regarding cultural resources, hosted by the National Conference of State Legislatures. **http://www.ncsl.org/programs/arts/statehist_intro.htm**

Continued

Sampling and Statistics in Archaeology

❑ Warming up to these subjects can be a difficult (but necessary) part of your training as an archaeologist. There are many books on sampling and statistics in general, far fewer that introduce the subjects with an archaeological or anthropological perspective.

- Drennan, Robert D. 1996. *Statistics for Archaeologists: A Commonsense Approach.* New York: Plenum Press.
- Madrigal, Lorena. 1998. *Statistics for Anthropology.* Cambridge, U.K.: Cambridge University Press.
- Thomas, David Hurst. 1986. *Refiguring Anthropology: First Principles of Probability and Statistics.* Prospect Heights, IL: Waveland Press.

CHAPTER 4

Background Research

For ask now of the days that are past, which were before you.

Deuteronomy 4:32

Those working on field crews, or students new to archaeology will probably not be responsible for collecting or evaluating the background information necessary to initiate a project. However, understanding how background research plugs into other aspects of an archaeological investigation will help you to better understand the decisions made that affect the conduct of fieldwork. And the proper background information will improve your ability to recognize archaeological evidence in the field. The general types of activities common to most endeavors are reviewed here.

A project's research design should include a clear statement of the purpose of fieldwork and the range of activities that will be performed. Research issues will vary from project to project and require that specific types of background information be compiled. Some degree of background information is in-hand at the time that a research design is prepared. Other background research and compilations of information may take place concurrently with field work and other project tasks, and can vary depending on whether a site survey, testing, or an intensive site study is underway.

In general, background research establishes what is already known about a project area or research topic, including the nature and location of any known archaeological sites, existing artifact collections, and the prehistoric and historic period developments of the general area. Background research provides a context in which the potential significance of any newly discovered sites can be evaluated. It also provides an environmental context for archaeological studies, and serves to focus fieldwork. For example, environmental data is used to identify areas with the potential to contain buried and stratified sedimentary sequences.

CULTURAL AND HISTORICAL BACKGROUND

Any field project will include a compilation and review of archaeological research that has already been performed in the specific area or general region, or that may be of special relevance to the area. Compiling this background will entail visiting or contacting a variety of repositories.

Public libraries are most useful for locating published histories and historic maps of the local area and county. These histories and maps can provide information about native and other inhabitants of an area, and the possible location of archaeological sites of both the prehistoric and historic periods. When dealing with urban areas, this type of research can be complex and needs to be carefully staged (Staski 1982). The place names used on historic and modern maps can provide some fairly important leads to the prehistoric cultural resources of an area (Figures 4.1, 4.2). Similar and additional references will also be found in the state library, and the holdings of local, county, and state historical societies. If you are unaware of the existence or locations of these repositories, check out the phone book and search the Internet.

Most states have archaeological societies (often organized into regional chapters) that count professionals, amateurs, artifact collectors, and the interested public among their members. Members of these societies may have firsthand knowledge of artifact collections and archaeological sites located within a project area. State societies often publish their own journals and newsletters, and hold regular meetings where local archaeological and

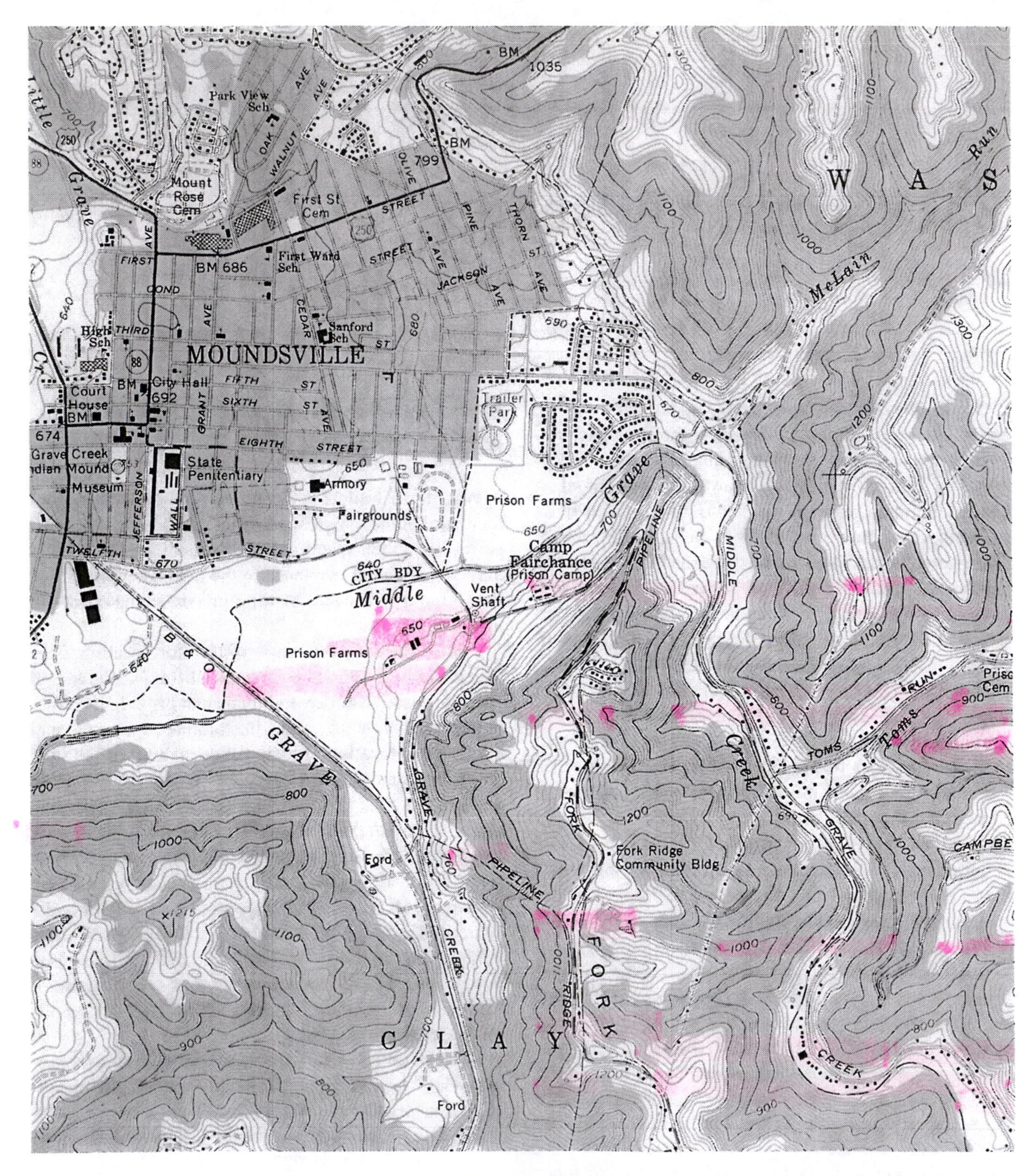

FIGURE 4.1.

A portion of the USGS Moundsville, West Virginia - Ohio 7.5 minute quadrangle, 1976. Scale 1:24,000. The name "Moundsville" is an obvious clue to the presence of Native American features. The stream name, "Grave Creek," is also suggestive. The town of Moundsville is located along a section of the Ohio River Valley where burial mounds of the prehistoric Adena culture are clustered. Grave Creek mound, located at Moundsville, is believed to be one of the largest Adena mounds (Dragoo 1963).

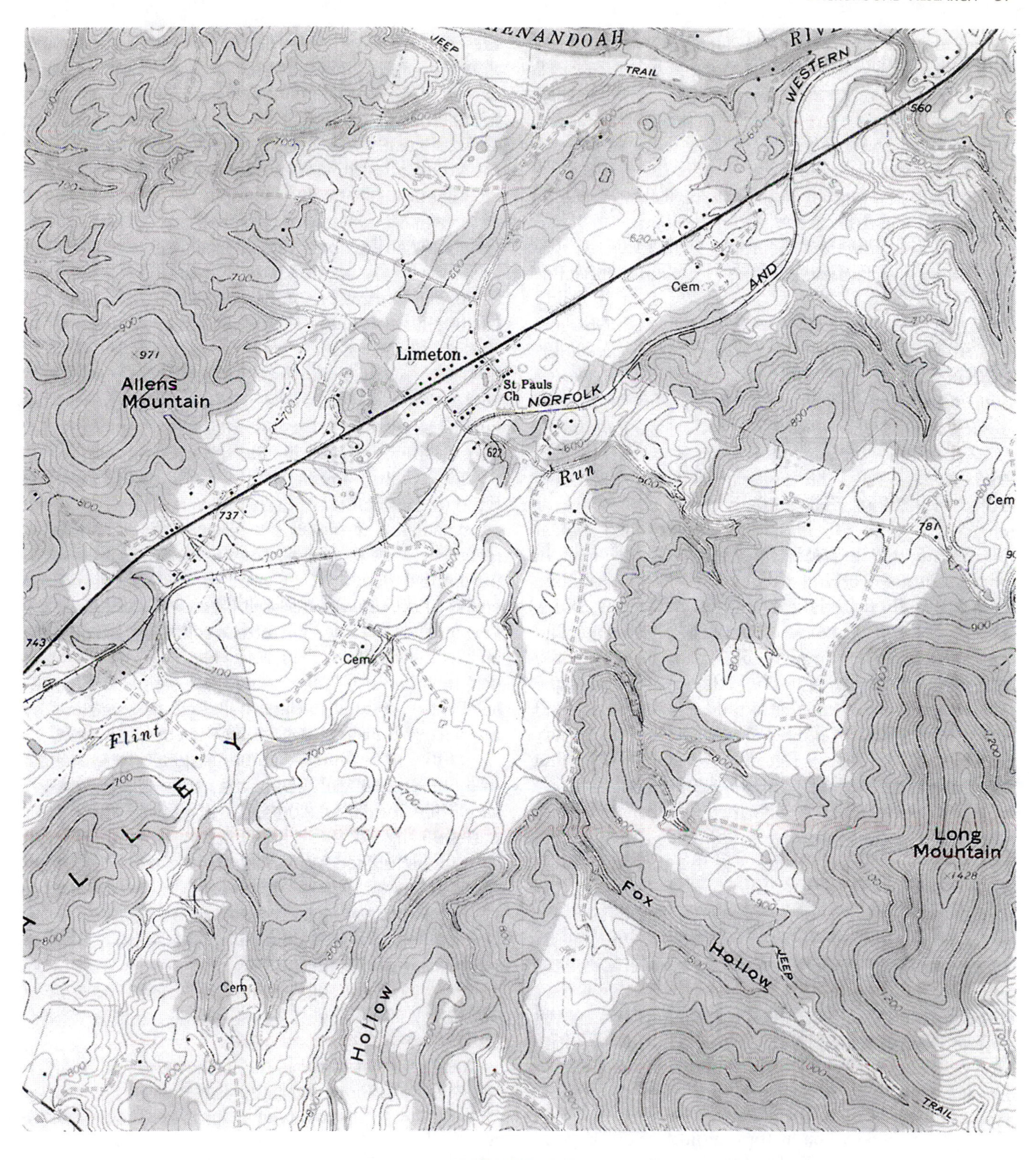

FIGURE 4.2.

A portion of the USGS Bentonville, Virginia 7.5 minute quadrangle, 1972. Scale 1:24,000.The stream named "Flint Run" that cuts diagonally across the map drains an area where deposits of jasper are a part of the bedrock. Jasper is a flintlike, cryptocrystalline type of rock favored by prehistoric artisans for the manufacturing of chipped stone tools. Indian jasper quarries and related workshops, some over 10,000 years old, have been located and studied by archaeologists in the Flint Run area (Gardner 1983). The town name, Limeton, is taken from the local limestone bedrock, which was burned in kilns by historic farmers of the area to produce lime for use on agricultural fields.

historical research is discussed. This is the first, but not only place to begin the search for published research applicable to a project area. State journals of archaeology are not typically well represented in local libraries, and even college and university libraries may not possess complete series. To help in the search, find a relatively recent issue of a journal and check the inside covers for publication and distribution information.

There are numerous journals that focus on the archaeology of particular regions of the country. For the archaeology of Native Americans a brief listing would include: *Northeastern Anthropology, Archaeology of Eastern North America, Southeastern Archaeology, Midcontinental Journal of Archaeology, Plains Anthropologist, Tebiwa, Journal of California and Great Basin Anthropology, The Masterkey (Southwest), The Kiva (Southwest),* and the *Southwestern Journal of Anthropology.* Area museums, universities, and scholarly organizations may support publication series of relevance. There are numerous national and topical journals in which research pertinent to the area in which you are working may be published. Searching them for useful material can be time consuming, and is usually prioritized below examinations of state and regional journals. Of course, there may be booklength treatments of relevance to the geographic area or topic that you are exploring that would turn up in general searches of library or publisher's catalogs. Because any publication will have a relevant bibliography, finding one useful reference will always lead to others.

Where a site survey is concerned, the focus is usually on references that will convey the scope of prehistoric and historic developments, or that might contain information about the nature and location of sites. For historically oriented projects on well-defined properties, like a farmstead or an urban area, it may be appropriate to examine the chain-of-title (list of all property owners, past and present) and legal documents related to former property owners. These documents can contain textual as well as graphic references to structures and activities once performed on a property that would be useful in the location and identification of archaeological deposits. This type of background research becomes even more critical when historic projects move beyond the site discovery phase of fieldwork. Beyond site discovery, background research becomes increasingly more topical and focused for both historic and prehistoric projects.

The office of each State Historic Preservation Officer (SHPO; see Appendix 3) typically maintains or has access to files documenting known archaeological sites and other cultural resources. Site locations are shown on 7.5 or 15 minute, United States Geological Survey (USGS) topographic maps (see chapter 6), as will National Register properties and districts, and archaeological sites of the historic period. In some cases, data relevant to National Register properties may be organized separately. Some SHPO offices also plot all areas that have been subjected to archaeological survey, regardless of whether or not the work resulted in the discovery of sites. This provides a visual record of the extent of research that has been carried out in a geographic area, and is important in the development of site prediction models and other spatially oriented research. Site data can be housed in a state office of archaeology or state archaeological survey, one of a number of state government programs or agencies that may be working under the umbrella of the SHPO.

The inventory forms used by states outline minimal types of information to be recorded when archaeological or cultural resources are discovered or "officially" recorded. This, of course, conditions the type of information that should be collected in the field, during the course of background research for an archaeological survey, or when interviewing artifact collectors and amateur archaeologists about their activities. The forms capture data important for both research purposes and the management of cultural resources, and are accompanied by manuals with instructions for how to complete them.

In order to encourage nonprofessionals to record the location of archaeological sites, some states employ simplified forms. Of course, it is okay to leave categories on any site forms blank or note that you, the person completing the form, simply don't know. Having a location and an inkling of what may be found there is an important start. States sponsor occasional workshops where professionals work with anyone wishing to record a site. Who is filling out an inventory form, and why they are completing one account for a lot of variability in the quality of information on file, regardless of how forms are designed.

Access to site data, particularly locations, is privileged information and usually only available to professionals and those involved in planning and management. The locations of sacred sites and some cultural resources on Indian lands may be kept confidential from cultural resource management professionals.

Forms vary from state to state, but many common themes are apparent. The assignment of the three-part Smithsonian site designation (see chapter 1) is the conclusion of the registration process, e.g., 11CW170 is the Riverton Site, the 170th site recorded in Crawford

County, Illinois (Winters 1969). In the meantime, the name(s) of the locality as used by collectors, or in the field by investigators should be listed. If the site was discovered as part of a cultural resource management study it is important to list any identification numbers used by the government agency or SHPO for tracking the project. The names and identification numbers are ways to link relevant materials (artifacts, field records, reports, review documents, and publications) that may be dispersed in a variety of locations.

Where is it? This obvious first concern requires listing information about the location of a site or resource. This includes political divisions of space (e.g., county, township, or other municipality) and details that enable the locality to be accurately portrayed on existing maps and relocated in the field by others. The names and addresses of property owners are recorded, as should be their attitude about the archaeological resources on their property and the possibility of further fieldwork. As noted above, USGS maps are used to record site locations and are the sources of some of the required political information. How to record site locations on USGS maps is reviewed in chapter 6. Being able to read and use maps is a skill required for more than just finding your way around in the field.

The type of site involved (open, rockshelter, cave, underwater) is listed along with any information available on its archaeological context (surface, buried, stratified, combination of surface and buried, integrity or degree of disturbance) and the size of the deposit. How the site was found or investigated (field methods) and by whom is recorded. Noting whether it was tested or not with excavations is important. If a site was discovered and recorded as part of a professional investigation, references to manuscripts and related publications will be included.

Ideally, copies of maps, field notes, field photos, drawings or photos of artifacts and features, artifact and feature descriptions and inventories will be a part of a site's record. Determining the possible age and cultural or ethnic affiliation of a site relies heavily on evaluations of the style or types of artifacts recovered when excavations are not involved. Artifact inventories, descriptions, and illustrations are especially important when the artifacts themselves are in private hands or located in a number of different places. A site may have been collected or worked by a number of individuals, and personal collections of artifacts can be damaged, lost, or dispersed for a variety of reasons.

Most inventory forms are designed to capture a wide range of environmental data. These include the site's association with a geographic or physiographic province, landform/topography, soil type, vegetation, bedrock or surficial geology, and surface water. Descriptions of current and past land use are also of importance. The role of environmental data in archaeological studies was highlighted in chapters 1 and 2. The formulation of models that predict site locations relies on this information for seeking correlations between attributes of the environment and the distribution of material evidence of cultural activities across the landscape.

Persons filling out forms are usually asked to provide opinions about any potential threats to a site or resource, the potential it might have for further research, and whether it might be worthy of nomination to the National Register of Historic Places. This type of information is useful to cultural resource managers in the early planning stages of federal or state undertakings.

In sum, if you are going into the field with the intention of finding sites, the categories on site inventory forms outline the types of data that minimally should be collected. If you are reviewing completed forms for an area where you will be working, the forms provide information about—

- ☑ who is, or has been active in the area, and how you can find these individuals (establishes who should be interviewed or consulted for additional background)
- ☑ the range of things that have been found (conditions your ability to recognize evidence and link the area with existing descriptive and interpretative literature)
- ☑ how and where sites can occur (provides archaeological and environmental contextual data, conditions field methods, your ability to recognize evidence, and interpretations of deposits)
- ☑ references to useful manuscripts, reports, and publications (can have an impact on the recognition of evidence in the field, field methods, and interpretations of deposits)
- ☑ what others have considered to be significant in terms of the National Register (conditions your interpretation of sites/deposits)

Besides site data, the SHPO or related office will have collections of the technical reports produced as the result of cultural resource management studies. These reports are an incredibly valuable resource for anyone performing background research. Even when a CRM project fails to result in the discovery of any new archaeological sites or cultural resources, the technical report typically contains a summary (and bibliography) of previous research completed in an area, and a compilation of relevant environmental data. Many states also have formal "state plans" for the management of cultural resources. The plans contain syntheses of regional prehistory and history that identify and prioritize important research questions and data needs. The SHPO or related office often will maintain a library of archaeological and environmental publications and manuscripts for the benefit of its own staff who review and evaluate projects, and outside researchers and cultural resource management personnel who use the facility.

The office of a Tribal Historic Preservation Officer can contain the same general range of information that SHPOs collect and maintain. If a project involves Indian lands, the holdings of both the SHPO and the THPO should be consulted, since they are not simply duplicates of one another. Other state and federal agencies that are frequently involved in cultural resource management studies may maintain their own files of site data that can include things not reflected in the holdings of the SHPO. Regional offices of the United States Army Corps of Engineers, the Bureau of Land Management, Forest Service, and the Fish and Wildlife Service should be consulted about the nature of their holdings. Museums, state-based departments of transportation, and colleges and universities with anthropology departments may also maintain site files and collections of unpublished technical reports, as well as artifact collections.

ENVIRONMENTAL DATA

Background on the geology, soils, vegetation, and topography of an area will increase your ability to recognize archaeological evidence in the field. In compiling this information you are creating a standard for what is natural. In the field, anything that is observed that doesn't fit these standards may be an artifact or product of human endeavor. Understanding the nature of the environment is of practical concern to how field strategies are designed—where are you going to look for archaeological deposits and how are you going to look for them. It is also of theoretical concern since we cannot fully understand the actions of humans if we don't understand the natural world that they inhabited.

It will be much easier to find texts dealing with environmental data in any library than it will archaeological ones. Most states have a Geological Survey that publishes and sells reports (and catalogs of published reports) that include maps and descriptions of what is being mapped. Search for titles like *The Geology of ______* (state), *The Geology of ______ County*, *Water Resources of ______ County*, or *Mineral Resources of ____ County*. The USGS also has a publication series that can include references important to your project area or topic under consideration. State departments of conservation and environmental resources are organizations to check for publications related to vegetation, streams, floodplains and wetlands. The former Soil Conservation Service (SCS), now part of the Natural Resources Conservation Service (NRCS) of the United States Department of Agriculture publishes and sells countywide soil surveys that include descriptions and maps. Because of the importance of geology, topography, vegetation, and surface water to the formation and nature of soils, soil surveys also contain valuable information about these other aspects of the existing environment. There are both regional districts and county-based offices of the SCS or the Soil Survey Division of the NRCS that will be listed in your phone book.

There are professional journal series focusing on geology, soils, sediments, geomorphology, ecology, and paleoenvironments. For the purposes of getting started in the field, texts with basic descriptions of area geology and vegetation, soil surveys, and available maps will provide the background that you need. Even though these books and maps may have been produced for public consumption, it may require extra work on your part to learn some of the basic vocabulary and concepts related to each field.

Geological maps depict the bedrock and surficial geology of an area, and are published in a variety of scales. Small-scale maps (those showing a relatively small area) offer the most detail. The most detailed map that you will probably be able to find of an area will be scaled at 1:24,000, the same scale as a USGS 7.5 minute quadrangle. This means that a single unit of measurement shown on the map is 24,000 times bigger in the "real" world. In other words, one inch on the USGS map corresponds with 2,000 feet on the ground (24,000 inches equals 2,000 feet). Geologic maps show the presumed areal extent of particular formations. This doesn't mean that actual bedrock will be exposed everywhere that the map indicates that a certain formation should exist. Bedrock is rarely exposed in coastal plains, where sediments and stream-deposited gravel make up the surficial geology. In areas that have been glaciated, tills composed of sediments and gravel transported into the area by the ice may mask bedrock. A review of geologic maps and relevant literature can aid in—

- ☑ the generation of expectations about what types of rocks could be encountered in a project area as a result of natural processes, enabling out-of-place materials (artifacts) to be more easily recognized
- ☑ an understanding of geology's influence on topography, which can ultimately improve your ability to recognize human-made landscapes
- ☑ the identification of rock types (lithic resources) that might have been useful to the prehistoric and historic inhabitants of an area
- ☑ the identification of possible sources of useful rock within and adjacent to a project area
- ☑ the identification of locations where quarries, rockshelters, or caves might exist within a project area

During the analysis of field data, the "lithic landscape" provided by geologic mapping is one context in which archaeological evidence of the procurement of resources, manufacturing, use, and distribution can be evaluated.

Since the maps portray formations, locating sources of rock types that might have been useful for the prehistoric and historic inhabitants of an area can require additional research. Flint, chert, jasper, and quartz, rock types frequently used by native artisans for the fashioning of chipped stone tools, occur within formations and will rarely be shown as specific units on a geologic map. On the other hand, sources of quartzite, rhyolite, metahyolite, and obsidian, also employed in the production of chipped stone tools, can be massive and are mapped as distinctive units on maps. The scale of the natural processes responsible for the origins of a particular type of rock usually determine whether they will be included in general geological mapping. Those resulting from large-scale processes will generally be mapped. The degree of detail on a map can also depend on whether the map was compiled during the past 20 years or earlier in the century. Maps resulting from specialized or focused studies may provide additional detail or deal with smaller-scale phenomena, although coverage may only involve a small geographic area.

The summary descriptions of formations included in the key to geologic maps must be supplemented by reference to other published materials. This heightens the chance of encountering mentions of materials of interest to an archaeologist. It is also useful to learn as much as possible about the characteristics of the lithic resources that might be found in a project area. How do the physical characteristics of a rock type influence the ways in which it might have been quarried, used, and transformed by humans? How do we distinguish naturally modified rock from culturally modified rock? Some types of rock used in the production of tools and implements can weather rather dramatically, making it difficult, if not impossible, for the uninformed person to recognize them as artifacts. Does a particular type of rock have attributes that allow it to be distinguished from similar rock types? Being able to link artifacts discarded at a site with the source of the raw material from which they are made (referred to as origin, or provenance studies) has incredible interpretive value.

Locating outcrops or exposures of rock can be accomplished using aerial photographs and topographic maps in conjunction with geologic maps. A topographic map will enable you to identify areas with extremely steep slopes. Comparing these slope areas with the geologic map of the area can suggest where outcrops may occur. The creation of the soils maps found in published soil surveys involves the use of aerial photos providing comprehensive coverage of large geographic areas. The photos are generally kept on file in the appropriate local office and can be used with permission. Some soils maps specifically locate outcrops, and as these maps are often based on an aerial photograph, outcrop areas may be visible even if not explicitly identified. Gravel deposits are associated with the stream channels, floodplains, and stream terraces of an area. A map of the surficial or Quaternary geology of an area may be necessary to locate ancient gravel deposits that occur well away from current streams.

Figure 4.3 (refer to Colorplates 2-3) of this book is a section of a geologic map of southeastern Pennsylvania included in a publication on the water resources of Bucks County (Hall 1973). Individual formations are shown as areas of distinctive color with associated letter codes. The level of detail that it shows is comparable to that of the geologic map of the entire state, published as a stand-alone map. The summary key that accompanies the map is sufficient to allow for the identification of formations that contain lithic resources of potential use to the historic and prehistoric inhabitants of the region: quartzite for chipped-stone tools; diabase, gneiss and granite for ground-stone tools like axes and celts; massive sandstones, diabase, gneiss, granite, limestone, and marble for historic building materials. Areas covered by Quaternary (the period of the last 2–3 million years) sediments and gravels, and glacially transported material (Illinoian and Jerseyan drift) can be noted. While it can be assumed that the Quaternary and glacial deposits will be exposed at the surface, the map really doesn't tell us where outcrops of the rocks that we are interested in might be located. Streams draining areas where bedrock of interest is located can have gravels in their channels and along their banks representing this bedrock.

Seeking out more detailed descriptions of the individual formations noted on the map, we would learn that there are in fact other types of rocks or lithic resources of interest in this same area of southeastern Pennsylvania. For example, the colors of the quartzites from different formations would be noted. We would learn that in the Lockatong Formation there are extensive areas of argillite, some of the limestones of the region contain dark-colored cherts, and jasper is found in association with the Hardyston Formation. Each of these rock types were employed in the chipped stone technologies of the region's Native Americans. Again, if these materials are in the bedrock, it is possible that they are also in the gravel deposits of the streams that drain the area. We might also learn of the rapid rate at which argillite can weather (Figure 4.4). Without this knowledge, it would be very easy to dismiss broken tools and chipping debris of this material as "non-artifacts" or products of nature.

Geologic fieldwork and mapping of the relatively small area encompassed by a single USGS 7.5 minute quadrangle have been completed for portions of many states. Figures 4.5 (Colorplate 1) and 4.6 are examples relevant to our Pennsylvania case study. Both show more detail than the previous map and have the advantage of being presented in conjunction with topography. This makes it easier to see the relationship between geology and landforms, and locate possible outcrops. In turn, mapping focused strictly on surficial deposits adds to the detail captured on Figure 4.6. We also learn in the descriptions of surficial deposits about their stratigraphy and that associated gravels include useful quartz, quartzite, and argillite.

FIGURE 4.4.

Weathered artifacts made from argillite and argillaceous material. When fresh, argillite is relatively hard and breaks with a conchoidal fracture, making it useful for the fashioning of chipped-stone tools. When left exposed at the surface, surfaces of artifacts can weather fairly rapidly, obscuring obvious flake scars and signs of modification by humans (also see Figure 2.3).

Although the examples above have focused on a single geographic area, the lessons they provide are applicable across the country. When compiling geological background for an archaeological project, consult a variety of sources and maps. Don't rely strictly on the characterization of formations that accompany the maps. This kind of background research can't be successful unless you already have some background, in this case a general knowledge of the geological resources exploited by the peoples of the past, and a basic familiarity with the language and scope of the field of geology.

KEY FOR FIGURE 4.5

(Refer to Colorplate 1)

Qal (pale yellow areas):	Alluvium; deposits of clay, silt, sand, and gravel along streams.
Qwo (yellow areas):	Outwash deposits; stratified sand and gravel deposits of fluvio-glacial origin, probably of Wisconsin age.
Trb (blue areas):	Member of the Brunswick formation, which includes red to reddish-brown and gray to greenish-gray clay-shale, mud-shale, silt-shale, mudstone, argillite, siltstone, and sandstone. Trb grades into limestone fanglomerate.
Trbl (blue areas):	Member of the Brunswick formation, which includes red to reddish-brown and gray to greenish-gray clay-shale, mud-shale, silt-shale, mudstone, argillite, siltstone, and sandstone. Includes quartzite fanglomerate.
Trbh (green areas):	Member of the Brunswick formation, which includes red to reddish-brown and gray to greenish-gray clay-shale, mud-shale, silt-shale, mudstone, argillite, siltstone, and sandstone. Trbh is adjacent to diabase intrusions.
Trd (orange areas):	Diabase, igneous sheets intrusive into the Brunswick formation. Unit is medium to coarse grained, dark gray to nearly black, has typical diabase texture, and is composed largely of grayish-green calcic plagioclase and green augite.

Abstracted from Drake et al. (1967).

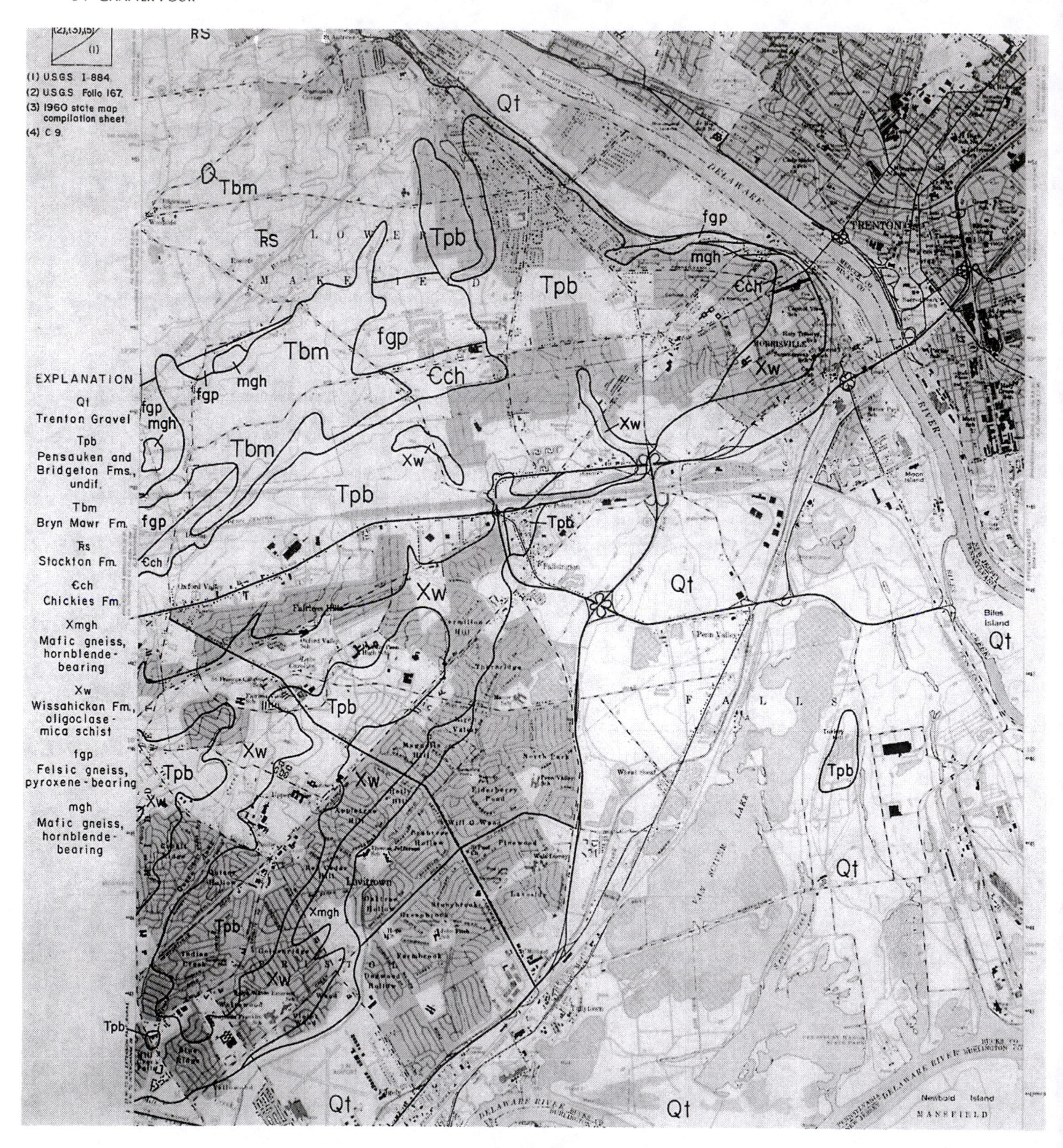

FIGURE 4.6.

Geologic map of the Trenton West, Pennsylvania and New Jersey quadrangle, 1978 (reduced). Individual formations are shown as areas bounded by a heavy black line with an identifying sequence of letters. Like Figure 4.5, this map shows the heightened detail that comes with mapping geological formations and deposits at the geographic scale of the 7.5-minute quadrangle (see Berg and Dodge 1981).

KEY FOR FIGURE 4.7

(Refer to Colorplate 4)

Qal (yellow areas on map) ALLUVIUM (HOLOCENE AND PLEISTOCENE)—Floodplain and channel deposits; characteristically very organic rich dark-colored silt and clay. Light-colored sands in Burlington at the mouth of Assunpink Creek. Tidal influence is strong near mouths of larger creeks and deposits are mostly muck. Several *C14* age determinations show that deposits range from Pleistocene through Holocene. Thickness unknown.

Qgwl (light brown areas on map) GRAYWACKE 1 (PLEISTOCENE)—Predominantly interstratified sand and gravelly sands which fine down valley. Surface altitudes of deposits south of Trenton range from 15 to 30 ft (4.5 to 9 meters) above sea level, and as high as 40 ft (12 meters) northwest of Trenton. Upper 6 ft (1.8 meters) of deposits typically are thin bedded, horizontally bedded, pale-red (10 R 6/2), pale-yellowish-brown (10 YR 6/2), or grayish-orange (10 YR 7/4) silt to very fine sand, commonly with thin gravel layer in base. Lower beds are commonly light-medium-gray (N 6) interstratified sand and gravelly sand as much as 40 ft (12 meters) thick. Gravel abundant and unweathered; boulders as much as 6 ft (1.8 meters) maximum dimension common. Layering typically includes horizontally stratified gravelly sand and cross-stratified sand. Sand contains dark rock fragments as a major constituent and may be compositionally considered a high-rank graywacke. In contrast with older higher deposits, gravels at this level contain limestone pebbles and cobbles. Leaching is, however, typical in upper 6 to 8 ft (1.8 to 2.4 meters) but below this depth pebbles are commonly coated or, rarely, cemented by calcium carbonate. Other types of gravel include quartzite, quartz, and red Triassic sandstone and shale. The age of Graywacke 1 is conjectural because no datable materials have been found. Based on field relationships up the Delaware River valley, these deposits could possibly be as young as Wisconsinan (Plum Point, about 24,000 radiocarbon years), or as old as Sangamon. We favor a Sangamon age because of the high elevation of these beds at Trenton (as high as 40 ft or about 12.2 meters) whereas the Wisconsinan-Holocene beds appear to lie at lower elevations in this part of the valley.

Qgw2 (dark brown areas on map) GRAYWACKE 2 (PLEISTOCENE)—Similar in texture and composition to Graywacke 1; that is, largely of interstratified sand and gravelly sand, but at higher elevations. Surfaces of terraces extensively dissected but range in altitude from 40 ft (12 meters) near Bristol to 70 ft (21 meters) beneath and northwest of the city of Trenton. Upper beds typically thin (1 in, (2.54 cm) to thick (5 ft (1.5 meters) horizontally stratified silt and very fine sand similar in color to upper beds of Graywacke 1. Lower beds thickest, as much as 30 ft (9 meters) and consist of interstratified, poorly crossbedded sand and horizontally bedded gravelly sand. Beds are also distinctly redder than Graywacke 1 and range from pale red (10 R 6/2) to grayish red (10 R 4/2) in part. More oxidized gravels (moderate-reddish orange, 10 R 6/6) as much as 20 ft (6 meters) thick are locally interstratified with these gravels particularly at base. Gravel is very abundant and many rock types are represented: quartzite, vein quartz, crystalline rocks, and Triassic sandstone, argillite, and shale. Feldspar-bearing gravel appears more weathered than that in Graywacke 1, particularly in the basal reddish-orange beds. Sands have generally the same graywacke composition as those in Graywacke 1 except in lower reddish-orange beds, which have much more feldspar than rock fragments and, therefore, are an arkose.

Tar1 (pink areas on map) ARKOSE 1 (PLIOCENE(?) AND MIOCENE(?))—Largely cross-stratified, fine to very coarse sand and horizontally bedded gravelly sand, typically stained by iron oxides throughout, although mostly dark reddish brown (10 YR 3/4) in upper 15 ft (4.5 meters). Beds below are commonly moderate orange pink (5 YR 8/4). Composition variable, but is largely a feldspathic quartz sand (subarkose to arkose); rock fragments are subordinate; and glauconite is locally a major constituent. Feldspar principally microcline but contains significant concentrations of plagioclase. Many of the plagioclase grains are skeletal, suggesting extensive postdepositional weathering of these beds. Gravel ranges in size to some rare boulders as much as 5 ft (1.5 meters) maximum dimension. Many rock types are represented: quartz, quartzite, red Triassic sandstone and shale, and a large number of crystalline rock types; feldspar-bearing gravels are largely saprolitized. Arkose in this area appears to be a series of degrading channels entering the Trenton area from the northeast and having surface altitudes ranging from a high of 80 ft (24 meters) to a low of 40 ft (12 meters). Base of channels follows same pattern, being roughly 40 ft (12 meters) above sea level where the channels have surface altitudes near 80 ft (24 meters), to nearly sea level (as at Rancocas) where surface elevations are about 40 ft (12 meters). The southernmost channel-system deposits are extensively iron-oxide cemented as much as 20 ft (6 meters) particularly in base. Deposits are as much as 40 ft (12 meters) thick in this area. The age of Arkose 1, or the Pensauken of some authors, is unknown because of a lack of distinct flora or fauna. Some have considered this unit to be Illinoisan because of its weathered character, but there is absolutely no evidence, except intuition, to support this age assignment. This unit could be as young as Sangamon or as old as late Miocene. A Tertiary age is favored because of the pervasive weathering throughout all of this unit.

Continued

KEY FOR FIGURE 4.7 (CONTINUED)

Tar2 (orange areas on map) ARKOSE 2 (PLIOCENE(?) AND MIOCENE(?))—Mostly sand, locally containing gravel, especially at base. Deposits similar to Arkose 1. Moderate-reddish-brown (10 R 4/6) near surface to pale-yellowish-orange or nearly white in lower beds. Well cross-stratified, having small to large co-sets; crossbeds mostly trough type, but planar beds common. Gravel mostly quartz and quartzite, but many other rock types present, notably crystalline rocks and Triassic red sandstone and shale. Crystalline clasts typically thoroughly saprolitized. Sand, chiefly quartz and feldspar, ranging in composition from protoquartzite to arkose; feldspar predominantly microcline. A soil zone as much as 15 ft (4.5 meters) thick is widely developed on the more sandy beds. Soil essentially a laterite with gibbsite and halloysite as primary weathering minerals. Deposits lie at distinctly higher altitudes than Arkose 1. As shown in cross section *A-A'*, these arkosic sands cover much of the main valley. Generally, the channels lie at higher altitudes (as much as 150 ft (45 meters) above sea level) and channel fills are thicker along the eastern side of the valley. Near the valley center, the upper surface of the deposits declines to near 100 ft (30 meters) and along the western edge rises again to nearly 130 ft (39.5 meters). The base of the deposits, for the most part, ranges between 65–75 ft (19.8–22.8 meters), except at the valley margins where the base is considerably higher. Section *B-B'* is drawn south of Trenton where the younger Quaternary units have entrenched through and lie below the general level of Arkose 2. Most of Arkose 2 in this general region apparently was stripped away and only occurs on hills having surface altitudes near 100 ft (30 meters). These hills are typically widely scattered, particularly east of the river. Larger, less dissected masses are present along the western edge of the deposit where the arkose overlies the crystalline rocks of the Piedmont. In this area, particularly near Silver Lake, surface altitudes of the arkose are as much as 160 ft (48.6 meters) and the base as low as 100 ft (30 meters). Down valley at the western edge of the map, the top of the arkose is a maximum of 120 ft (36 meters) and the base near 90 ft (27 meters). South of the river and along the western edge, the base of the arkose is at nearly 75 ft (22.8 meters). The overall distribution of the arkose in the map area suggests a series of degrading channels. Unit is as much as 70 ft (21 meters) thick. Arkose 2, based on stratigraphic relationships, is older than Arkose 1. This unit was transported into the main valley from the northeast, perhaps representing deposits of the ancestral Hudson. Indeed, this unit appears to be the oldest deposit within the main valley. No datable material has been found in this unit, so again its age is conjectural. Poorly preserved leaf imprints have been found at one locality near New Brunswick and have been interpreted to indicate a warm climate. The depth of weathering could be interpreted as indicating great antiquity, probably indicating a Tertiary age.

The more that you know about what might be beneath the ground before you delve into it, the better your chances of success as a field archaeologist. Published soil surveys contain a wealth of relevant information. As with geological source material, the more training or education that you have in the subject, the more you will be able to accomplish with this type of background information (see chapter 7).

A variety of data are used to produce the descriptions and maps that are part of every soil survey. Geologic maps, aerial photos, topographic maps, observations of current land use, and the analysis of profiles and soil samples derived from fieldwork are employed. Like geological maps, soil maps are abstractions to a certain degree. It is not possible to dig a hole in every small landscape within a geographic area to determine the type of soil present there. Although the preparation of a soil survey involves a good deal of fieldwork, the preparation of maps relies heavily on an understanding of the relationship between the characteristics of a soil and geology, topography, surface water and drainage characteristics, and the natural processes that can shape a landscape. Models of these relationships, standards of soil classification, and field data are used to identify areas where distinctive soils should exist, even if fieldwork did not occur in each individual area. The scale at which these relationships are modeled for generating maps is generally no smaller than an area of 6–7 acres in extent. This doesn't detract from the usefulness of soil surveys for archaeological research, but it does mean that you may eventually encounter variation in the field that does not conform with data in a published soil survey.

Soil data and maps are organized into soil series and phases of individual series. Soils that have profiles that are similar to one another make up a soil series. Patterned variations in the soils grouped as a series are the basis for defining soil phases. In many cases this variation results from a soil's position on a slope and how this effects soil formation, weathering, and erosion. Individual descriptions of soil series and phases include the characteristics (color, texture, structure, pH) and thickness of individual layers or strata within a profile, overall stratigraphy, the types of landscapes and geology or parent material on which the soil occurs, drainage characteristics, and how related landscapes are currently used, or could be used. You will learn more about how to read and understand soil profile descriptions in chapter 7. The maps in a soil survey show the geographic distribution of phases of individual soil series. Scaled aerial photographs are used as the base for individual soil maps. A review of published soil surveys can aid in

- the generation of expectations about what types of sediments, soils, and stratigraphy could be encountered in a project area as a result of natural processes, enabling human impact on the landscape and features to be more easily recognized
- identifying the depths to which test excavations will have to proceed in the search for archaeological deposits in specific areas or on specific landscapes
- identifying areas where examination of surfaces could reveal traces of archaeological deposits
- using stratigraphy and soil weathering characteristics to estimate the age of strata and horizons
- identifying potential logistical constraints for excavations and other fieldwork
- the compilation of information on land use history, and how this might affect field methods and the integrity of any archaeological deposits that might be found on specific landscapes
- identifying locations of possible sources of clay for use in the manufacturing of ceramics and brick
- identifying the natural processes that have been active in shaping the landscapes on which you are working
- reconstructing paleoenvironments

As you can see from this list, the thoughtful use of a soil survey can have a dramatic impact on the basic organization of fieldwork and budgeting time and funds for a project, or preparing a grant proposal or competitive bid for a cultural resource management study. Knowing the depth that soils can attain on a specific landscape will allow the experienced archaeologist to estimate how long it will take to complete an excavation of a given size. The depth to which an excavation must be taken in turn conditions the size of the unit that is opened (see chapter 9). An archaeologist with appropriate training, or a project's soil scientist or geomorphologist, can also estimate the gross age of a soil based upon physical characteristics that reflect the weathering of sediments over time, like the number and nature of "B" horizons and soil structure (see chapter 7). Statements about the relative age of soils may be included in a soil survey. The description of a series may note that the soil is developed from recent alluvial (water-deposited) sediments, while another may be old alluvium. Some soils may contain a well-dated horizon or stratum, owing to its distinctive character and involvement in previous archaeological or paleoenvironmental research. In all of these cases, information about the potential age of a soil or a stratigraphic layer helps an archaeologist decide how deep to dig. If the cultural prehistory of your area stretches back 12,000 years, then you need to examine 12,000 years worth of soil if you are searching for sites. If the soil column of the landscape on which you are working is 15 feet thick and 40,000 years old, why dig through the whole sequence?

How "heavy" are the soils to be excavated, that is, what percentage of clay do they contain? The clay fraction of a soil can affect the rate at which excavated matrix can be dry screened. The higher the clay fraction, the longer it will take. There may be so much clay in a soil that it might be more efficient to use water screening (see chapter 9). What are the drainage characteristics of a soil? If a soil is poorly drained, thought must be given to when (the season) excavations might best be undertaken. If there isn't enough flexibility in a project's schedule to accommodate seasonal changes in rainfall and climate, then it might become necessary to plan for the use of pumps or other equipment to control water in excavations.

Descriptions will also note whether a soil is one that is typically plowed, and since soil maps are based on aerial photos, the extent of agricultural fields easily can be seen. Together with information about the potential age of a soil, an archaeologist can decide how much of a survey project will involve surface inspections and how reliable these inspections might be in providing evidence of the full range of occupations in the area through time. The type of slope that a soil occupies is part of the standard information provided in a soil survey. With the exception of caves, rockshelters, and cliff dwellings, archaeological sites rarely occur on slopes greater than 12 percent. If you are engaged in an archaeological survey, these data can be used to exclude some project areas from all but a cursory examination.

How much rock can be expected to be found in the ground as a result of natural processes, and of what type, are noted in the description of soil series and phases. This may help you to decide whether the things being found in excavations are artifacts. Features may be visible as soil anomalies with distinctive colors that are in contrast to the surrounding matrix (see chapter 2). But they also can be textural anomalies with no visible color differences that allow you to distinguish them from the surrounding matrix. Knowing what textural variations you can expect within and between strata as a result of natural processes (see chapter 7) is necessary if you are going to effectively recognize textural anomalies. Soil descriptions provide you with some of this background. The pH of soils is also noted so you can get a good idea about the potential for organic artifacts and ecofacts to be preserved.

Soil maps can depict environmental details not found on other types of maps. Rock outcrops, the location of gravel deposits, springs and small streams not shown on USGS topographic maps may be shown. Since soil is a function of geology or parent material, topography, climate, vegetation, organisms, and the passage of time, it reflects the environments and natural processes of the past. With the appropriate training, soils and stratigraphic data can be helpful in your effort to reconstruct paleoenvironments.

An example of a soil map is shown in Figure 4.8. It depicts a 2 x 3-mile portion of the Yamhill River Valley in Oregon. Table 4.1 organizes the soil phases on the map by landform, and Table 4.2 gives examples of descriptions of individual phases. Current land use is easily seen on the aerial photo that serves as the base of the map. There are agricultural fields throughout the area interspersed with woodlots (areas with dark shading). A large orchard or tree plantation is visible in the northcentral portion of the map as a patterned distribution of trees. If we were planning an archaeological survey, the extent of agricultural land dictates that systematic surface examinations would be a major part of the field strategy. We might also assume that any shallowly buried sites have been disturbed to some degree by plowing. The soil descriptions for these agricultural areas indicate that the typical uppermost strata of a profile is a plowzone or "Ap" horizon (see Table 4.2). The planting and maintenance of orchards or tree nurseries can have an even greater impact on archaeological deposits (see below). A number of the soils listed in Table 4.1 occupy slopes greater than 12 percent and can probably be eliminated from any detailed consideration, streamlining the scope of the fieldwork we're planning. Areas of shale rock land are 50–75 percent rock outcrop (see Table 4.2).

FIGURE 4.8.

A portion of Sheet 48 from the *Soil Survey of the Yamhill Area, Oregon* and accompanying key (Otte et al. 1974). Scale 1:20,000. The map is oriented north-south. Areas occupied by a specific type of soil are bounded by a black, heavy line and accompanied by an identifying letter code. The map base is a scaled aerial photograph of the area. These types of maps provide information valuable for planning archaeological surveys, understanding site formation processes, and reconstructing the environment. See next page for accompanying key.

CONVENTIONAL SIGNS

WORKS AND STRUCTURES

Highways and roads

- Dual
- Good motor
- Poor motor
- Trail

Highway markers

- National Interstate
- U. S.
- State or county

Railroads

- Single track
- Multiple track
- Abandoned

Bridges and crossings

- Road
- Trail
- Railroad
- Ferry (FY)
- Ford (FORD)
- Grade
- R. R. over
- R. R. under

Tunnel

Buildings

- School
- Church

Mine and quarry

Gravel pit

Power line

Pipeline

Cemetery

Dams

Levee

Tanks

Well, oil or gas

Forest fire or lookout station

Windmill

BOUNDARIES

- National or state
- County
- Minor civil division
- Reservation
- Land grant
- Small park, cemetery, airport
- Land survey division corners

DRAINAGE

Streams, double-line

- Perennial
- Intermittent

Streams, single-line

- Perennial
- Intermittent
  - Crossable with tillage implements
  - Not crossable with tillage implements
  - Unclassified

Canals and ditches

Lakes and ponds

- Perennial (water, w)
- Intermittent (int)

Spring

Marsh or swamp

Wet spot

Drainage end or alluvial fan

RELIEF

Escarpments

- Bedrock
- Other

Prominent peak

Depressions (Large, Small)

- Crossable with tillage implements
- Not crossable with tillage implements
- Contains water most of the time

SOIL SURVEY DATA

Soil boundary

- and symbol (Dx)

Gravel

Stoniness

- Stony
- Very stony

Rock outcrops

Chert fragments

Clay spot

Sand spot

Gumbo or scabby spot

Made land

Severely eroded spot

Blowout, wind erosion

Gully

Slide or slip

FIGURE 4.8.

Continued

TABLE 4.1

MAPPED SOILS ORGANIZED BY LANDFORM

ALLUVIAL, BOTTOMS, AREAS ALONG STREAMS

Ch — Chehalis silty clay loam, 0–3% slopes
Ck — Chehalis silty clay loam, overflow areas
We — Wapato silty clay loam, 0–3% slopes

ALLUVIAL FANS

CeC — Chehalem silty clay loam, 3–12% slopes
Co — Cove silty clay loam, 2–7% slopes
Cs – Cove silty clay loam, thick surface, 0–2% slopes

STREAM TERRACES

Am — Amity silt loam, 0–2% slopes
CaD — Carlton silt loam, 12–20% slopes
Dc — Dayton silt loam, thick surface, 0–2% slopes
HeB — Hazelair silty clay loam, acid variant, 2–7% slopes
Te — Terrace escarpments, 20–40% slopes

HILLS, UPLAND SLOPES, RIDGETOPS

HcD — Hazelair silty clay loam, 7–20% slopes
JrD — Jory clay loam, 12–20% slopes
JrE — Jory clay loam, 20–30% slopes
JrF — Jory clay loam, 30–60% slopes
NeB — Nekia clay loam, 2–7% slopes
PaD — Panther silty clay loam, 4–20% slopes
PcC — Peavine silty clay loam, 2–12% slopes
PcE — Peavine silty clay loam, 20–30% slopes
PCF — Peavine silty clay loam, 30–60% slopes
PeD — Peavine silty clay loam, moderately shallow, 7–20% slopes
SH —Shale rock land
SL —Stony Land
StD — Steiwer silty clay loam, 5–20% slopes
StE — Steiwer silty clay loam, 20–30% slopes
WeD — Willakenzie silty clay loam, 12–20% slopes
WeE — Willakenzie silty clay loam, 20–30% slopes
WkB — Willakenzie silty clay loam, shallow, 2–7% slopes
WkD — Willakenzie silty clay loam, shallow, 7–20% slopes
YaE — Yamhill silt loam, 20–30% slopes
YaF — Yamhill silt loam, 30–50% slopes
YhB — Yamhill silt loam, moderately shallow, 2–7% slopes
YhD — Yamhill silt loam, moderately shallow, 7–20% slopes

TABLE 4.2

EXAMPLES OF SOIL DESCRIPTIONS

COVE SERIES

The Cove series consists of poorly drained, nearly level or gently sloping soils that formed in recent alluvium on bottom lands and fans. The areas are smooth, and some are slightly concave. Depth to clay generally ranges from 7 to 24 inches, but some areas have clay throughout. Elevations range from 30 to 300 feet. Annual precipitation is 40 to 60 inches, average annual air temperature is 53 degrees F., and the frost free season is 165 to 210 days. In areas that are not cultivated, the vegetation is ash and willow, sedges, cattails, and grasses. Cove soils are associated with McBee and Chehalis soils. In a representative profile, the surface layer is very dark gray silty clay loam about 8 inches thick. The subsoil is mottled very dark gray, very firm clay about 33 inches thick. It is underlain by mottled dark grayish brown clay that extends to a depth of more than 60 inches. Cove soils are used for hay and pasture and occasionally for spring grain and field corn. They are also used for wildlife habitat.

TABLE 4.2

EXAMPLES OF SOIL DESCRIPTIONS (Continued)

Cove silty clay loam (Cn) 0 to 2 percent slopes. This soil is on nearly level or slightly concave bottom lands. Representative profile about 50 feet north of the Amity-Bellevue road and 25 feet west of ditch; NE 1/4 NW 1/4 SE 1/4 sec. 25 T. 5 W:

Ap–0 to 8 inches, very dark gray (10YR 3/1) silty clay loam, dark gray (10YR 4/1) when dry; many fine, distinct, yellowish brown and dark reddish-brown mottles; moderate, fine, subangular blocky structure; firm, hard, sticky, plastic; common fine roots; common line and very fine pores; medium acid (pH 5.6); abrupt, smooth boundary. (7 to 16 inches thick)

B21g–8 to 16 inches, very dark gray (10YR 3/1) clay, dark gray (N 4/0) when dry; many fine, distinct, dark yellowish-brown and dark reddish brown mottles; weak, medium, prismatic structure that parts to moderate, fine, subangular blocky; very firm, very hard, very sticky, very plastic; few fine roots; many very fine pores; slightly acid (pH 6.1); clear smooth boundary. (7 to 10 inches thick)

B22g–16 to 41 inches, very dark gray (N 3/0) clay, dark gray (N 4/0) when dry; common, fine, faint, dark yellowish-brown mottles; moderate, medium and coarse prismatic structure; very firm, very hard, very sticky, very plastic; few fine roots; common very fine pores; occasional slickensides that do not intersect; few scattered fragments of igneous rock and sedimentary rock; slightly acid (pH 6.4); clear smooth boundary (15 to 26 inches thick).

Cg–41 to 60 inches, dark grayish brown (2.5Y 4/2) clay, gray (N 5/0) when dry; many fine, dark reddish brown and dark yellowish-brown mottles; massive, very firm, very hard, very sticky, very plastic; very few roots; common very fine pores; slightly acid (pH 6.4).

Depth to clay is 7 to 16 inches. Moist values are 2 or 3 and chromas are 1 or less throughout, except that in the A and C horizons chromas range to 2. Dry values are 4 to 5. Hues are predominately 10YR in the A horizon but range to 2.5Y or neutral; hues are generally 2.5Y or neutral in the B horizon but range to 10YR. Structure is moderate or strong in the solum. Mottling is distinct or prominent within a depth of 20 inches. A few fine rock fragments occur in places. Included with this soil in mapping are areas of Wapato and Labish soils. These areas are less than an acre in size and make up less than 5 percent of the total acreage. This Cove soil is very slowly permeable, and roots seldom penetrate below a depth of 16 inches. It is difficult to cultivate except in summer. The available water capacity is 4 to 6 inches. Surface runoff is very slow, and ponding often occurs during winter. Overflow is common during winter. The erosion hazard is slight. Fertility is low. Most of the acreage is cultivated. Hay and pasture plants are the principal crops. Spring grain is grown occasionally. Drainage by ditches and surface smoothing extend the season of use. Capability unit IVw-2; not placed in a woodland group; wildlife group 2.

Cove silty clay loam, thick surface (Cs) 0 to 2 percent slopes. The profile of this soil is similar to that of Cove silty clay loam, except that the depth to clay ranges from 16 to 24 inches. Texture in the surface layer ranges from silty clay loam to silty clay. Available water capacity is 6 to 7.5 inches, and the fertility is moderately low. Hay and pasture plants are the principal crops, but spring grain and vegetable crops are also grown. Capability unit IIIw-2; not placed in a woodland group; wildlife group 2.

Cove silty clay loam, fan (Co) 2 to 7 percent slope. This soil is on alluvial fans. It has a profile similar to that of Cove silty clay loam. The lower part of the subsoil is 20 to 40 percent strongly weathered siltstone, shale, and a few basalt fragments. This soil is associated with the somewhat poorly drained Chehalem soils. Included with this soil in mapping are areas less than an acre in size of similar soils that have a surface horizon as much as 24 inches thick over the clay subsoil. Also included are areas of a Chehalem soil that are as much as an acre in size. Hay, pasture plants, and occasionally small grain are grown. This soil has a high water table caused by seepage from adjacent hills during winter and spring. Capability unit IVw-1; not placed in a woodland group; wildlife group 2.

TABLE 4.2

EXAMPLES OF SOIL DESCRIPTIONS (Continued)

PEAVINE SERIES

The Peavine series consists of well drained soils that formed over sedimentary rock on low hills and foot slopes of the Coast Range. These soils have 2 to 60 percent slopes. Elevations range from 400 to 1200 feet. Annual precipitation is 55 to 70 inches, average air temperature is 51 degrees F., and the frost free period is 165 to 210 days. In areas that are not cultivated, the vegetation is Douglas-fir, bigleaf maple, oak, and poison-oak. Peavine soils are associated with Willakenzie, Melby, Ead, Panther, Nekia, Jory, and Olyis soils. In a representative profile, the surface layer is very dark brown and dark-brown silty clay loam about 10 inches thick. The subsoil is dark reddish-brown and yellowish-red, firm silty clay about 26 inches thick. It is underlain by fractured strongly weathered shale at a depth of about 36 inches. Peavine soils are used mainly for orchards, small grains, hay, pasture, and timber. They are also used for water supply, wildlife habitat, recreation and home sites.

Peavine silty clay loam (PcC) 2 to 12 percent slopes. This gently sloping to strongly sloping soil is on ridge tops, on side slopes of low hills, and on foot slopes of the Coast Range. Slopes are commonly more than 5 percent. Representative profile about 50 yards north on field road from county road along ridge top in the southeastern corner of Moores Valley; SW1/4 SW1/4 SE1/4 sec. 10 T. 3 S., R., 5W.:

A1–0 to 4 inches, very dark brown and dark brown (7.5YR 2/2 and 7.5YR 3/2) silty clay loam, dark brown (7.5YR 4/2 and 4/3) when dry; moderate, fine, subangular blocky structure; friable, slightly hard, sticky, plastic; many fine and very fine roots; common very fine tubular pores; few fine fragments of shale; medium acid (pH 5.6); clear, smooth boundary. (3 to 7 inches thick)

A3–4 to 10 inches, dark brown (7.5YR 3/2) silty clay loam, brown (7.5YR 4/4) when dry; moderate, fine, subangular blocky structure; firm, hard, sticky, plastic; many fine and fine roots; many very fine tubular pores; few fine fragments of shale; very strongly acid (pH 5.0); clear, wavy boundary. (0 to 8 inches thick)

B1t–10 to 15 inches, dark reddish-brown (5YR 3/4) silty clay, yellowish-red (5YR 4/6) when dry; moderate, fine, subangular blocky structure; firm, hard, very sticky, very plastic; many fine and medium roots; few fine tubular pores; few thin clay films on ped surfaces and in pores; very strongly acid (pH 4.6); clear, wary boundary. (0 to 8 inches thick)

B2t–15 to 26 inches, yellowish-red (5YR 4/6) silty clay, yellowish red (5YR 5/6) when dry; moderate, medium, subangular blocky structure; firm, hard, very sticky, very plastic; many medium roots; many very fine tubular pores; common thin clay films on ped surfaces and in pores; many very fine fragments of shale; very strongly acid (pH 4.5); clear, wavy boundary. (8 to 15 inches thick)

C1–36 to 49 inches, light yellowish-brown to brownish-yellow fractured, strongly weathered shale with yellowish-red (5YR 4/6) silty clay loam in fractures, yellowish-red (5YR 5/6) when dry; massive, firm, very hard, sticky, plastic; common fine roots in fractures; common very fine pores in soil material; many thin clay films coat the shale; very strongly acid (pH 4.5); gradual, wavy boundary. (4 to 10 inches thick)

C2–49 to 64 inches, light yellowish-brown to brownish-yellow fractured, strongly weathered shale with light yellowish-brown (10YR 6/4) to brownish-yellow (10YR 6/6) silty clay loam in fractures, yellow (10YR 7/6) when dry; massive, few medium roots in fractures; many moderately thick, yellowish-red (5YR 4/6) clay films coat the shale; extremely acid (pH 4.3); gradual, wavy boundary. (0 to 20 inches thick)

R–64 to 84 inches, light yellowish-brown (10YR 6/4) to brownish-yellow (10YR 6/6) very hard fractured bedrock.

TABLE 4.2
EXAMPLES OF SOIL DESCRIPTIONS (Continued)

Included with this soil in mapping are areas of Panther. Willakenzie, Jory, Nekia, and more steeply sloping Peavine soils. Included areas are less than 2 acres in size in cultivated areas and less than 5 acres in size in woodland areas. They occupy less than 10 percent of the total acreage. This Peavine soil has moderately slow permeability. Roots can penetrate to bedrock. Tilth is moderate, and the soil can be cultivated most of the year except during winter and early in spring. Surface runoff is slow to medium, and erosion is a moderate hazard in unprotected areas during rainy periods. Fertility is moderate. Less than half the acreage is cultivated. Orchard trees, small grain, hay, and pasture plants are grown. Douglas-fir grows in wooded areas. This soil may slide and slump on the steeper slopes during periods of sustained rain. Capability unit IIIe-2; woodland group 2c1; wildlife group 3.

Shale Rock Land (SH) Shale rock land is 50 to 75 percent rock outcrops. The rest is well-drained soils that are too variable to identify and map separately. The soils are less than 20 inches deep over siltstone, sandstone, and shale. They are strongly sloping to very steep. In areas that are not cultivated, the vegetation is oak, poison-oak, grasses, and some Douglas-fir. The soils are loam to clay in texture and contain *few* to many sedimentary rock fragments. In some cultivated fields, these soils are very severely eroded. Moist hues are 5YR to 10YR, and values and chromas are 2 through 4. Included in mapping are areas of soils that are deeper than 20 inches. These included areas are less than an acre in size and occupy less than 10 percent of the total acreage. The soils of this mapping unit have moderate to slow permeability. Roots penetrate to a depth of 20 inches. Tilth is poor to moderate. The available water capacity is 2 to 5 inches. Surface runoff is medium to rapid, and erosion is a severe hazard in unprotected areas during heavy rains. Fertility is low. Less than a third of the acreage is cultivated. Hay and pasture plants are the principal crops. Small grain can be grown where slopes are favorable and rock outcrops are of limited extent. Uncultivated areas are in natural oak-grass and pasture. Capability unit, VIe-4; not placed in a woodland group; wildlife group 6.

STEIWER SERIES

The Steiwer series consists of well-drained soils that formed from old alluvium and colluvium. These soils have 5 to 50 percent slopes. Depth to bedrock is 20 to 40 inches. Elevations range from 250 to 650 feet. Annual precipitation is 40 to 50 inches, average annual air temperature is 53 degrees F., and the frost-free period is 165 to 210 days. In areas that are not cultivated, the vegetation is grasses, oak, and poison-oak. Steiwer soils are associated with moderately fine textured Willakenzie, Hazelair and Yamhill soils and Shale rock land. In a representative profile, the surface layer is dark brown silty clay loam about 6 inches thick. The subsoil is clay loam that is dark brown in the upper part, dark yellowish-brown in the lower part, and about 21 inches duck. Shale that has sandstone lenses is at a depth of about 27 inches. Steiwer soils are used primarily for grain, hay, and pasture. They are also used for wildlife habitat, recreation, water supply, and home sites.

Steiwer silty clay loam (StD) 5 to 20 percent slopes. This soil is on low hills. Representative profile in a pasture northeast of Sheridan, 1000 feet east and 540 feet north of the Sheridan City dump; NW1/4SW 1/4SW1/4 sec. 25, T. 5S., R. 6W:

Ap–0 to 6 inches, dark-brown (10YR 3/3) silty clay loam, brown (10YR 5/3) when dry; moderate, fine, subangular blocky structure that parts to moderate, fine, granular; firm, hard, sticky, plastic; many very fine roots; common very fine tubular and irregular pores; few, very fine, weathered fragments of shale; strongly acid (pH 5.4); abrupt, smooth boundary. (4 to 8 inches thick)

B1–6 to 10 inches, dark-brown (10YR 3/3) clay loam, brown (10YR 5/3) when dry; moderate, very fine, subangular blocky structure; friable, hard, sticky, plastic; many very fine roots; many

TABLE 4.2

EXAMPLES OF SOIL DESCRIPTIONS (Continued)

very fine tubular pores; many, very fine, weathered fragments of shale; strongly acid (pH 5.3); clear, wavy boundary. (3 to 8 inches thick)

B21–10 to 19 inches, dark-brown (10YR 3/3) clay loam, brown (10YR 5/3) when dry; moderate, very fine, subangular blocky structure; friable, hard, sticky, plastic; common fine roots; many very and very fine pores; few, fine, strongly weathered fragments of shale; strongly acid (pH 5.3); gradual, wavy boundary. (6 to 15 inches thick)

B22–19 to 27 inches, dark yellowish-brown (10YR 3/4) clay loam, brown (10YR 5/3) when dry; moderate, fine, subangular blocky structure; firm, hard, sticky, plastic; few fine roots; many very fine pores; 15 percent very fine, strongly weathered fragments of shale; strongly acid (pH 5.2); abrupt, irregular boundary. (6 to 12 inches thick)

IIC–27 inches, strong-brown (7.5YR 5/6) to reddish-yellow(7.5YR 6/6) variegated shale that has sandstone lenses; reddish-brown clay coatings in the fractures; very strongly acid (pH 4.9).

The A horizon ranges in texture from silt loam to silty clay loam or clay loam. Moist values are 2 and 3, and chromas are 2 or 3. The B2 horizon is silty clay loam or clay loam. Moist values are 3 or 4, and dry values are 5 or 6; chromas are 3 or 4. Hues range from 10YR to 7.5YR. Content of siltstone and shale fragments ranges from a few in the upper part of the solum to as much as 30 percent in the lower part. These fragments are weathered to some degree. Depth to bedrock ranges from 20 to 40 inches, but is commonly 24 to 32 inches. Included with this soil in mapping are areas of Yamhill and Willakenzie soils, more steeply sloping Steiwer soils and Shale rock land. Also included are areas of Stony land. The included areas are less than 2 acres in size and occupy as much as 10 percent of some mapped areas. This soil has a moderately slow permeability. Roots can penetrate to depths of 20 to 40 inches. The available water capacity is 3.5 to 8 inches. Surface runoff is medium and erosion hazard is severe in unprotected areas during rainy periods. Fertility is low.

Less than half the acreage is cultivated. Grain, hay and pasture plants are the principal crops. A few prune orchards are in production. Oak grassland pasture is the principal use. Capability unit IVe-1; not placed in a woodland group; wildlife group 3.

Wapato silty clay loam (Wc) 0 to 3 percent. This soil is in low lying areas along streams. It has smooth topography and is subject to short periods of overflowing and ponding. Representative profile in a field of vetch along the South Yamhill River, 50 feet south of abandoned county road; NW1/4NW1/4NE1/4 sec. 23, T.5 s., r. 5 W:

Ap–0 to 9 inches, very dark grayish-brown (10YR 3/2) silty clay loam, dark grayish-brown (10YR 4/2) when dry; very dark brown (10YR 2/2) coatings on peds; weak and moderate, fine, subangular blocky structure; firm, slightly hard, slightly sticky, plastic; many very fine pores; many fine roots; slightly acid (pH 6.4); abrupt, smooth boundary. (6 to 9 inches thick)

A1–9 to 16 inches, very dark grayish-brown (10YR 3/2) silty clay loam, very dark grayish-brown (10YR 4/2) when dry; very dark brown coatings on ped surfaces; many, fine, distinct, reddish-brown (5YR 3/2) mottles; few, fine, black stains and coatings; moderate, fine, subangular blocky structure; friable, hard, slightly sticky, plastic; many very fine and common fine pores; common fine roots; slightly acid (pH 6.2); gradual, smooth boundary. (6 to 10 inches thick)

B21g–16 to 22 inches, dark grayish-brown (10YR 4/2) silty clay loam, grayish-brown (10YR 5/2) when dry; many, fine, distinct, dark reddish-brown (5YR 3/2) mottles; few, fine, black stains; moderate, medium and fine, subangular blocky structure; friable, hard, sticky, plastic; many very fine and few fine pores; few fine roots; medium acid (pH 5.8); clear, smooth boundary. (5 to 8 inches thick)

TABLE 4.2

EXAMPLES OF SOIL DESCRIPTIONS (Continued)

B22g–22 to 32 inches, dark grayish-brown to grayish-brown (10YR 4/2–5/2) silty clay loam; many, fine, distinct, dark reddish-brown (5YR 4/4) mottles; common, fine, black stains; moderate, medium, fine, subangular blocky structure; firm, hard, sticky, plastic; many very fine and few fine pores; medium acid (pH 5.8); clear, smooth boundary. (9 to 22 inches thick)

B3g–32 to 60 inches, grayish-brown (10YR 5/2) silty clay, light gray (10YR 7/2) when dry; many, fine, distinct, dark-brown (7.5YR 4/4) mottles; common, medium and fine, black stains; weak, subangular blocky structure; firm, very sticky, plastic; few fine pores; medium acid (pH 5.6)

The A horizon has moist values of 2 and 3, dry values of 4 or 5, and chromas of 2 or 3. In places, distinct mottles occur throughout the A horizon or only in the lower part. The A horizon is dominantly silty clay loam but is silt loam in places. The B2 horizon has moist values of 4 and 5 and chromas of 1 and 2; hue is 10YR to 2.5Y, and in places it is 5Y. Mottles are distinct to prominent. Texture is dominantly silty clay loam, but ranges to a silty clay in the lower part below a depth of 30 inches. A few waterworn pebbles are embedded in the solum. In places the solum is underlain by stratified layers that contain pebbles and stones below a depth of 40 inches. Included with this soil in mapping are areas of Cove and Chehalis soils, narrow, steeper sided drainage ways, and in the Sheridan area, some unnamed gravelly soils. These included areas are as much as an acre in size and make up less than 5 percent of the total acreage. This Wapato soil is moderately slowly permeable to roots and water. During winter and early spring, a temporary water table restricts root growth. The available water capacity is 10 to 12 inches. Tilth is moderately good, but seedbed preparation can be difficult if the soil is worked when too wet or too dry. Surface runoff is slow, and water ponds for short periods during winter. The erosion hazard is slight. This soil is subject to occasional to frequent overflow. Fertility is moderate. Most of the acreage has been cleared for cultivation. Small grain, hay and pasture plants are the principal crops. Corn, other late planted vegetable crops, and grass and legumes for seed are also important. Drainage by either open ditches or tile is needed in order to lower the water table in spring. Because of the low lying position of the soil, drainage outlets are often difficult to establish. Capability unit IIIw-5; not placed in a woodland group; wildlife group 2.

The above is abstracted from Otte et al. (1974).

Intermittent streams are shown that do not appear on the USGS quadrangle of the same area, crucial information for models used to predict site locations. Rock outcrops and gravel deposits are scattered from west to east across the central portions of the area (see the "o" and "v" symbols on the map and accompanying key). Soil profiles are shallow in these areas, as they are in zones mapped as shale rock land or stony land. Depending on the degree of surface exposures, test excavations may not be necessary to search for archaeological deposits in these areas. On the other hand, areas of shale rock land and stony land are rarely plowed, so even shallowly buried archaeological deposits, if they exist, might be relatively undisturbed.

The depth of excavations needed to search for the widest possible range of prehistoric archaeological deposits will vary from landform to landform, as can be surmised from the examples of soil profiles in Table 4.2. Excavation units measuring from 1.5 x 1.5 feet to 5 x 5 feet on a side could be used to examine these deposits. Dry screening of excavated matrix will be relatively slow since many of the area's soils are silty clay loams. The pH of the soils ranges from 4.9 (strongly acid) to 6.4 (slightly acid), so organic artifacts will not typically be preserved. The southern third of the area's landscapes have been shaped by alluvial processes, i.e., flowing water transporting and depositing sediments. Since landscapes affected by streams are some of the most dynamic, we might expect to find the most well-preserved and stratified archaeological deposits in these areas. The Cove and Wapato soils are poorly drained on a seasonal basis, and excavations would best be carried out in the summer season.

Don't be daunted by the complex detail in the soil descriptions and their interpretation. As an archaeologist you will spend your life in the field dealing with material evidence embedded in some type of sedimentary or soil matrix. Doesn't it make sense to learn as much as you can about sediment and soils?

Topographic maps, those that depict the surface features of the earth, are another easily obtained and rich source of environmental and other background data. In addition to depicting elevations and the rise and fall of the landscape, they show current cultural features (buildings, roads, railroads, airports, etc), open versus forested land, streams, and surface water. Federal and state governments are the largest source of them. The USGS publishes the greatest variety with scales ranging from that of the always-important 7.5 minute quadrangle at 1:24,000, capturing an area of approximately 7 x 9 miles, to those illustrating large regions at 1:250,000. State topographic or geological surveys may publish countywide topographic maps at scales of 1:50,000 to 1:62,500. The smallest scale topographic maps are produced as part of construction or engineering projects that alter the land in some way, like highways or residential developments. These are of particular importance to archaeologists involved in cultural resource management studies as they are often studying the selfsame areas. Like soil maps, topographic maps aid in the identification of landforms that can be linked with specific environmental processes, and land use with different potentials for where and how archaeological deposits might occur.

Aerial photos are also a source of environmental as well as cultural data. I've already noted some of their uses above. County or regional offices of the SCS/NRCS keep series on file owing to their importance in preparing soil maps and managing conservation. The National Aerial Photography Program of the USGS is another good source of photos. Crop marks, patterned growth in vegetation, and patterned stains in surface soils hinting at subsurface features and archaeological deposits are often most obvious from the air. Aerial photos may also be available from a state's geological survey.

Background information on the forests and vegetation of an area can be obtained from a variety of published texts and maps. The scales of published maps vary greatly. In any case, mapped data, like those for geology and soils, are generalized models of what actually exists or would be anticipated to exist on a specific landscape. Vegetation maps are often published by a state's department of natural resources. Information on vegetation also is typically included in published soil surveys. The preparation of Environmental Impact Statements (EIS) are often part of legally mandated projects requiring cultural resource studies. If so, the company with which you are working will have access to the detailed vegetation and other environmental data compiled as part of the EIS.

LAND USE HISTORY

How a landscape has been used in the recent past can be important to know when planning and conducting an archaeological survey. What have historic occupants done to the land that might affect the way that archaeological sites are searched for? How might they have altered the integrity of an existing archaeological deposit? Working out the land use history of a property or project area makes use of the wide variety of information collected during background research and interviews with landowners. Historic maps, aerial photos, soil and topographic maps are ready sources of information on this topic. Staski (1982) assesses the type of subsurface disturbance and potential archaeological remains associated with different types of structures commonly found in urban areas.

Looking at dated series of topographic maps and aerial photos can illustrate how a landscape has been transformed over time. They can indicate how the extent of open, farmed, and forested land has changed, where and how development has progressed. While older aerial photos are often kept on file, it can be more difficult to locate older versions of USGS topographic maps for a given area. When existing maps are updated, older versions are often liquidated. Libraries that include maps in their holdings are probably the best place to look first for these older maps. The National Oceanic and Atmospheric Administration (NOAA) is the source of topographic maps of coastal areas that also chart water depth. Similar maps have been produced since the late 19th century and as a group provide one means of examining recent environmental change along shorelines related to sea level rise, stream dynamics, and tides.

Care must be taken in correlating historic maps that lack latitude or longitude references with more modern maps. Use as a comparative reference point something that is least likely to be misrepresented or overgeneralized on both maps. The alignment of roads can change, the location of some structures can be generalized.

FIGURE 4.9.

A pit is being dug around this large tree so that its root ball may be isolated and wrapped in burlap, prior to being removed from the nursery. This typical activity can have a dramatic impact on archaeological deposits should they exist on the same landscape.

FIGURE 4.10.

The linear bare patches in this field are where sod has been cut and stripped from a larger field of sod. Sod farming results in the gradual loss of sediment from the landscape being farmed. This practice may expose artifacts at the surface and result in their displacement.

Focus on some cultural or natural feature that might not have been altered substantially like bridges or the junction of major streams.

Recognizing where farms once existed or now occur shouldn't be too difficult. Plowing associated with farms creates extensive surface exposures for examination even as it mixes and churns any shallowly buried or layered sedimentary and archaeological deposits. Ponds may be dug and the course of small streams altered. Small quarries may have been opened. These are landscape alterations that usually can be discerned from maps and aerial photos. Other important alterations will not be so apparent, and should be the focus of interviews with property owners.

Orchards, nurseries, sod farms, and areas that have been repeatedly logged should be identified since all can have an impact on subsurface deposits and archaeological sites (Figures 4.9 and 4.10). The planting and removal of trees will disturb deposits well below the level that a plow will. Where nurseries are involved, it is not unusual to encounter subsurface disturbances 3–5 feet deep associated with the removal of more mature trees. Old or dead orchard and nursery stock may be removed with bulldozers, compounding the degree of subsurface disturbance. All of these activities can result in artifacts being brought to the surface where they might be found during a surface survey. It can make the initial discovery of sites easier, but presents problems with determining the size of an archaeological deposit and the degree to which it has been disturbed.

Logging practices today are more environmentally friendly than they were in the past. The creation of roads, landings where logs are stockpiled for transport, and the use of heavy machinery like skidders, forms ruts and exposes once-vegetated surfaces to erosion. This process may ultimately reveal artifacts and archaeological sites but disturb deposits in the process. Historic background research may reveal areas that have been repeatedly logged. Plantations of trees will be visible on modern topographic maps and air photos, fingerprinting areas that have been intensively logged. Wood and paper companies will maintain records dealing with their logging activities as will any government agency involved in forestry.

If you are participating in an archaeological survey that encompasses residential areas you should be aware of the earth-moving and grading that is typically associated with their construction. The topography of a development may be graded with heavy machinery to improve drainage relative to where roads, utilities, and houses will be located (Figures 4.11 and 4.12). This may expose archaeological sites or bring them nearer an existing surface in some cases, and in other cases result in their burial under fill. Once construction begins, the topsoil or "A" horizon of an area is stripped and stockpiled for later use (Figure 4.13). Any artifacts in this soil are displaced from their original context, and artifacts and features in the subsoil may be exposed. When construction is complete, the topsoil is used in the final grading and landscaping of the development.

FIGURE 4.11.

Grading of the existing topography for practical and aesthetic purposes is often associated with the construction of housing. In this photo, fill is being placed over an existing surface. The truck is parked on the original surface. Placed fill and the new surface being created are visible to the left.

FIGURE 4.12.

Road construction in general involves alteration of the existing landscape. Here the topsoil has been stripped and stockpiled in advance of additional grading. The flagged stakes in the picture are used to guide subsequent cutting and filling operations.

Any artifacts in the topsoil get redeposited in a variety of new contexts. An archaeological survey of such an area will discover artifacts and features. The challenge is to determine which represent intact deposits. Being able to recognize the stratigraphic fingerprint of cutting and filling associated with grading will also influence the strategies used in test excavations. Municipal authorities that approve construction plans often keep relevant maps on file, so it may be possible to examine

FIGURE 4.13.

Standard construction practices can have a dramatic impact on landscapes and the archaeological deposits that they may contain. Here you see a portion of a housing development under construction. The topsoil or "A" horizon has been mechanically stripped from the area and stockpiled adjacent to the house being built. Exposed subsoil is evident in the foreground of the picture.

"before" and "after" topographic maps of a landscape to see where the greatest impacts have occurred.

Most types of road construction involve the alteration of the landscape, as does the building of above-ground utilities and rights-of-way. As in the examples discussed above, grading can result in the exposure, displacement, or burial of archaeological deposits. The construction of below-ground utilities (e.g., gas, oil, water, and sewer pipelines) involves the digging of deep, if not broad trenches. This can disturb buried archaeological deposits. It can also result in bringing once-buried artifacts to the surface and redepositing them there. Once-buried sediments and the artifacts associated with them may be employed in regrading surfaces once construction trenches are backfilled. Government and utility authorities maintain plans of this construction that may be useful in planning an archaeological survey. Excavations or subsurface testing within a utility right-of-way or any public right-of-way have the potential to encounter buried utilities such as water, gas, oil, or electrical pipes and lines. Most states have toll-free telephone numbers that allow anyone planning any type of excavation to arrange for a utility markout before any digging is initiated.

Existing industries will be easy to spot on the landscape and their potential impact on the environment and archaeological deposits assessed. Historic background research will reveal former industries of relevance to a project. In addition to a consideration of landscape alteration, archaeologists must be aware of hazardous or toxic wastes that might be associated with an industry and the landscape it occupies. These pose a health threat to crews working in the field at all levels of archaeological investigation. There are formal procedures and regulations that guide working in such areas once they have been identified, and archaeologists do work in them (see chapter 5).

In archaeological excavations you will be searching for floral and faunal remains that represent the environments of the past, as well as the subsistence, crafts, and medicinal practices of humans. Identifying the current flora and fauna of an area provides a comparative base for comprehending the nature and degree of environmental change over time. This comparative base is also useful in deciding whether microforms of ecofacts are in fact directly associated with an archaeological deposit. Are the seeds recovered from an excavation level ancient, or are they modern ones moved through the ground by the action of groundwater or the activities of insects and animals? Is the pollen in the same level recent or ancient? Knowing what can be found on a landscape now will figure in the resolution of these questions.

In planning an archaeological survey, don't automatically forsake an area because construction or other landscape-altering activities have occurred. Be aware of the problems and prospects that come along with them. Significant portions of sites can be preserved even in the midst of profound disturbances (Figure 4.14).

FIGURE 4.14.

Significant archaeological deposits can be found even in highly developed areas like downtown Manhattan in New York City. The layered logs and timbers seen here are the remains of late-18th-century wharves exposed at the Assay Site.

LANDOWNER AND INFORMANT INTERVIEWS

Landowners are interviewed prior to fieldwork if for no other reason than to obtain permission to trespass and work on their property. Of course there is much more to be learned from such encounters, and it's a great chance to communicate the workings of archaeology to the public. The same is true of interviews with artifact collectors or other knowledgeable individuals that you may contact. You will learn and you will teach. There needs to be a give and take in any interview. If you are not already aware of who owns a property, they can be identified by visiting the appropriate municipal authority and examining tax maps for the area. It's a fairly speedy process and the documents are open to the public. Owners may not reside on the property and you may end up speaking with them and a resident tenant. The names and addresses of collectors will be amassed in the course of reviewing site data at the SHPO and consulting with other archaeologists who work in the same region, as will information about other persons with expertises you wish to tap (e.g., local historians, geologists, mineral collectors, librarians).

Initial contact to schedule an interview is best made over the phone or in person. When private property is involved in a cultural resource management investigation, landowners may be notified through the mail by the responsible government department that a project is being planned or in progress, asking for their cooperation. This doesn't absolve the archaeologist from her responsibility to make personal contact. Again, there is too much to be learned from this interaction. I've known too many landowners who simply chuck any unrecognized mail into the garbage without ever opening it.

Don't misrepresent yourself, those for whom you work, or the project in which you are engaged. Be sincere, be yourself, and be clear about what you hope to do. Remember that you are asking favors of a complete stranger. In all of the years that I've conducted these types of interviews I have never had a landowner ask me to sign a liability waiver or provide proof that I was covered by insurance should something happen to me while working on their property. Regardless, you should be aware of whether you are covered for liability and injury while working on a project.

Sooner or later you will encounter hard feelings and high emotion working in cultural resource management. People may be faced with relocation or the condemnation and taking of their property in order that a federal undertaking go forward. They might be upset by the fact that they may soon be living next to an 8-lane highway or a sewage treatment plant. You may be the first person involved in any way with such projects who comes face-to-face with the landowner. It doesn't matter that you are an archaeologist and are only concerned with a tiny piece of a much bigger endeavor. Get ready to be cussed, hear complaints, questions, and things best reserved for the lawyers and bureaucrats managing the project. Be patient, answer honestly the questions that you can, and make your role in the overall process clear. Be prepared to hear the word "no" once in a while. I'm not going to go into details about how to establish rapport with a person during an interview. Everyone's personality is different and the same strategy will not work for everyone. Explore some of the texts that I've recommended at the end of this chapter.

Once you have explained what you would like to do and why, and received the hoped for "yes," be sure to review property boundaries with the landowner. You and your colleagues don't want to inadvertently trespass on the land of others. Find out who the neighbors are. One cooperative landowner can often smooth the way in dealing with other landowners. Find out how long they have owned the property, since this will have an impact on the other types of questions that you might ask them. Remember that your goal is to learn about the presence of artifacts and potential archaeological sites on the property, supplement background information on local history, and learn about land use so that its potential effect on archaeological deposits can be taken into account during the conduct of fieldwork. Following are examples of relevant queries.

☑ Have any artifacts ever been found on the property or have they ever seen things that they consider to be unusual or out of the ordinary? Other than objects, this might include strange coloration in surface soils, unusual topography, or patches of ground where the drainage is radically different from that of surrounding areas. You might want to qualify what you mean by an artifact and give some examples. Nonprofessionals may have more limited views of what the term denotes.

☑ Do they allow people to collect artifacts on their property and are they aware of any individuals who do so with positive results?

☑ What kinds of things have been found?

☑ How old are standing structures on the property?

☑ Are there any old buildings or ruins?

☑ Have they ever researched or seen a chain of title for the property?

☑ Do they have any old maps or plats of the property? In short, you want to find out what they know about the cultural history of their land.

☑ Land use history is equally important. How is the property currently used?

☑ What sort of improvements have they made that involved altering the landscape?

☑ Are there, or have there ever been any quarries or mines on the property?

☑ Is household garbage or other debris disposed of on the property? Where?

☑ How long have individual dumps been in use?

☑ Has the land ever been farmed and are they aware of any old orchards or tree nurseries?

If the property is an active farm, find out about the general cultivation methods employed. There is more than one way to plow/till the land and each has different implications for how archaeological deposits might be affected. Have any agricultural fields been tiled? Tiling is a means of improving poorly drained ground. It involves digging a system of deep ditches across a field, lining them with drainage tile, then covering them over. The excavation of the ditches can bring more deeply buried artifacts to the surface or disturb intact features. During an interview, property owners should be asked about the construction of drainage systems and if they have any plan maps of them. Poorly drained ground may also be in-filled with sediments from other locations. Learn about the location of any underground pipelines for irrigation systems.

What types of crops are grown? Later, you might also want to learn about the range of chemical fertilizers typically used on the fields. This would be important if soil chemistry is being used to unravel the cultural and natural stratigraphy of excavations to search for features or identify activity areas. What types of animals are kept? As you will see in chapter 8, the activity of animals on the landscape can lead to the exposure of surfaces, the exhumation of sediments, and the discovery of artifacts. One type of farm animal that has a nearly systematic effect on subsurface deposits are pigs. Their constant rooting in the ground can turn a once barren field rocky. Their delving may reveal shallowly buried artifact deposits. Fields where pigs have been penned for prolonged periods are definitely worth a look.

For all of the reasons I've discussed, it might be useful to interview previous property owners. Current landowners may not have been in possession of the property for very long and won't be able to answer your questions. Previous owners may be associated with specific land uses that the present owner is not, e.g., the previous owner farmed the land, the current owner does not. Previous owners may be part of a family that has owned and used the land for generations, and have intimate knowledge of its history.

Your review of site record forms will apprise you of some of the information that you will want to get from the artifact collectors or amateur archaeologists that you interview. Other concerns will be site or project specific, or influenced by your personal interests and expertise. The attitude of the collector will have a big impact on the direction that your interview takes. Some are willing to have their collections examined but are unwilling or cautious about sharing the exact location of sites. There are useful things that can be learned from looking at a collection even if you don't know the precise location of the site that produced it. Obviously, the level of effort that you expend in examining a collection and interviewing a collector will depend on whether they provide information about the location of sites. It is a good practice to field check the location of sites with your informant. This may enable you to place the location more accurately on available maps, and it may stir your informant's memory about additional details.

Finds made on public property by individuals can be problematic. It is illegal for private individuals to remove artifacts from state, federal, or Indian lands without appropriate permits. Some artifact collectors, and certainly a large percentage of the general public may be unaware of this in their zeal to alert authorities about finding an artifact or possible site. Some artifact collectors will offer to provide information about sites on state or federal lands if they can remain anonymous in the documentation that is created. The same might hold true for sites that have been discovered on private property by individuals. In some states, formal permission from a private landowner is necessary in order for an individual to excavate and remove artifacts from a property (e.g., Oregon; see Gorospe 1985). It is your responsibility to make your informant aware if such a situation exists. How you proceed from there is the subject of debate and many ethical dilemmas.

I recommend traveling in pairs to any interview regardless of whether you are comfortable with the interview procedure or not. Any untoward behavior is less likely to be initiated when three people are involved in a meeting rather than two. It also makes it easier to find your way around what might be unfamiliar geographic territory. I find it difficult to maintain a running conversation and keep detailed notes about the important things being said. Two people can juggle the situation much more efficiently. A two-person interview team is especially important when interviewing collectors and examining collections. The details will be flying! Someone can be taking copious notes or perhaps shooting pictures while the other asks questions, responds, and generally keeps things moving. If all parties concerned are comfortable, a portable tape recorder may ease the situation. When using a tape recorder, transcribe the useful material on the tapes as soon as possible after the interview. This creates a duplicate and perhaps more durable record of the information you've collected, and makes it easier to integrate into subsequent reports.

LOGISTICAL MATTERS

As I noted at the outset of the chapter, as someone just getting started in archaeology you will probably not be responsible for collecting or evaluating background information. If access to such information is not provided by those in charge of the project, ask for it. Minimally, you should be aware of the geology, topography, and soils of an area since this background directly impacts your ability to recognize evidence in the field. You can prepare yourself by reading about the history and prehistory of an area on your own. Some of this information will already have been collected by those responsible for crafting a project's research design. If part of a cultural resource management study, additional and more focused background research may be done by project specialists or staff while fieldwork is in progress.

Familiarize yourself (or again, defer to your supervisor or instructor) with the laws and procedures relevant to the type of fieldwork in which you will be engaged. If your fieldwork is part of a cultural resource management study find out which government agencies or departments are involved, and what stage of compliance the project has reached. What permits, if any, are required in order for the archaeological investigation to proceed? Archaeological fieldwork on federally or state-owned land usually requires some type of permit issued by a government agency. What will be the procedure if human remains are encountered in the field?

Get a sense of how the overall project is organized in terms of staff and individual responsibilities. Be clear about what is expected of you in the field and don't be afraid to ask questions. There are no such things as stupid questions, just stupid answers. Make sure that your supervisor or instructor is aware of your level of skill. Gearing yourself up for fieldwork, dressing properly, and health and safety concerns are the subject of the next chapter.

APPLYING YOUR KNOWLEDGE

1. Obtain a site or cultural resource inventory form, or site record form, from your state office of archaeology or SHPO. Some states provide these online through the Internet (e.g., North Carolina, Wyoming). The state government pages of your phone book will help you locate your SHPO. Listings are also provided in Appendix 3. Review the categories of information that the form requires and the rationale for each.

2. Find and read summaries of the prehistory and history of an area or region with which you are familiar. Obtain a geological map covering this area. Using the map, identify formations or rock types within formations that contain materials known to have been used by the historic and prehistoric inhabitants of the area. Try to locate sources of these rocks in the field and collect hand samples of each. An easy way to get a quick look at an area's geology is to find a stream that drains it and examine the types of rock in its gravel bars and shoreline deposits. When identifying rocks in stream deposits it is always best to examine a freshly broken surface. The colors and textures of the waterworn surfaces or cortex of a rock can be misleading. You will need a rock hammer, or have to be comfortable using a "hammerstone" in the manner of native artisans to break the rock. There are field guides to rock identification that can help you familiarize yourself with what your are seeing, or you might visit a museum of natural history and browse their geological displays.

3. Buy the USGS 7.5 minute quadrangle for the area in which you live. Visit your local or other library and locate a 19th-century map or historical atlas that covers your area. Compare what each map shows for the same geographic area. What has changed over time? Are there houses or buildings standing today that are in the same location as 19th-century structures? Are there any roads that retain their historic alignments? Are the courses of rivers and streams the same on both maps? Have any place names changed or been dropped from use since the 19th century?

4. Obtain the soil survey for the area in which you live and identify the sheet or sheets that correspond with the topographic map used in exercise #3. In what ways do the maps complement one another? Other than soils, are there cultural or natural features shown on the soil map that do not appear on the topographic maps? Identify soil phases that occupy slopes of 12 percent or greater. Locate these same areas on the topographic map to get a better sense of how topographic maps depict the rise and fall of the landscape.

5. Your job is to interview an artifact collector who is active in the area in which you will be doing fieldwork. The collector has an extensive collection of 5,000 pieces from area sites that includes artifacts that he has collected on his own and purchased from others. He is willing to talk to you but is uncertain at the outset whether he will tell you the exact location of each site. He is also aware that some of his activities are illegal; he has collected artifacts on state and federal lands. He has two objects that were surface collected but which the existing literature maintains are found only in Native American burials of the early historic period. He also has human teeth from an excavated pit feature that also contained stone spearheads with styles typically of the time from 4,000 BC to 2,500 BC. Outline your goals, interview strategy, and the topics and questions that you would like to discuss with him. If you are taking part in an archaeological field school or a class on archaeological methods, try role-playing with a classmate who will take the part of the collector. In preparation for this exercise, you might want to review the ethical discussions in chapter 1, and those of cultural resource management law in chapter 3 and Appendix 2.

DIG DEEPER

Background Research

❑ These references will give you an idea of what it is like to do some types of background research and apply it to the investigation of a particular issue or problem.

- Barber, Russel J. 1994. *Doing Historical Archaeology: Exercises Using Documentary, Oral, and Material Evidence.* Upper Saddle River, NJ: Prentice-Hall, Inc.
- Barber, Russell J. and Frances F. Berdan. 1998. *The Emperor's Mirror: Understanding Cultures through Primary Sources.* Tucson: University of Arizona Press.

Environmental Data:

❑ Remember that when using Internet resources, addresses can change. If any of the addresses listed below fail to work, use a search engine and the key words in the website of interest.

❑ The website maintained by the United States Geological Survey is a good place to search for geologic and topographic maps and aerial photos of interest (National Aerial Photography Program). It includes lisitings of state geological surveys (addresses and phone numbers) and available state-based geological maps. Begin at **http://www.usgs.gov/** or go straight to USGS Publications and Data Products at **http://www.usgs.gov/pubprod/** Some states provide access to topographic maps online at no cost. The Topozone, produced by Maps a la carte, Inc. (**www.Topozone.com**) is a free online source of topographic maps, including 7.5 minute quadrangles, for the entire United States. The DeLorme Mapping Company, Freeport, Maine, sells digitized 7.5 minute topographic maps by state on CD ROM with many useful software features. Also check the Yellow Pages of your phone book under "maps" for suppliers of hard copies of maps.

❑ The Natural Resources Conservation Service of the United States Department of Agriculture: This agency has a website that provides access to plant, soils and hydrologic data. Go to **http://www.nrcs.usda.gov/** and select "Technical Services" then "Data Resources." It includes lisitings of published soil surveys by state and county through its Soil Survey Division at **http://www.statlab.iastate.edu/soils/soildiv/sslists/sslisthome.html**

Conducting Personal Interviews:

❑ Although these references were not written with archaeologists in mind and include more detail than you will need, they provide a comprehensive look at what is involved in doing informant interviews. Getting used to doing interviews can be an uncomfortable experience (it was for me) and the more preparation you do, the better.

- Ives, Edward D. 1995. *The Tape Recorded Interview: A Manual for Field Workers in Folklore and Oral History.* Knoxville: University of Tennessee Press.
- Kutsche, Paul. 1998. *Field Ethnography: A Manual for Doing Cultural Anthropology.* Upper Saddle River, NJ: Prentice-Hall, Inc.

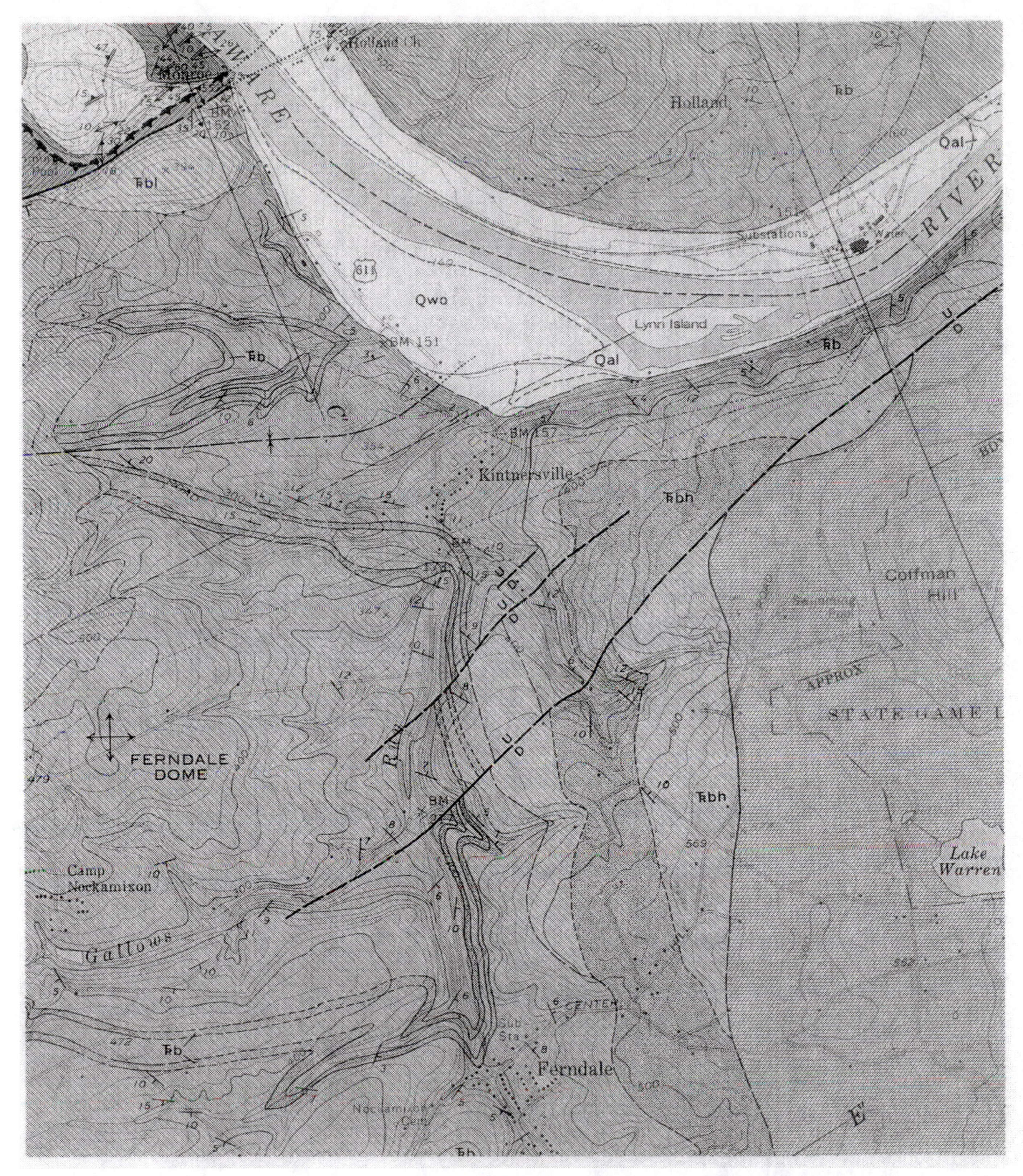

FIGURE 4.5.

A portion of the geologic map of the Riegelsville, Pennsylvania and New Jersey quadrangle (Drake et al. 1967).This map reflects more detailed geologic fieldwork than what is shown in Figure 4.3, and is executed at a smaller scale (1:24,000). Since a USGS topographic quadrangle was used as a base, the relationship between geological formations/deposits with topography can be seen. Unfortunately, not all areas in Pennsylvania or other states have been mapped at this level. Refer to the key on page 63.

FIGURE 4.3.

A portion of a Geologic Map of Southeastern Pennsylvania (Stose and Jonas, 1973) and accompanying key. Scale 1:380,160.
Colorplate 2

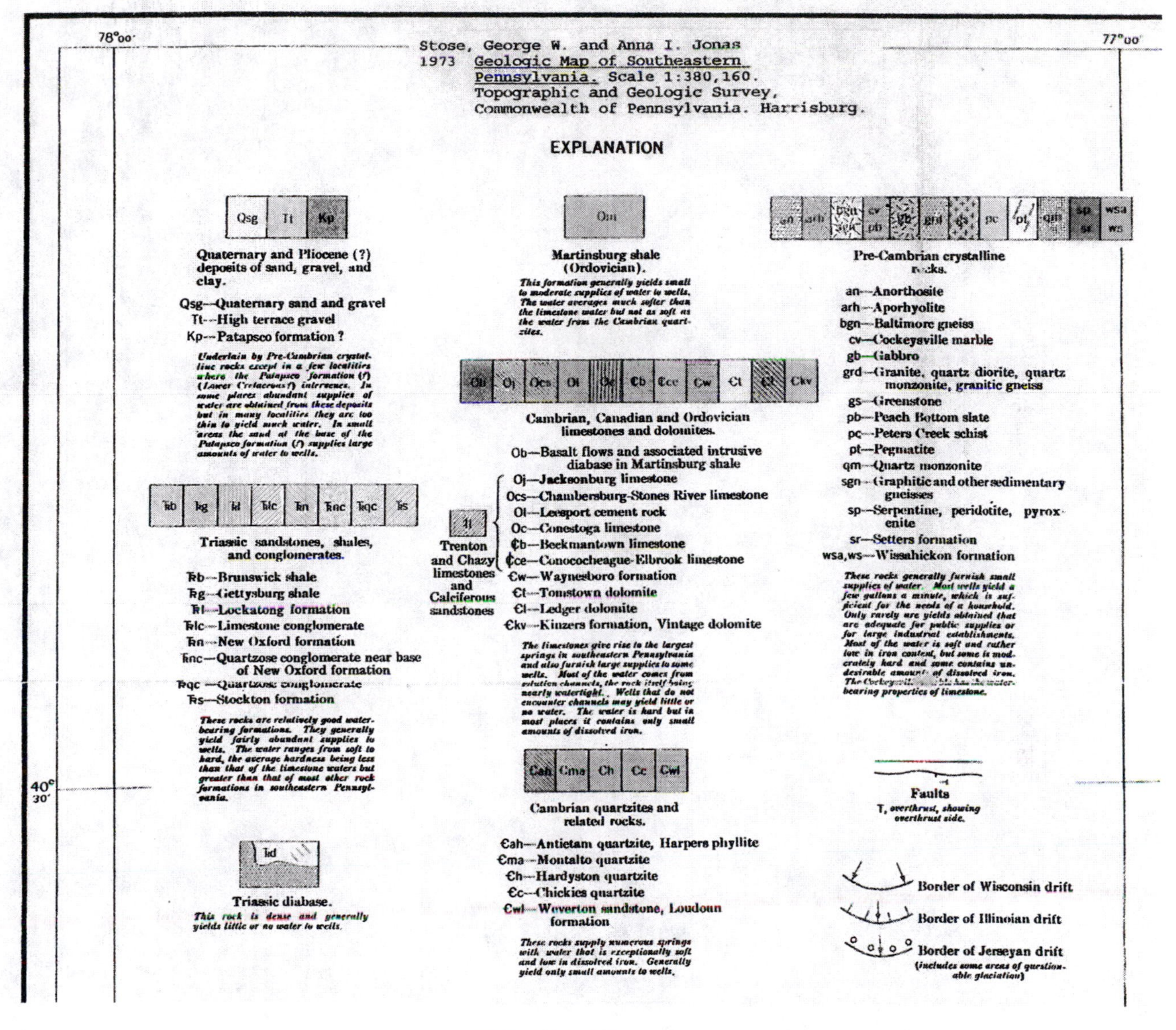

FIGURE 4.3.

Continued

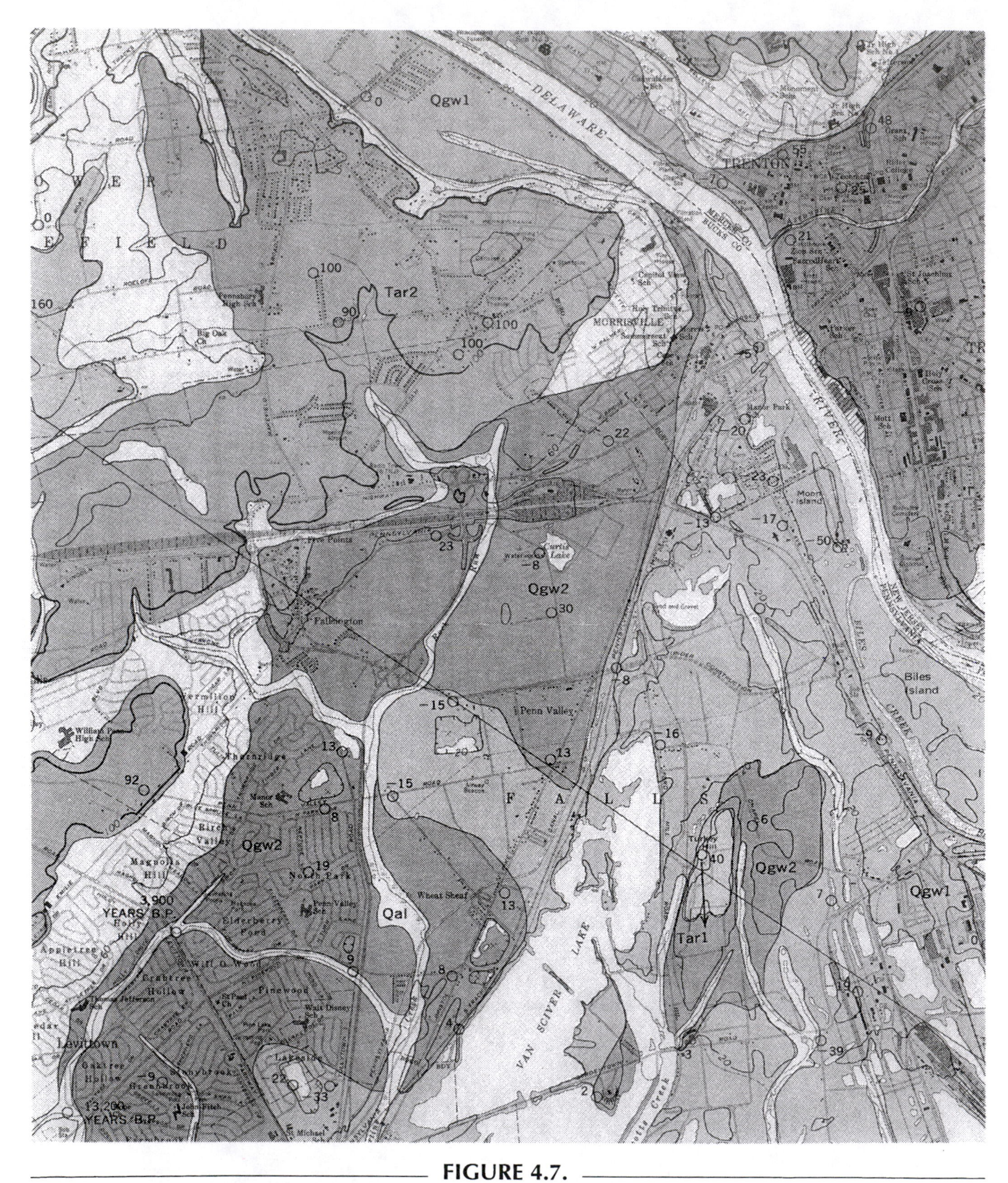

FIGURE 4.7.

A portion of a Geologic Map of the Surficial Deposits in the Trenton Area, New Jersey and Pennsylvania and accompanying key. Scale 1:48,000 (Owens and Minard, 1975). This map segment overlaps the coverage provided by Figure 4.6. Since its focus is strictly surficial deposits, it provides detail not replicated by the mapping of Quaternary deposits shown on Figure 4.6. Refer to the key on pages 65-66.

Colorplate 4

CONTROL DATA AND MONUMENTS

Aerial photograph roll and frame number* 3-20

Horizontal control

- Third order or better, permanent mark — Neace
- With third order or better elevation — BM 45.1; Pike BM 45.1
- Checked spot elevation — 19.5
- Coincident with section corner — Cactus
- Unmonumented*

Vertical control

- Third order or better, with tablet — BM × 16.3
- Third order or better, recoverable mark — × 120.0
- Bench mark at found section corner — BM 18.6
- Spot elevation — × 5.3

Boundary monument

- With tablet — BM 21.6
- Without tablet — 171.3
- With number and elevation — 67 301.1

U.S. mineral or location monument

CONTOURS

Topographic

- Intermediate
- Index
- Supplementary
- Depression
- Cut; fill

Bathymetric

- Intermediate
- Index
- Primary
- Index Primary
- Supplementary

BOUNDARIES

- National
- State or territorial
- County or equivalent
- Civil township or equivalent
- Incorporated city or equivalent
- Park, reservation, or monument
- Small park

*Provisional Edition maps only

Provisional Edition maps were established to expedite completion of the remaining large scale topographic quadrangles of the conterminous United States. They contain essentially the same level of information as the standard series maps. This series can be easily recognized by the title "Provisional Edition" in the lower right hand corner.

LAND SURVEY SYSTEMS

U.S. Public Land Survey System

- Township or range line
  - Location doubtful
- Section line
  - Location doubtful

Found section corner; found closing corner

Witness corner; meander corner — WC; MC

Other land surveys

- Township or range line
- Section line

Land grant or mining claim; monument

Fence line

SURFACE FEATURES

- Levee — Levee
- Sand or mud area, dunes, or shifting sand — Sand
- Intricate surface area — Strip Mine
- Gravel beach or glacial moraine — Gravel
- Tailings pond — Tailings Pond

MINES AND CAVES

- Quarry or open pit mine
- Gravel, sand, clay, or borrow pit
- Mine tunnel or cave entrance
- Prospect; mine shaft
- Mine dump — Mine dump
- Tailings — Tailings

VEGETATION

- Woods
- Scrub
- Orchard
- Vineyard
- Mangrove

GLACIERS AND PERMANENT SNOWFIELDS

- Contours and limits
- Form lines

MARINE SHORELINE

Topographic maps

- Approximate mean high water
- Indefinite or unsurveyed

Topographic-bathymetric maps

- Mean high water
- Apparent (edge of vegetation)

FIGURE 6.1 TOPOGRAPHIC MAP SYMBOLS.

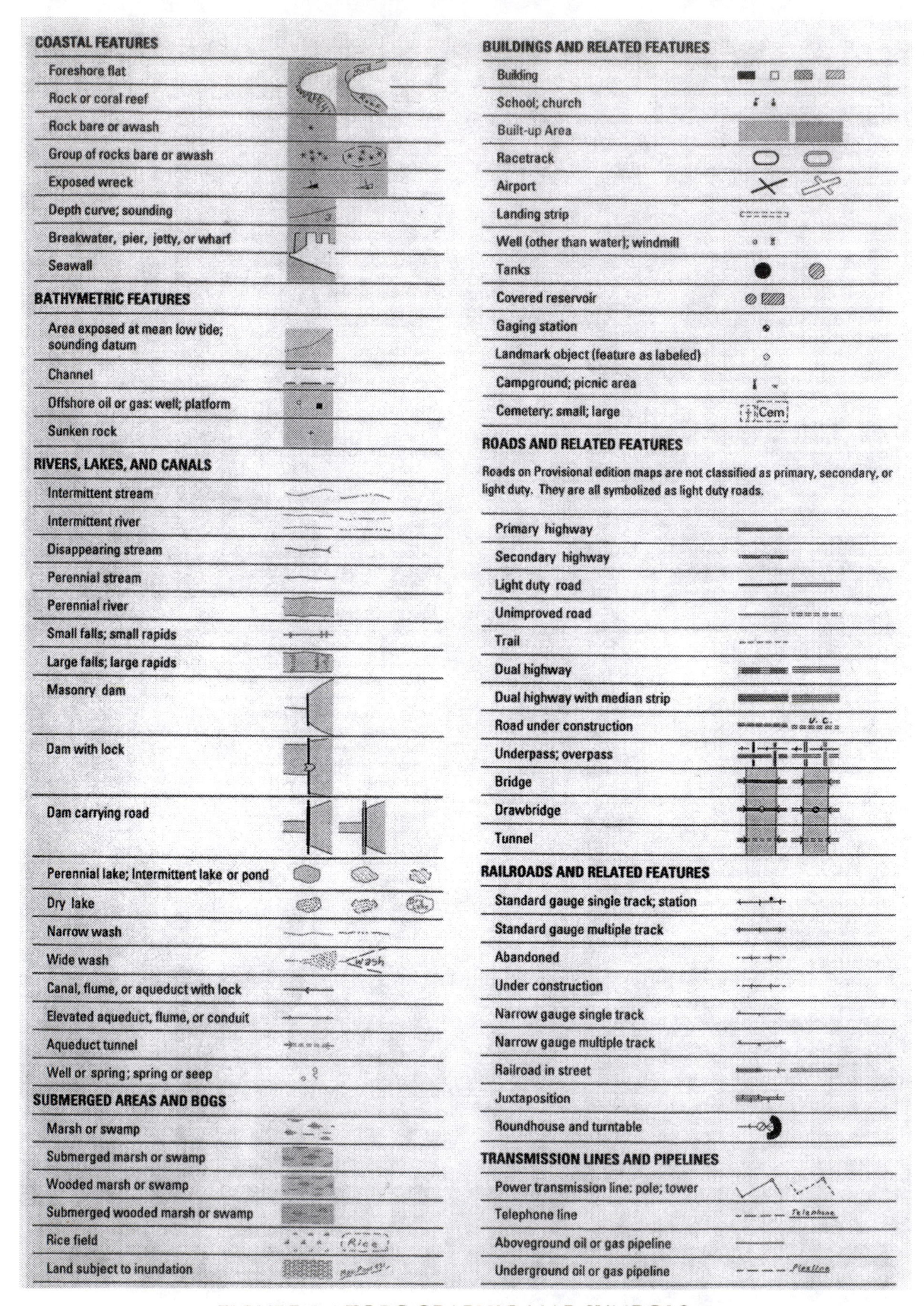

FIGURE 6.1 TOPOGRAPHIC MAP SYMBOLS.

Continued.

Colorplate 6

CHAPTER 5

Preparing for the Field

"A sign," he said, "a sign."
"It is this," I answered, producing from beneath the folds of my roquelaire a trowel.

Edgar Allan Poe, The Cask of Amontillado

This chapter has two goals. The first is to give advice about how you can outfit yourself for the field to prepare for the technical tasks that you will be performing and the context in which you will be performing them. How you gear up for the field will vary depending on the type of fieldwork being planned (e.g., survey, testing, intensive excavations), the type of archaeological resources involved (e.g., open sites, caves and rockshelters, underwater sites, structural and industrial ruins), and where the work takes place (e.g., woodlands, desert, mountains).

The second goal is to make you aware of the *potential* hazards associated with fieldwork and how they can be avoided or minimized. This discussion of hazards is not meant to discourage your participation in fieldwork. Our daily lives are full of potential hazards, some of which we prepare for and others that we deal with as they arise. Some situations are more hazardous and worthy of forethought and planning than others. Some we take for granted because of the situation or culture in which we were raised. So it is with fieldwork. The nature of hazards associated with a project will vary along the lines of the variables I noted regarding gearing up for the field. What should you know, who should you contact, and what should you do before beginning work in an area? How to prepare for potential hazards can be simply a matter of common sense, or a matter of legal procedure, as in many cultural resource management projects. I want to raise your consciousness about the nature of fieldwork. Thinking ahead will make your field experience more enjoyable and productive.

FIELD KITS

The tools required to carry out archaeological fieldwork usually are provided by the persons or organization running the investigation or project. Eventually you will want to put together your own field kit. Items that will be useful in a variety of situations are illustrated in Figure 5.1, a look at my personal field kit:

- ☑ Trowel
- ☑ Ice Pick
- ☑ Knife
- ☑ Root Clippers
- ☑ Work Gloves
- ☑ Compass
- ☑ Rulers/Tapes
- ☑ Plumb Bob
- ☑ Line Level
- ☑ Calculator
- ☑ 10X Hand Lens
- ☑ Field Guide to soils
- ☑ Personal Provisions
- ☑ Fieldbook

TROWEL

Most people prefer to work with a pointing trowel with a blade that is 4.5″ to 5″ long when new. Be aware that pointing trowels come in greater blade lengths. The Marshalltown brand has always been popular with archaeologists. The blade and shank of the trowel are a single piece of forged metal. Any brand of trowel that

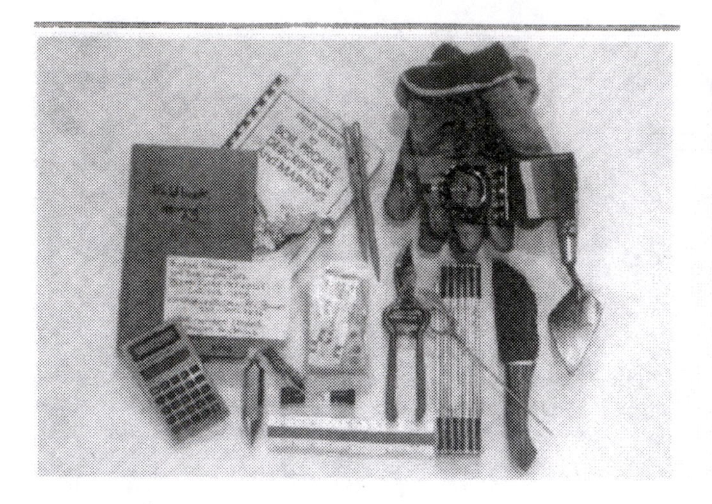

FIGURE 5.1.

Some essential items in a personal toolkit: a trowel, line level and string, small folding or sheath knife, folding rule or tape, compass, calculator, notebook, pencils, scaled ruler, plumb bob, ice pick or thin probe, work gloves, hand lens, guide for recording and describing soils and profiles, personal first aid kit and data card.

FIGURES 5.2, 5.3.

The trowel sharpening process. The secret to this operation is positioning the trowel on a flat surface allowing the blade to be sharpened to hang slightly over the edge of the anchoring surface. You want to work the trowel blade to a sharp, not rounded margin. The initial effort may take you 10-15 minutes, daily maintenance less so.

you buy should have this feature. Those that have a shank that is spot welded or soldered to the blade simply will not last. There are some interesting regional preferences in trowels (Bleed 2000).

The blades of your trowel should be sharpened on a bevel from the top side since the trowel is used both to move and "cut" through sediments. Having a sharpened trowel makes this easier and enables you to effectively deal with small roots without recourse to clippers or pruning shears. Cutting sediments of a profile with a sharpened trowel also helps to bring up the color and textural qualities of the sediments that you are examining. Use a milled, common bastard file to put a beveled edge on your trowel (see Figures 5.2 and 5.3). Maintaining this edge will be a daily chore with the file. Over time the size and shape of your trowel blade will change as a result of this practice. Notice how asymmetrical the blade of my trowel is. This is from resharpening it more frequently on one side versus another. I'm right-handed so one margin of the trowel blade experiences heavier use than the other.

Many excavators mark their trowels and other tools in some way. This helps to distinguish your tools from those of others that invariably get laid around excavations during the course of the workday. A dab of bright, dayglo paint on the butt end of your trowel will also make it easier to find if mislaid. You will soon learn how easy it is to lose tools in the field if you are not attentive. Attention to your tools is another way of demonstrating that you are thinking about what you are doing.

Some archaeologists carry a collection of trowels with different sizes and shapes of blades (Figure 5.4). These can make it easier to expose and work around artifacts that are left in place on excavation floors for photography and mapping as a group (Figure 5.5). Others may use metal or wooden or bamboo spoons of varying sizes, or dental tools. The Ingalls pick (see Figure 5.4) is an implement that is a reasonable substitute for a trowel in many circumstances. It has a fine "cutting" margin that can be used like the blade of a trowel. The handle and the orientation of the blade make it possible to do heavier work with the pick, like breaking up a stiff sediment or matrix, than what is possible with a trowel. The length of the handle, however, means that it is not as easily packed and transported as a trowel.

ICE PICK

A household ice pick or short length of stiff wire can be a useful probe when excavating a level within a unit. It can help you to quickly (and relatively gently) locate larger artifacts before they are exposed with trowel or shovel. It also provides a means of "feeling" textural changes in sediments, useful for describing sediments and defining stratigraphic boundaries (see chapter 7). A short brush will also be useful for cleaning objects on the floors of excavations. While a brush may also be useful for moving sediments across the floor of an excavation, this action blurs the color and texture of the matrix to your eye.

KNIFE

A knife has always been my idea of a do-everything tool, and today you can get knives that are one component of a multipurpose tool, like the popular "Leatherman." Folding knives have the advantage of being more portable than sheath knives. Folding knives with blades that can be locked are preferred, since they will never accidentally close and cut you while in use. Sheath knives are great for heavier types of cutting and light chopping, and are useful in assessing the compaction and texture of sediments in a profile (see chapter 7).

FIGURE 5.5.

Different sizes and shapes of trowels or other implements can be useful when artifacts are exposed and left in place on the floor of an excavation for subsequent mapping and photography. The excavator in this photo has the difficult job of exposing the closely packed rock of a prehistoric Native American hearth without displacing any individual pieces.

FIGURE 5.4.

Varieties of trowels and the Ingalls pick, sometimes used as an alternative to the trowel.

ROOT CLIPPERS

There are numerous implements useful for dealing with roots both large and small encountered in excavations. I carry a pair of sturdy, sharp clippers primarily for use in cleaning the profiles of excavation walls for drawing, photography and analysis, and not for handling all situations requiring the removal of roots. A knife may do double duty here as well, but may be more difficult to use.

Work Gloves

Work gloves are a necessary item for many reasons. Accidents involving the hand tools that you will be using can result in cuts and scrapes, and gloves can help to minimize or prevent some of these incidents. Gloves also provide a degree of protection from the development of blisters. The oils of poison ivy and poison oak that cause rashes are associated with the roots of the plants as well as the leaves, and may adhere to sediments adjacent to the roots. You may not only be digging through sediments in which there are the roots of these plants, but you will also be screening the excavated matrix. Wearing gloves will help lessen your contact with these oils.

Compass

Given the importance of context and maps, it is hard to imagine going out into the field without a compass. There are many useful types. The one that I've pictured is an inexpensive Silva Ranger with a see-through base that has horizontal scales. The scales that you get on the compass base can vary. Try to select one that is 1:24,000, since this matches the scale of the 7.5 minute USGS maps commonly used in the field. In addition to the standard functions of a compass that help you find your way, locate yourself on a map, aid you in map drawing and the layout of excavation grids and units (see chapter 6), this version also has a feature that allows you to read the percent slope of a landscape or rock outcrop.

Rulers

Folding rules and tapes are used in laying out excavation units, mapping the location of artifacts and features within units, and recording profiles of excavations or other exposures. Mine are scaled both in metric and engineer-scaled feet so that I can be flexible depending on the measurement system being used in the field or employed on available maps. Double-scaled rules will be more expensive to purchase. I also carry a small, straight-edged ruler, scaled in a variety of ways, to assist with the drawing of maps and sketches in my fieldbook.

Plumb Bob, Line Level and Calculator

Two implements useful in mapping artifacts and features within an excavation unit are the plumb bob and a line level. One method of mapping the horizontal location of objects involves using a plumb bob in conjunction with tapes (see chapters 6 and 9). A plumb bob is also handy for keeping the walls of an excavation straight, or "plumb" (see chapter 9). Of course, it is a necessary component of surveying equipment like a transit. The line level is suspended on a string and is used in recording depths and recording stratigraphy (see chapters 7 and 9). A small calculator will also come in handy for a variety of in-field calculations, most involving mapping (see chapter 6).

Field Guide

Although I carry a field guide to describing soils, what I use most often from it are a series of charts that enable you to determine sediment texture, assess soil structure, and mottling (see chapter 7). These can be reduced through photocopying to make them more portable, and laminated to prolong their life in the field. A Munsell color chart is also necessary for descriptive purposes (see chapter 7), but its cost is prohibitive for the personal field kit.

10X Hand Lens

A 10X hand lens is sufficient for field examination of the sediments in an archaeological deposit and to check for obvious wear or edge damage on the margins of artifacts.

Personal Provisions

Many of you may already carry a card that provides critical information for those who might attend you in an emergency. You should always have one when in the field. It should include your name, address, and phone number; age; if you take medication on a regular basis, the medication involved and what it is supposed to treat; whether you are allergic to any medications; whether you wear contact lenses; health insurance provider and policy number; name, address, and phone number of a person to call in case of an emergency, and their relationship to you. Medical alert tags, bracelets, or necklaces should be worn by those who regularly take a medication or have an allergy to medications that might be administered in emergency situations. I also carry some personal first-aid items like Band-Aids, handi-wipes, moleskin for blisters, aspirin/tylenol, and the occasional non-prescription pill for hay fever. On a seasonal basis I add sunscreen and bug repellent.

Fieldbook

The final, and most important item in my kit is the fieldbook used for keeping a daily log of field activities, observations, interpretations, interviews, and "things-to-do." I consider it the most important item because without notes, you're just another collector not really thinking about what you are doing. Engineering, surveying, and forestry supply companies sell notebooks that are especially suited for use in the field and are favored by many archaeologists. They are relatively small in size (7.5″ x 4.5″) and easy to pack. They have hard covers and the pages are perfect-bound so you won't lose them no matter how roughly you treat the book. The pages are treated to resist water and won't "melt" if you get caught in the rain. A popular item with some archaeologists is an implement that combines a clipboard with a shallow box capable of holding data sheets, notebook, paper, and pencils (Figure 5.6). It not only protects paperwork from the elements but provides a firm surface on which to write and draw. Its bulk is a potential drawback.

Your own additions or deletions to the brief inventory that I have reviewed will depend on the type of fieldwork in which you are engaged and where the fieldwork takes place. Not illustrated is the sturdy backpack or rucksack that you will need to lug these things around along with your lunch, snacks, and fluids. Make sure that you select one that is comfortable to wear and can take a beating, i.e., put up with repeated cycles of getting wet, dirty, and cleaned. Even if you're only hiking short distances during the day, you might want to consider getting a pack with some type of internal frame or lower back support.

FIGURE 5.6.

An example of a "portable desk" used by many archaeologists in the field. It is a combination clipboard-box for storing notes and data sheets, pencils/pens, and implements handy in sketching maps. It helps to protect the all-important paperwork in the field and provides a firm surface on which to write.

DRESSING FOR THE FIELD

It's impossible to describe a dress code that is appropriate for all types of field work, in all places, and at all times. Shorts and T-shirts may be fine for surface survey in open areas or desertlike conditions but are inappropriate for surface survey in overgrown fields or woods. Steel-toed boots are smart footwear when working around heavy machinery being used in an excavation, but are not necessary at other times.

CLOTHING

Safety and relative comfort are primary considerations when dressing for the field, with safety coming first. Any considerations must take into account the activities that will be performed in the field and the field environment. Dressing also can be a matter of legally mandated safety requirements depending upon who is sponsoring the project, or where it is being carried out. Cultural resource management firms may have internal policies about dress that relate to safety, comfort, and being perceived as a professional by outsiders. The sensitivities of the people in the communities where you are working must also be considered.

Aspects of the natural environment in which you are working that deserve consideration when planning your dress include intense exposure to the sun, dust, and airborne particulate matter, exposure to extremes in temperature, poisonous and irritating plants, briars and thick undergrowth, biting insects and some types of wildlife, like snakes. Fieldwork usually always involves a good deal of walking and carrying burdens, lifting, and lots of squatting. Choose durable and comfortably fitting clothing with these activities in mind. Your project director or field supervisor should provide guidelines for you to follow before going to the field. Again, the key here is dressing appropriately for planned or anticipated activities, and the environment in which they will occur (see Figures 5.7–5.15).

Layering your clothing may be necessary when working in areas where the temperature or your level of activity (and perspiration) fluctuates dramatically during the day. In cases where you're consistently getting soaked with perspiration, crossing streams, or dealing with unpredictable wind-chill factors, it might be wise to tote an extra T-shirt, pair of socks, or sweater. Sooner or later rain gear will come in handy. How often you pack your rain gear along with your daily necessities will depend on the weather and your particular field situation.

FOOTWEAR

What constitutes appropriate footwear can vary from situation to situation. In general, ankle-height boots are a wise choice for footwear. They provide support for the hiking you may be doing over all sorts of different terrain, and are protection from abrasion, accidental cuts, and many types of impacts that you might encounter in the field or on site. They have a firm sole that will stand up to the rigors of shoveling while protecting your foot. Smooth-soled shoes are preferred for working on the floor of excavations, especially during mapping procedures, because they cause the least damage or disturbance to the matrix on which you are standing and squatting. Finding a smooth-soled shoe that is also a good hiking boot may not always be possible, however. Steel-toed work boots provide necessary protection for your feet when working around heavy equipment. Leather gators may be worn by those working in areas where there is the danger of snake bites.

FIGURE 5.7.

Dressing appropriately for all conditions to be encountered in the field can be a complicated matter. These field technicians are making their daily crossing to Hendricks Island, Bucks County, Pennsylvania. There's always the potential to get wet! Upon landing, they will hike 1/4 to 1/2 mile with their gear to the sites where they are excavating. In the open areas around the excavations shorts would be great to beat the summer heat. But to get that far on the island, excavators have to traverse woodlands in which poison ivy, greenbriars, and ticks are not unknown. Lightweight long pants and shirts are a healthy and comfortable compromise. In the afternoon they will be returning to the canoe landing with their gear and the artifacts recovered during the day. Sturdy and comfortable packs are a must. Footwear must be suitable for hiking and working in excavations.

It's tough to envision a field situation where bare feet or sandals would be appropriate. Bare feet used to be the norm when mapping cleaned excavation floors. But where there are hand tools, shovels, sifting screens, or artifacts there is the potential for injury to unprotected feet.

HEADGEAR

Indiana Jones has nothing on real archaeologists when it comes to stylish headgear. Brimmed hats or caps are smart choices for working in the sun. Hard hats are necessary when working around heavy machinery and when using certain power tools, like chainsaws. They should also be worn when working in caves or whenever there is a danger of falling objects, like in structural ruins or deep excavations. Bandanas are a versatile article of clothing that may be used in combination with headgear for safety and comfort in the field. I have seen field workers anchor them under their hats allowing them to fall over their neck, like a kepi, to shield them from the sun. Bandanas may be wetted and worn on the head, under a hat, or around the neck to keep cool. If the water is cold, so much the better. Two students in one of my field schools used to soak clean bandanas in water and then put them in the freezer. They would carry them to the field in the small coolers in which they kept their lunch and would use them on their neck and head during the day as needed. Of course, using even a small cooler is only going to be convenient when you are working on a single excavation and are not required to do much walking or hiking on a daily basis.

PERSONAL PROTECTIVE EQUIPMENT

Personal protective equipment (PPE), an official term, has the defined purpose of shielding or isolating individuals from chemical, physical, and biological hazards that may be encountered while on the job (OSHA 2001: Part 1910 Subpart I). OSHA regulations regarding PPE

Once I was working on a farm owned by a family of old-order Amish. I had just finished speaking with the head of the family, explaining the purpose of my work and what I would like to do on the property. It was July and the weather was steaming. He was dressed in black pants and a long-sleeved white shirt buttoned up to his neck. I was in cutoffs and a T-shirt. Just looking at him made me feel hotter than I already was. Throughout my first day there I kept noticing a group of older girls peeking around the corner of the barn at me and laughing. One of the men later confided in me that because of the way that I was dressed, I was basically "naked" in their eyes, and the girls were alternately being embarrassed and getting a chuckle out of it. As a courtesy, and in deference to their sensibilities, I began wearing long pants and a more modest shirt, and praying for a break in the weather.

are directly applicable to ***employees;*** students and volunteers on projects fall into a grey area. However, the directors of any project have an obligation to look out for the welfare of the people under their supervision.

Many of the things that I have mentioned above are considered to be PPE under specific circumstances: gloves and proper clothing when sharpening tools or using tools that are sharp or have been sharpened; hard hats where appropriate; suitable footwear; articles of blaze-orange clothing when in the woods during hunting season, or when working in low visibility or high traffic areas; respirators and filter masks when working under dusty conditions, in caves or other environments where there is a potential hazard from airborne bacteria (see below). Archaeologists even work on sites that are contaminated with various levels of toxic waste and spend the day in "moon suits" (Tyvek suits)! I suspect that you won't find an archaeological field school that has to contend with toxic waste, but some people's first experience with fieldwork may be as fledgling employees of a cultural resource firm where anything can happen.

FIGURE 5.8.

In this scene, excavators are wearing shorts and T-shirts to cope with the summer weather, but retain ankle-height work boots for safety.

PERSONAL APPEARANCE

How people perceive you when you are in the field and how you would like to be perceived (or how your supervisors would like you to be perceived) are issues to consider when dressing. It's rare that you are in the field on your own. Usually you are representing someone, be it your school or university, those sponsoring or funding the work, a company, or an archaeological society. Each of these organizations has a concern that you take your work seriously, and that you are taken seriously by property owners, onlookers, visitors, and local residents. There is no professional standard of dress for an archaeologist, but your appearance (including personal grooming) can communicate your attitude about the work that you are doing and its importance. It is also a way of signaling your regard for and consideration of the community in which you are working.

The typical landowner doesn't want to see raggedy-looking youths or "hippies" roaming their property. It can be a tough line to walk. Archaeologists work outside doing physical labor and get very dirty in the process, so why be concerned about how we look? On the other hand, we're engaged in a scientific pursuit and anything that underscores our professional intentions and garners the respect of the public is worth doing. The organization that is sponsoring the project in which you are participating, or the company for which you may be working, may have a dress code that melds concerns with how staff are perceived with health and safety concerns. Some may even provide shirts or caps with organization/company logos on them to be worn in the field. This helps the public and landowners to

FIGURE 5.9.

This crew has paused at the end of the day for a portrait showing off their individual approaches to dressing for winter weather in the Middle Atlantic Region.

FIGURE 5.11.

Most of the crew here are wearing vests of blaze orange. If you are in the field during hunting season some article of your clothing, whether it's a cap, shirt, vest, or jacket, should be blaze orange as a safety precaution. OSHA standards require the wearing of blaze-orange attire by any employee working in the woods or adjacent to roads. (Please refer to the back cover).

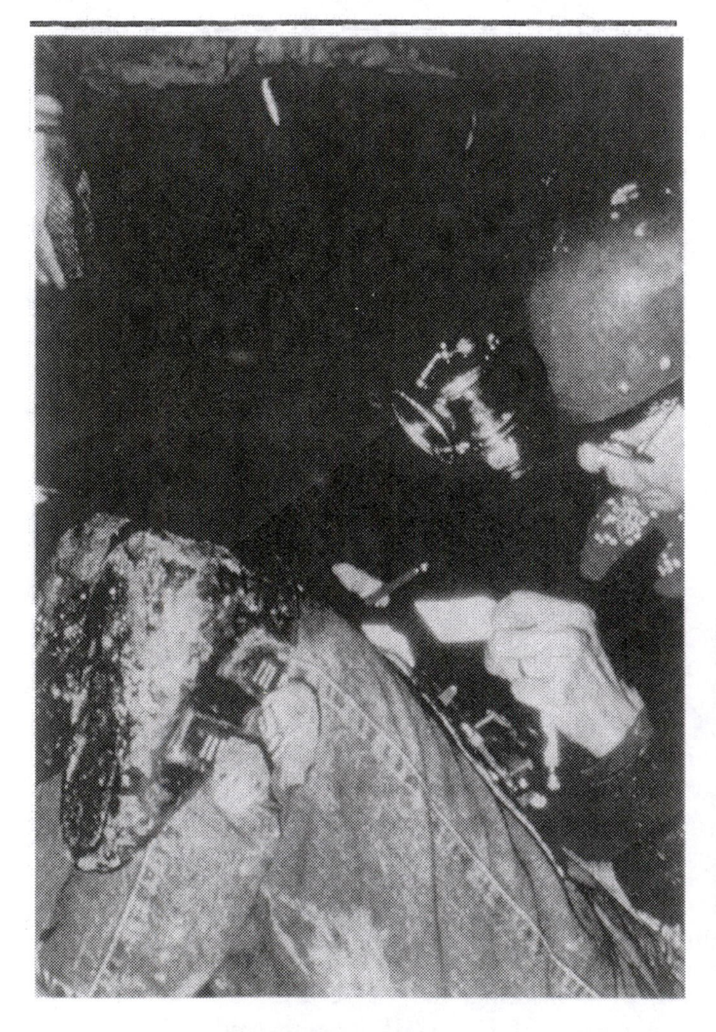

FIGURE 5.10.

An archaeologist outfitted for working in a cave—Dr. Patty Jo Watson in Hour Glass Cave, Colorado.

readily identify the members of a field team. Think twice about whether any graphics or text on your clothing might be offensive, or whether your attire itself might be considered to be immodest. The review of health and safety issues below have implications for how to dress for the field and reinforce some of the suggestions that I have made.

HEALTH AND SAFETY CONCERNS

The people that take you into the field need to be aware of situations that can impact the health and safety of their crew and plan accordingly. Greater personal awareness of these issues makes everyone's job easier. Potential hazards will vary depending on the type of fieldwork, the type of archaeological resources involved, and the environment in which the work takes place. Remember that there is a distinction (albeit subtle at times) between what teachers and supervisors need to do for the students and volunteers in their care, and the legal obligations of employers to their employees.

PERSONAL SAFETY

The Occupational Safety & Health Administration (OSHA), a federal organization whose main goal is to protect employees in any work environment, has set a number of standards and regulations that should become as familiar to the field archaeologist as his/her personal excavation tools (e.g., Akerson 1995). The standards are organized with relevance to the construction industry, since archaeology is not specifically addressed and construction jobs involve working outside, using hand tools and machinery, digging and working in excavations. OSHA standards may be embedded or enhanced in state- or agency-specific plans regulating health and safety. In addition, the insurance carried by the school or organization sponsoring an archaeological field project may have requirements regarding the health and safety of participants.

If you are an employee, your employer should have and implement a health and safety plan. This plan should be in a written format and be made available to all employees. Aspects of the plan will be site specific, varying with the scope of a project. It must identify potential hazards, including hazardous materials used on the job site (material safety data sheet), and outline emergency procedures and other measures to be taken to avoid or mitigate hazards. The plan should include phone numbers and the locations of accessible emergency and ambulance services, hospitals, police, relevant travel routes, and required PPE. There must be an on-site person who is able to identify existing and predictable hazards in the surroundings, or working conditions which are unsanitary, hazardous, or dangerous to employees, and who has authorization to take prompt corrective measures to eliminate them. This requires specific training in hazard analysis and the use of protective systems per OSHA standards.

In addition to the health and safety plan, your company may train you or others in a number of on-site safety measures which may include items such as basic first aid and CPR, hazard communication, respiratory protection, personal protective equipment, and working safely in excavations and confined spaces. The list of potential hazards on a project needs to be reviewed with crew on a periodic basis.

Some of the OSHA standards are less than specific. For example:

> *The employer must determine which employees are exposed to possible head injury hazards, and assure that they wear appropriate head protection. Where employees are not exposed to possible head injuries, head protection is not required by OSHA standards. . . . OSHA has no exhaustive guidelines for determining whether head protection is required, this must be done on a case-by-case basis, depending on the specific operation, worksite, potential hazards, and other circumstances (29 CFR1926.100; 1910.132; 1910.135; OSHA Standards Interpretation and Compliance Letters, Clarification on standards for head protection online at:* ***http://www.osha-slc.gov/OshDoc/Interp_data/I19830823A.htm>****).*

If the regulations for personal protection remain slightly ambiguous, it is to give the employer the freedom to implement these standards as they apply to each job situation. You may find though that an employer may require you to wear some of the personal protective safety measures at all times, just to ensure that you are always in compliance with OSHA safety measures. It's simply a matter of good planning for field schools and other projects that don't have to adhere to OSHA standards to create their own health and safety plans.

Table 5.1 lists some potential hazards associated with archaeological fieldwork. The frequency with which some of these might present themselves will vary by

FIGURE 5.12.

Hard hats are required headgear when working around heavy machinery or when falling objects can be a hazard, such as when excavating in deep units.

geographic region. For example, field crews working in the Southwest need to be concerned about airborne maladies like valley fever and hantavirus while those working in the East rarely give these a thought.

Conscientious field supervisors typically collect personal information pertinent to the health of their crew. This information could be important if you are injured or have some physical problem while in the field that requires medical attention, as I noted in my admonition to carry a personal data card. If you have first-aid or lifesaving training, advise your supervisor. It is important that someone on the field crew be sufficiently skilled to render basic first aid if needed, and a well-stocked first-aid kit should always be on site or with the crew.

Fieldwork can involve a lot of walking, squatting, lifting, and labor involving movements and muscles that you might not ordinarily use. Stretching and warming up with light exercise at the beginning of the day is a good idea. The old standbys, situps/crunches and pushups are great for toning up the abdominal, lower back, shoulder, and arm muscles that you will be using. If you have little or no experience with doing physical labor outdoors, talk with someone who does. Seek advice from the people running the field project. If you have any doubt about your ability to perform, check with your doctor or schedule your yearly physical in advance of the field season. Make sure that your last tetanus shot is still valid. Cuts are one of the most frequent accidents to occur in the field, and this precaution will help to forestall any complications from infection.

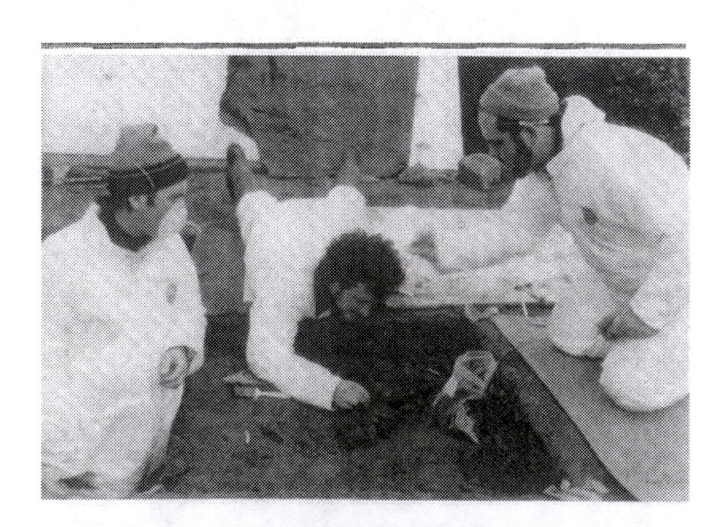

FIGURE 5.13.

Even when associated with hazardous waste, the examination of an archaeological site may be unavoidable. The crew shown in this picture is working on a site with level D+ waste requiring the use of Tyvek suits and booties.

It is a good idea to leave an itinerary with a responsible party or landowner if you are going into remote areas on a daily basis or for prolonged periods. Packing walkie-talkies, cell phones, or flares may be a good idea under these circumstances. In extremely remote wilderness areas carrying firearms may be appropriate. Proceed with caution here, and make sure that you understand local and state gun laws. No one should be carrying a gun who is not licensed or trained in its safe use.

Quite a bit of an archaeologist's field time can be spent within an excavation or trench. The OSHA regulations for trench safety are many, and it is important that everyone become familiar with the most basic of safety measures. Face it, soil is heavy. Even a small amount collapsing from the wall of a unit can cause severe injury. Some OSHA standards present problems given the excavation techniques used by archaeologists, and the types of observations that need to be made in excavations. These issues are addressed in chapter 9. In general, an excavation less than 5 feet deep in a stable soil or matrix requires no unusual measures. Be wary of getting into anything deeper without proper precautions.

Living in the field as part of a field school or project can be a landmark experience. Rules for daily chores and guidelines for general behavior and the resolution of conflicts are unavoidable in such situations, as are concerns with health and safety. You will get many chances to hone your social skills, as well as make new friends and develop relationships worth maintaining. Sometimes living in the field can test your mental health, since obtaining privacy on a regular basis after the work day is over may be difficult. It can be tough to escape an unappreciated social scene when there is no place to escape to. Self-discipline involving things like educational reading and personal study, or even eating and sleeping habits, can be hard to maintain in group settings.

Some possible field ailments are listed in Appendix 4 along with summaries of their symptoms and treatment. Familiarize yourself with ones that are appropriate to your circumstances and watch your fellow field technicians. You are part of a team. Help one another. The Center for Disease Control can be consulted for conditions specific to geographic regions. Should you have to see a doctor, be sure to advise her where you have been working and what you have been doing. It is always a good idea to be clear about the scope of your personal health insurance and to learn about any coverage that may extend to you as a result of who is sponsoring the field project.

TABLE 5.1.
POTENTIAL FIELD HAZARDS

Existing Environment
Insects bites (allergic reactions to stings, Lyme Disease, Rocky Mountain Spotted Fever), animal bites (snakes, dogs, rodents, rabid animals, livestock), other livestock-related traumas
Poisonous plants (poison ivy, poison oak, poison sumac)
Extremes in climate, climate-related dehydration, water intoxication, exposures, frostbite, hypothermia, sunburn, sun poisoning, heat cramps, heat exhaustion, skin cancer
Excessive pollen, dust, and associated airborne diseases in some regions and specific environments (e.g., hantavirus, valley fever)
Pre-existing hazardous wastes, current use of potentially hazardous chemicals employed in farming, horticulture, and industry
Hazards related to underground utilities
Areas open during hunting seasons, game lands, areas where visibility is poor (e.g., dense woods/undergrowth), areas of vehicular traffic
Personal allergies (pollen, dust, poison ivy/oak, bee or other insect stings) or pre-existing medical conditions
Work Environment
Cuts, abrasions and bruises related to the use of tools employed in brush clearing during survey and excavation, and to the use of hand tools and excavating equipment
Falls related to hiking/walking, getting in and out of excavations
Hazards related to working around heavy, earth-moving equipment
Cave-ins of excavations, rock shelters and cave ceilings and walls
Fire and burns from equipment used to heat excavation shelters in the winter
Muscle and back strain/physical stress from labor performed
Remotedness of field site(s) and access to emergency and medical services
Method of transportation to the field and in the field
Living in the Field
Hazards associated with type and location of accommodations
Who is sharing accommodations and personal hygiene
Method of transportation to and from living accommodations
Cuts and abrasions from using camping-related equipment
Fire and burns from heating and cooking equipment
Mental health (isolation from friends and family; awkward social situations)
Assault, sexual harassment

Abstracted from Flanagan 1995; Howell 1990.

PUBLIC SAFETY

You also have to think about how your work may create potential hazards for the public, livestock, and wildlife. Excavations left open overnight or for prolonged periods are the greatest concern. If you are working in a remote location, flagging surrounding an excavation may be sufficient together with whatever covering is being used to protect the excavation itself. It may be necessary to erect some type of temporary fencing (e.g., snow fence) with caution signs in areas where excavations are more accessible to the public. Where there is a chance of livestock wandering into excavations, stouter fencing with barbed wire may be necessary.

As I look back over this discussion, I wonder how many readers have been made nervous about the potential hazards of working in the field. Some people take certain things for granted and deal with them out-of-hand. Having been raised in what was once a rural area, I grew up avoiding certain plants, animals, and insects and rarely gave them a second thought in adulthood. I was taught basic first-aid and survival skills. Those with no woodscraft need to consider such things a bit more deliberately at first. In turn, I had many lessons to learn about power tools and heavy machinery. It is the same way with fieldwork. You train yourself to recognize potential hazards and how to avoid or mitigate them. You develop good habits. You never become complacent, but daily life in the field quickly ceases to be a white-knuckled adventure in paranoia.

FIGURE 5.14.

Field technicians at work in a hazardous situation. In a trench over their heads, they should be wearing hard hats, have ready access to ladders, and the trench itself should be shored and/or banked to some degree.

FIELD ETIQUETTE

It is only natural that where health, safety, and group cooperation are concerned there will be guidelines for acceptable behavior. Every field session should begin with a review of standards and the expectations of those supervising the fieldwork. Of course, obscene language or behavior, sexual harassment, fighting, or any threatening behavior are not tolerated at any time.

Find out about the organizational structure of the field team and how you fit into the big picture. What's the chain of command? To whom do you address your questions about field procedures or things that you don't understand? Who do you see about replacing your broken trowel or getting a blank data sheet for your new excavation level? Some field schools or projects may be small and intimate enough that you won't need to ask these questions. As a rooky in the field your biggest concern should be what is expected of you, how closely your work is monitored, and guidelines for knowing when it's time to ask for help.

Your field orientation should include a review of the rhythm of the day—when will the work day start and end, are breaks scheduled or taken on an as-needed basis, when is lunch, will there be lectures, reviews of latest finds and the progress of work? If you are excavating, breaks, smoking, drinking, and eating should take place outside of your unit.

Whether you are involved with site survey or excavation, you will have some responsibilities for collecting artifacts, seeing that they are bagged with appropriate provenience data, making notes and/or filling out data sheets. You will undoubtedly be accountable for the field equipment that you use. Keep your equipment organized and know where it is throughout the day. You are in the field to observe, record, and collect data, and it is easy to get distracted or overwhelmed at times. Developing good equipment habits is one less thing that you will have to worry about. Reserve some time at the end of the day to clean anything that needs it (e.g., trowels, shovels, buckets, screens). Some tools may be issued and turned in on a daily basis to an

FIGURE 5.15.

Field technicians working in a deep trench with appropriate safety measures.

equipment manager. Don't take the attitude that they are someone else's property and the equipment manager will take care of things. Respect for your tools, even if they are borrowed, is respect for your job.

Staying organized and keeping your work area clean are signs that you are thinking about what you are doing and the importance of your task. Take care about resting on or against the walls of excavations, or standing too close to their top edges. You don't want to find yourself sliding into a hole as a result of a cave-in, nor do you want to disturb the contexts that you and your colleagues are taking such care to examine. Leaning shovels or other tools against the walls of excavations should also be avoided.

Part of the joy of fieldwork is sharing your finds and observations with the members of your crew. Don't presume that it is okay to step down into someone's excavation, open their artifact bags for a quick look, or scan their notebooks or data sheets. Ask for permission. Your field supervisors should also do so as a courtesy. Also, check with your field supervisor about photographs that you might want to take of sites, excavations, or artifacts. Usually this is just a formality. These same courtesies should be practiced when you or your crew visit other field projects. Remember, you are learning to behave as a professional.

Occasionally you may get visitors in the field. The best thing that you can do is be sincere about communicating what you are doing and why you are doing it. Be enthusiastic. It is your chance to educate the public, a process that will yield dividends far into the future. You will probably be asked the same battery of questions a thousand times by a thousand different people—how do you know where to dig, why do archaeologists work in square or rectangular holes, how do you know how deep to dig, what's the best thing that you ever found? Visitors may also have their own interpretations, strong at times, about what you are finding and may deliberately draw you into a debate. Explain yourself as best as you can, be courteous, and enjoy a conversation with a person who has taken the time to come out into the field. Everyone can learn from such experiences.

Some questions may be asked about relatively sensitive information such as how much the project is costing, who is paying for it, how much money you are making as a field technician, and things of this nature. Your crew chief or supervisor should make it clear well before you encounter any visitors what information, if any, is considered to be sensitive, and which questions should be referred to them. This is especially important when the press is on-site and what you say might end up broadcast on the six o'clock news or in the newspaper.

Having Native Americans or members of descendant communities on-site or in the field with you can be an exhilarating experience. Members of descendant communities may even be included in project teams. Too often we think of the the public, and descendant communities in particular, as passive consumers of the information that archaeology generates. We get wrapped up in taking our notes, measurements, and photos, and trying to be objective, forgetting the potential value of other perspectives on what we do, the different interests that people have in the past, and how archaeological data of the past may be used by nonprofessionals. The best thing that you can do is interact and communicate with people in a respectful manner. This, at least, we owe them. We are studying their ancestors and like other members of the public, they ultimately support what we do through the federal and state governments.

APPLYING YOUR KNOWLEDGE

1. Generate a list of potential hazards for the environmental area in which you will be working, and the type of project in which you will be participating. What steps can you take personally to prepare for these hazards?

DIG DEEPER

Equipment

❑ Your local hardware store can provide a number of the items that you might want to include in your personal field kit. In addition, there are companies that cater to the equipment needs of professionals involved with surveying, forestry, agriculture, ecology, and engineering. Check out Forestry Suppliers, Inc., P.O. Box 8397, Jackson, Mississippi 39284-8397 or online at **fsi@forestry-suppliers.com**> ARCHMAT is a company that deals consistently with archaeologists and is located at P.O. Box 1418, Merrimack, New Hampshire 03054-1418 or online at **ARCHMAT@aol.com**>

❑ **http://home.houston.rr.com/bbeck/abbie/archsup.html** provides links to a variety of sources for useful field supplies. Silva, Suunto, and Brunton are major brands of compasses and sighting instruments. Silva is a division of Johnson Outdoors, Inc., P.O. Box 1604, Binghamton, New York 13902-1604. The Brunton Company is located at 620 East Monroe Avenue, Riverton, Wyoming 82501. Search the Internet for Silva, Suunto, or Brunton compasses to locate suppliers or regional outlets.

❑ Remember that when using Internet resources, addresses can change. If any of the addresses listed here fail to work, use a search engine and the key words in the website of interest.

Health and Safety Issues

- Flanagan, Joseph. 1995. What You Don't Know Can Hurt You. *Federal Archaeology* 8(2):10–13. A brief article summarizing the range of hazards that archaeologists might face in the field.
- Howell, Nancy. 1990. *Surviving Fieldwork: A Report of the Advisory Panel on Health and Safety in Fieldwork, American Anthropological Association.* Washington, D.C.: American Anthropological Association.

❑ A wide-ranging study of the health and safety problems faced by anthropologists working in the United States and abroad. This book includes a discussion of medical problems that anthropologists have had. It embraces and goes well beyond the difficulties that archaeologists face. Really interesting reading.

- Niquette, Charles M. 1997. Hard hat archaeology. *Society for American Archaeology Newsletter* 15(3):15–17. A brief but illuminating discussion of what it means to comply with OSHA regulations.
- Poirier, David A. and Kenneth L. Feder (eds). 2000. *Dangerous Places: Health, Safety, and Archaeology.* Westport, Connecticut: Bergin and Garvey, Greenwood Publishing Group.

Continued

❑ A collection of articles covering a wide variety of the potential hazards encountered by archaeologists in the field.

- Center for Disease Control , Division of Parasitic Disease—provides link to summary information about various diseases. **http://www.cdc.gov/ncidod/dpd/dpd.htm>**
- Center for Disease Control, Division of Bacterial and Mycotic Diseases—provides links to summary information about various diseases. **http://www.cdc.gov/ncidod/dbmd/diseaseinfo/**
- Gets you to the regulations section of the Occupational Safety & Health Administration (OSHA) of the United States Department of Labor. Many of the activities related to archaeology are covered under the Construction Industry. **http://www.osha-slc.gov/OshStd_toc/OSHA_Std_toc.html>**

Working with Others

❑ Watkins, Joe, K. Anne Pyburn, and Pam Cressey. 2000. Community relations: What the practicing archaeologist needs to know to work effectively with local and/or decsendant communites. In *Teaching Archaeology in the Twenty-First Century,* edited by Susan J. Bender and George S. Smith, pp.73–81. Society for American Archaeology, Washington, D.C. A good place to start getting a perspective on working with Native Americans and other descendant groups.

Chapter 6

Maps, Surveying, and Mapmaking

"By indirections find directions out."

William Shakespeare Hamlet

As an archaeologist you will constantly be using and making maps. Maps provide an approximation of the physical nature of the area in which you will be working. Because of this, they are helpful in planning the practical aspects of fieldwork. For example, what's the best way to gain access to a project area—which roads should be used and what are they like, how far will crew have to walk with equipment, what is the nature of streams to be navigated, how much of the area is woods versus open field? Maps provide a general idea of what the natural landscape should look like, what types of vegetation, rocks and soils should occur. These data aid you in the recognition of anomalies that may be artifacts or the material result of some type of human behavior. They also help the educated archaeologist determine general areas where archaeological deposits may occur near existing surfaces, where they may be deeply buried, or where caves or rockshelters might exist.

Once in the field, a person skilled in reading a topographic map and armed with a compass will be able to tell where they are at all times, and be able to use the map to locate specific areas or features. Topographic maps are essential for finding your way in areas well removed from roads. The use of a Global Positioning System (GPS) instrument can make orienteering and navigation in the field even easier. Of course you will use maps to record the locations of your excavations and sites, and compile data critical to various analyses. You will draw maps of individual sites showing topography, standing structures or ruins, other surface features, the layout of excavation units, and the position of artifacts within these units. You will create maps as a means of analyzing the spatial distribution of artifacts and features. The importance of maps is such that I haven't been able to avoid talking about them in the chapters leading up to this one. The message here is learn to love maps.

The types of maps most frequently used by archaeologists are geological, soil, and topographic maps. Of this select group, topographic maps are probably the most important because of their impact on the planning and execution of fieldwork. Chapter 4 reviewed the nature and potential uses of geological and soil maps and where one might obtain them, and included an introduction to topographic maps. Here I want to focus additional attention on topographic maps.

UNDERSTANDING AND READING TOPOGRAPHIC MAPS

Topographic maps, in addition to depicting elevations and the rise and fall of the landscape, show existing cultural features, open versus forested land, streams, and surface water. Topography literally refers to both the natural and artificial surface features of a region. As I noted in chapter 4, the federal and state governments are the best, or most readily available sources for topographic maps of a variety of scales. Some states have made maps accessible online via the Internet. Digital versions of 7.5 minute topographic maps by state are available on CD-ROM from the DeLorme Mapping Company, Freeport, Maine.

MAP SCALES

A comparison of the maps shown in Figures 6.1–6.3 illustrates the relative usefulness of maps of differing scales. These figures show an area about 6 1/4 inches square from each map. Obviously, the amount of information that can be shown on a map is limited by the scale of the map. Scale is a relative concept when comparing maps, and refers to the degree to which a map represents a geographic reality. Small-scale maps show a more extensive geographic area than large-scale maps; a small-scale map has a smaller ***representative fraction*** (the relationship between map area and actual geographic area, e.g., 1:250,000) than a large-scale map (e.g., 1:24,000). Small-scale maps therefore cannot depict as much detail as large-scale maps.

The small-scale map, 1:250,000, shown in Figure 6.1 depicts an extensive geographic region (one unit of linear measurement on the map represents 250,000 real-world units). The rise and fall of the landscape (elevations) are depicted using a large contour interval. This means that landscapes are represented in a highly generalized fashion relative to larger scale maps. Roads, place names, and developed areas are the primary cultural features shown. These are not useful maps in the field since smaller landscapes and a variety of cultural features are simply invisible owing to the map's scale. Maps of this scale are useful for evaluating the physical context of a project area in terms of the broader region. Trends in regional geography and physiography are easily seen.

Maps scaled at 1:150,000 are published privately for a variety of states in an *Atlas and Gazetteer* series by the DeLorme Mapping Company. While still of a relatively small scale, there is more detail in topography compared to maps at the 1:250,000 scale. These maps are useful for planning travel routes and for locating convenient cities or large towns with services (e.g., hospitals, food, lodging, hardware) that might be important during fieldwork.

The 1:62,500 scale map (see Figure 6.2) shows topography in greater detail and still covers a fairly extensive area. This map scale is often chosen to create topographic maps on a county-by-county basis within individual states. Scales can range to as large as 1:50,000 for such coverage. The contour interval is smaller, providing for a more realistic depiction of the landscape. Road systems are shown in greater detail and some structures are also depicted. I prefer this type of map for making travel and logistical plans related to fieldwork. It retains sufficient topographic detail to be also useful as a base for recording known site locations in evaluations of settlement patterns.

The USGS 7.5 minute quadrangle, topographic series (see Figure 6.3) with a scale of 1:24,000, is the most important published topographic map for fieldwork for much of the United States. Natural and cultural features are shown in the greatest detail, although a relatively small geographic area is depicted. Except for densely settled or developed areas, all structures are shown. Coverage at this scale is available for the coterminous United States with the exception of Alaska. Only portions of Alaska are covered by 1:24,000 maps—most of the maps of Alaska have a scale of 1:63,360. Hawaii has complete map coverage at the 1:24,000 scale, as do the United States Virgin Islands. Puerto Rico's map coverage is at a scale of 1:20,000.

On any individual 7.5 minute map an actual area of approximately 7 miles by 9 miles is reduced to approximately 18 inches by 23 inches. The overall map sheet measures about 22 inches by 27 inches. To find the 7.5 minute topographic series map (or maps) for a particular area you will need to consult a map index, usually available at the store or the online source you are using. Map indexes can always be obtained from the USGS. Indexes are organized by state and show the area of the state gridded into 7.5 minute quadrangles, and other smaller scale maps if applicable. Each quadrangle is named, most often after a large town, city, or outstanding cultural or natural feature that falls within its boundaries. The date that an individual map was last field checked, revised, or inspected will also be shown. Major towns and cities are shown on the index map to allow you to orient yourself and choose the appropriate quadrangle. Increments of latitude and longitude also appear around the edges of the index map to facilitate map selection. I'll highlight some important features of a typical 7.5 minute topographic map. Be aware that some maps may contain more information than others, and the location of this information on the map may vary, depending on the age of the map.

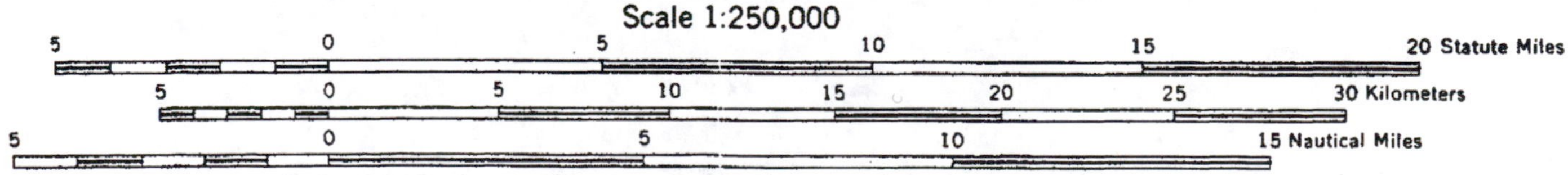

FIGURE 6.1.

Portion of the Baltimore 1:250,000 scale topographic map including portions of Maryland, Pennsylvania, Virginia, and West Virginia, USGS, 1978. The black arrow indicates the area of Marley, Maryland in the central portion of the graphic. Refer to the topographic map symbols on Colorplates 5 and 6.

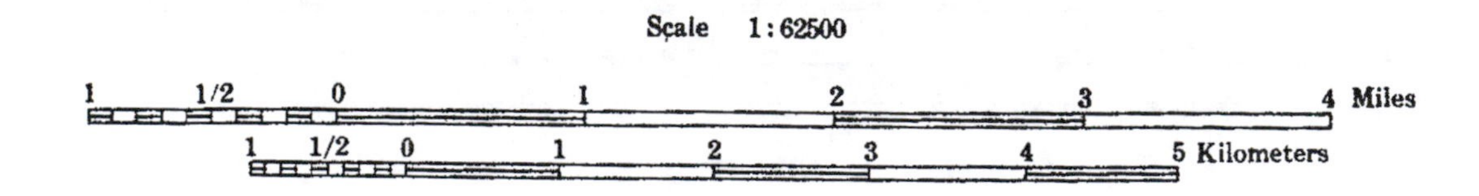

FIGURE 6.2.

Portion of the topographic map of Anne Arundel County, Maryland Geological Survey, 1978. Scale 1:62,500. Area shown overlaps that depicted in Figure 6.1. The black arrow indicates the developed area of Marley, Maryland on the left side of the graphic.

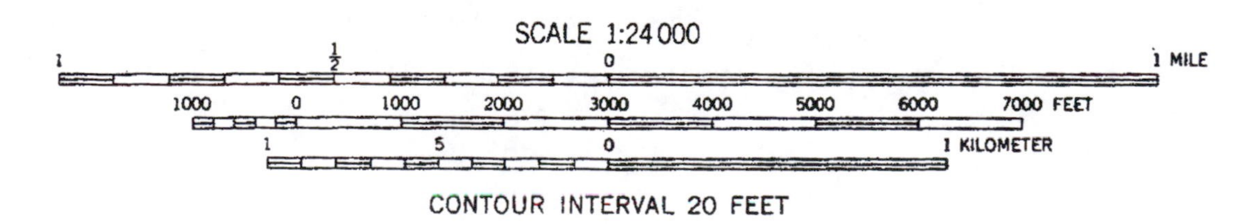

FIGURE 6.3.

Portion of the Curtis Bay, Maryland, 7.5 minute quadrangle, USGS, 1957. Scale 1:24,000. Area shown overlaps that depicted in Figures 6.1 and 6.2. Marley, Maryland is obvious on the map.

INFORMATION AROUND THE MAP FRAME

A variety of information is shown around the margins of the topographic map, as well as on the map proper. The name of the map and the state or states it encompasses are shown on the upper and lower right-hand corners of the map sheet (Figure 6.4). The 7.5 minute quadrangle, although it depicts a relatively small area, can encompass portions of different states. The Flaming Gorge quadrangle includes portions of the states of Utah and Wyoming. Map indexes for each state would include the Flaming Gorge quadrangle. Beneath the title in the lower right corner of the sheet is the date that the map was last field checked, compared with aerial photographs of the same area, or revised. In some cases two dates may be shown, one in black ink and one in purple ink. The black date refers to when the originally published map was field checked. The purple date refers to when the original map was revised using information obtained from aerial photographs. When a purple date appears on a map, all features that were added to the map as a result of the aerial photo revisions will also appear in purple. The location of the quadrangle within larger state boundaries is shown to the left of this title block. Above the title block, a key for the various roads depicted on the map is provided. A variety of publication, accuracy, and coordinate/grid system

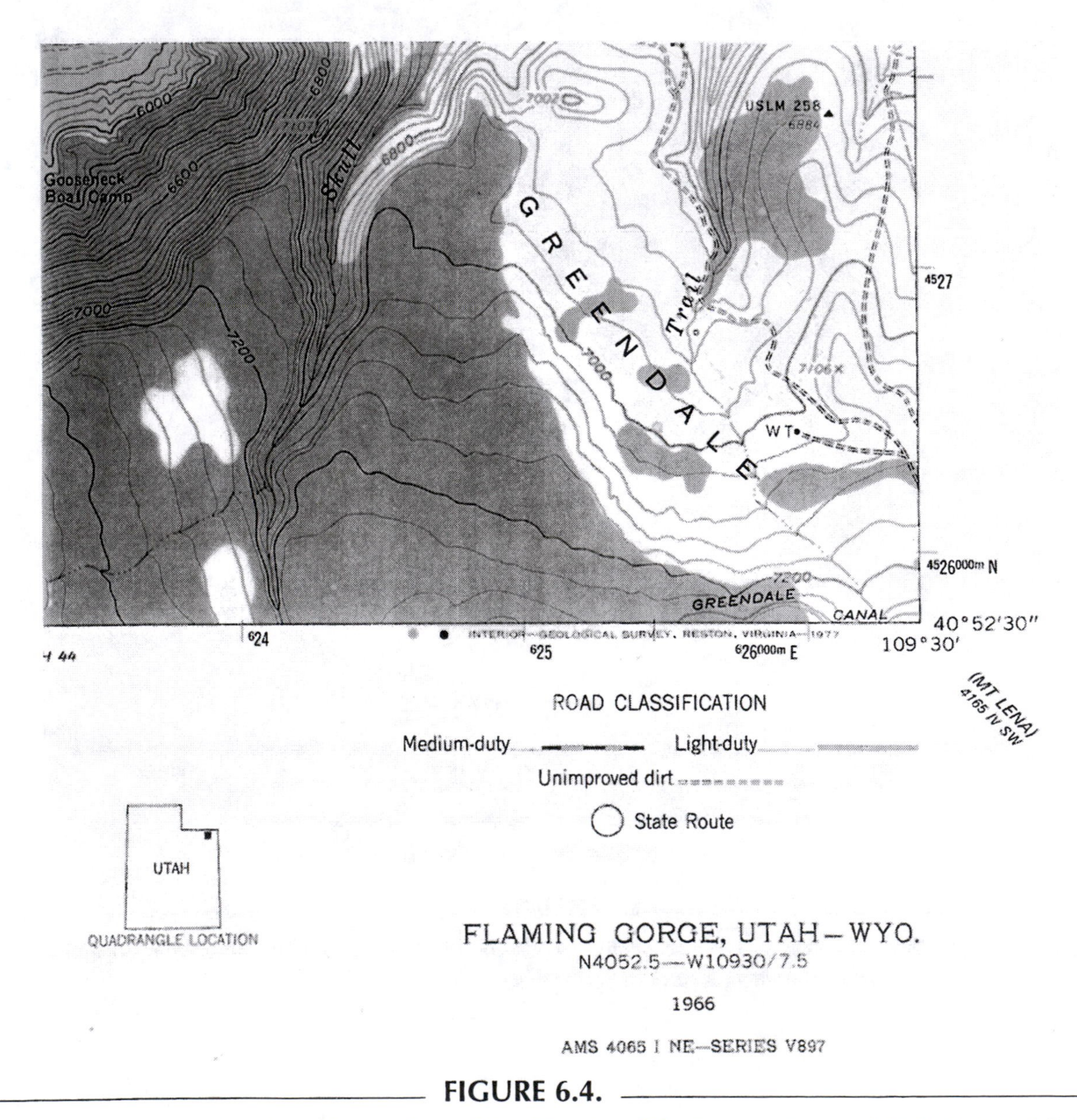

FIGURE 6.4.

Lower right (southeastern) portion of the Flaming Gorge, Utah-Wyoming, 7.5 minute quadrangle, USGS, 1966, highlighting the map title block.

data are shown in the bottom left corner of the map sheet, opposite the lower title block (Figure 6.5).

At each of the four corners of the map, and at the center of each side of the map, are printed the names of adjacent quadrangles. The names appear in capital letters within parentheses. An individual quadrangle is flanked by eight additional maps. Mt. Lena is the name of the map that joins the lower right (or southeastern) corner of the Flaming Gorge quadrangle (see Figure 6.4), and Rileyville is the name of the map that joins the southwestern corner of the Strasburg quadrangle (see Figure 6.5).

At the bottom center of each map are bar scales for measuring horizontal distance (Figure 6.6). In addition, the contour interval is described. Contour intervals appear as a series of brown lines on the map and are a convention for representing the rise and fall of the landscape (a three-dimensional entity) in two dimensions. The contour interval is the change in elevation between two adjacent contour lines. In the case of Figure 6.6, the change in elevation between two adjacent contour lines is 20 feet. We'll discuss contour lines in more detail later.

To the left of the bar scales are a series of lines/arrows showing the relationship between ***magnetic north*** (MN), the ***"true" north*** line represented by a star, and grid north (GN) as used in the ***Universal Transverse Mercator*** (UTM) grid system (see Figure 6.6). The true north line is drawn with reference to the position of the earth's north pole. The vertical lines framing the map sheet are aligned along this true north direction. The north and south poles are the opposing ends of the diameter line that represents the axis about which the sphere of the earth rotates. They are

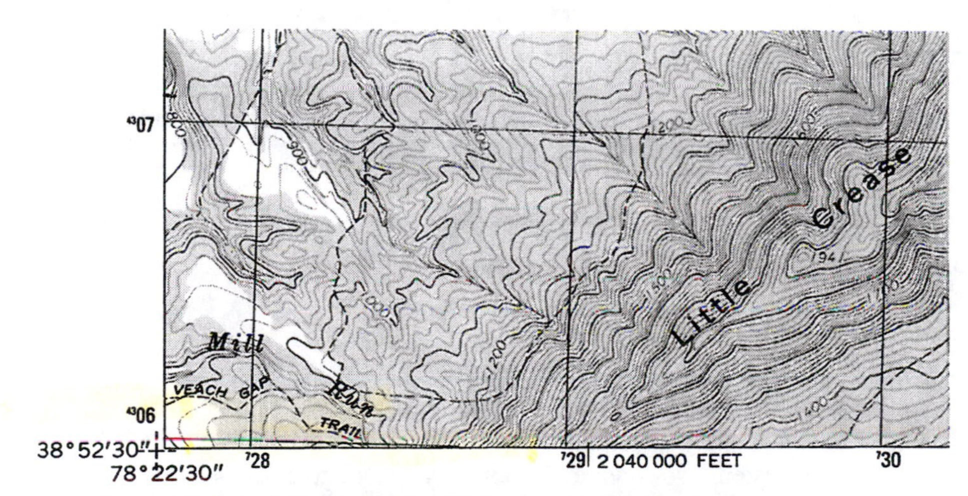

FIGURE 6.5.

Lower left portion (southwestern) of the Strasburg, Virginia 7.5 minute quadrangle, USGS, 1994, highlighting publication and accuracy data.

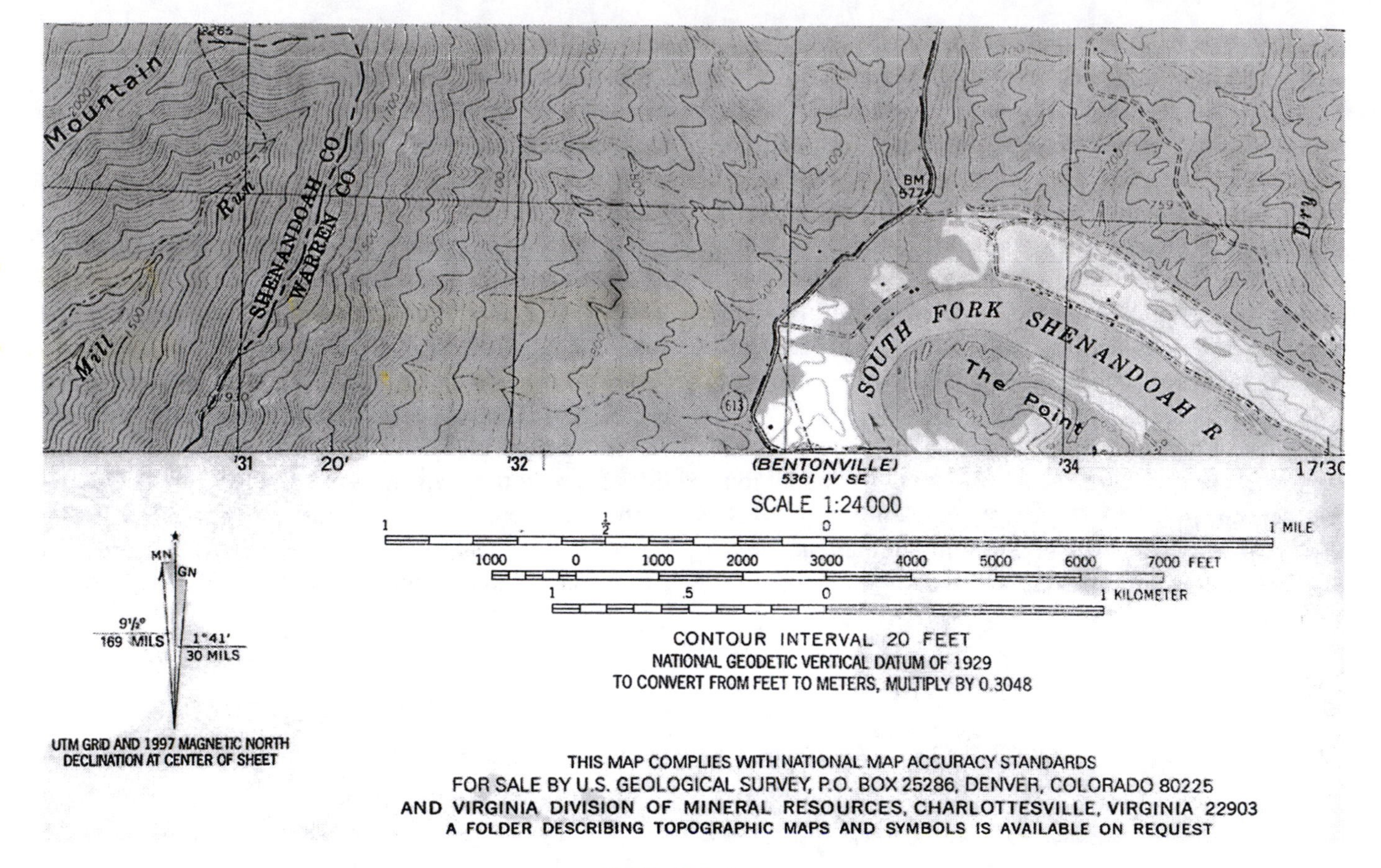

FIGURE 6.6.

Bottom center portion of the Strasburg, Virginia 7.5 minute quadrangle, USGS, 1994, showing bar scales for linear measurements. Scale 1:24,000.

geographically fixed points. Maps are therefore oriented to a standard that theoretically does not change, making them part of the same organized system worldwide. The up and down orientation of a map is north-south, the left to right orientation is west-east.

Magnetic north and south do not align precisely with the geographic north and south poles of the earth. The magnetic field of the earth is not uniform and has changed over time; how magnetic north is recorded varies depending on your location on the earth's surface. The UTM system is a metric grid that covers the earth with the exception of the polar regions. Its coordinate system allows specific points on the earth's surface to be located with precision. Latitude and longitude are used in a similar fashion for locating points.

The variance between magnetic north and true north, and true north and the UTM grid orientation, is graphically displayed in the diagram of three lines seen in Figure 6.6. This deviance from true north is referred to as ***magnetic declination*** and is measured in degrees and minutes. For the area shown on the Strasburg quadrangle, magnetic north differs from true north by nine and one half degrees, the UTM grid from true north by one degree and 41 minutes.

Using a topographic map in the field to find your way around necessitates orienting the map to some fixed reference. Magnetic north as determined by a compass is the most easily established referent. But, since it differs from the true north on which the topographic map is based, we must compensate for this difference by figuring the declination into any readings or calculations. Many compasses are designed to be adjusted for the declination of a given area, avoiding the need for mental gymnastics.

Coordinate and Grid Systems

I've mentioned several terms that bear further discussion in order to understand other features of the 7.5 minute quadrangle, and how these maps are used by archaeologists: latitude, longitude, degrees and minutes, and the UTM grid. The north and south poles, the equator, latitude, and longitude are used by mapmakers (cartographers) to develop a geographic grid for the earth and determine the precise location of features on its surface. The equator is the great circle of the earth created by a plane perpendicular to the rotational axis of the earth, and equidistant from the north and south poles (Figure 6.7). Lines, or parallels of ***latitude*** represent the east-west component of the geographic grid. They are oriented parallel to the equator and progress, and are measured, to the north and south from there. All portions of lines of latitude are parallel to one another. That's where the phrase parallels of latitude comes from. The north-south component of the grid system is lines, or meridians of ***longitude.*** These lines run from the north pole to the south pole and cross all lines of latitude at right angles. These lines are parallel to one another only in certain places, primarily near the equator. You can see how individual lines begin to converge as the poles are approached (see Figure 6.7). Lines of longitude progress, and are measured, to the west or to the east from the ***prime meridian.*** The prime meridian is a line of longitude chosen as a standard referent for all others. It is the one that passes through the Royal Observatory at Greenwich, England.

Latitude and longitude are components of a ***geographic coordinate system*** (grid) that covers the surface of the earth and can be used to describe specific locations, i.e., where a line of latitude intersects a line of longitude. Figure 6.7 only shows a few examples of lines of latitude and longitude. Obviously, many more exist in the conceptual sense or we would not be able to use them to describe any point on the earth's surface. A topographic quadrangle is bracketed by two lines of latitude and two lines of longitude. Since lines of longitude are not parallel except near the equator, the quadrangles actually represent slight trapezoids rather than rectangles. However, this distortion is generally not evident on individual maps because of the relatively small areas that they cover.

Circles, or arcs representing portions of circles, are commonly divided into ***degrees*** and subdivisions of degrees. A single degree represents 1/360 of a circle, or more precisely, 1/360 of the angle that the radius of a circle describes in a full revolution (Figure 6.8). Therefore, a complete circle represents 360 degrees. A degree can be subdivided into 60 ***minutes,*** and a minute into 60 ***seconds.*** Remember that these terms are referring to portions of angles. These divisions of a circle will be familiar to anyone who has ever looked at the face of a compass or used a protractor. The symbol for degree(s) is the familiar small circle positioned to the

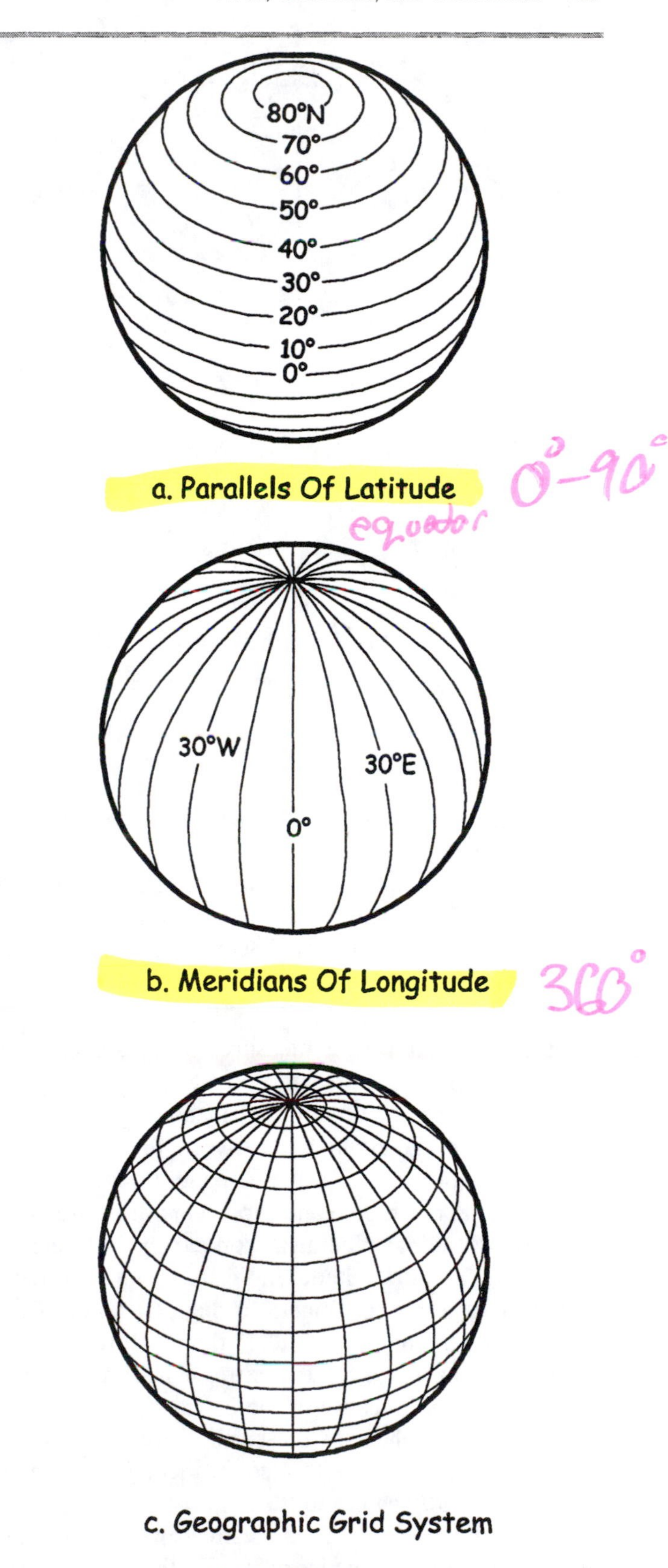

FIGURE 6.7.

The geographic grid system of the earth involving the poles, the equator, parallels of latitude, and meridians of longitude.

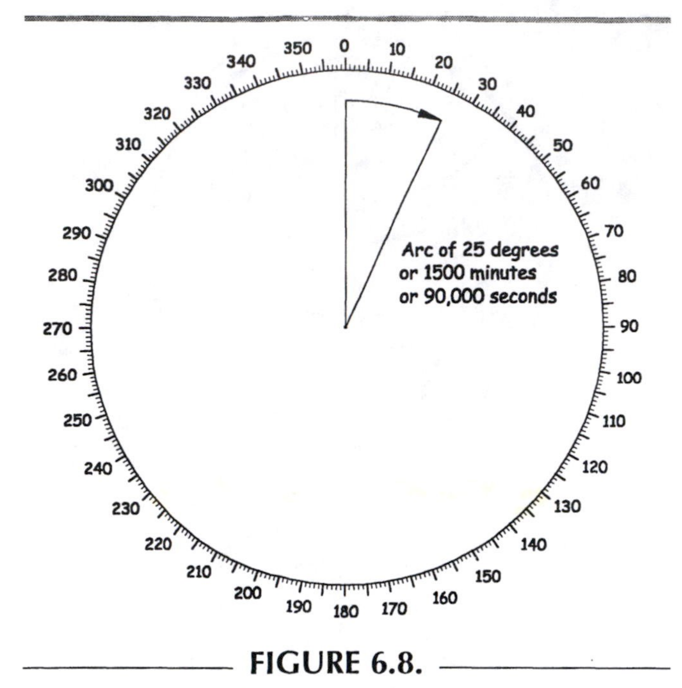

FIGURE 6.8.

A circle and its divisions into degrees, minutes, and seconds.

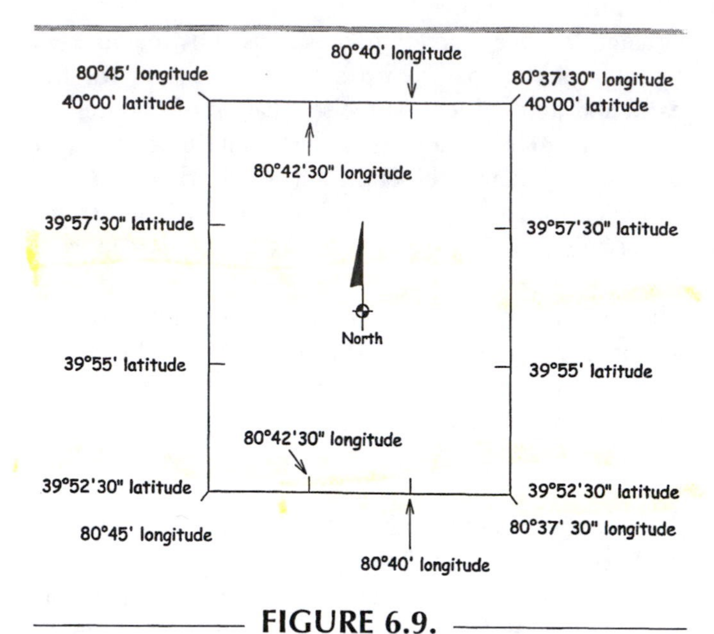

FIGURE 6.9.

Schematic of the Moundsville, West Virginia-Ohio 7.5 minute quadrangle, USGS, 1976, showing latitude and longitude reference points. The positioning of these reference points is typical of most maps of this scale.

upper right of a numeral. Minutes are shown by a single stroke positioned at the upper right of a numeral, e.g., 25′ (read twenty-five minutes), and seconds are designated by two strokes to the upper right of a numeral, e.g., 30″ (read 30 seconds). Seventy degrees, twenty-five minutes, and thirty seconds would be expressed as 70°25′30″.

Lines of latitude are measured or designated by degrees north or south of the equator, so 20 degrees north latitude (or N20° latitude) means that you have a line of latitude, running east-west, that is situated 10 degrees north of the equator. Ninety degrees north latitude is the north pole and represents the extreme in northern latitudinal readings. Conversely, 40 degrees south latitude (S40° latitude) indicates a line of latitude, running east-west, positioned 40 degrees south of the equator. Ninety degrees south latitude represents the south pole and represents the extreme in southern latitudinal readings. Because of the orientation of lines of latitude on the sphere of the earth, only 180 degrees of a circle's arc are needed to describe them (90 degrees to the north and 90 degrees to the south).

In turn, lines of longitude are measured by degrees east or west of the prime meridian. To the east they can progress to 180 degrees, and to the west they can progress 180 degrees until both converge again on the prime meridian. Because of their orientation on the sphere of the earth their description employs the entire 360 degrees of a circle (180 degrees east, plus 180 degrees west).

Figures 6.9 and 6.10 depict the latitude and longitude reference points on a typical 7.5 minute quadrangle. At the upper right or northeastern corner of the map is the latitude (40°00′) and longitude (80°37′30″) for that point or location. Two other readings for latitude are shown as you move down the right or eastern margin of the map. Figure 6.10 only shows the first of these. The reading of 57′30″ along the edge of the map is adjacent to a short black, vertical line, or tick mark, that runs from the map margin to the left. All reference points for latitude and longitude are designated by similar tick marks. The reading of 57′30″ is actually shorthand for 39°57′30″ of latitude, since we have moved south of our initial point of 40°00′ and are headed towards our southernmost point at 39°52′30″ (compare Figures 6.9 and 6.10). This shorthand is employed between all of the corners of the map.

The longitude reference points work in the same way. We begin with a reading of 80°37′30″ in the upper right or northeastern corner of the map, and moving to the left the first appropriate tick mark encountered has a reading of 40′ (compare Figures 6.9 and 6.10). This is shorthand for 80°40′ since we are moving left or west from our starting point of 80°37′30″ longitude headed

FIGURE 6.10.

Portion of the Moundsville, West Virginia-Ohio 7.5 minute quadrangle, USGS, 1976. Latitude and longitude degree, minute and second readings are arranged along the edges of the map as indicated by the arrows (arrows added, not part of original map).

for the upper left or northwestern corner of the map at 80°45′ longitude.

Now you can appreciate why this topographic map is called a 7.5 minute quadrangle. The northern margin of the map is 40°00′ of latitude and the southern margin 39°52′30″ of latitude. The difference between these readings is 7.5 minutes or 7′30″. In turn, there are 6.5 minutes, or 6′30″, between the lines of longitude that frame the eastern and western margins of the map.

Like lines of latitude and longitude, UTM is a grid and coordinate system used in mapping. Representing the curved surface of the earth on the flat, two-dimensional sheet of a map invariably involves some degree of distortion, regardless of what type of system is used. In attempts to deal with such distortion, the UTM system employs north-south strips of the earth's surface in its map projections. The north-south strips are relatively narrow, comprising 6 degrees of longitude (Figure 6.11). Sixty of these strips, or ***UTM zones,*** are used to portray the earth's surface from the equator north to 84° latitude and south to 80° latitude. Each UTM zone is given an identifying number. Polar regions are not mapped in the UTM system.

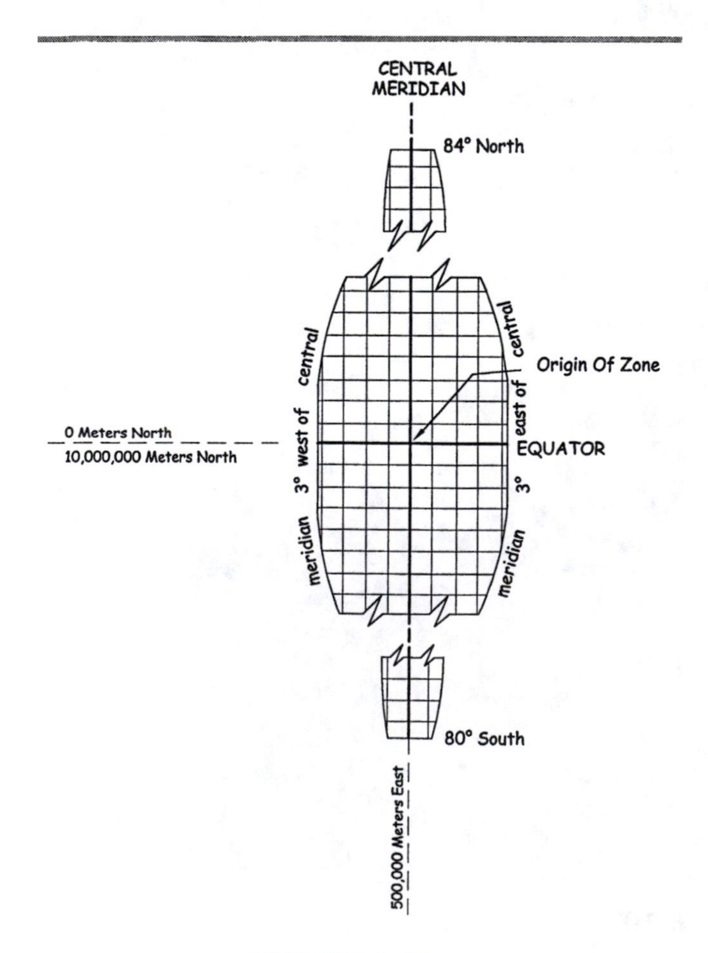

FIGURE 6.11.

Example of a UTM (Universal Transverse Mercator) strip or zone.

The grid lines of each UTM zone are parallel to the equator and the prime meridian, which are taken as the base points for the system. Each UTM zone has a ***central meridian*** or central vertical grid line. The central horizontal line of each zone corresponds with the earth's equator. Horizontal grid lines within a zone are measured and designated in meters from the central horizontal line representing the equator. Moving north, measurements start from 0 meter and become progressively higher. Moving south from this line measurements begin at 10,000,000 meters and progressively decrease. What this reflects is a measurement of a longitudinal distance around the earth beginning at the equator, moving north to the pole, and then wrapping around the opposing side of the earth back to the starting point at the equator. Vertical lines within a zone are measured with reference to an imaginary point that lies beyond the zone itself, 500,000 meters west of its central meridian. Thus all measurements of vertical lines within the zone are made moving east (to the right) from this reference point.

The UTM grid lines differ from the grid established by lines of latitude and longitude because of their parallel and perpendicular quality. Remember that lines of latitude are always parallel to one another while lines of longitude are not always parallel to one another. If you look closely at the UTM grid lines projected upwards from the map edges in Figures 6.5 and 6.6, you will see that they are skewed relative to the lines of latitude and longitude that frame the edges of the map. This reflects the noted differences between the UTM and latitude and longitude grid systems.

The UTM zone that a 7.5 minute quadrangle falls within is noted in the text appearing in the lower left hand corner of the map (see Figure 6.5). On maps that don't have a UTM grid projected over them, light blue tick marks around the border of the map are used to designate UTM grid points (see Figure 6.10). The tick marks are short straight lines that originate at the black line that defines the map edge, and proceed a short distance away from the map edge. This is in the opposite direction of the black tick marks used to denote points of latitude and longitude. Next to the tick marks are numbers designating a measurement in meters (Figure 6.12). The tick marks are usually placed every 1,000 meters as will be noted in the text in the lower left-hand corner of the 7.5 minute quadrangle (for example, see Figure 6.5).

Complete meter designations for a UTM grid point are shown typically only near the southeastern (lower right-hand) corner of a 7.5 minute map. As you can see in Figure 6.12, the first UTM grid point moving north from the map's corner is 4,526,000 meters N ("northing"); this point is 4,526,000 meters north of the equator in the grid system of this particular UTM zone. The first UTM tick mark encountered moving west (to the left) from the corner is 615,000 meters E ("easting"); this point is 615,000 meters east of the imaginary line outside of the UTM zone used as a reference point for the designation of all of the zone's vertical grid lines. To avoid confusion, all UTM designations of single points (a place where a horizontal and vertical grid line intersect) are expressed as a northing and easting. Tick marks on the east and west (left and right) sides of the map refer to northing measurements and the horizontal lines of the UTM grid. Tick marks on the north and south (top and

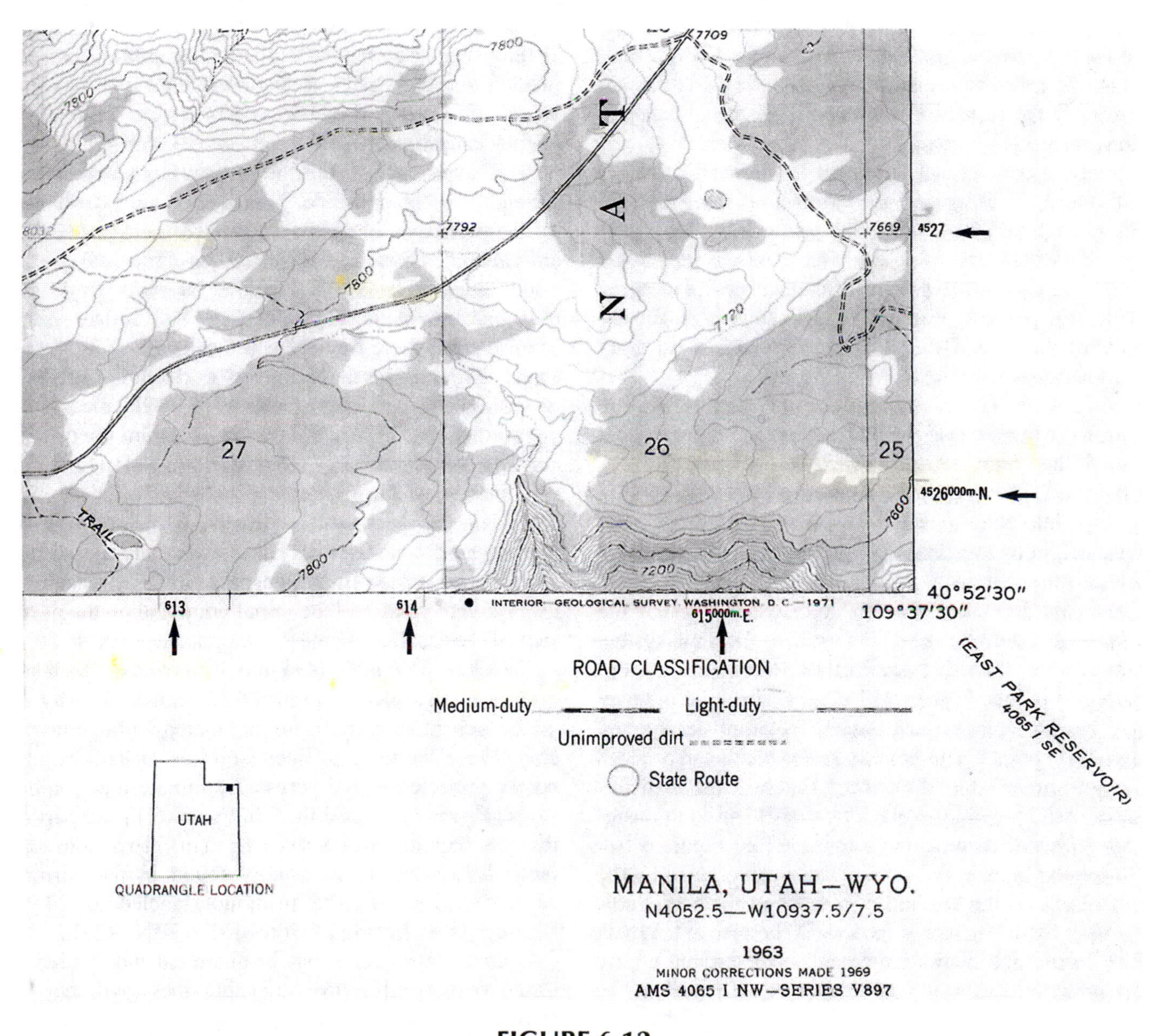

FIGURE 6.12.

Portion of the Manila Utah-Wyoming 7.5-minute quadrangle, USGS, 1963/1969. The full "northing" (4,526,000 meters N) and "easting" (615,000 meters E) measurements of points on the UTM grid are shown just to the left, and just above the southeastern corner of the map. Measurements at other UTM grid points at intervals of 1,000 meters are expressed in a numerical shorthand. UTM grid points are indicated by arrows (arrows added, not part of original map).

bottom) sides of the map refer to easting measurements and the vertical lines of the UTM grid.

A numerical shorthand or abbreviation is used for the other UTM tick marks around the border of the 7.5 minute map. For example, the first grid point north of 4,526,000 meters N is designated as 4527 which actually means 4,527,000 meters N. Remember that designated UTM grid points or tick marks generally are spaced at intervals of 1,000 meters (see Figure 6.12). To the left or west of 615,000 meters E is 614, which translates as 614,000 meters E. In short, the last three zeroes are not shown in these designations. Occasionally a 7.5 minute map will overlap two UTM zones. This will be made apparent in the text in the lower left corner of the map and in the complete meter designations for grid tick marks.

Only more recently published maps will project UTM grid lines across the surface of the map (see Figures 6.5 and 6.6). When the grid is projected it will be done with black lines; blue tick marks at the margin of the map are not employed. The meter designations of each grid point appear where grid lines intersect the edge of the map. Other grids may be drawn on topographic maps using red lines.

Each state has its own grid system that is tied into national geodetic standards or reference points. This is called the ***State Plane Coordinate System*** (SPCS). Like the UTM system, this statewide mapping grid is divided into zones, each with a central meridian and a false origin or imaginary point that falls outside of the zone to the southwest, and from which distance measurements are taken for the north-south lines of the zone's grid. Unlike the UTM system, the state system employs the English system of measurement to designate grid points. Figure 6.13 is an example of the map tick marks and designations in feet that accompany these grid points. The tick marks are black and proceed away from the edge of the map. This orientation differentiates them from the black tick marks used to designate points of latitude and longitude (see Figure 6.10). Horizontal lines in the state grid are represented by the tick marks on the left and right sides of the map, vertical lines by tick marks at the top and bottom of the map. Few of the tick marks are actually labeled, but all are placed at intervals of 10,000 feet, as indicated in the text that appears in the lower left corner of the map sheet (see Figure 6.13; also see Figure 6.5). If a map straddles two different zones, distinctive tick marks are used to denote the points in each zone.

A final grid system that you will encounter on some 7.5 minute maps is that of the ***United States Public Land Survey System*** (USPLSS). Its use began in the late 18th century and continues today with elaborations. The colonial government was interested in providing a systematic survey of lands north and west of the Ohio River, relatively new portions of the U.S. where settlements were expanding. ***Principal meridians*** (north-south lines) and horizontal base lines were selected to establish subsequent grids (Figure 6.14). The intersection of a principal meridian with its associated horizontal baseline has a latitude and longitude designation, and serves as the ***initial point*** of the survey for laying out the remainder of the grid and the designation of units within the system. The federal Bureau of Land Management (BLM), not the USGS, is responsible for maintaining records of surveying related to this system (Bureau of Land Management 1973; Thompson 1988:81).

The largest unit within the system is a square measuring six miles on a side. The squares are established by the placement of east-west lines (***townships***) at six-mile intervals from the horizontal baseline, and north-south lines (***ranges***) at six-mile intervals from the principal meridian (Figure 6.15). The square units formed from these lines also are referred to as townships. Moving north from the horizontal baseline, township lines are designated as T.1N, T.2N, etc.; moving south T.1S, T.2S, etc. Moving east from the principal meridian, range lines are designated as R.1E, R.2E, etc.; moving to the west of the meridian R.1W, R.2W, and so on. In combination, a township line and a range line are used to designate a particular square measuring six miles on a side (the township). So the first square unit (township) east of the initial point and on the north side of the baseline would be designated T.1N, R.1E.

Each township is divided into 36 ***sections,*** which are each one mile square. Figure 6.15 shows the way in which individual sections are numbered within a township. These are the grid lines (solid or dashed) prominently projected in red across 7.5 minute topographic maps (Figure 6.16; grid lines in Figure 6.12 are part of the U.S. Public Land Survey System). Township and range designations are usually found in the corners of the map sheet. The grid unit labeled as 24 in Figure 6.16 is section 24 of township T3N, R22E.

Sections themselves may be quartered and divided in other ways (see Figure 6.15), but these will not be shown on the 7.5 minute topographic map. Some sections can be irregularly shaped because of the difficulty of overlaying a two-dimensional system of square grids over the curved surface of the earth. Subdivisions (***fractional lots***) of irregularly shaped sections can themselves be irregular in shape. These are numbered

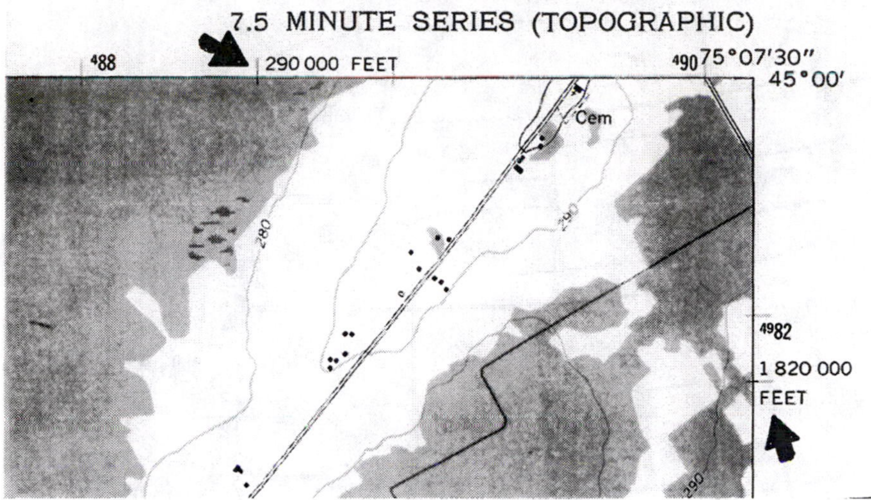

FIGURE 6.13.

Portions of the Morrisburg, Ontario-New York 7.5 minute quadrangle, USGS, 1964. Grid points in the State Plane Coordinate System are indicated by black tick marks adjacent to designations in feet (see arrows). Other black tick marks showing points in this grid system are placed along the edge of the map at intervals of 10,000 feet as indicated by the text found in the lower left corner of the map sheet.

sequentially by the surveyors who lay them out (Napton and Greathouse 1997:196).

The U.S. Public Land Survey System allows you to define progressively smaller plots of land. The grid system is an actual one, unlike the geographic coordinate or UTM systems; it has been laid out by surveyors with the corners of sections marked by monuments anchored in the ground. The monuments are capped with a brass plate that provides information on the location that it represents. Where the monuments are known to still exist, the corners are designated with "+" at the intersection of grid lines on the topographic map. The corners of some quarter sections may also be marked with monuments. Surveying related to the Public Land System is ongoing and all relevant information may not be shown on the edition of the map that you are using. Regional offices of the BLM or USGS may be able to provide you with additional information.

For archaeologists, the monuments of the grid system are important locational references to which artifact finds and sites can be tied. Principal meridian, township, range, section, and subdivisions of sections may be used on some archaeological site forms to record the so-called legal location of the site. Remember that not all states have been mapped using this system. The system generally excludes the area covered by the 18th-century colonies/states and West Virginia, Tennessee, Kentucky, Texas, and Hawaii

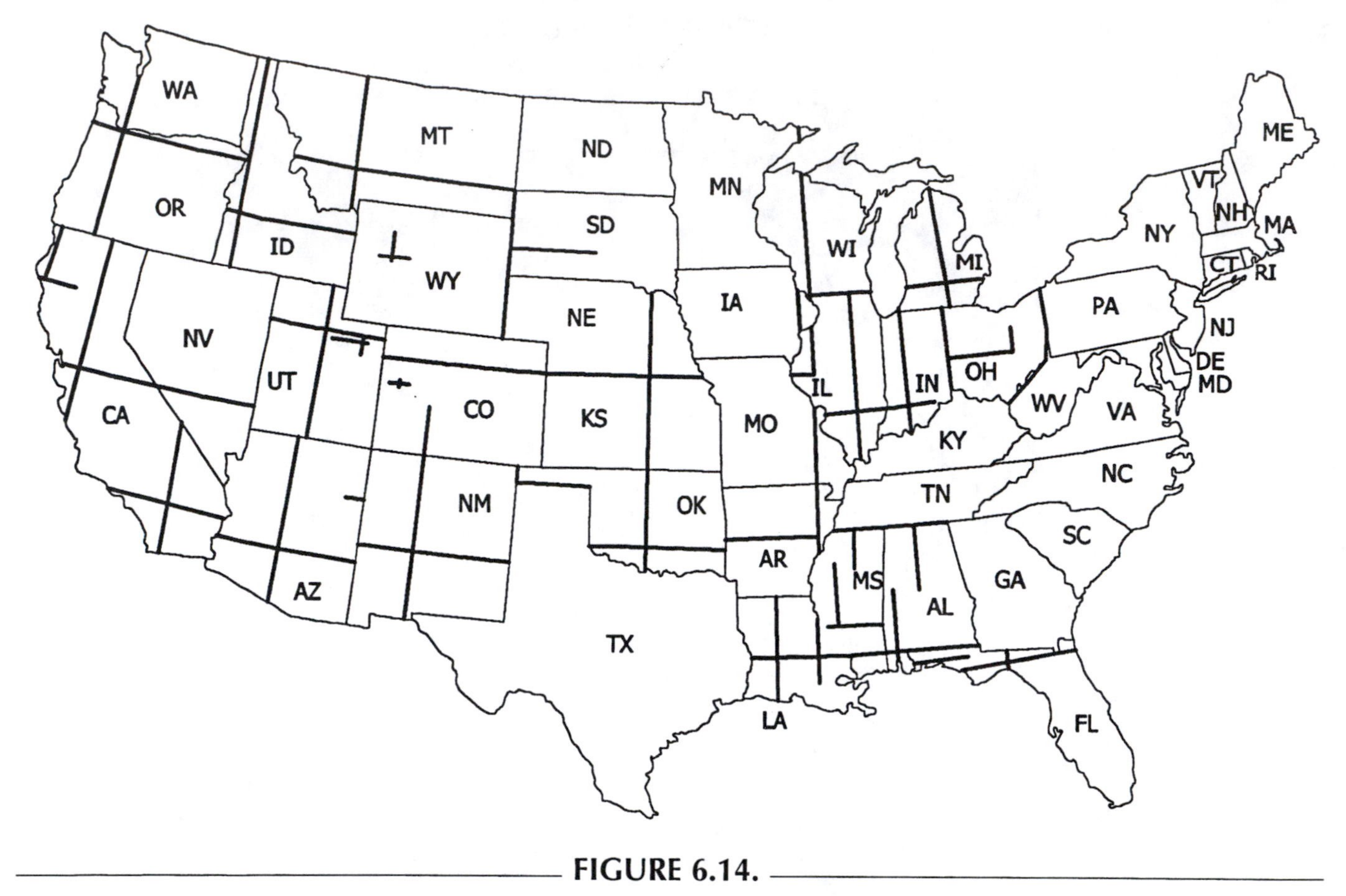

FIGURE 6.14.

Meridians and base lines used to establish the Public Land Survey in portions of the United States. The darker, heavier lines represent the base lines and meridians, lighter lines are state boundaries. Derived from Thompson 1988: Figure 88.

(cf. Napton and Greathouse 1997: Figure 9.9; Thompson 1988: Figure 88).

We've now reviewed four different coordinate or grid systems that are expressed in some way on 7.5 minute topographic maps: the geographic coordinate system (latitude and longitude), the UTM system, the state plane coordinate system, and the U.S. Public Land Survey System. No one system serves all of the myriad needs of archaeologists, although some are used more consistently than others. Latitude and longitude are used to identify site locations in the field, for plotting site locations on maps to accompany archaeological site forms, and to some extent for finding your way around in the field and relocating sites using a GPS instrument. The UTM system has long been used to identify/plot the location of sites on maps submitted with archaeological site forms. In this one respect it is quicker and easier to employ than latitude and longitude. The SPCS is little used by archaeologists, even though it is associated with locational monuments in the field that are useful for surveying. The USPLSS, like the GCS and UTM systems, can be used to describe the location of an archaeological site and plot it on a 7.5 minute topographic map. It is also important for archaeologists in the field because of the numerous monuments that exist and their systematic placement, allowing sites and surveying data to be tied accurately into the existing system and making the relocation of sites in the field easier.

TOPOGRAPHIC MAP SYMBOLS

Within the boundaries of the 7.5 minute topographic map are a variety of symbols and colors used to convey information. Table 6.1 (in the color section) provides a key to the various symbols that can be found on maps. Most of the terms associated with these symbols are self-evident. However, the more background that you have in geology or geography the greater your comfort level in translating map symbols and terms. For example, what is a glacial moraine? What's the difference between an intermittent stream and a perennial stream? In general, the following associations of map color and type of mapped feature can be noted:

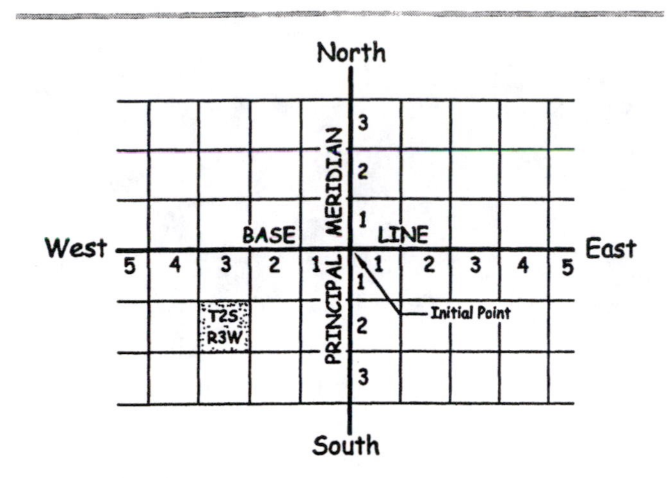

Township Grid With Location Of Township 2 South, Range 3 West

6	5	4	3	2	1
7	8	9	10	11	12
18	17	16	15	Section 14	13
19	20	21	22	23	24
30	29	28	27	26	25
31	32	33	34	35	36

Township 2 South, Range 3 West With Location Of Section 14

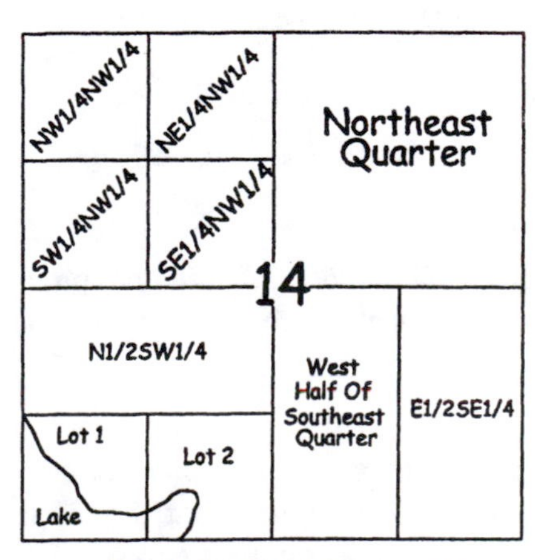

Possible Subdivisions Of Section 14

FIGURE 6.15.

Township, range, and section divisions of the United States Public Land Survey System. Redrawn from Thompson 1988: Figure 87.

- **blue**—water features like streams, rivers, ponds, lakes, bays. Also used as shading in depicting glaciers and snowfields, and land that may be periodically inundated.
- **black**—cultural features like roads, buildings, and some boundary lines. UTM grids when lines are projected across a map.
- **brown**—contour lines representing the configuration of the surface of the land. Also used as shading in depicting deserts and dry washes (channels in which water flows during storm events), intricate surfaces, tailings from excavations (e.g., ponds, mines), and areas of sand, dunes, or mud.
- **gray**—unsurveyed areas in Alaska.
- **green**—vegetation cover like forest, scrub, orchards, nurseries, vineyards.
- **red**—major roads. The grid lines and corners of the USPLSS projected across a map and other features of land surveying systems.
- **pink**—shading used for urban areas.
- **purple**—any map features added as a result of the evaluation of aerial photos taken after the original production of a map.

ELEVATION

Contour lines are one of the most important features of the 7.5 minute topographic map. These brown lines are the mapmaker's abstraction of elevation and the rise and fall of the landscape. Using contour lines is a two-dimensional way of providing a perception of three dimensions (Figure 6.17). Contour lines are a summary or approximation of the intricacies of a landscape, providing us with a sense of what we would see if we were actually on the ground in a specific place. An individual contour line connects areas on the landscape with the same elevation, without crossing over lines representing higher or lower elevations. Contour lines never intersect or cross over one another. Contour lines also may be used to depict the configuration of surfaces that are now underwater. The contour lines will be brown if the land has been recently inundated, such as in the case of reservoirs or dams. Blue contour lines are used to show more typical underwater landscapes.

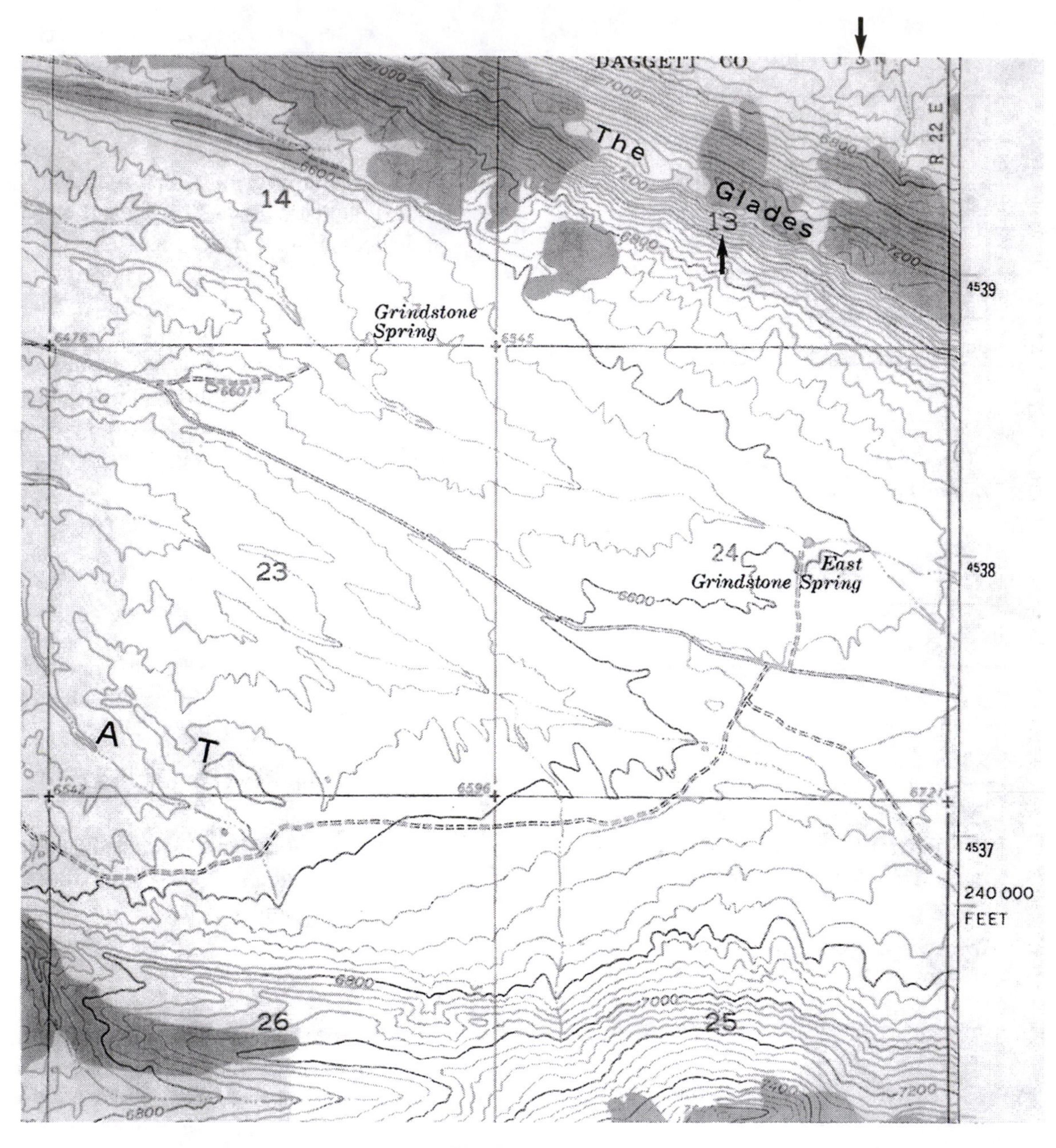

FIGURE 6.16.

Northeastern portion of the Dutch John, Utah-Wyoming 7.5-minute quadrangle, USGS, 1966. The grid lines projected across the map are sectional boundaries of the U.S. Public Land Survey system. Designations for the township and range of this portion of the grid are indicated by the black arrows in the graphic.

Figure 6.18 illustrates that contour lines are a summary or approximation of trends on the earth's surface. Twelve actual elevations (above sea level) are shown for a portion of a hypothetical landscape. Using these data, I've chosen to draw the contour line representing 25 feet. Only one of the actual readings is 25 feet, but there are a number of places where the elevation of 25 feet is assumed to exist. For example, we assume that somewhere on the slope between the actual readings of 20 feet and 30 feet is an elevation of 25 feet. To draw the contour line I've assumed that the slope between these two known points is consistent, so 25 feet in elevation must exist halfway between them. The remainder of the contour line is drawn using the same relationship between change in elevation between known points, and the distance between these points. This is a simple example. The contour lines on USGS topographic maps are actually constructed using a

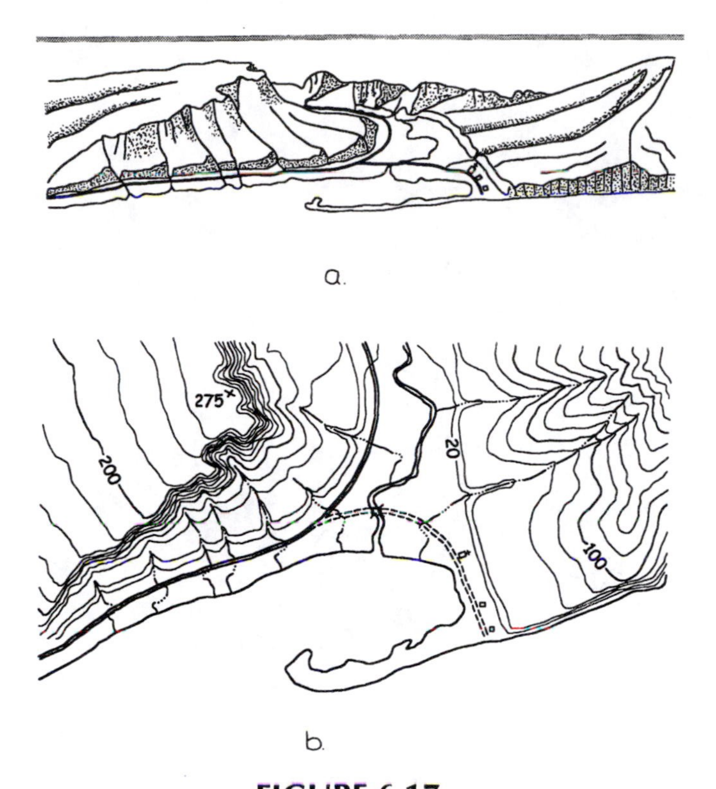

FIGURE 6.17.

An example of a three-dimensional landscape (a), expressed as a topographic map employing contour lines (b). Adapted from Thompson 1988: Figure 22.

combination of aerial photographs, the surveying of control points on the ground, and an electronic, computer-assisted image correlator system.

Contour line elevations are expressed in feet for most of the maps covering the United States, although some have been prepared in the metric system. Not all contour lines are labeled as to the elevation that they represent. Every fifth contour line (***guide contour***) is thicker and slightly darker in color than the others. Select guide contours are labeled with an elevation making it possible for the map reader to determine the elevation represented by adjacent contour lines (Figure 6.19). The standardized difference in elevation between individual contour lines is called the ***contour interval.*** Standard contour intervals are 5, 10, 20, and 40 feet, or 1, 2, 5, 10, and 20 meters when the metric system is employed. Other intervals are used where the terrain cannot be adequately depicted using the standard ones.

Other map features provide additional information about elevations. Benchmarks are locations where an exact elevation has been determined. These are marked with a black "X" and the abbreviation "BM" with an associated elevation reading on topographic maps (see Figure 6.19). In the field, benchmarks usually consist of a metal pipe and concrete monument set in the ground so that its top is at ground level (Figure 6.20). The top of the monument is capped with a brass plate inscribed with information about its location and elevation. As noted previously, the corners of sections in the USPLSS are monumented and correlated with an exact horizontal and vertical (elevation) measurement. Elevations are noted adjacent to the red crosses that denote section corners. Spot elevations are denoted by a small "x" and associated elevation reading (see Figure 6.19). These are generally not monumented, but are associated with some distinctive feature on the landscape like a road intersection or the summit of a hill or mountain. Spot elevations where the value is shown in black have been field checked, while those printed in brown have not (Thompson 1988: Figure 21).

The distance between contour lines reflects how gradually or how quickly the ground surface is sloping. Contour lines relatively close together represent a steeper slope than those that are spaced farther apart; elevation is increasing over a relatively short, horizontal distance. You can see in Figure 6.19 that the landscapes adjacent to the Green River are steep in contrast to those located farther to the south. Figure 6.21 is a topographic profile for a portion of this area (Line A-B on Figure 6.19), and is another way to visualize the earth's surface as represented by contour lines. The profile is constructed using elevation and distance values shown on the map. Climbing up a long slope with an angle of 37°, like those shown, would be quite a scramble.

The slope of a surface or landscape can be evaluated without having to draw a topographic profile. First measure the distance over the area in which you are interested, then determine the amount that elevation has changed over this distance by seeing how many contour lines were crossed. These two values can then be used to construct a slope triangle, the angle of which can be measured with a protractor. Alternatively you can calculate the gradient of the slope using the same two values, distance, and the change in elevation that occurs over that distance. Gradient is the rate of regular ascent or descent of a slope or line, and is usually expressed as a percent. Calculating gradient involves dividing the "rise" (the change in elevation) by the "run" (the distance over the landscape of concern). The gradient of the slope in Figure 6.21 is nearly 80 percent.

The mapmaking process requires making decisions about which elevations to depict and the contour interval to employ. These decisions are sensitive to the scale of the map (see Figures 6.1–6.3) and the nature of the

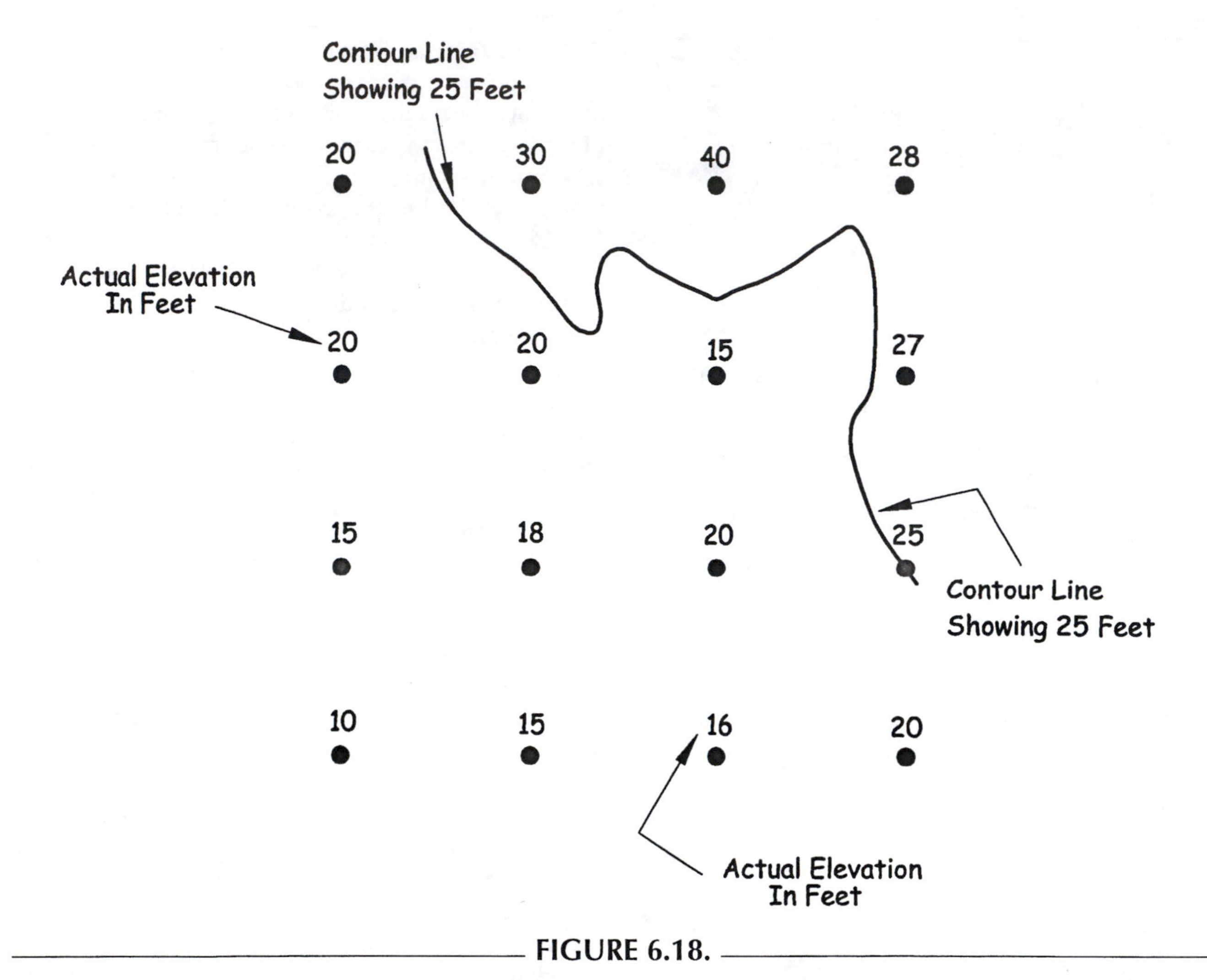

FIGURE 6.18.

The construction of a contour line (25 feet in elevation) from a series of real elevations on a landscape.

landscape to be depicted. For terrain that is steeply sloping, it may be graphically impossible to show contour intervals for every 10 feet of elevation, like the portions of the Flaming Gorge map in the examples above. If we were to attempt to map the same area with an interval of 10 feet, the contour lines would be so closely spaced or on top of one another that they would not be distinguishable as individual lines. In turn, a contour interval of 10 feet may be inadequate to reveal some landscape features on terrain that is flat to undulating. Even on 7.5 minute maps, it is not possible to use a contour interval that will reveal all individual landscapes, or aspects of the landscape that might be of interest to archaeologists.

In using topographic maps where streams are mapped, "V"s made by the contour lines crossing the streams will always point upstream or upgrade (obvious examples can be seen in Figures 6.10 and 6.19). In areas where contour lines are a patterned series of "V"s but no stream is present, a gully, ravine, or valley with sloping sides and an obvious grade running the length of the landform is depicted. Again, the "V"s are pointing upgrade or upslope.

Archaeologists working on cultural resource management investigations may be able to take advantage of detailed topographic maps made especially for their project area. Since CRM projects are often related to some type of construction or land modification, the civil engineers, developers, or agencies proposing the construction will need detailed maps for project planning, design, and engineering. In these cases, topography is mapped with much more detail and using smaller contour intervals than what would be available on the appropriate 7.5 minute USGS quadrangle. Floodplains and wetlands may also be delimited with greater detail. In the field, the archaeologist can use the benchmarks and survey stations established by the map surveyors as points of reference for their own mapping efforts.

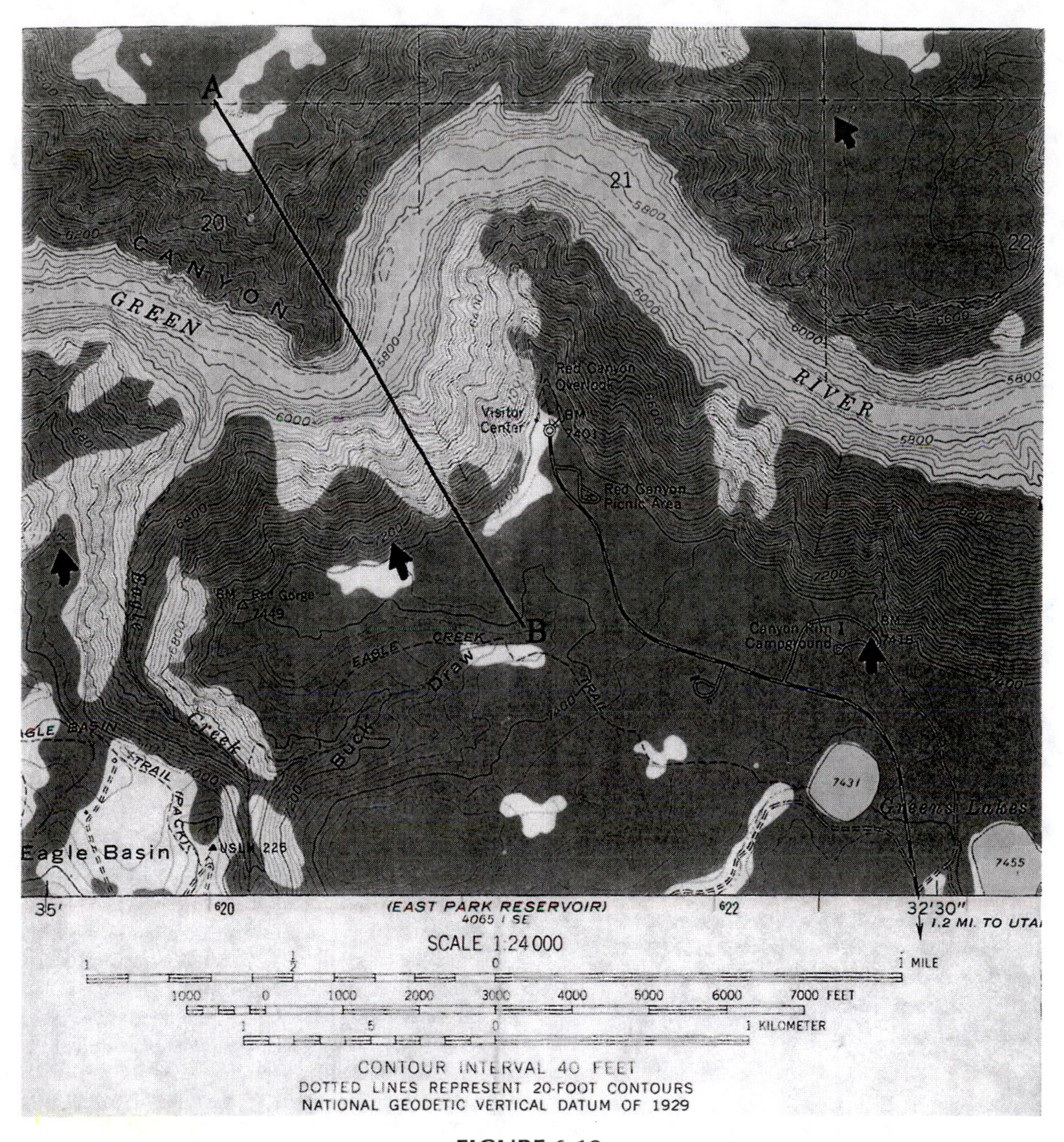

FIGURE 6.19.

Portion of the Flaming Gorge, Utah-Wyoming, 7.5 minute quadrangle, USGS, 1966. Examples of guide contours, benchmarks, section corners, and spot elevations are indicated by the arrows. The line A-B is used to generate the topographic profile shown in Figure 6.21.

INSTRUMENTS

The magnetic compass is the most basic and economical instrument used to determine direction, measure angles, and do rough surveying in the field. There are many different types of hand-held compasses with a variety of features, and with widely varying prices. I'll review some different types of compasses.

All compasses, of course, enable you to determine the direction of magnetic north and the other cardinal directions using a magnetized needle on a pivot (electronic/digital compasses are an exception).

FIGURE 6.20.

A type of metal pipe and concrete monument used for establishing elevation benchmarks in the field. The monument will be buried in the ground with its top flush, or slightly above ground surface. This example is approximately two feet long.

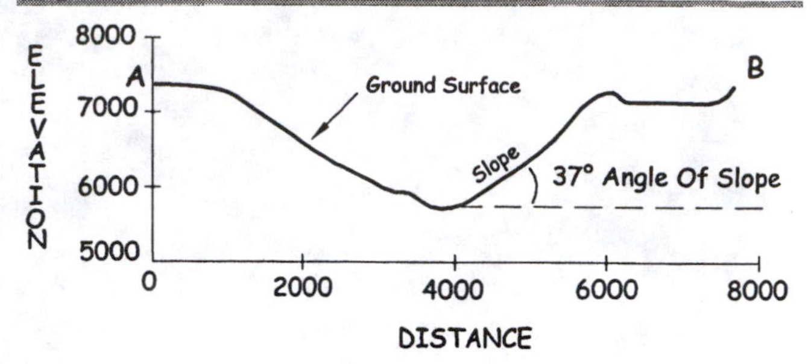

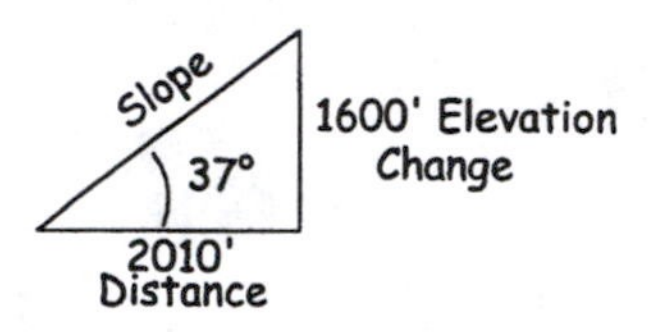

Calculating Gradient = $\frac{\text{Rise (1600')}}{\text{Run (2010')}}$ = 0.796 = 79.6% Slope

FIGURE 6.21.

Topographic profile for line A-B from a portion of the Flaming Gorge, Utah-Wyoming, 7.5 minute quadrangle, USGS, 1966. Methods for determining the angle of slope and gradient or percent of slope for a portion of the profile also are shown.

Because the effects of magnetism vary with proximity to the earth's poles, the dip of the compass needle varies from an ideal horizontal plane depending on whether you are in the northern or southern hemisphere. Compasses are made for specific use in either the northern or southern hemispheres with needles counter-weighted on one end or the other. International models are available that will work in either hemisphere. When selecting a compass, look for one with a housing that is filled, or damped, with some type of nonfreezing liquid. This decreases the amount of back and forth swing that the needle will go through before aligning with north, i.e., you're ready to use the compass much sooner than would be possible with a compass that is not damped with liquid. The nonfreezing liquid allows you to use the compass in a variety of weather conditions. The red-colored portion of the needle in compasses indicates north.

All compasses employ the 360 degree divisions of a circle and their faces, dials, or housings can be graduated in a number of different ways to reflect this (Figure 6.22). Graduations that proceed clockwise from 0 through 360 degrees are referred to as ***azimuth,*** or provide azimuth readings. Graduations that divide the compass face into quadrants based on the four cardinal directions and the 90 degrees that each contains are referred to as ***bearings.*** The compasses on mechanical transits are graduated in this way. Bearings are expressed relative to the northern or southern hemispheres of the compass. The term bearing is also used to refer to the direction or the degree reading from one point or object to another, which in many cases means taking a compass reading from the point where you are located to a distant point or object upon which you sight.

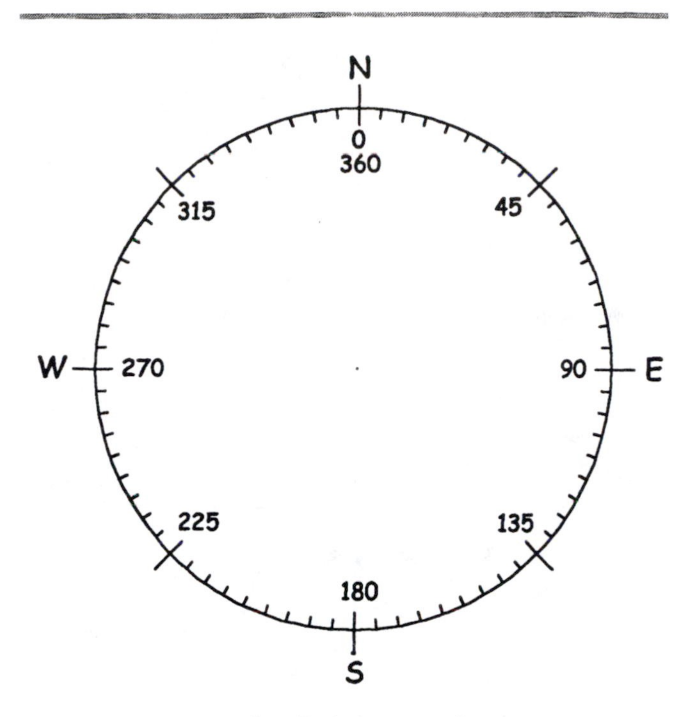

a. Azimuth divisions of a compass

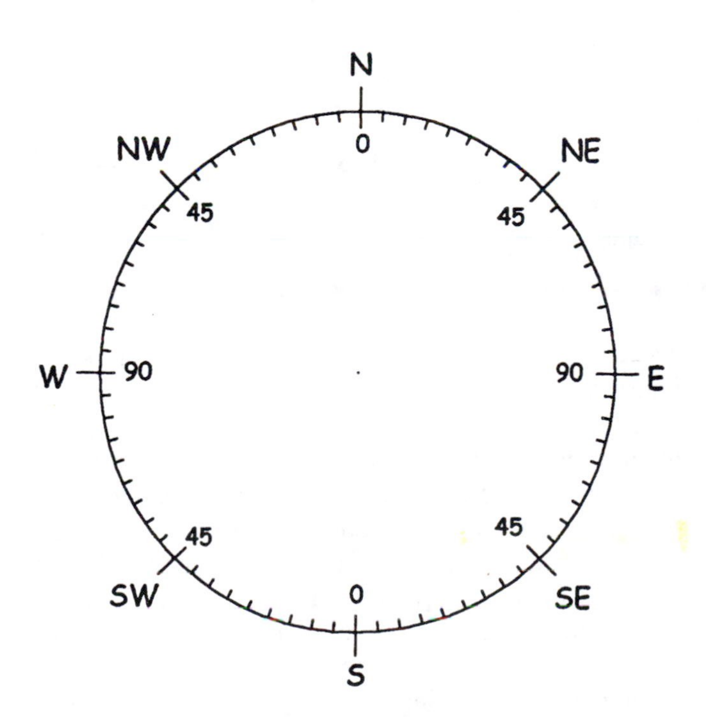

b. Quadrant and bearing divisions of a compass

FIGURE 6.22.

Compass graduations.

Azimuth and bearing provide two different ways to refer to angles and direction. With the compass oriented to magnetic north, an azimuth of 45 degrees can also be expressed as N45E (read North, 45 degrees East). One might also call this due Northeast. An azimuth of 215 degrees would be S45W (read South, 45 degrees West) or due Southwest.

Hand-bearing or ***sighting compasses*** (Figure 6.23) allow bearings to be taken from the position where you are standing to any point upon which you sight. The compass's graduated dial is free floating and is always fixed with 0 at the north arrow. More expensive models allow readings as precise as 15–30 minutes of arc. Because the graduated dial or housing cannot be rotated, it is difficult to use these compasses with maps to plot a series of bearings. Also, this type of compass generally cannot be adjusted for magnetic declination, so you will have to do some math to correct for this when using your bearings to plot map positions. Electronic/digital versions of this type of compass are available and are easy to read. They are battery operated, which may be a concern for users who spend prolonged periods in the field.

Baseplate or protractor compasses (Figures 6.24a, b) are some of the most widely used by archaeologists because of their utility with topographic maps and ***orienteering*** (using maps to find your way or locate

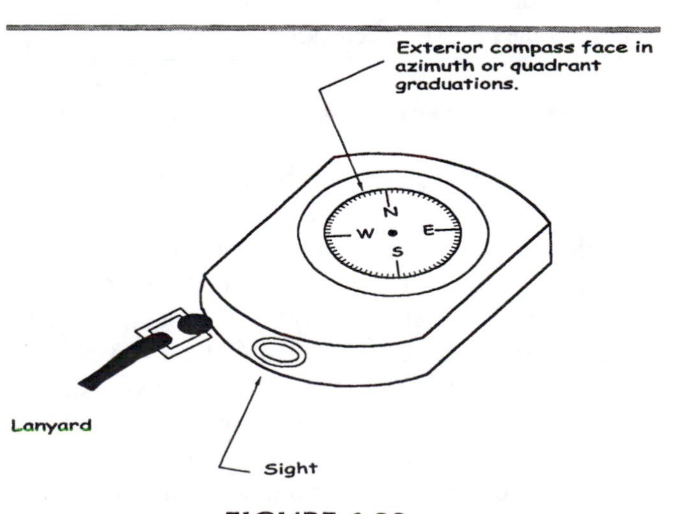

FIGURE 6.23.

Example of a hand-bearing or sighting compass. The compass face and graduations are visible on the surface of the instrument. For taking accurate bearings to a specific point, the user looks through the sight, aligns on the target or objective, and reads the angle on the scales visible through the sight.

FIGURE 6.24A.

An example of a baseplate compass, the *Silva Ranger Type 15*. The compass is shown open. When closed neither the mirror nor the face of the compass are exposed.

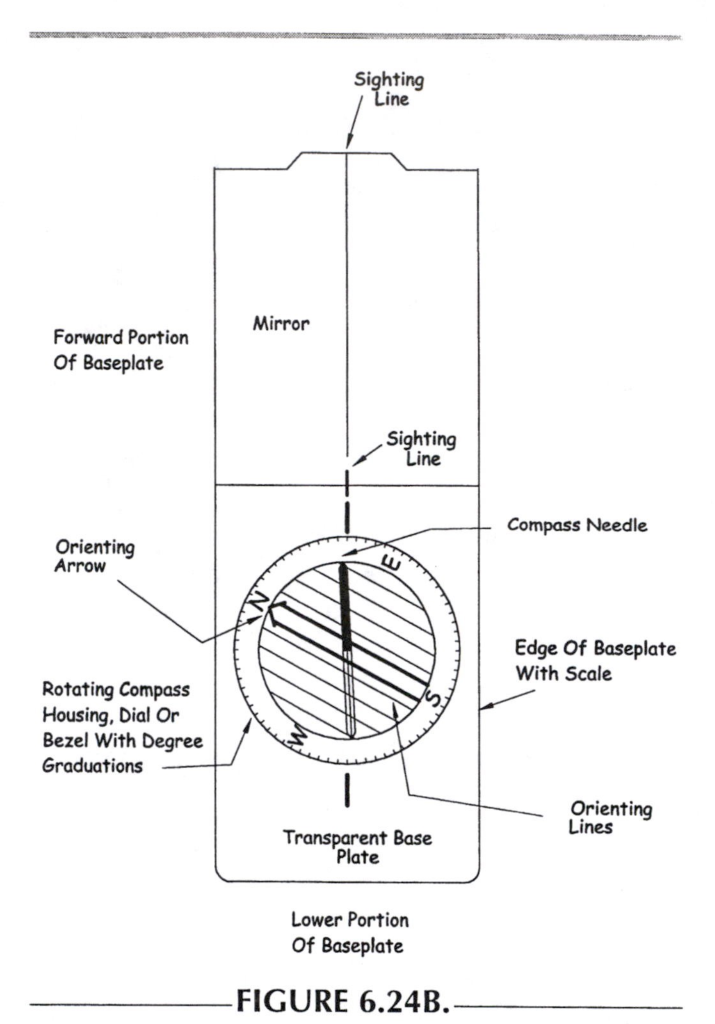

FIGURE 6.24B.

A schematic of a baseplate compass identifying significant features.

things). Baseplate compasses consist of a compass with a rotating housing, dial, or bezel mounted on a see-through rectangular base that includes map scales for measuring distance. In fact, the baseplate compass was designed to be used specifically with topographic maps and orienteering (Jacobson 1999; Kjellstrom 1994). In addition to the normal functions of a compass, it can be used as a protractor with a map to determine bearings to positions without orienting the map to north, something not possible with a hand-bearing compass. Because the graduated housing or dial can be rotated, bearings or lines of travel can be set or fixed, requiring only that the compass's north arrow be aligned when the compass is in use. This is not possible with a hand-bearing compass. Baseplate compasses can be adjusted for magnetic declination. Baseplate compasses, however, generally provide readings to single degrees. Many are equipped with mirror sights that allow the compass to be sighted and read when held at eye level, theoretically improving the accuracy of the reading. The accuracy of a reading using any type of compass depends on holding the instrument level so that the north arrow may swing freely and be accurately aligned. Some baseplate compasses are available with built-in bubble levels.

Pocket transits include a magnetic compass and are designed to read both horizontal and vertical angles. Remember that a basic compass only reads horizontal angles. Measuring vertical angles is important in surveying and mapmaking. The angle of a slope can be used in a trigonometric formula to reduce a distance over the slope to a horizontal measurement for the purpose of creating a map. A simplified ***clinometer,*** an instrument for measuring slopes or angles of elevation, also can be found on some types of baseplate compasses. Pocket transits have bubble levels that allow them to be used with greater accuracy when reading horizontal or vertical angles. In addition, many are made to attach to a staff or tripod for even greater stability and accuracy in use. The Brunton Company of Wyoming is probably best known among archaeologists and geologists for their pocket transits. These versatile instruments can perform many of the operations of a transit or theodolite while being small and lightweight, making them useful in situations where transporting a transit or theodolite would be impractical.

Once you have experience in using a compass and see the range of situations in which it can operate, choosing a specific model for your field kit will be easier. Any decent compass that you purchase will come with detailed instructions for its use. In my experience, a good baseplate compass seems to be the most versatile for archaeologists for the money invested. A pocket transit is more useful when working in remote areas where it may not be practical to use a transit, theodolite, or other type of surveying equipment for detailed mapping and the construction of surface collection or excavation grids.

The ***Global Positioning System*** (GPS) is a satellite-assisted means of telling you where you are on the surface of the earth and, depending on the sophistication of the equipment that you are using, how you arrived at your position. It can also generate information about distance and altitude, again depending on the type of equipment used. A series of satellites orbiting the earth are equipped with solar-powered radio transmitters. Ground-based stations keep track of the position of these satellites. Because of their orbits, a person with a GPS receiver has access to several of these satellites at any given time. When the GPS receiver is activated, it receives signals from four or more satellites indicating their position and the time of signal transmission. The GPS receiver uses this information to triangulate your position on the surface, which is expressed in latitude and longitude or UTM coordinates. This information is displayed on the LCD screen of the GPS receiver. Depending on the type of GPS used and the procedures applied in data collection, the accuracy of readings can vary widely. There are stand-alone, hand-held GPS receivers (Figure 6.25) and backpack-mounted versions with external antennas. Readings within 5–30 meters of the actual location are typical of intermediate models of equipment, while accuracy below 5 meters and to the level of centimeters and millimeters is possible with more sophisticated equipment and processing procedures. The use of a remote antenna (one that is not built into a hand-held GPS receiver) and processing software can improve the accuracy of readings. Be aware that there is a tremendous amount of variability in GPS equipment now on the market, from systems that an outdoors person or field technician might be able to afford to expensive, land-surveying-quality systems.

Most GPS receivers allow the storage of readings so that you can literally reconstruct your movements over the landscape by plotting the individual latitude and longitude coordinates or UTM readings on a map. Bearings can also be generated. Some GPS models can be used as a navigation system in the ways that a compass can be used, providing digital and graphic displays of cardinal directions, bearings, latitude and longitude, and changes in the same as you move across the landscape. Processing software can make GPS units compatible with digitized maps and mapping programs so

FIGURE 6.25.

An example of a hand-held GPS receiver.

that the readings accumulated in the field can be downloaded into a computer for easier plotting and manipulation. There are a variety of models and features, and ancillary equipment that extends the performance and capabilities of GPS receivers. Catalogs published by suppliers will give you the best idea of the range of what is available. A base station, like Trimble's Community Base Station, is an integrated GPS receiver and software system that operates on a personal computer and combines the various functions that I've outlined.

There are a variety of surveying instruments that allow you to determine direction, take bearings, measure angles, distances, and elevations, and use these data to create maps and locate your position upon the surface of the earth. An older and once traditional means of making maps for archaeological purposes employs a plane table and alidade. A ***plane table*** is basically a sturdy drawing board mounted on a tripod that can be leveled. The drawing board is rectangular and because of the way in which it is mounted to the tripod, the edges of the board can be aligned in any direction selected. Graph paper for drawing maps is pinned or taped to the surface of the board. An ***alidade*** is a telescope mounted on a straight-edged base that parallels the line of sight of the scope (Figure 6.26). This sits on the plane table, but is not attached to it. The straight-edged base of the scope contains a bubble level. When the scope is placed on the plane table, it can be used to determine when the table has been adjusted to a level position. Cross hairs within the scope are used for reading a ***stadia rod*** (Figure 6.27). A stadia is a graduated rod (in feet or meters) that is used for indirectly measuring distance and differences in elevation. The example on the left in Figure 6.27 is graduated in feet, tenths of feet, and hundredths of feet. A third decimal place can be estimated depending on the degree to which a black or while graduation is split. The larger numbers

FIGURE 6.26.

An older traditional-styled plane table and alidade.

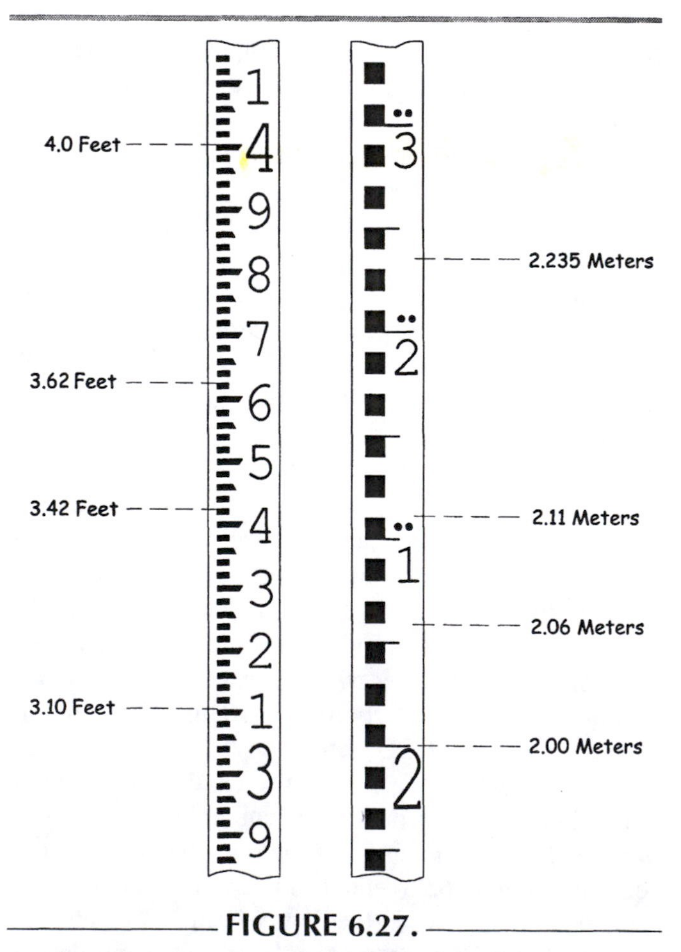

FIGURE 6.27.

Sections of stadia rods useful for distances of up to 400 feet or about 120 meters. Beyond this distance stadia with larger and heavier graduations, and different patterns of graduations, are needed for effective sighting and calculations (Anderson and Mikhail 1998:195). Examples are shown of how rod graduations are read.

designating feet in Figure 6.27 (3 and 4) are typically shown in red on a stadia rod. The rod section on the right in Figure 6.27 allows readings of meters, decimeters, and centimeters with estimations of millimeters based upon the degree to which a black or white graduation is split. The large number 2 on this rod indicates the meter, usually shown in red on an actual rod. The two dots placed above every demarcated decimeter (1, 2, and 3) means that a reading is 2 meters + the relevant decimeter, centimeter, and millimeter value. The numbers of dots change as the meter changes.

Any instrument used to read a stadia rod has to have the capability to be leveled, or have a means of reading the vertical angle at which the scope is inclined when used in a position that is not level. A traditional alidade has no components for reading horizontal angles. Distances to mapped points from the plane table, and the direction of the mapped point relative to alignment and location of the plane table are directly plotted on the graph paper attached to the plane table. To measure the angle between two points, plot their locations on your plane table map then use a protractor to measure the angle of the arc between the points.

With a plane table and alidade you literally create a map as you do the surveying in the field. When the surveying is done, the map is done. You can tell right away if mistakes have been made in taking or plotting readings, or if additional detail is needed, because you have both the map and the mapped phenomena (e.g., excavation grids, units, terrain) there in front of you. Other types of surveying instruments require that you first take whatever readings are appropriate and then plot them in a separate operation. In other words, you don't draw the map as you go. The eventual plotting of readings to create a map may or may not be done in the field, depending on circumstances. Any mistakes made in the surveying will only be evident as a result of drawing the map. On the other hand, maps created using a plane table and alidade are not as accurate as those compiled from readings taken with other types of surveying instruments like a transit or theodolite. A plane table and alidade also are more cumbersome to use when mapping large areas, and are inappropriate for certain types of surveying, like extending lines over great distances, or running extended traverses to relate different locations to one another.

FIGURE 6.28.

A traditional type of mechanical transit.

The mechanical or engineer's ***transit*** (Figure 6.28) is a precision surveying instrument used for turning and measuring horizontal and vertical angles, extending lines, and measuring distance and differences in elevation with the use of a stadia rod. It consists of a telescope mounted on a vertical framework, which is attached to a series of horizontally oriented, circular plates. The entire instrument is then attached to a tripod. The scope can be rotated horizontally, either clockwise or counterclockwise, in conjunction with a graduated horizontal circle/scale that allows the angle of any particular rotation to be determined. The horizontally rotating scope and graduated circle can be used as a protractor in the same way as a baseplate compass. The difference is that the transit performs this function with a much greater degree of accuracy. The telescope of the transit also can be rotated vertically, independent of the base of the instrument, and has a graduated arc for reading vertical angles to within 1 minute of arc. Internally, the scope contains horizontal cross hairs for reading a stadia rod and a vertical cross hair for general alignment on a target.

The horizontal graduated circle of a transit is scaled from 0 to 360 degrees, exhibiting markings for degrees and half degrees (30 minutes). Inset along the inside of this graduated circle is a ***vernier scale.*** It consists of divisions of the smallest increment on the primary scale of the transit and provides a refinement of the reading of an angle (minutes and seconds depending on the type of vernier). Horizontal angles typically can be read to 1 minute of arc. In general, a vernier is an auxiliary scale that allows more precise readings of the smallest division of an instrument's main scale. Some stadia rods are also equipped with vernier scales. Electronic transits provide angle measurements as a digital display on a small LCD screen attached to the instrument, rather than require you to read them off mechanical scales.

Every transit has some means for leveling the horizontal plates of the instrument, either in the form of bubble levels and screws that enable portions of the base plate to be raised or lowered, or with screws and an optical sight. The baseplate must be leveled for any type of operation. The transit's telescope also has its own bubble level and adjustment knobs for leveling. When the scope is leveled, reading distance and elevation off a stadia rod is a fairly straightforward process. When the scope is used in an inclined position when reading a stadia rod, the vertical angle must be determined and plugged into various formulas for determining distance and differences in elevation. There are knobs for adjustment and manipulation of levels and scales located in various positions on the transit. These can vary slightly in location depending on the model of the instrument.

Many transits are equipped with a magnetic compass and related graduated circle, usually organized by quadrants rather than azimuth. Although it is not handheld, it can be used for the same purposes as a sighting or hand-bearing compass. Because the transit's compass is situated on a plate that can be leveled and can be sighted with the telescope, it is more accurate than a hand-held compass. Of course, the transit is more cumbersome and time consuming to use.

A ***plumb bob*** is a pointed metal weight suspended from a string used to project the horizontal location of a point from one elevation to another (Anderson and Mikhail 1998). It can be suspended from the midpoint of the base of a transit attached to a tripod to orient the instrument directly above a fixed point on the ground. Some transits have an optical "plummet" that enables the instrument to be set up without the use of a plumb bob.

A transit and ***theodolite*** (Figure 6.29) share many of the same basic features. A theodolite, however, provides more accurate measurement of angles, generally approaching 1 second of arc. Scales are enclosed, and on older, more traditional models those scales are viewed through an optical train. On newer electronic models there is no need to read scales. Readings are provided in digital displays on a small LCD screen mounted on the instrument. Electronic theodolites can be coupled with computers/data recorders in the field to store all measurements taken. Theodolites are more compact and lighter in weight than transits, and generally do not include a magnetic compass. Both the base and telescope levels are of high sensitivity. Instruments of more recent age are positioned using an optical plummet rather than a plumb bob.

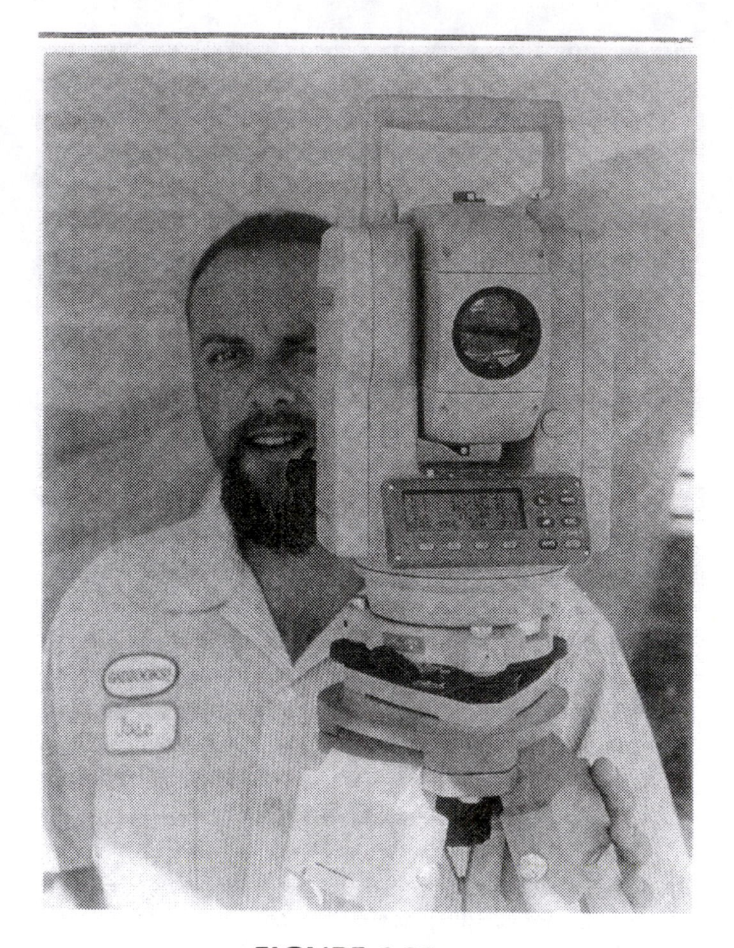

FIGURE 6.29.

A total station combining a theodolite with an EDM.

An ***electronic distance measurement*** (EDM) instrument can be attached to a theodolite to provide readings without the need to read a stadia rod. A target consisting of one or more prisms is used in place of a stadia rod. The theodolite is sighted on the prism and an electronically generated beam of light is emitted from the EDM. The prism reflects the beam back to the theodolite/EDM. Through the phase (wave form of the beam) comparison of the generated signal and the reflected signal, the EDM is able to calculate a distance. ***Total stations*** combine an electronic theodolite and EDM in a single housing (see Figure 6.29). They provide a level of accuracy for surveying operations performed by archaeologists that is much higher than traditionally attained (Rick 1996).

A variety of instruments are available that are used only to measure differences in elevation between points. These are called ***levels.*** They are mounted on a tripod, have some type of mechanism for leveling the instrument on a horizontal plane, and are used in conjunction with a stadia rod. Elevations are only one type of surveying data that archaeologists record in the field, and since all transits and theodolites can perform this and other functions, a level is not a critical piece of field equipment.

An ***Abney level*** differs from the levels just described in that it can also function as a clinometer, an instrument used to measure the angle of slopes. It is a hand-held, small telescope-like device, typically with 5X magnification. A vertical arc with scales for reading angles and percent of slope is mounted on the scope. With the help of a bubble level on the scope, the instrument measures the angle between the horizontal plane and the slope of the surface between two points (where you are standing with the instrument and the point toward which you are sighting). Knowing the angle of a slope, regardless of whether it is determined with an Abney level, clinometer, transit, or theodolite is important for reducing the distance along a slope to a horizontal distance used in mapping.

Clinometers are a standard feature on pocket transits and some types of baseplate compasses. More sophisticated clinometers are stand-alone, hand-held instruments that are generally easier to read than those built into compasses and pocket transits. These clinometers are of the same general size and shape as the hand-bearing compass shown in Figure 6.23. These clinometers also contain multiple scales so that both angles and percent of slope can be determined. Cosines to 45 degrees are listed on the exterior of many models, which enables in-field calculations to be done without the need for a calculator with trigonometric functions. ***Cosine*** (abbreviated COS) is the trigonometric relationship between the size of an angle and the sides of a right triangle, and is a component in the formula for reducing a slope measurement to a distance on a horizontal plane. Clinometers with a ***secant scale*** (secant is the reciprocal of the cosine) preclude the need for a trigonometric calculator, but some basic arithmetic is still needed to figure out the reduction of a slope measurement to a horizontal distance.

The basic principles of surveying involve the measurement of angles, distances, and differences in elevation in order to create maps, extend lines, or relate locations to one another. These principles do not change. However, the specific details of how an instrument is set up, manipulated, and read do. You can refer to any number of texts for detailed descriptions of the surveying instruments that I've summarized and their operation (e.g., Anderson and Mikhail 1998; Bettes 1992; Hogg 1980). The best way to learn is to actually use an instrument, something that should be part of your formal training as an archaeologist. Once you have experience on one type of instrument, learning how to handle a different model will be relatively easy. GPS units, electronic theodolites, EDMs, and total stations can simplify and speed up many surveying tasks in the field while providing incredible accuracy. However, they all have one potentially fatal flaw—they are powered by batteries. When the batteries fail, the equipment cannot be operated, or some operations and calculations have to be performed mechanically or by hand, as with older styles of instruments. Don't feel cheated if you learn the basics of surveying on an older style of transit or theodolite. The basic principles of surveying can be translated to any instrument, and you may end up with a greater appreciation of skills that people trained on electronic devices take for granted.

In the sections that follow I'll discuss a number of general principles guiding the use of surveying instruments and the basic tasks for which archaeologists use them. I'll also offer some comments on the utility of different surveying instruments for specific tasks. Deciding which instrument to use for a given operation depends on the nature of the operation itself, the level of accuracy that you need, field conditions, and in some cases the funding available to meet equipment needs. I'll also review ways to accomplish some tasks without the use of surveying equipment, instead relying on the use of measuring tapes and some basic geometry and trigonometry. The techniques employed to make maps do not vary as much as do the instruments that can be used to make them (Spier 1970:6). While all of the operations that I describe are effective, some are simplified versions of more rigorous and varied approaches taken by professional surveyors. Taking a basic surveying course through a department of engineering at a college or university will broaden your understanding and improve your skills.

MEASURING DISTANCE

Determining distance, either roughly or precisely, will be a part of most surveying operations in the field. There are a variety of ways to estimate or determine distance, each with different levels of accuracy, and each useful in different types of situations. Your ***measured pace*** will be useful for roughly estimating distances when navigating or orienteering with a compass, or making sketch maps of sites, excavations, and terrain. To calculate your measured pace, you will need a tape, preferably 200 feet, or 50 or 100 meters long. Find a flat open space where the tape can be fully extended in a straight line. Decide whether you are going to calculate the length of a single step/pace or a double step/pace. Repeatedly walk the length of the outstretched tape counting your paces as you go. Don't exaggerate your stride, walk normally. Divide the distance that you have walked by the number of paces/steps that you have taken. Do this several times to see how your pace might vary and to determine an average pace.

For long distances it is a good idea to know the length of time that it takes you to walk a half mile to a mile. Walk a measured course along a road at a leisurely pace and time your progress. If possible, walk and time a measured course through the woods or on a trail for comparison. Relatively small and inexpensive electronic ***pedometers*** that attach to your belt or waistband can keep track of paces walked, distances, and elapsed time. These can be very handy when traversing long distances and keeping a mental count of paces is bothersome. Like other electronic equipment, it is important to keep batteries charged and to have spares on hand.

Uneven ground and slopes will decrease the accuracy of pacing as a way of estimating distance. Remember that on a map, terrain is rendered in two dimensions (imagine that you are in the sky looking down at a landscape and have no depth perception). Thus a slope is reduced to a horizontal distance (***slope reduction***) that won't equate with the actual distance that you covered walking across it. This is illustrated in Figure 6.30 with a cross section of two different slopes. The distance across each slope is shown, 304.1 feet (slope *a*) and 360.6 feet (slope *b*), respectively. Although these distances are very different, each slope can be reduced to a horizontal distance of 300 feet, which would be the measurement used on a map. Contour lines are the convention used for showing differences in slope or ground surface over horizontal distance, and allow the map reader to recognize differences between them.

Employing tapes is obviously a more accurate means of measuring distance, although more time consuming than pacing. Open reel-style tapes made with a reinforced synthetic fabric like dacron or fiberglass are the easiest to use (Figure 6.31). They are flexible, sturdy, and easy to clean. They are generally available in lengths up to 300 feet and 100 meters. Open and enclosed reel tapes of steel and chrome or nylon clad steel are harder to keep clean and can be difficult to handle. They are not as flexible as fabric or fiberglass tapes, they tend to break more frequently because of the stress on the metal, and can inadvertently scrape and cut your hands if not handled attentively. But steel tapes are considered to be more accurate than other types because they aren't subject to the same degree of distortion through prolonged use as fabric or fiberglass tapes. For the purposes of archaeological surveying and mapping, the differences in accuracy between these types of tapes should be of little concern.

Smaller pocket reel/case tapes (steel) and folding rules are useful for measurements within excavation

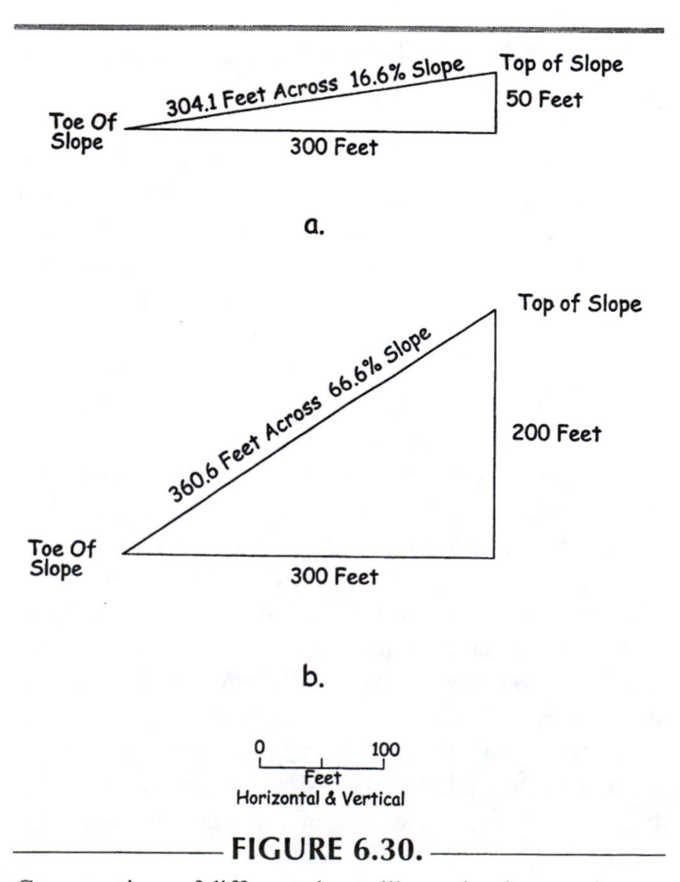

FIGURE 6.30.

Cross sections of different slopes illustrating how each reduces to the same horizontal distance (300 feet) for the purpose of mapping.

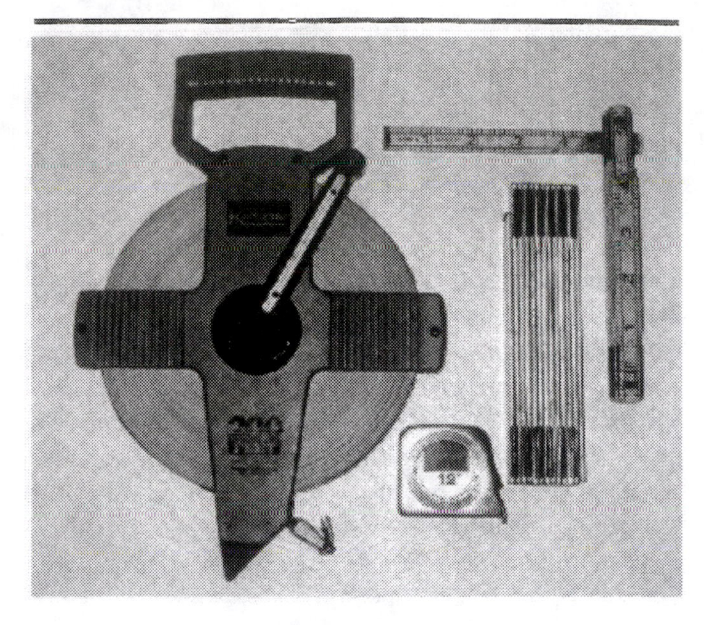

FIGURE 6.31.

Types of tapes (large reel, pocket reel/case, folding rule; English and metric).

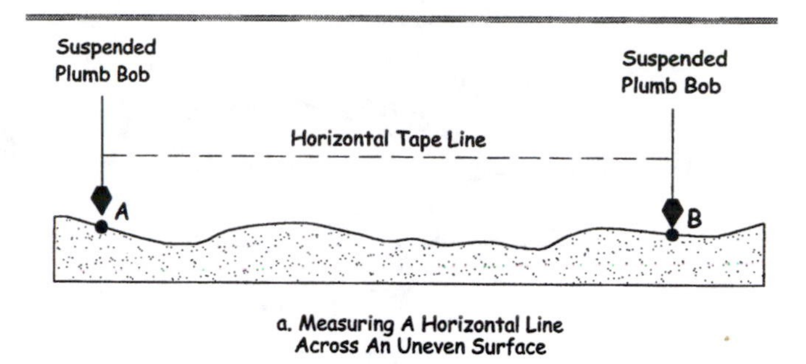

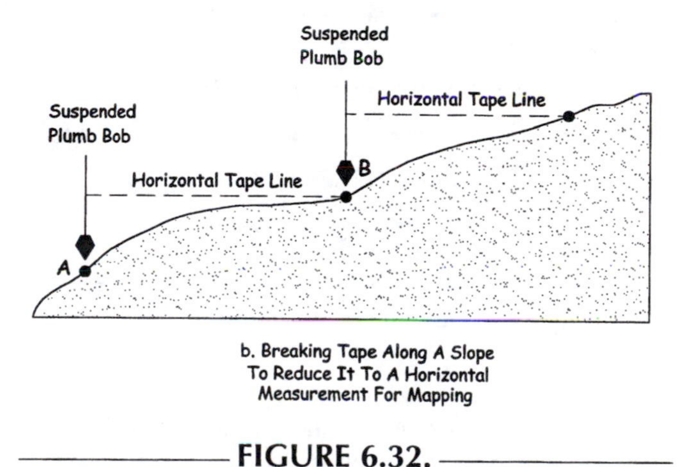

FIGURE 6.32.

Methods for taping distance over uneven terrain and slopes.

units and for preparing maps of excavation floors and artifact locations. I prefer tapes with engineer-scaled feet on one side and metric divisions on the other. They allow you to function in a greater variety of situations. In cultural resource management studies, archaeologists often have access to detailed engineering maps or survey data prepared in the English system.

Some methods for taping distances over uneven terrain and slopes are illustrated in Figure 6.32. Where the slope is not severe it will be possible for a tape to be pulled and held horizontally between two points at a moderate distance (Figure 6.32a). Each person holding an end of the tape must use a plumb bob to orient the tape above their respective points on the ground, since the unevenness of the surface makes it impossible to get a horizontal reading taping along the ground's surface. The height at which the tape is held will vary depending on the degree of slope between the points, other obstructions, and the convenience of the surveyors. The key is to pull the tape taut and hold it in a horizontal position. Deviations will introduce error into the measurement. The longer the distance, the more difficult it will be to hold the tape taut and horizontal. Personal trial and error will give you a sense of the limits of this technique.

Breaking tape uses the same basic approach of a tape oriented horizontally and plumbed over two points, but with the slope broken down into manageable pieces for horizontal measurement (Figure 6.32b). In this case, only one person need use a plumb bob to orient the tape over a point while the other person holds the tape at ground level. This operation is repeated until all portions of the slope have been measured. It is important that each time a setup is moved it is anchored on the last point of the previous setup. In our example this means that the person who held the tape at point A will reposition themselves at point B, while the other person establishes the new point C. Marking points used in taping with surveying pins, or simply nails stuck into the ground, will ensure greater accuracy in this operation.

Another means of reducing a slope to a horizontal measurement can be accomplished with an Abney level or clinometer and a tape measurement along the slope itself. The Abney level or clinometer is used to determine the angle of the slope. This angle and the taped measurement of the slope are then plugged into a trigonometric formula to reduce the slope to a horizontal measurement. This tactic makes use of the relationships

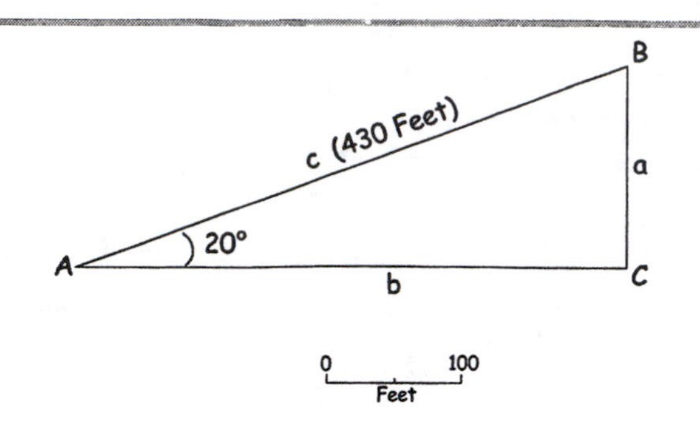

A, B and C are angles of the right triangle; a, b and c are the sides (lengths) of the right triangle.

To solve for b (length), use the trigonometric formula: COS. A = $\frac{b}{c}$

In this example, A = 20 degrees; COS. 20 degrees = .93969; c = 430 feet

Then: .93969 = $\frac{b}{430}$ Balance the equation: .93969 x 430 = b

Result: b = 404.07 feet = the slope reduction of c

FIGURE 6.33.

Calculating a slope reduction.

between the size of an angle and the sides of a right triangle. Figure 6.33 provides an example. Our slope is represented by side c of the triangle. Using a clinometer, we've determined that the angle of the slope is 20 degrees. The distance along the surface of the slope has been taped as 430 feet. What we want to find out is the value of side b of the triangle, which is the horizontal reduction of the slope. Plugging the known angle and known distance into the formula, we see that b, the slope reduction, is 404.07 feet. Determining the cosine of a slope angle will require that you have a calculator with trigonometric functions, or refer to trigonometric tables that list the cosine values of specific angles. The calculator is easier to transport, and will be useful in other field operations.

Distance also can be determined using an alidade, transit, or theodolite in combination with a stadia rod. Regardless of the specific instrument used, it must be set up on or over a point from which measurements are to be taken. The instrument must be leveled, including the scope, or the scope must be equipped with scales for reading vertical angles. Looking through the scope of the instrument, you will see three horizontal cross hairs and a single vertical cross hair. The topmost and lowest horizontal cross hairs (***stadia hairs***) are critical to determining distance in this operation. With a level scope, the distance from the point over which the instrument is setup to the point where the stadia rod is positioned is calculated by the following procedure. First, read the stadia rod where the topmost cross hair intersects the graduations of the rod. Take an additional reading, where the lowest cross hair intersects the rod. Then subtract the lower reading from the upper reading. Finally, multiply this result by 100, or move the decimal point two places to the left. What you have actually done is multiplied the ***stadia interval*** (the difference between the upper and lower cross hair readings) by the ***stadia factor,*** which is usually 100 (Anderson and Mikhail 1998:351–353). Since you will probably be using an instrument with an internally focusing telescope, it will not be necessary to add the distance from the center of the instrument (scope) to its principal focus to your result. The formula for determining the distance with a level scope is

$$\textit{Distance} = Ks + C$$

where K = stadia factor (usually 100), s = stadia interval (top stadia hair minus bottom stadia hair), and C = distance from the center of the instrument (scope) to its principal focus, or

$$\textit{Distance} = (100 \times \textit{stadia interval}) + 0$$

Distance can also be measured with a stadia rod and an instrument when the scope is not level. ***Inclined sights*** for measuring distance are often necessary on slopes or in situations where the height of the instrument is lower in elevation than the surfaces towards which you are shooting. The distance along the slope itself can be calculated, as well as the horizontal distance to which it can be reduced. The horizontal distance is what we are most interested in for mapping purposes. In this operation it is calculated as

$$\textit{Horizontal distance} = Ks \cos^2(\textit{angle}) + C \cos(\textit{angle})$$

where K = stadia factor (usually 100), s = stadia interval (top stadia hair minus bottom stadia hair), $\cos^2$ is the cosine of the angle (at which the instrument scope is inclined from the horizontal) squared, and C = distance from the center of the instrument (scope) to its principal focus, or

$$\textit{Horizontal distance} = (100 \times \textit{stadia interval}) \times (\textit{cosine of angle squared}) + 0$$

Distances can be measured using an instrument and stadia rod more quickly and more efficiently in a greater variety of situations than through taping. When dealing with uneven terrain, slopes, and great distances, using the stadia rod is probably a more accurate means of measuring distance. However, there are circumstances when using tapes in conjunction with an instrument may be advisable, such as the creation of detailed excavation plans or plans of complex features.

Theodolites equipped with an EDM provide an even greater measure of accuracy and efficiency than using a stadia rod. Again, the instrument must be set up over a point from which measurements are desired. Measuring distance with an EDM does require placing a reflective target at each point to which a distance measurement is to be taken, in the same way that a stadia rod must be positioned. Readings and conversions are instantly available as a digital display on an LCD screen and can be stored electronically.

A GPS unit also can generate distance measurements. The accuracy of the measurement will depend on the type of equipment being used. With a hand-held receiver this operation first requires that the GPS unit be activated and initialized at a starting point, and then left on as you move to other points. Taking multiple GPS readings at each point and then averaging these readings will increase the accuracy of the operation, even if you are using less expensive equipment. Post-field correction of data using GPS-related computer software will also help to reduce errors. Nearness to buildings and operating under a heavy tree canopy can present problems with satellite reception.

LOCATING AREAS

With a 7.5 minute quadrangle map, many locations will be relatively easy to find in the field because of their association with, or nearness to recognizable natural or cultural features shown on the map. In these cases it is a straightforward process to plot directions and scale distances off one or more mapped landmarks. Not all fieldwork takes place in remote areas that are difficult to find.

In conjunction with a topographic map, a baseplate compass is useful for general navigation into harder-to-reach areas where fieldwork is to take place. Let's say that your destination is the northwest corner of Deep Lake, shown on the hypothetical topographic map of Figure 6.34. In order to plot a compass bearing to follow towards your destination, you must first select a starting point on the map that should be easy to locate in the field, or that you have already located in the field. I've

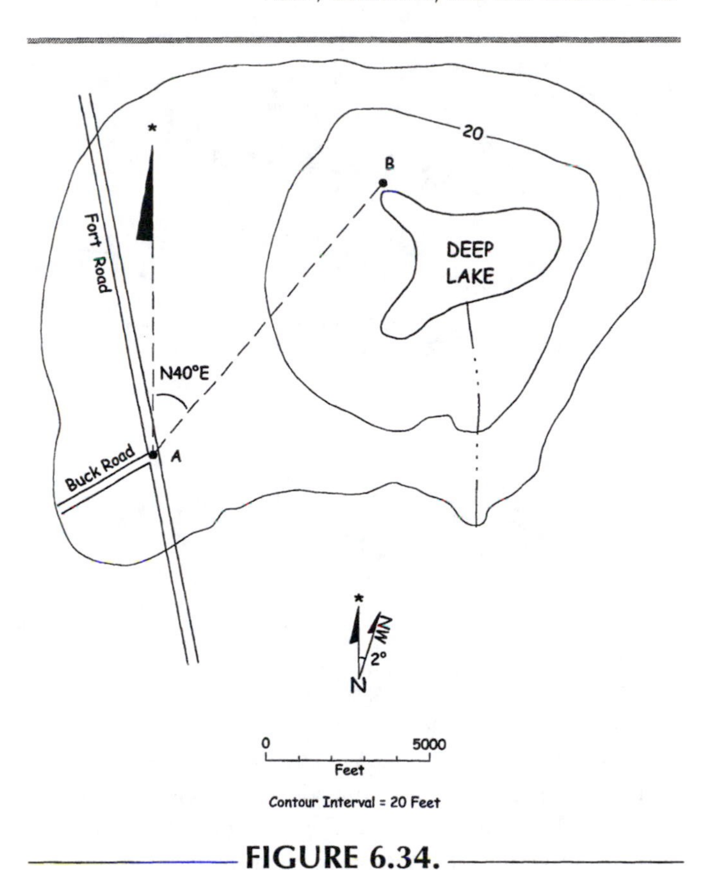

FIGURE 6.34.

Portion of a topographic map showing a starting point (A) and intended destination (B). The direction of travel determined off true north, the orientation of the map, is North 40 degrees East.

chosen the intersection of Buck and Fort roads as an easy to locate starting point. Take the compass and place the lower portion of either the left or right edge of the baseplate on the starting point on the map (Point A). Align the forward portion of the same side of the baseplate on your destination (Point B on the map). You've now created a straight line between your starting point and your destination, one that is also parallel to the line of sight of the compass (Figure 6.35).

To determine the compass bearing of this line of travel, rotate the compass housing so that the orienting arrow and orienting lines parallel the direction of true north on the map. Don't let the baseplate move from its Point A to Point B orientation during this operation, and don't be concerned about what the magnetized compass needle may be doing. The degree graduation on the compass housing that is now adjacent to the sighting line on the forward portion of the baseplate is your compass bearing relative to true north. In this example, the bearing would be North 40 degrees East, or an azimuth of 40 degrees (see Figure 6.35). However,

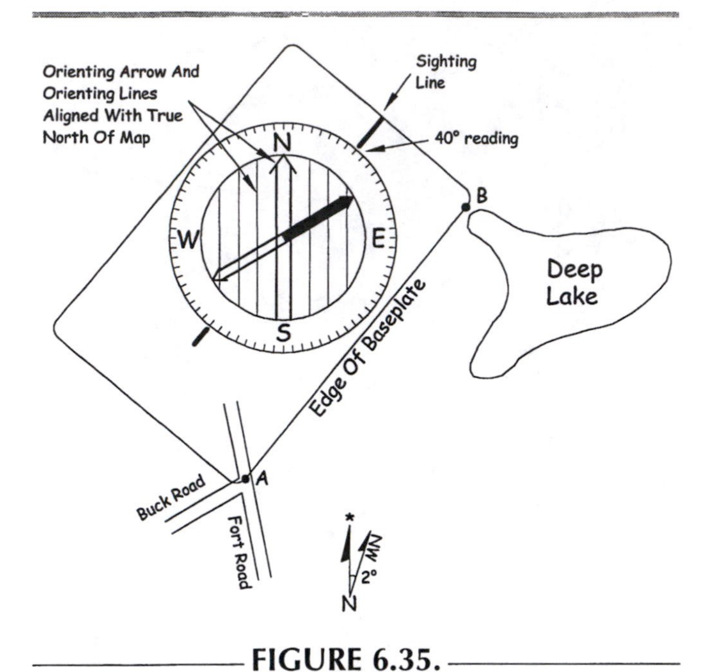

FIGURE 6.35.

Positioning of baseplate compass to determine compass bearing from starting point (A) to destination (B). Following the alignment of the edge of the baseplate with the starting and end points, the orienting arrow and orienting lines of the compass have been rotated until they parallel the true north orientation of the map. The compass bearing from A to B is read where the compass graduation is adjacent to the compass sighting line.

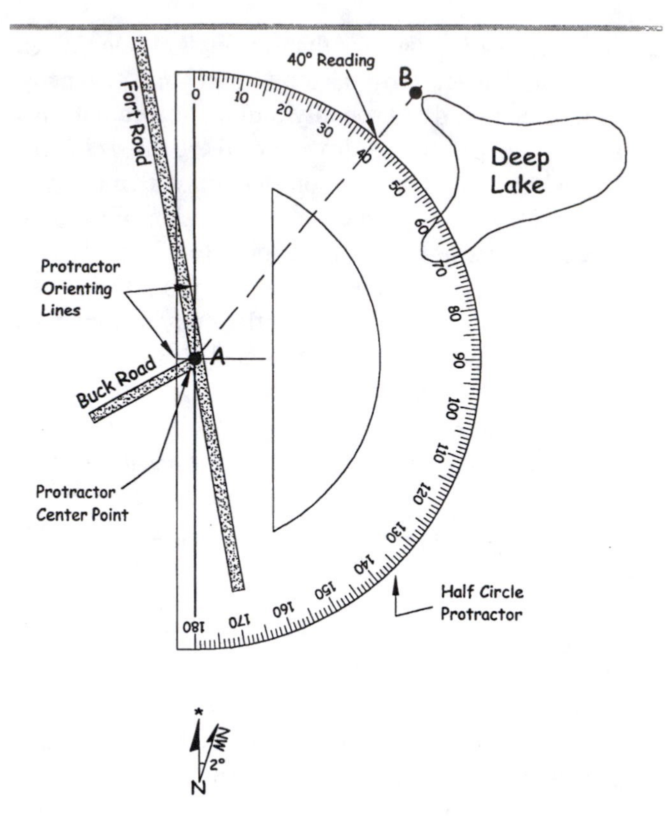

FIGURE 6.36.

Using a protractor to determine the bearing from starting point (A) to destination (B). The protractor's center point is aligned over Point A and its orienting line with true north as shown on the map. The angle reading is taken where the line of travel intersects the arc of the protractor.

you must take the magnetic declination of your area into consideration because when it is time to orient the compass using the magnetic needle and begin your journey, the orientation will be to magnetic north rather than true north. In our example, the declination is 2 degrees to the east of true north, so the compass bearing to be followed in the field would be North 38 degrees East, or an azimuth of 38 degrees. If you have mechanically adjusted your compass to the magnetic declination of the area, this computation won't be necessary.

Determining the bearing from A to B also can be accomplished with a protractor (Figure 6.36). The center point of the protractor is placed over Point A and its 0–180 degree orienting line aligned with true north on the map. The angle reading is taken where the graduated arc of the protractor intersects the line of travel between points A and B, in this case, 40 degrees. You will still need the compass to navigate to your destination.

With the bearing computed, you are ready to use the compass to navigate from the starting point to the northwest corner of Deep Lake. You are standing at the starting point. The compass dial remains turned so that 40 degrees (or 38 degrees if you have not adjusted the compass for the magnetic declination) is aligned with the sight line on the forward portion of the baseplate. Holding the compass in your hand at waist height, and keeping it as level as possible, turn your body (or the baseplate of the compass) until the magnetic compass needle is aligned with the orienting arrow. The sighting line of the compass will now be pointing in your direction of travel, North 40 degrees East. Scaling the distance from the map, you know that the northwestern corner of the lake is about 11,000 feet, or over two miles away. You can then estimate the number of paces it will take you to reach this point were you able to walk in a straight line, or estimate the amount of time walking such a distance would require.

As you walk you will need to periodically use the compass to keep yourself heading in the right direction. Again, the compass dial remains turned so that 40 degrees is adjacent to the sighting line. Maneuver yourself or the baseplate of the compass, as you did at the starting point, so that the compass needle aligns with the orienting arrow. As before, your line of travel will be in the direction in which the compass's sighting line is pointing. Rarely will you be able to walk straight towards a target or destination without confronting obstacles. Any time that you check your bearing with the compass, try to pick a landmark in the distance that is in the line of travel. This could be a distinctive looking tree, rock outcropping, etc. Visually focusing on such a landmark will make your walk easier, especially when maneuvering around obstacles, and reduce the number of times that you will have to refer to the compass. Keeping track of the distance you have covered will be more problematic depending on how many detours you are forced to make away from the line of travel.

Over relatively short distances like that of the Deep Lake example, ending up in the vicinity of your intended destination shouldn't be a problem. However, the greater the distance traveled, and the greater the number of detours around obstacles, the greater the potential for error in your line of travel. For example, one degree of compass error over a mile course means that you will miss your target by 92 feet, at 4 degrees, by 368 feet (Jacobson 1999:29). Recognizing mapped landmarks and reading the topography will help you to stay on course.

Obviously a transit or theodolite could be set up at the starting point and a 40-degree angle turned off magnetic north to establish a line of travel to reach the northwestern corner of the lake. This would provide you with a very accurate line of sight and help you to pick out visual landmarks to guide your travel. But setting up the transit or theodolite and transporting it to multiple positions where additional readings might be required to keep you on track would be very time consuming and cumbersome. It is unlikely that you would need to use these surveying instruments to find your way into or relocate a general area.

With a programmable GPS model, you can enter the latitude and longitude of your destination, initialize or orient the GPS at the selected starting point, and allow the GPS to keep you on course with continuous readings of direction and the rough distance covered displayed on the LCD screen of the instrument. GPS-related software will allow you to graphically plot this course on a personal computer.

Finding your way to some locations may not be as easy as plotting and following a single bearing, but will require multiple bearings followed for variable distances. Graphically, this would look like a series of straight lines connecting points at which a particular line of travel is initiated and terminated. This is what surveyors call a ***traverse,*** with the points at the established ends of individual lines being ***traverse stations*** or ***traverse points,*** although the terms are used most frequently in land surveying operations and not simple navigating or orienteering. Traverses may be necessary to get around major obstacles or simply to follow the easiest or most convenient route to your destination. Plotting a course of bearings and distances will, of course, depend on the quality of the map that you are using. For example, are there streams that can't be crossed without the use of watercraft, cliff faces or slopes that can't be climbed, gorges or ravines that can't be negotiated? The procedure for plotting a series of bearings and distances to arrive at a destination is simply a reiteration of the

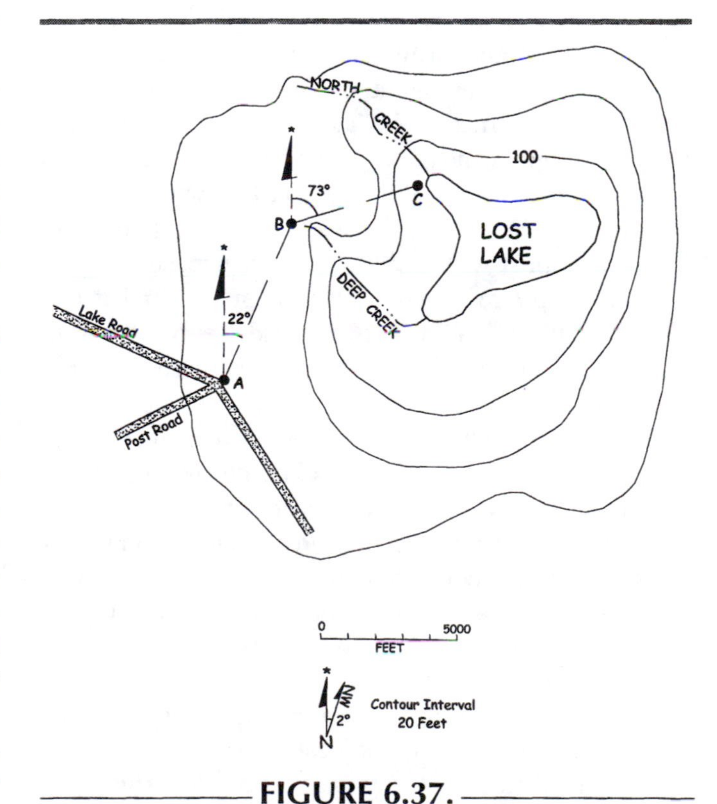

FIGURE 6.37.

Portion of a topographic map showing two distinctive bearings and distances to be traveled from starting point A at the intersection of Lake and Post roads, to the final destination at point C at the northwestern corner of Lost Lake.

operation described in the Deep Lake example. The difference is that there is more than one "starting point" from which bearings are taken and distances measured.

In the example shown in Figure 6.37 our goal is to reach the northwestern corner of Lost Lake. The most accessible and easily located starting point for the trip is the intersection of the Lake and Post roads (point A). Assessing the topographic map we see that we can't walk straight to our destination (point C) because of the course of Deep Creek, and our unwillingness to walk over a mile carrying a canoe. A bearing is plotted (North 22 degrees East or an azimuth of 22 degrees) using the procedures discussed above, and the distance scaled (6,400 feet) to the head of Deep Creek (point B). Because we've chosen an endpoint for our first line of travel that coincides with a natural landmark that should be easily distinguished, paying strict attention to the distance traveled along the bearing once in the field won't be critical. Using the head of Deep Creek as a new starting position (point B), a new bearing is plotted (North 73 degrees East or an azimuth of 73 degrees) and distance scaled (5,000 feet) to the final destination at Lost Lake (point C).

Because of the availability of recognizable landmarks in this example, you could find your way to the northwestern corner of the lake employing only casual use of a compass and paying little heed to distance. You know from looking at the map that following bearings ranging anywhere from North 22 degrees East to North 70 degrees East will enable you to intersect some portion of Deep Creek. Once the stream is found it is a simple matter to follow it to its head, get on its northern side, and follow its course to where it empties into Lost Lake. The shoreline of Lost Lake could then be followed north to the mouth of North Creek, an obvious landmark in proximity to our final destination (point C). While this is not the most efficient route to travel, especially if you are lugging field equipment, it emphasizes the fact that competent map reading and the use of obvious natural and human-made landmarks make it possible to navigate to many locations without any elaborate measurements and calculations.

Precision in locating an area on a map, or providing a technical description of its location, is most important when artifact find spots, archaeological deposits, sites, or excavations are involved. The goal is to be accurate enough that anyone could use your map plots or directions to relocate a position. But first you have to determine where you are, or where the phenomenon of interest is. This task can be accomplished in a number of ways.

In some cases, the location to be determined may be associated with distinctive landmarks or landscapes already appearing on published maps. For example, if an archaeological site were discovered on the western side of where North Creek empties into Lost Lake in Figure 6.37, its location would be easy to designate on the map without recourse to other operations. Latitude and longitude, or UTM coordinates for the site could be determined from the map itself, or produced while on-site with a GPS. With these data anyone could find their way back to the site in any number of different ways.

GPS readings make the documentation of a location easy, even when there are no distinctive mapped natural or cultural landmarks/features with which it can be associated. The longitude, latitude, or UTM coordinates that the GPS generates can be used to plot the position on a map. Without a GPS you must tie the location that you wish to document into a landmark or feature that appears on a published map. This will require determining bearings or azimuths, and measuring distances from the location that you wish to document to a mapped landmark or feature. The landmark or feature could be a road intersection, bridge, stream junction, benchmark, survey monument, or prominent building; in other words, something that appears on the map that is likely to be relatively permanent on the landscape.

DOCUMENTING SITE LOCATIONS

You've discovered the structural ruins of an 18th-century cabin in a forested area (Figure 6.38). While there are distinctive aspects of the topography immediately surrounding the site, none are of a sufficient scale to be shown on the 7.5 minute quadrangle for the area. The nearest, relatively permanent landmark shown on the map is a road intersection located to the southwest. To tie or reference the site location to this landmark, you will have to use a compass or other surveying instrument to run traverse lines from the site to the intersection. How would you do this with a baseplate compass?

To make the traverse as accurate as possible you want to have a clear line of sight between each traverse point, i.e., between points A and B, between points B and C, between points C and D. You have a general idea of where you are, or should have, because you've been using a map and compass to find your way around during the day, leading up to the discovery of the 18th-century ruins. You know that you need to head in a southwesterly direction to encounter the road

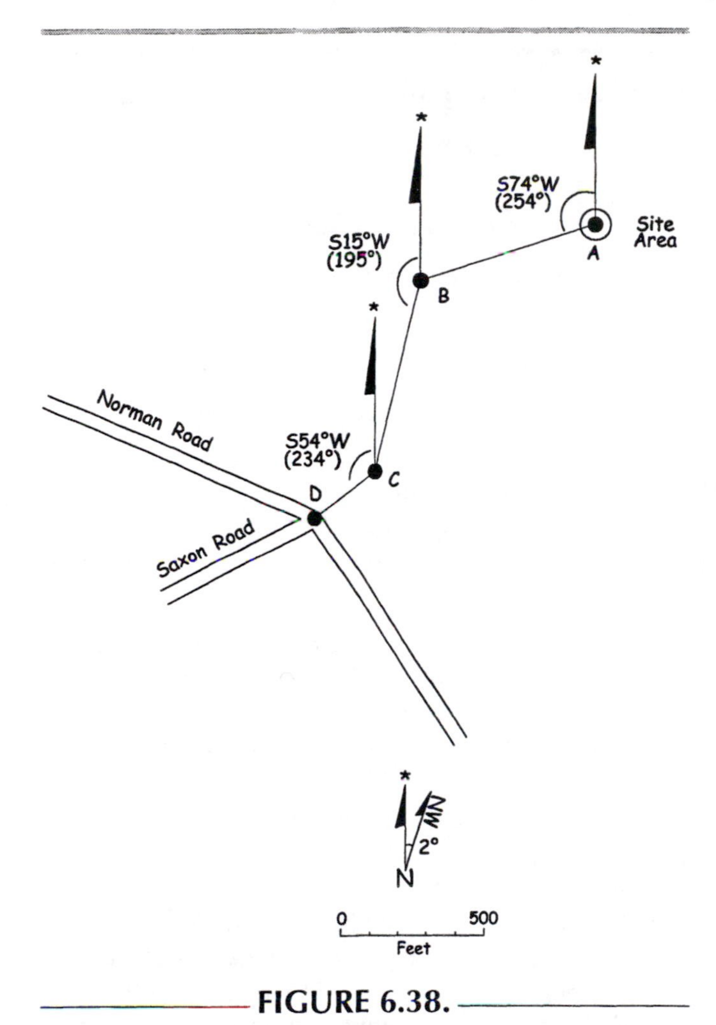

FIGURE 6.38.

Compass traverse lines referencing a site location to a landmark (road intersection) depicted on a published map.

intersection. But because you can't determine your precise location on the topographic map, you can't simply plot a bearing from where you are to the road intersection and follow it in the way that was used in the Deep Lake example. Standing on-site at traverse point A, you pick the longest line of sight that you can get in a southwesterly direction. The end of this sight will be traverse point B. Position a crew member at point B and plant a surveyor's pin or appropriate marker in the ground. Your markers are temporary; once the traverse is complete you can remove them. Blazes put on trees, stumps, or rocks could also serve as traverse points. The crew member gives you a sight at point B by holding a range pole or suspended plumb bob over the point, or you sight directly to the blaze on an appropriate marker. Standing at point A with your compass, hold it level either at waist height or to your eye and point the compass sights at B. Without moving the compass, rotate its housing or dial until the orienting arrow is in alignment with the compass needle. If you are using the compass at eye level, tilt the mirrored cover toward you until you can see the reflection of the compass face before you rotate the housing. When you can clearly see the compass face in the mirror, rotate the housing to align the orienting arrow and compass needle.

Once the orienting arrow and needle are aligned the bearing can be read. The degree graduation on the compass housing that is now adjacent to the forward sighting line is your compass bearing relative to true north, assuming that you've adjusted the compass for local declination: an azimuth of 254 degrees or South 74 degrees West (see Figure 6.38). This is the same procedure that you would use to take a bearing in the field to any target or objective. If possible, tape the distance between traverse points A and B. Pacing will introduce a higher degree of error into the operation. Whether the distance is paced or taped, it will be necessary for someone to remain at point A with the compass to insure that a straight line is followed to point B by those doing the pacing or taping.

The entire procedure is then repeated from point B to C, and so on, until the road intersection is reached. Beginning with point B, backsights or back bearings are also taken. This means that when you occupy point B, not only are you taking a forward bearing to point C, but you are also taking an additional bearing sighting back to point A. Back bearings provide a check on the accuracy of your original forward bearings and the error inherent in the overall operation. Table 6.2 lists the forward and back bearings for the lines in the compass traverse. I've paired the shots that refer to the same traverse line. Since the arc from one end of a straight line to the other end (a semicircle) equals 180 degrees, the difference between the back bearing and the previous forward bearing should be 180 degrees. For example, the difference in the angle between the shot from A to B (254 degrees) and the shot from B to A (74 degrees) is 180 degrees. I've constructed an ideal situation with these bearings. In reality, you will probably detect some form of error in your traverse work when using a compass. Any detected error can be divided equally between the bearings taken and used to compute corrected bearings. With a baseplate compass you should be suspicious of errors exceeding 1.5 degrees and reshoot the traverse line. Probable maximum errors should be less than one degree if you are using a compass with optical sights or a pocket transit (Jacobson 1999:31). Back bearings also provide the necessary

TABLE 6.2

Bearings For Compass Traverse, Figure 6.38

Station		Bearing	Azimuth
At	To		
A	B	South 74 degrees West	254 degrees
B	A	North 74 degrees East	74 degrees
B	C	South 15 degrees West	195 degrees
C	B	North 15 degrees East	15 degrees
C	D	South 54 degrees West	234 degrees
D	C	North 54 degrees East	54 degrees

directions for anyone beginning at the road intersection to follow your traverse back to the site area.

An alternative to the traverse operation just described would be to pick a few convenient bearings, beginning from point A, and physically clear the line of sight. The goal is to maximize the length of each traverse line and reduce the number of lines needed to reach the intersection. The fewer bearings that you have to follow in the compass traverse, the smaller the degree of potential error in your operation. The tradeoff is the amount of time and effort expended in clearing the traverse lines.

Using a transit or theodolite in the above operations increases the accuracy of the measurements. Angles can be read with greater precision and distance measurements easily obtained using a stadia rod or EDM. The downside is that the bulky equipment has to be transported by hand during the day's movements through the field. A transit/theodolite traverse, turning angles to the right, is shown in Figure 6.39. When using either of these instruments it is necessary to establish traverse points over which the instrument can be set up, unlike the compass traverse where a blazed tree or boulder might be used, or a cleared patch of ground where you might have inscribed an "X" or imbedded a stick. Establish a traverse point A in some portion of the site area. Choose a traverse point B and set the instrument up over this point, leveling it. Determine the magnetic bearing of the line BA. If you are using a traditional engineer's transit, or a theodolite equipped with a tube or trough compass, this is done by freeing the compass needle and allowing it to settle on north, then adjusting the horizontal circle/scale of the instrument so that "0" is aligned with the compass needle. Now turn the instrument to sight on point A and read the angle on the horizontal circle/scale. This will be the azimuth of line BA relative to true north. If you are using a theodolite that has no mounted compass, then determine the bearing using a compass. Measure the distance between B and A using a stadia rod or EDM. If the distance is taped, use the instrument to keep the persons doing the measuring on line.

With the bearing read, sight the instrument on point A with the horizontal circle/scale set to "0". Then turn the instrument to the right (clockwise) until you can sight the point that you selected to be C. Read the angle (121 degrees) on the horizontal circle/scale. Now set up the instrument over point C, sight back to point B with the horizontal circle/scale set to "0". Turn the angle to the right until you sight point D. Read the measured angle that this represents (219 degrees). Also record the magnetic bearing for line CB. Measure the distance between the points. Finally, with either the instrument or a compass, occupy point D and determine the bearing for line DC.

The azimuth for lines CB and DC can be calculated independently of taking a magnetic bearing at each traverse point. It does require, however, that you know the bearing/azimuth of the initial line BA. The calculation is as follows:

74 degrees *(initial azimuth, line BA)*
+121 degrees (first angle to the right)
195 degrees
- 180 degrees
15 degrees *(azimuth of line CB)*
+219 degrees (2nd angle to the right)
234 degrees
- 180 degrees
54 degrees *(azimuth of line DC)*

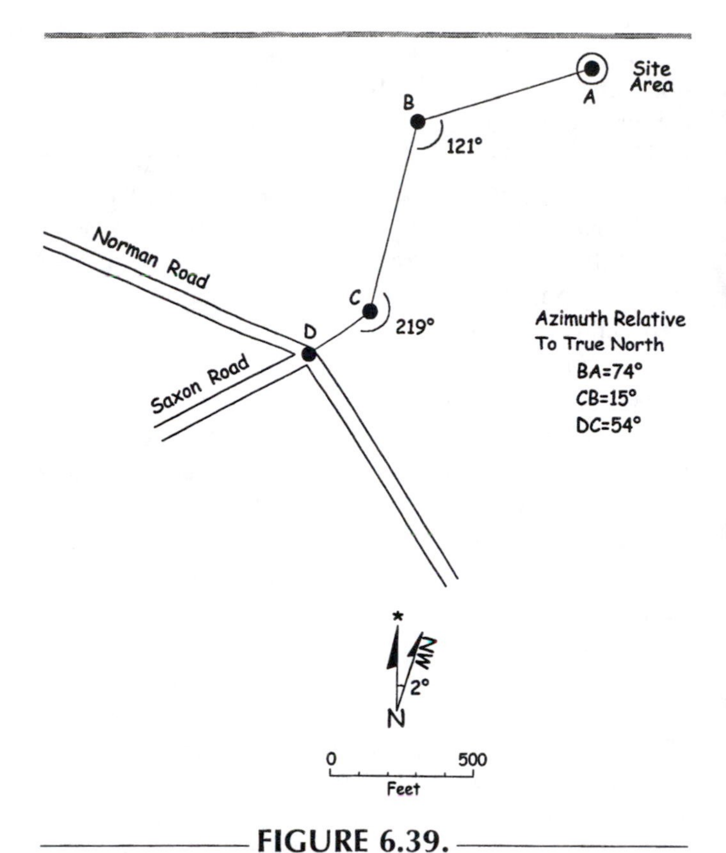

FIGURE 6.39.

Transit/theodolite traverse, turning angles to the right, used to tie a site location into a landmark (road intersection) depicted on a published map.

FIGURE 6.40.

A ball type of marker employed in an electronic marker system used by some archaeologists for relocating sites. The ball is buried, and its depth and position relative to other site features recorded.

This calculation can be used as a cross check with the magnetic bearings/azimuths taken at each traverse point.

The traverses described above are examples of an ***open traverse.*** This means that only one end point of the traverse has a fixed or known horizontal location (the road intersection in my examples). In a ***closed traverse*** both ends of the traverse are known points, or a traverse may begin at a known point and loop around to eventually finish at the same point. Closed traverses are more accurate because they provide a computational cross check on directions and distances not possible with an open traverse (see Anderson and Mikhail 1998:385–490 for a detailed discussion of different types of traverses).

With your survey data, regardless of whether it was derived from a GPS, compass, transit or theodolite traverse, you should be able to plot the position of point A on a USGS 7.5 minute quadrangle. When you compare the various methods I've described for documenting or referencing a location, your appreciation for the use of GPS in archaeology should be even greater than when we started.

While the methods I've outlined should be sufficient to relocate an area or site, area-specific or site-specific sketches, maps, and verbal descriptions generated by you, the archaeologist, are critical for understanding the location in more detail. It's simply not good enough to be able to say, "well, here we are". What is the size and shape of the site area or cultural features that you are attempting to document? Where are the locations of any subsurface tests or excavations that have been completed? Where are the locations where samples of artifacts were collected from the surface? What and where are distinctive natural features in the site area that aren't shown on published maps, but could be used as internal landmarks for determining site boundaries or the location of test excavations? How do these locations relate to the initial point that you used in your traverse linking the area with a mapped landmark or feature? The creation of scaled plans and topographic maps is discussed later in this chapter.

One way that some archaeologists have used to enhance the accuracy of site or area relocation is through the use of buried ***electronic marker systems*** (Whitlam 1998). The technology involved is borrowed from the underground utilities industry where the ability to precisely relocate hard-to-find things on landscapes that may change over time is important. A marker system includes a small passive antenna encased in a durable, damage-resistant shell, which is buried (Figure 6.40), and a locator device that resembles a metal detector in appearance. Depending on the model, markers can be buried from 2 to 8 feet below the surface, keeping them out of reach of vandals, the curious passerby, and the myriad types of disturbances that can affect the surface of the earth. The position of the buried marker relative to other aspects of a site or locality is noted at the time of burial. The locator device works by sending out a pulse on a specific frequency that matches that of the buried marker. Each marker is preset to respond to a specific frequency. The locator is capable of detecting the buried marker if it passes within approximately one to three feet of it. Use of an electronic marker system doesn't preclude the need to document as accurately as possible the location of a site or area of interest. But it does provide a means of confirming that an exact position has been relocated and a key to orienting any spatial information generated by previous fieldwork on site.

SYSTEMS OF RECORDATION

Three different systems have been used to various degrees to record the location of archaeological sites when they are officially registered with the state (see chapter 4): latitude and longitude coordinates, the various subdivisions of the township and range system, and UTM coordinates. Latitude and longitude have been used sporadically in this country for systems of site recording, especially during earlier parts of the 20th century (e.g., Cross 1941; Skinner and Schrabisch 1913:42). Latitude and longitude coordinates for a site's location can be easily generated by a GPS when on site. Coordinates can also be calculated based upon a plotted location on a USGS 7.5 minute quadrangle. Latitude and longitude tick marks are spaced at 2.5 minute intervals along the edges of the map. By orienting an interpolation template on the control points that come closest to the site's location, coordinates for the site's position are obtained. Interpolation templates are available from the USGS or companies like Forestry Suppliers.

UTM coordinates are favored for recording site locations over latitude and longitude because of the ease in taking the measurements. UTM coordinates can be generated on site by many models of GPS. UTM coordinates also can be determined from a plotted location on a USGS 7.5 minute quadrangle, or any topographic map that includes UTM tick marks. First, determine the UTM grid zone in which your map is included. The grid zone is listed in the box of text in the lower left corner of the map (for example see Figure 6.5). As you will recall, UTM grid ticks (blue) are arranged along the borders of the map, generally spaced at intervals of 1,000 meters. Some maps project the UTM grid system across the entire map from these control ticks. If you are using a map with a projected UTM grid the next step in this procedure won't be necessary.

Select the most convenient UTM tick marks on flanking sides of the map and construct a box that encompasses the site area that you wish to record. Do this by using a straight edge and pencil to project lines across the map from UTM ticks at the map's border (Figure 6.41). The box will generally measure 1,000 meters on a side unless your area of interest occurs adjacent to one of the map's corners, where UTM tick marks may not be spaced at 1,000 meter intervals. In these cases the reconstructed box will not be square. In the example shown in Figure 6.41, the UTM tick marks used to project the grid are 4,972,000 meters (northing), 4,973,000 meters (northing), 501,000 meters (easting), and 502,000 meters (easting).

Now determine the UTM coordinates of the southwest or lower left corner of the grid box that you have created. Simply follow the projected lines that intersect at this point back to their origin at the map's edge. In our example, the coordinates are northing 4,972,000 meters and easting 501,000 meters.

To complete the calculation of the UTM coordinates that correspond with the center of the site area you will need a UTM coordinate grid/counter (Figure 6.42), available from the USGS or companies like Forestry Suppliers. Pick the scale on the coordinate counter that matches the scale of your map. In our example, the scale is 1:24,000, standard for a 7.5 minute quadrangle. Orient the UTM coordinate counter in the southwest corner of the grid box (Figure 6.43) and measure the number of meters from this point moving right to the location of the site. The measurement is 880 meters. Add this figure to the easting coordinate of the southwest corner: 501,000 + 880 = easting 501,880 meters. Use the vertical portion of the coordinate counter to measure the distance up or north from the base of the grid box. This measurement is 221 meters and is added to the original northing coordinate: 4,972,000 + 221 = northing 4,972,221 meters. To avoid confusion, always take UTM readings from west to east (left to right) for easting and south to north (bottom to top) for northing within the grid box. The UTM designation for the center of the

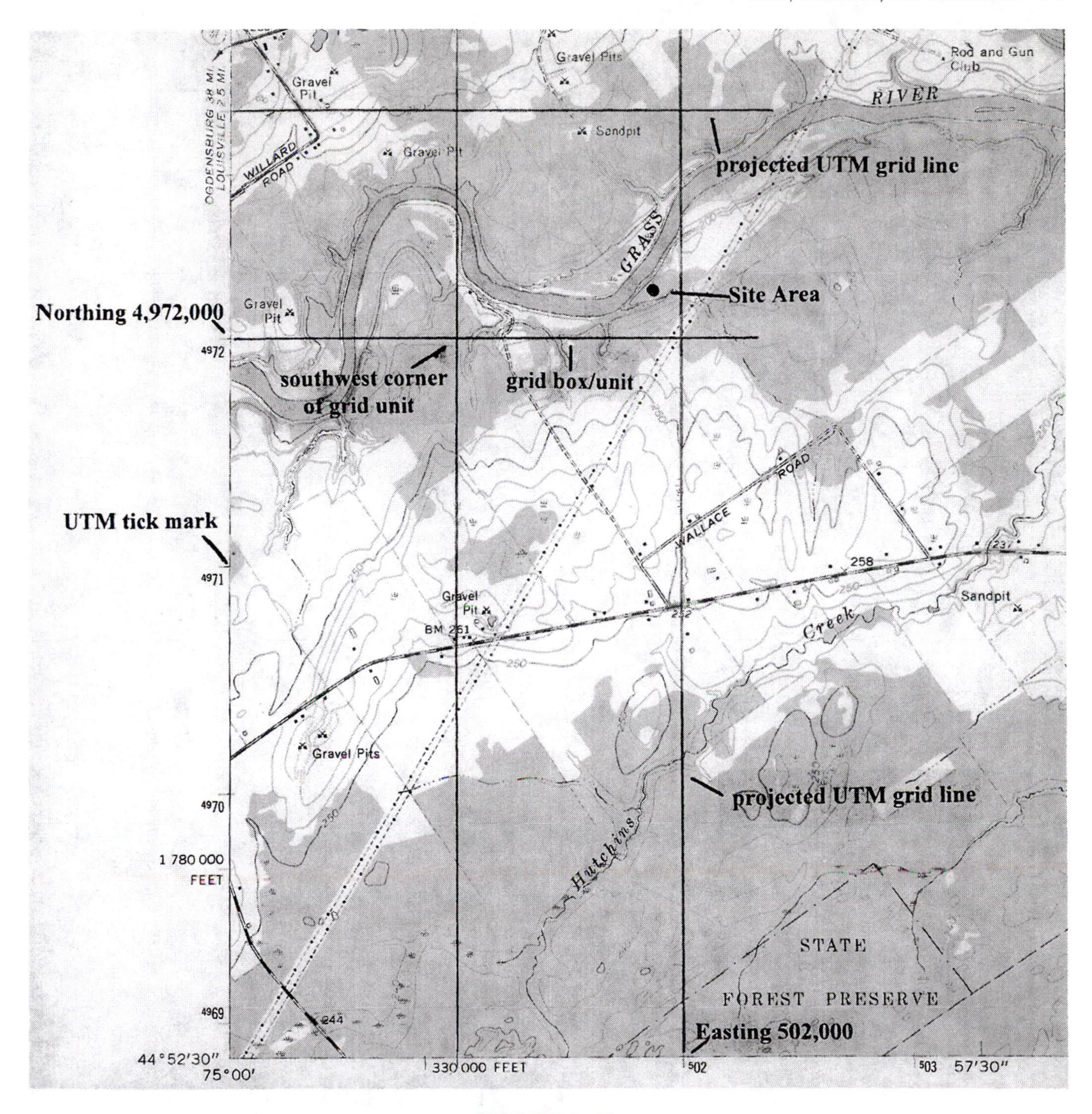

FIGURE 6.41.

Constructing a UTM grid unit encompassing a site area. The grid unit has been constructed by projecting lines out from UTM tick marks along the edge of the USGS 7.5 minute quadrangle (Massena, New York - Ontario, 1964).

site in our example is expressed as Zone 18, easting 501,880 meters, northing 4,972,221 meters. When a large site area is involved, UTM coordinates may be determined for various locations along its boundary, as well as its center.

States within the United States Public Land Survey System may use township, range, section, and section subdivisions to describe the "legal" location of a site. Regular sections (1 mile square) and irregular sections are depicted on 7.5 minute topographic maps. Regular

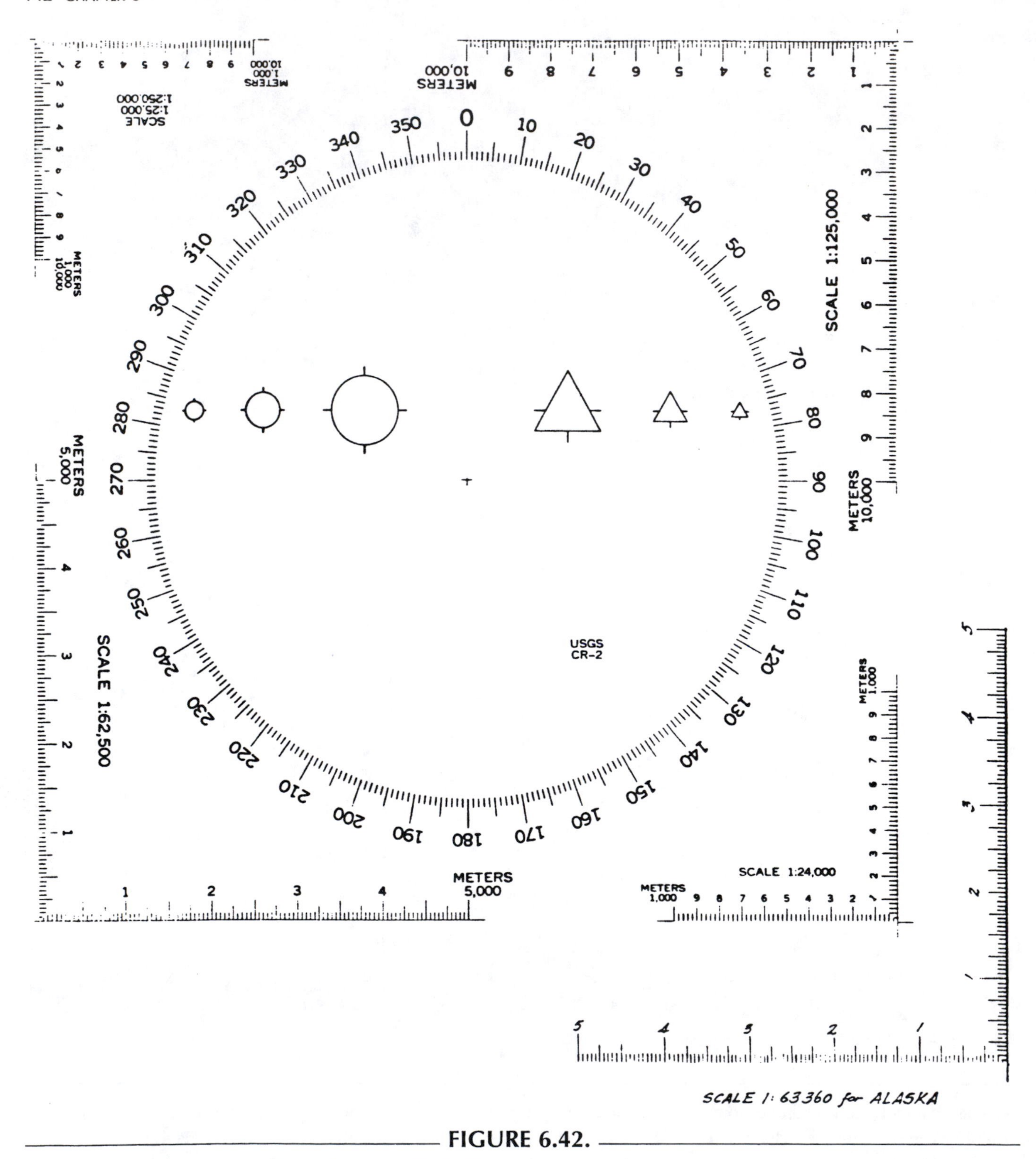

FIGURE 6.42.

A UTM coordinate counter, available from the USGS. The counter is a transparency on mylar so that it can be overlaid on topographic maps.

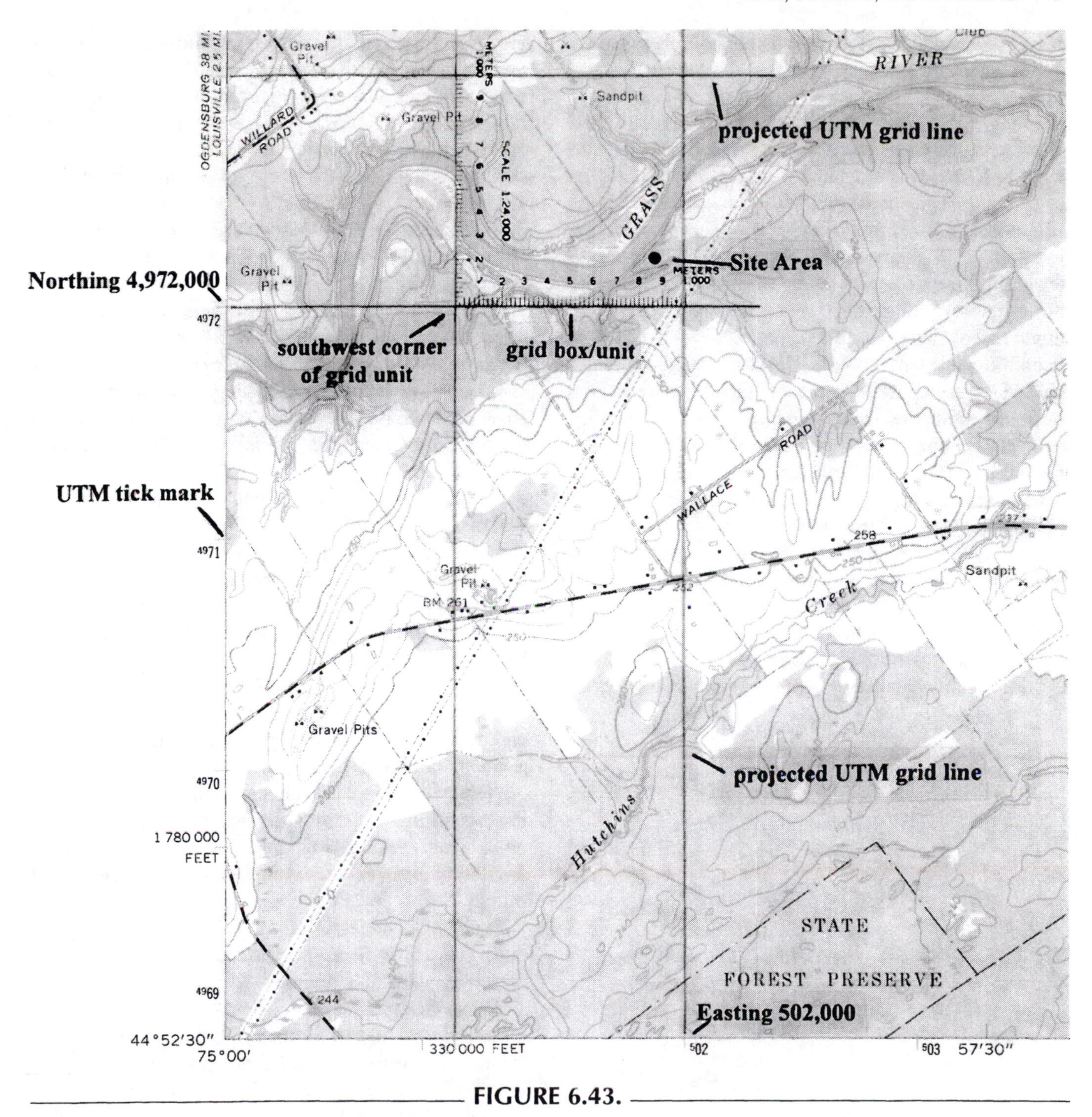

FIGURE 6.43.

UTM coordinate counter oriented within the grid unit shown in Figure 6.41. Measurements are taken from west to east, and south to north with the counter in order to locate the center of the archaeological site.

sections may be subdivided into quarters, the quarters into quarters, and so forth until an area of no more than 10 acres has been identified that encompasses the site (see Figure 6.15; Napton and Greathouse 1997:197). Transparent templates can be obtained to aid in the subdivision of regular sections.

LAYING OUT LINES OR TRANSECTS

You have already seen how useful laying out lines can be in finding and relocating positions in the field. Establishing or extending lines with a compass or surveying instruments serves a number of other purposes as well. During a site survey, areas are often tested with transects (linear alignments or distributions) of auger borings, shovel test pits, or small excavations. When working in wooded areas, a compass can be used to sight in the rough alignment of a transect to be cleared of vegetation prior to beginning any subsurface testing. Compass and tapes may also be suitable for laying out locations for subsurface testing along transects, depending on the nature of the terrain and the length of the transects. For transects that are more than a few hundred feet long, laying out the line with a transit or theodolite and stadia rod or EDM might be more advisable. The greater the number of lines that need to be laid out, and the more complex the relationship between the lines, the greater the chances for cumulative and sizable errors resulting from using a compass. Accuracy is important. You need data sufficient to plot your test locations on a map and enable someone to relocate the general area ***and*** specific subsurface tests, especially those that produced artifacts, features, or important information about stratigraphy.

Choosing the direction of any given line follows from the nature of the landscape and the work that you are doing. Your goal might be to systematically test an

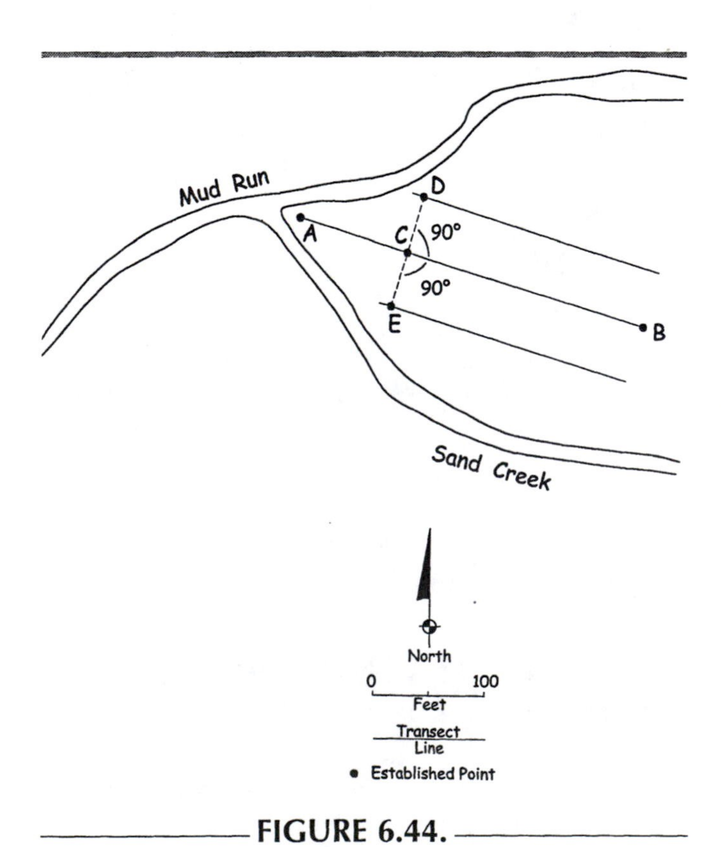

FIGURE 6.44.

Laying out a series of parallel lines/transects.

area with a series of shovel test excavations arranged in evenly spaced transects, so the lines that you establish must parallel one another (Figure 6.44). Once the initial line is established (AB), you have points of reference for locating and orienting subsequent lines. Occupy a point (C) along the initial line and set up your surveying instrument. Sight on either end of the established line AB (your choice, although the longer the sight, the better, like CB) with your horizontal circle/scale set to "0". This will be the point of reference off which you will turn angles to establish new lines. Watching your horizontal circle/scale, turn 90 degrees to the left. Measure the appropriate distance from the instrument's position, in this case 50 feet, along the 90-degree line. Set a new point (D). Rotate the instrument back to the "0" sight along the initial line (CB). Now turn 90 degrees to the right, measure 50 feet along this line, and set a point (E). You now have one point each set on two new transect lines.

Occupy one of the new points, E. From here, sight the instrument back to your original setup at C with your horizontal circle/scale set to "0". Turn the instrument 90 degrees to the left. This will be one portion of the new transect/line. Set a point at the end of the line. Now turn back to "0" (C) and continue to the right to 90 degrees. This will enable you to establish the other end of your new transect/line. Repeat this operation to establish other lines/transects.

New points can be established along any portion of a line by pulling a tape straight and taut between existing points, and setting new points at desired intervals. Doing this over uneven ground will require the use of plumb bobs by the crew handling the tape. Alternatively, the tape can be pulled from the instrument setup. The instrument is used to keep crew handling the tape and setting new points on line.

The same operation could be completed with a baseplate compass, or any compass with a rotating housing or dial. Of course it would not approach the same degree of accuracy as that achieved with a transit or theodolite. Stand at A and take the bearing of the initial line AB: South 74 degrees East or an azimuth of 106 degrees (Figure 6.44). Turning 90 degrees to the left of this bearing (towards D) would bring you to a bearing of North 16 degrees East or an azimuth of 16 degrees. While still standing at C, turn the dial of your compass so that the index or forward sight is aligned with 16 degrees. Now slowly turn your body until the compass needle aligns with the orienting arrow. The forward sight of the compass will now be pointed at D, having turned 90 degrees off the initial line.

ESTABLISHING GRIDS AND UNITS

Laying out a series of parallel lines off an initial baseline is one step away from creating a grid system. All that is needed is the creation of additional lines at right angles or 90 degrees. Grids are used in controlled surface collections of archaeological sites, as a way of subdividing an area for the purposes of sampling, as a way of organizing the spatial distribution of excavation units, and as a way of organizing elevation readings for the construction of contour maps. They are a basic way of subdividing space, controlling and documenting the context of things within that space. Grid systems can be very large, covering acres or even square miles of territory, or relatively small, accommodating a small number of excavation units. The size depends on the task at hand.

A grid can be constructed with a transit or theodolite by first establishing an initial baseline with an orientation and length of your choosing. Initially, set at least two fixed points along the baseline, preferably at either end. Determine the size of the intervals or units that will be employed in the grid. Let's say that the goal is to grid a plowed farm field into units 50 feet square (Figure 6.45) for a surface collection of artifacts (surface collections of all types will be discussed in detail in chapter 8). The field is oblong to rectangular, so we'll pick a baseline (AB) orientation that will run the length of the field and generally parallel the creek and wood's edge. This strategy will allow us to create large portions of the grid from fewer set ups of the surveying equipment. And because we're adjusting the orientation of the grid to the orientation of the landscape, we should end up with fewer partial or oddly shaped collection units.

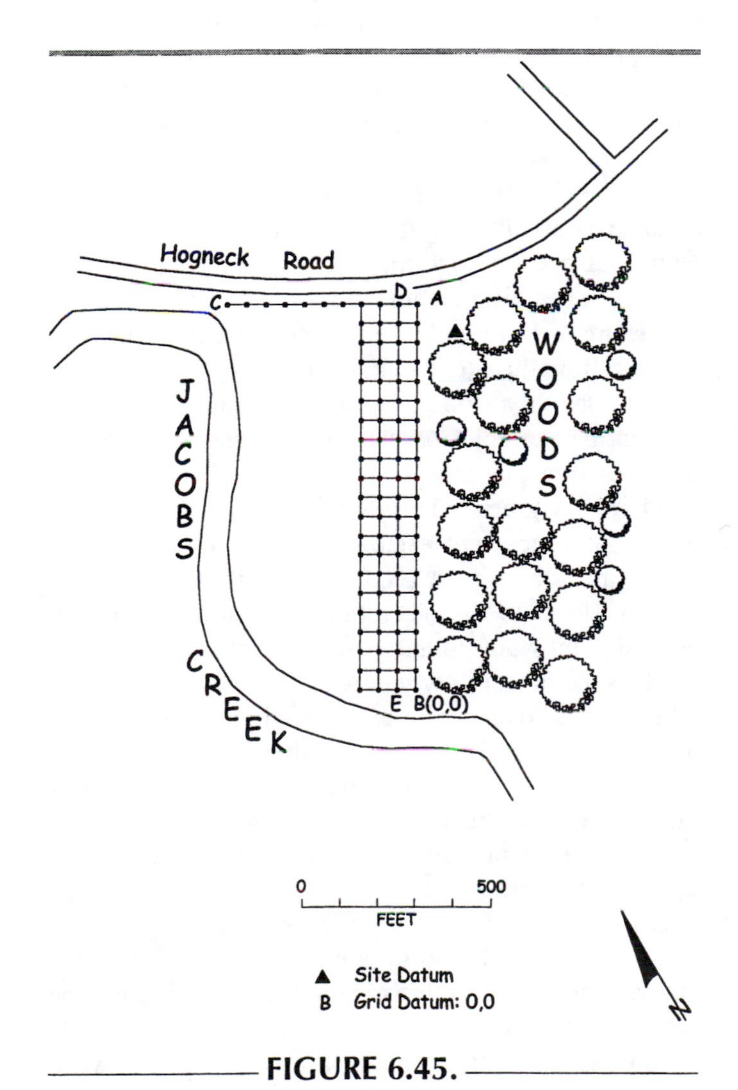

FIGURE 6.45.

Constructing a grid for a controlled surface collection.

Create the baseline AB by setting a fixed point at A and B. Set up your instrument over A and sight to B, setting the horizontal circle/scale to "0". Record the magnetic bearing of the line AB for future reference. Line AB will be the point of reference to which all other lines of the grid will be oriented. Using your instrument to keep crew on line, tape the distance between A and B, setting points every 50 feet. Turn 90 degrees to the right off line AB and set point C, establishing line AC. Using your instrument to keep crew on line, tape the distance between A and C, setting points every 50 feet. You can now occupy the individual points set along line AC to set the remainder of the grid.

For example, move your instrument to the first grid point (D) to the west of A. Sight the instrument on C with your horizontal circle/scale set to "0". C was selected for the sight rather than A because it is at a greater distance from the instrument; longer sights are preferred for establishing a reference prior to turning angles. Rotate the scope vertically so that you are now looking back at point A. This doesn't involve any movement of the horizontal circle/scale. If you are using a surveying instrument with a scope that can't be rotated vertically, then simply turn 180 degrees to get the back sight. In either case, the instrument's scope should be perfectly aligned on A without having to make any further adjustments of the horizontal circle/scale. If the alignment is not precise start again, this time by sighting on point A with the horizontal circle/scale set to "0". Vertically rotate the scope and look back at point C. The instrument should be in perfect alignment with C. If it is not in alignment, then there is something wrong with the way in which you set up your instrument over point D, or there is something wrong with the way in which D itself was established. Recheck your instrument setup and try again. If things still aren't working out, go back to point A and recheck the alignment of line AC and the points along it. Because multiple lines and points will be established as part of the grid, any initial error will be propagated through all subsequent operations and measurements. You need to identify and eliminate or minimize errors early in the operation.

Back to point D. Your instrument is set up there, you've sighted on C and the horizontal circle/scale is set to "0", you've back sighted to A and everything checks out. Now, using the horizontal circle/scale, turn 90 degrees to establish the line DE. Using your instrument to keep crew on line, tape the distance between D and E, setting points every 50 feet.

Repetitions of the operation described at each of the points along the line AC will complete the basic grid system. As you can see, occupying sequential points along line AC to set remaining portions of the grid requires setting up your instrument at fewer points than if your were working off line AB. The fewer setups, the smaller the potential for error.

The designation of individual units within the grid system is determined based upon their spatial relationship with a grid ***datum.*** Datum refers to a starting point, or point of origin to which other things are related, and is used in a number of different ways by archaeologists. For a grid system that employs measurements of distance and direction, the datum can be thought of as the 0,0 point. For the purpose of convenience, we will designate the direction of parallel lines AB, DE, etc., as grid north-south, and line AC and those that parallel it as grid east-west. Neither of these sets of lines corresponds exactly with the compass directions; however, using a directional referent will make labeling and referring to grid units easier. We'll use point B as the grid datum. With B as the starting point, all grid units will be situated to the north or west. The choice of B as the grid datum means that we don't have to complicate grid unit designations with considerations of units that might be located to the north and east, south and west, or south and east. The result will be fewer errors in filling out data forms, artifact bags, and labels. Whenever possible, try to pick a grid datum so that all of the system's units fall within a single quadrant, i.e., northeast, northwest, southeast, or southwest of the datum.

With B as the datum or 0,0 point of the system, other points on the grid can be labeled based on how far to the north, and far to the west they are from it. The point 50 feet north of B along line BA is North 50, West 0 (abbreviated N50W0). Subsequent points along line BA would have north readings that would change, but would retain West 0 designations. In contrast, the first point 50 feet to the west (left) of B would be North 0, West 50 (N0W50), the next N0W100, the next N0W150, and so forth. The first grid point north of (above) E would be N50W50.

Rather than list the four corners of a collection unit when referring to it in field notes or listing it on data forms or labels, the coordinates of a single corner are used as a reference. In American archaeology it is the tradition to refer to any unit in a grid, regardless of whether it is for surface collection or excavation, by its southwestern corner. For example, point E is the southwest corner of the unit N0W50. The unit north or above this would be N50W50.

Like any type of area-specific or site-specific map, a grid system must be tied into a point or landmark that has some permanence and can be relocated. Once relocated, this type of datum (site datum) is used to orient or reconstruct previous mapping efforts. In the example shown in Figure 6.45, an archaeologist might bury or anchor a metal pipe or a concrete monument in the ground just inside the woods east of point A and use it as a site datum. Placing the site datum in the field itself would subject it to disturbance from future agricultural activities. The distance and magnetic bearing from the site datum to A would be determined (Figure 6.46). The distance from the site datum to another grid point on the baseline should also be recorded. Grid point N900E0 is used in the example. Also record the angle created by the lines from the site datum to each of the two grid points (55 degrees, 11 minutes, 55 seconds). Do this with your instrument established over the site datum and sighted on point A (N1000E0) with the horizontal circle/scale set to "0". Then turn the instrument to sight on grid point N900E0 and read the angle.

The specifics of how you relate a grid system or any other mapped data to a site datum will vary from situation to situation. But you will always be asking and answering the same basic question: what measurements, angles, or bearings do I need to know to reconstruct previous mapping operations if I find the site datum and nothing else?

A final step is to reference the site datum to a landmark that can be found on a published map. The road intersection shown on the map in Figure 6.45 would be such a landmark. Relating the intersection to the site datum would require running a traverse line between the two points. Using a GPS unit to record the latitude, longitude and UTM coordinates of the site datum also is a good idea.

Small grids or rectilinear units also can be established through triangulation using tapes and working off two established points. ***Triangulation*** employs the known relationships between the angles and sides of a triangle to establish lines, locate positions, and determine unknown distances and angles. In establishing a rectilinear unit or grid, understanding the right triangle is key (Figure 6.47). Any rectilinear shape can be divided into a number of right triangles. As an example, let's suppose that you are in the field and want to lay out a square unit for excavation measuring five feet on a side (Figure 6.48). Start by establishing a line (AB) that is five feet long from end to end. The orientation of this line is up to you. The problem is that you simply can't measure five feet off each of these points to complete your unit. What will ensure that the unit is perfectly square and that the angles in each corner are 90 degrees? You might get lucky and after repeated trial and error orient the new points so that all sides of the unit they create measure exactly five feet on a side. This is where the Pythagorean theorem comes in handy. Any square is made up of two identical right triangles with a common hypotenuse. If you knew the hypotenuse of a right triangle that has two sides measuring five feet, you could triangulate the remaining points of your unit from your established points A and B and be confident in ending up with a square with internal angles of 90 degrees. Solve the equation for side "c", or the hypotenuse of a right triangle, and you're ready to go.

The actual task of triangulating the corners of the unit requires three people and two tapes: one person at point A holding the "0" end of a tape, a person at point B holding the "0" end of another tape, and a third person pulling the tapes taut and overlapping them to determine their intersection, and setting the new point. When all four corner points are set, check the measurements between each to insure that it is five feet. Retriangulate points C and D, if necessary, after making sure that the distance between A and B is exactly five feet. The triangulation process can be tricky on uneven or sloping ground and will require pulling the tapes in a

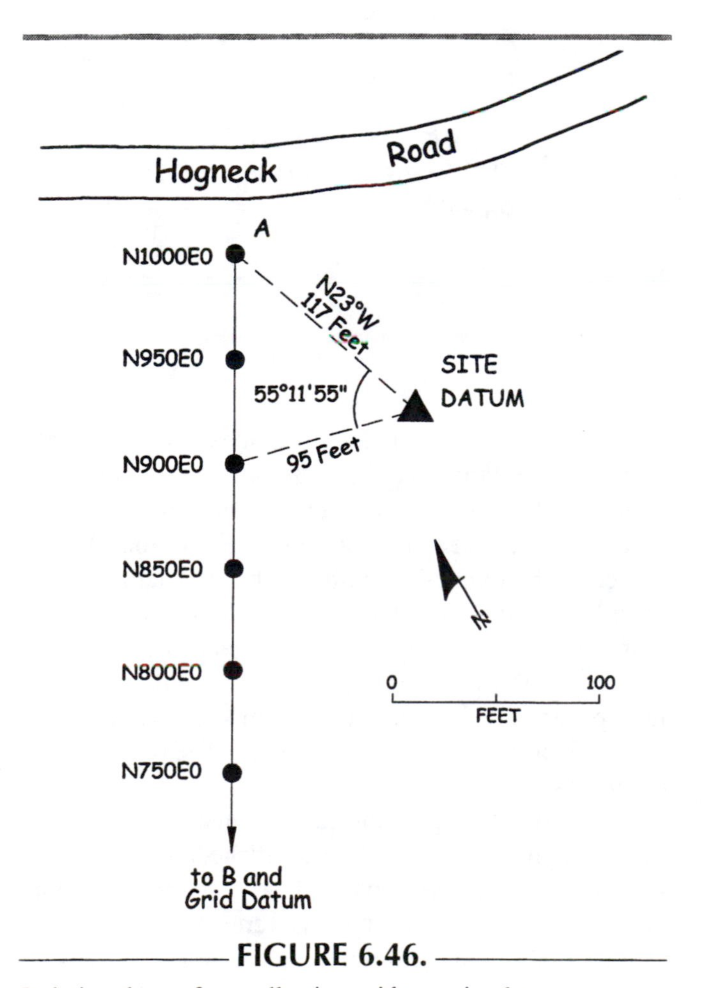

FIGURE 6.46.

Relating the surface collection grid to a site datum.

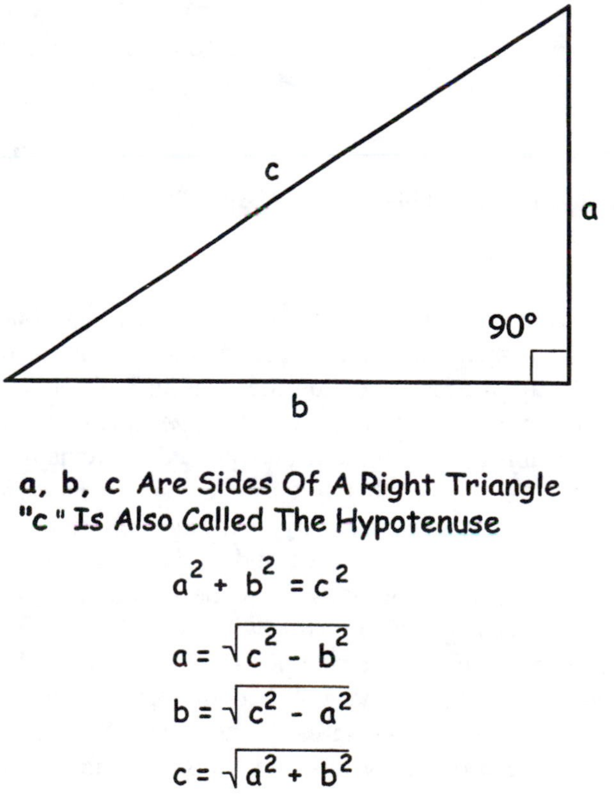

c
a
90°
b

a, b, c Are Sides Of A Right Triangle
"c" Is Also Called The Hypotenuse

$$a^2 + b^2 = c^2$$

$$a = \sqrt{c^2 - b^2}$$

$$b = \sqrt{c^2 - a^2}$$

$$c = \sqrt{a^2 + b^2}$$

FIGURE 6.47.

The right triangle and the Pythagorean theorem.

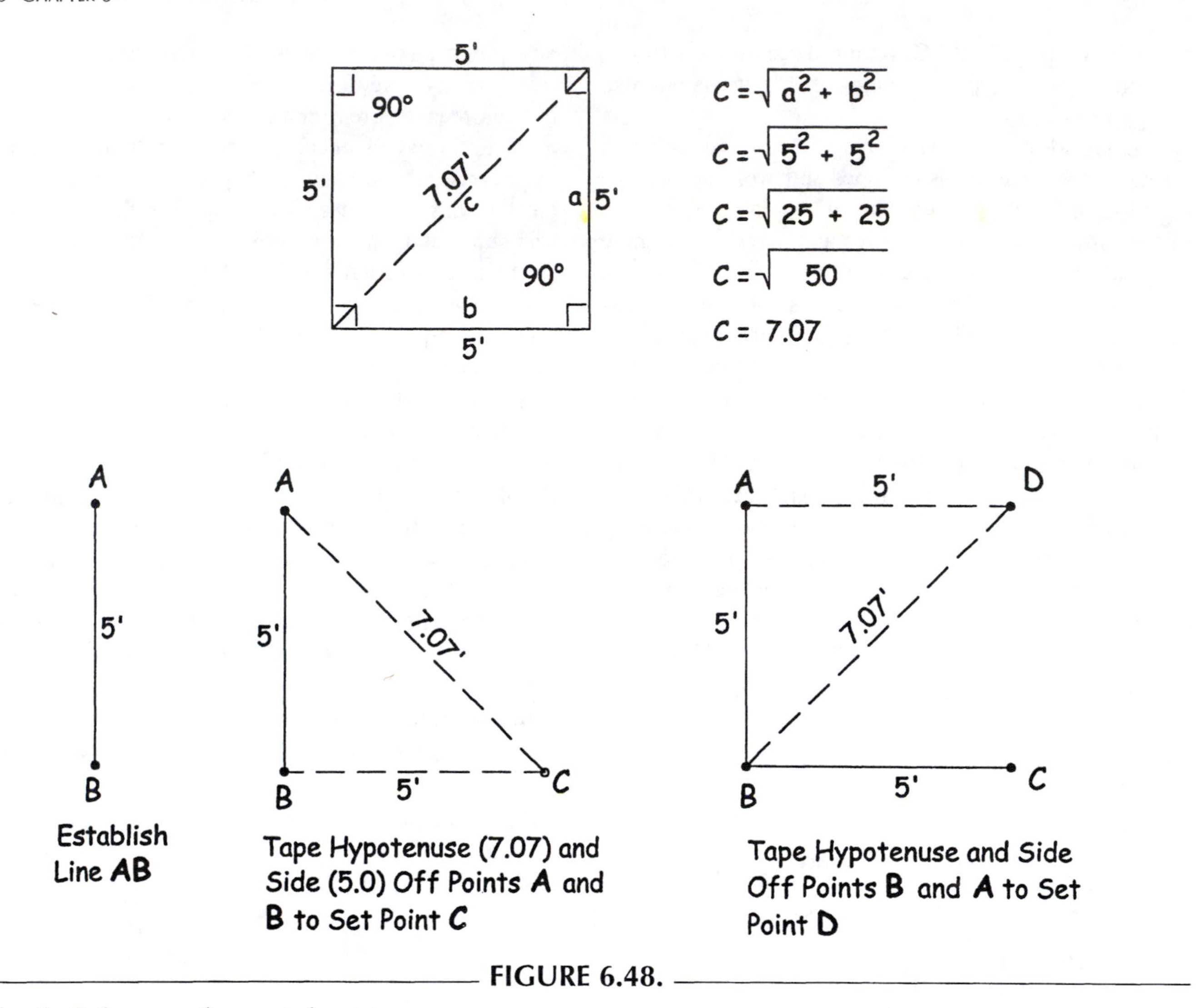

FIGURE 6.48.

Using the Pythagorean theorem to lay out a square.

horizontal line that is level and using a plumb bob to translate their point of intersection to the ground surface.

It doesn't matter if you are dealing with engineer-scaled feet or metric units, or what size square you want to establish, the formula and procedure remains the same. You know what you want the sides of your unit to measure; solve for the hypotenuse of the right triangle of which they are a part, and take it from there. The procedure and formula is the same if you want to lay out a rectangle. You still will be solving for the hypotenuse of a right triangle, but the adjacent sides of the triangle (those that will become part of your unit) will be of different lengths (Figure 6.49).

You can probably imagine ways in which triangulation with tapes off fixed points could be used to establish a grid system. For example, it would be fairly simple to triangulate the corners of a 30′ x 10′ rectangle, set points every five feet along the sides of the rectangle by pulling a tape between two corners, then play connect-a-dot with a tape between the points established along the rectangular frame (Figure 6.50). The grid could be expanded with additional triangulation. An alternative method (Figure 6.51) is to establish a single long line, pull a tape between the two endpoints and set intervening points at a regular interval, then progressively triangulate units off this initial line of fixed points. You can see that the possibilities are endless.

Progressive triangulation can propagate any errors in the initial lines and points established, and increase them as the system is expanded, so it is always important to double check your measurements and the final product (individual grid units) for accuracy. Extensive triangulation over uneven ground and slopes can be problematic, as I have already noted. Use a surveying instrument in these cases if at all possible.

Finally, laying out straight lines or traverses is useful for establishing controls or fixed points on which to base additional surveying operations, like the creation of plan maps and topographic maps.

CREATING PLAN MAPS

A plan map depicts spatial relationships between the things of interest, but does not include elevations or all aspects of the topography. The surveying operations used to collect data for constructing plan maps are

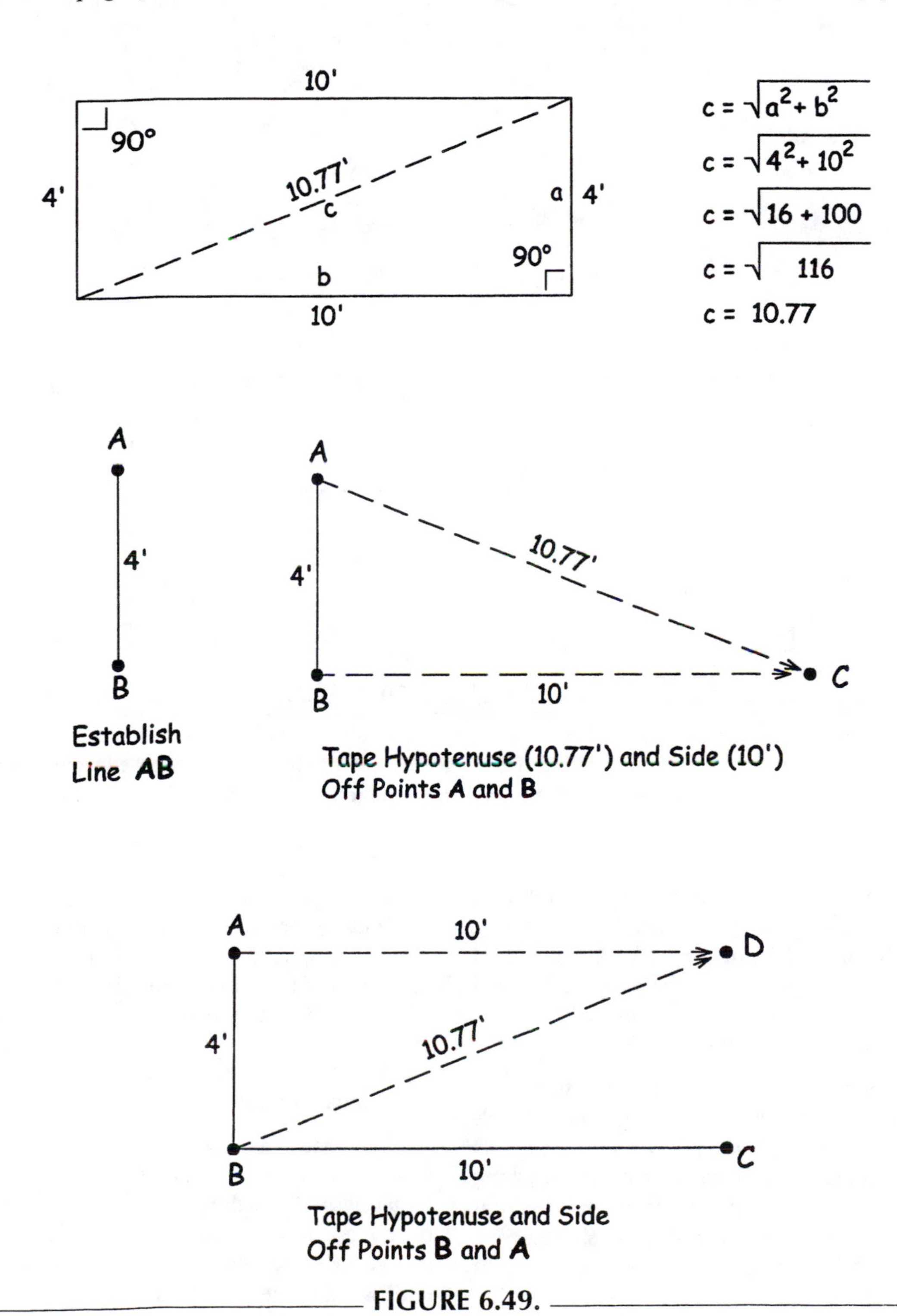

FIGURE 6.49.

Using the Pythagorean theorem to lay out a rectangle.

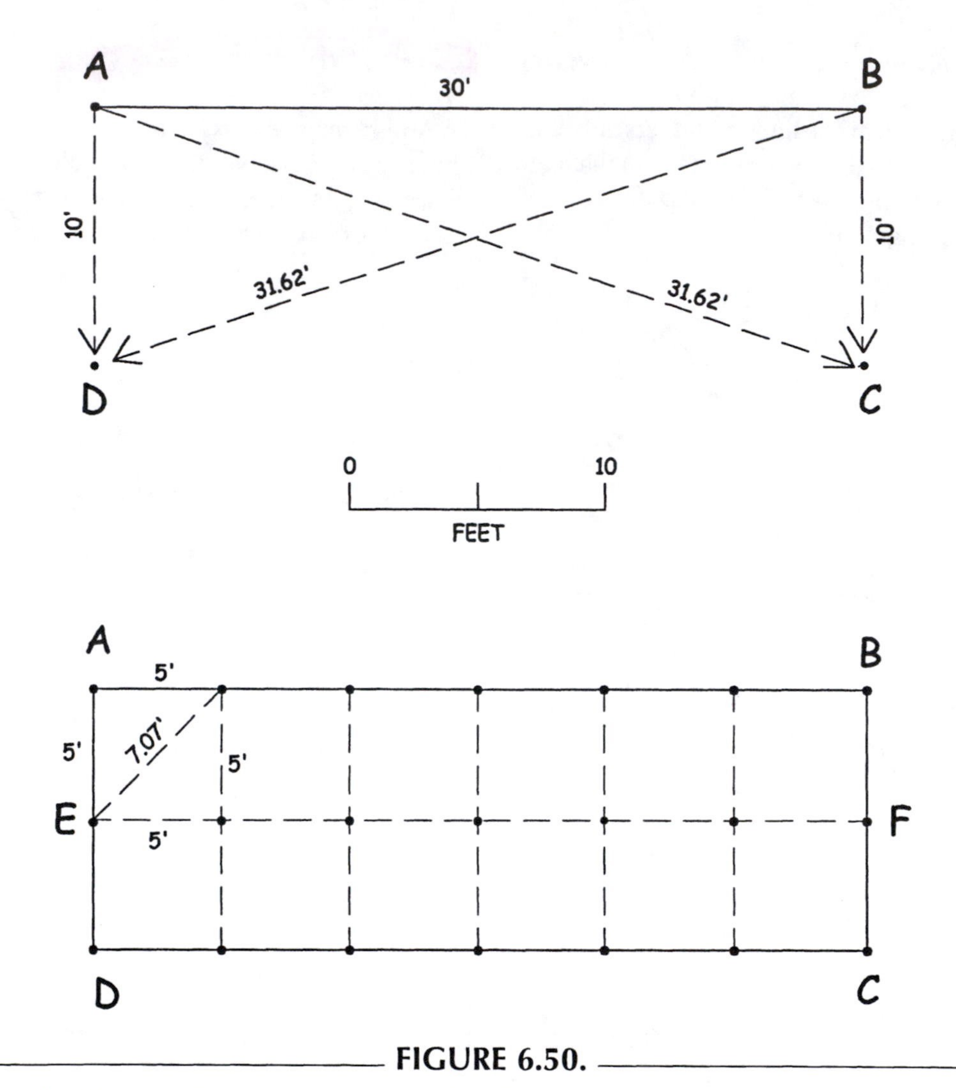

FIGURE 6.50.

Triangulating a small grid system. Triangulate the rectangle (ABCD). Pulling a tape between the corners of the rectangle set points every five feet. Pull a tape between points E and F and set points every five feet. Check the length of the sides and the diagonal (hypotenuse) of the 5′ ×5′ units established by this procedure for accuracy.

often done in conjunction with collecting data relevant to mapping topography or elevations. I'm discussing them separately for the sake of convenience and clarity. Archaeologists create plan maps for a variety of reasons, and the details included on a map vary with the intent of the map. Plan maps are made to:

- ☑ document where you searched for artifacts or archaeological deposits—location of surface collection grids, excavations or any subsurface tests performed (regardless of whether or not anything was found in them), and prominent aspects (natural or cultural) of the area that would enable someone to see how they influenced your testing strategies or to aid in relocation
- ☑ document site areas showing the distribution of finds or features on the surface, the location of any excavations or subsurface tests, and prominent aspects (natural or cultural) of the area that might aid in site interpretation and relocation
- ☑ document the location of artifacts, features, or anomalies within excavations

Plan maps can be constructed with surveying instruments, or through triangulation using only tapes. Regardless of whether you use a surveying instrument or triangulate with tapes, the mapping operation must start with the establishment of points of reference to which subsequent measurements of distances and

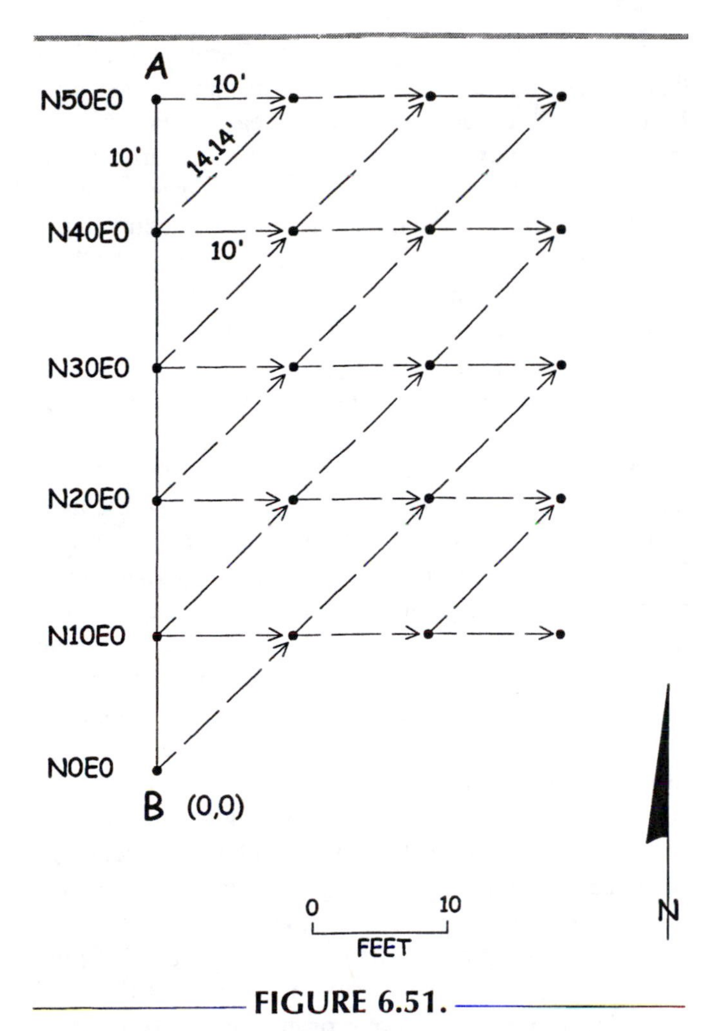

FIGURE 6.51.

Another method for creating a grid system using triangulation. Establish line AB. Pull a tape between the points and set intervening points at a regular interval. Progressively triangulate individual grid units off the fixed points of this initial line. Check the length of the sides and the diagonal (hypotenuse) of individual units for accuracy.

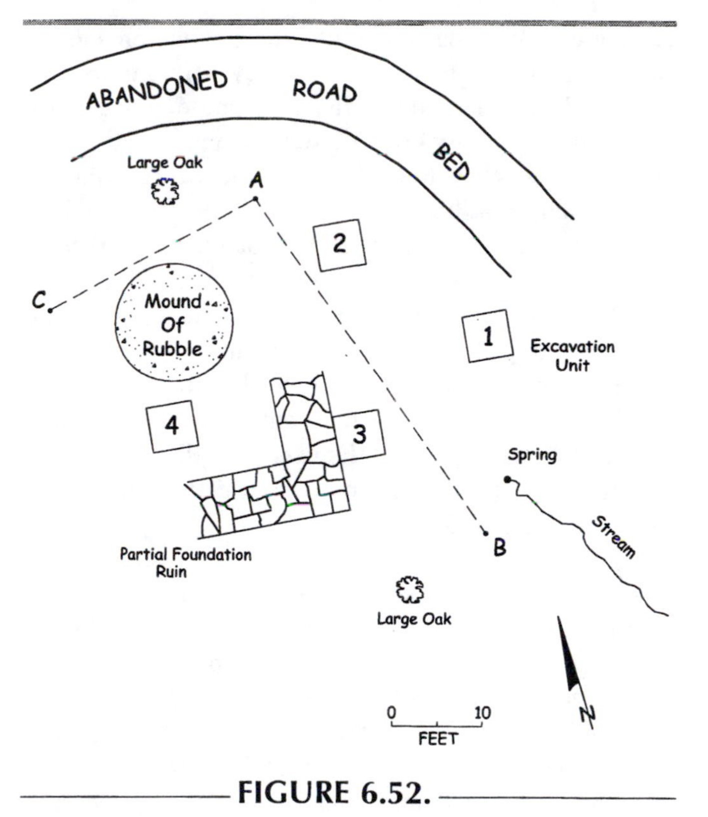

FIGURE 6.52.

Creating a plan map of a site area. Mapping takes place from reference points A, B, and C, chosen to provide a clear line of site to all of the natural and cultural features to be included in the plan. All measured angles and distances are related in some way to these reference points.

angles will be related. In some cases, two points (a single line) may suffice. In others, it may be necessary to run traverse lines with multiple points. The choice of initial reference points is an arbitrary decision on the part of the mapper. Position the points so that you can see the greatest number of things to be mapped from them. You also need to have a clear line of sight from one point to the next.

Consider the example of the site area shown in Figure 6.52. Points A and B are situated so that there is a clear line of sight between the two, and they provide access to all of the features to be mapped. The points themselves can be a surveying pin or a wooden stake in the ground with a nail or tack in it to serve as the precise point. For points that might have to be re-occupied over the course of a mapping operation, or that will be in use over a long period of fieldwork, it is best to use stakes. They won't be as easily displaced or disturbed as surveying pins or anything else that you might use to mark a point. Measure the distance between the points and the magnetic bearing of the line AB. Everything that we do from this stage forward will in some way be related back to this line and the points on it.

Set up your instrument over point A and sight to B with the horizontal circle/scale set to "0". B becomes the reference point off which angles will be turned to the natural and cultural features to be mapped, and A will be the point from which the distance to these features will be measured. To map the position of excavation unit #1, turn an angle off your reference point B to the northeast corner of the unit and measure the distance to it from A. Do the same for the other corners of the unit.

The number of shots that you need to take to be able to adequately render something on a map depends on its shape. Rectilinear forms like the excavation units or the foundation ruins are relatively straightforward; shoot all corners, endpoints, and angle points. Use your judgement to determine how many individual shots are necessary to depict curvilinear forms like the abandoned roadbed and the mound of rubble. Visualize the curving edge of the road or the mound as a series of dots connected by short, straight lines. Think of what you are doing as constructing a picture by connecting a series of dots. Some things, like the large oak trees, need not be recorded in excruciating detail. Turn the angle to the center of each tree and measure the distance to it. Estimate the diameter of the tree trunks and put them in your notes to be added to the map when drafted.

If you can't see everything that you want to map from your instrument setup at A, it will be necessary to relocate the instrument. Whenever you relocate, it must be to a point that is already mapped and related in some way to your existing mapping references. For example, you could occupy point B with your instrument, sight back to point A with the horizontal circle/scale set to "0", and turn new angles and measure distances to things like the large oak tree or the southern and eastern margins of the foundation ruin. To get sufficient shots to depict the horizontal extent of the mound of rubble, or to map excavation unit #4, you might have to set a new point (C), or run a traverse line with additional points. Pick a location for C that has a clear line of sight to A and also provides the needed views of the boundaries of the rubble mound and excavation unit #4. With the instrument set up over A, sight to point B with the horizontal circle/scale set to "0". Turn the angle off this point of reference to the position that you selected for point C. Record the angle and measure the distance from A to C. Now occupy C with the instrument and sight back to A with the horizontal circle/scale set to "0". Your new angles and distance measurements to the western edge of the rubble mound and unit #4 will be from these new points of reference.

You must keep detailed notes of this, or any other surveying operation that you perform. Given the necessary shifting to different reference points and the multiple measurements required to create a plan map, you can appreciate the potential for confusion without adequate notes. Make unscaled rough sketches of the area and things that you are mapping, and the location of reference points and instrument set ups. Describe your procedure. For every shot that you take, record where the instrument is located, what is being sighted, and the relevant angle and distance measurements (in many cases you will also be recording elevations). An example of notes related to the surveying for the plan map of our hypothetical site are given in Table 6.3.

Using a transit or theodolite provides the greatest degree of accuracy in this type of mapping operation. Their use has one disadvantage, however. You won't know if you have made errors in reading angles or distances until you plot the data to create the map. If possible, plot the points and draw the map before leaving the field. It might be difficult or impossible to return to the field when you've discovered mapping errors back in the office or laboratory. If you will be doing your plotting or map drawing in the lab or office, your notes,

TABLE 6.3

Example of Surveying Notes for Plan Map

Established reference points *A* and *B* for mapping site area containing structural ruins. *A* is located on the northern end of the site between unit 2 and the abandoned roadbed and is a wooden hub with nail. *B* is located on the southern end of the site between the spring and the large oak tree and is a wooden hub with nail. Distance between *A* and *B* is 43.5 feet (taped).

Set up over *A*, sighted to *B* for "0" reference point. Magnetic bearing of line *AB* is South 24 degrees East. All angles turned to the right. Distances determined with EDM.

AT	TO	ANGLE	DISTANCE
A	NE corner, unit 1	328 degrees, 15 minutes	29.70 feet
A	NW corner, unit 1	334 degrees, 30 minutes	25.95 feet
A	SE corner, unit 1	335 degrees	32.50 feet
A	SW corner, unit 1	341 degrees, 30 minutes	29.00 feet

and any electronically stored readings will be your only guide to unraveling any problems or inconsistencies that might be found. Keep good notes! A plane table and alidade allow you to create the map as shots are taken, but it does not provide the same degree of accuracy that can be achieved with a transit or theodolite.

If you are dealing with a relatively open area with terrain that isn't too uneven or sloping, you might consider gathering data for your plan map through triangulation using tapes. This process still involves the judicious placement of reference points or traverse lines to which subsequent measurements will be related. The method works because two straight lines can only intersect in a single point. If you know where each of the lines originates and the length of each line, their point of intersection can be determined. The points of intersection that we're interested in are positions in the connect-a-dot pictures that we want to draw of excavation units, roads, ruins, etc. Let's stick with the same site area that we examined above and consider, for example, how we might begin to map the abandoned roadbed (Figure 6.53).

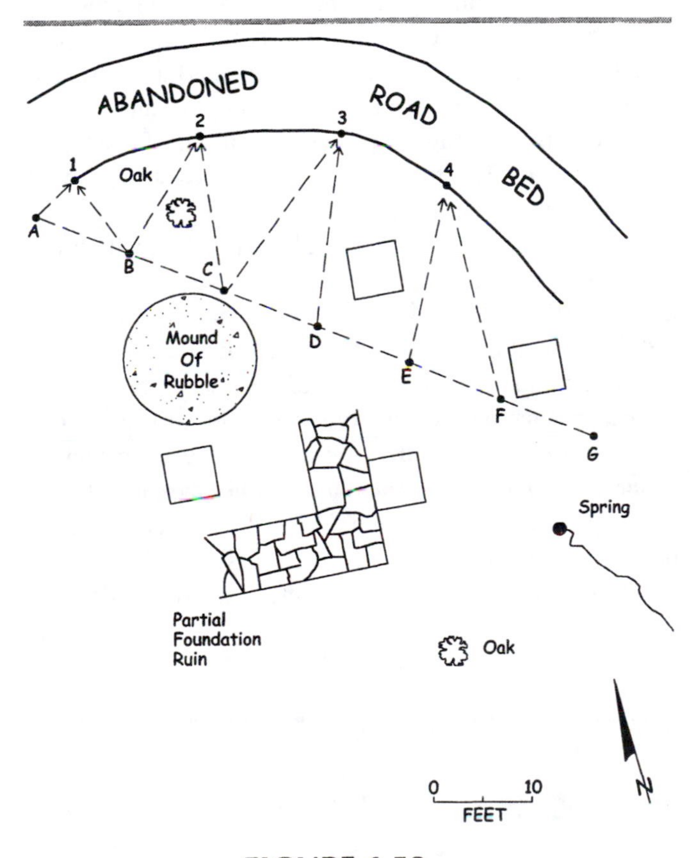

FIGURE 6.53.

Creating a plan map of a site area through triangulation. Distance measurements are taken to selected positions from reference points established along the taped line AG, chosen to provide easy access to the greatest number of natural and cultural features to be included in the plan. The distance to a selected position must be recorded from two reference points for triangulation. No angles need be measured in this mapping technique.

Mapping using triangulation requires using more reference points and taking more measurements than if you were using a surveying instrument. A line (AG) of reference points is established that provides access to the greatest number of features to be included in the map. The line is created by selecting its endpoints and pulling a tape between them to set additional points. The length of the line is known (60 feet), as is the distance of each of the points along it from either end. In this example the reference points are set every ten feet, but they can be set at any distance that makes triangulation of a particular point more convenient. Reference points need not be set at regular intervals, but you must know their distance from one end of the line or the other. Positions along the curve of the abandoned roadbed are selected for mapping and numbered for the purposes of keeping notes. As with any curvilinear form, the goal is to break the curve down into a series of connected straight lines. A distance must be taped from two reference points to the position along the curve to be mapped.

The remaining natural and cultural features of interest can be mapped in the same way. For each point needed to describe the feature's shape, you must have a distance measurement from two reference points. It may not be possible to complete the map with measurements off a single line of reference points. Any additional lines used will have to be triangulated off some portion of the original line AG, so that all measurements relate back to the same original standard, accurately preserving spatial relationships. Creating plan maps in this fashion may be necessary when working in remote areas where transporting heavy surveying gear is impractical (or when the battery pack of your total station dies!).

Plan maps are generally "floating," that is, they are internally consistent and accurate but are not inherently linked to something that you might find on a published topographic map. They should eventually, however, be linked to a site datum or some type of landmark or feature that can be relocated.

As a field technician, most of the mapping that you probably will do will be within excavation units, plotting the three-dimensional location of artifacts and features. Techniques used in this type of mapping are discussed in chapter 9 along with excavation techniques.

DETERMINING ELEVATION AND CREATING TOPOGRAPHIC MAPS

A transit, theodolite, or tripod-mounted level in combination with a stadia rod can be used to determine elevations, as can a total station and reflective prisms. Surveying for elevations is often combined with the operations for creating a plan map. The elevation of a point is its vertical distance above or below some arbitrarily defined standard or datum. One standard is sea level, and the elevations shown on published topographic maps are related to it. Archaeologists frequently establish an elevation datum or benchmark in the field, arbitrarily chosen to represent "0" elevation, to which all other elevations are related. These site- or area-specific elevation datums may then be tied into USGS, USPLSS, or other benchmarks or monuments whose elevation relative to sea level have been determined. This, in turn, allows the site- or area-specific elevations to be translated to elevations above sea level. Elevations may be determined for spot locations, for example, the elevation datums of individual excavation units used to record the vertical context of artifacts and features, or for the purposes of constructing a topographic map of the surface of an area. Techniques used for recording elevations within the context of individual excavation units are discussed in chapter 9.

Any operation to determine elevations requires working off a standard reference, either a point whose elevation has already been determined, or an arbitrarily chosen elevation datum or benchmark. An elevation datum or benchmark must be solid, firm, and hopefully removed from disturbances during the life span of its anticipated use. Concrete and pipe monuments capped with a brass plate on which a point or other data can be inscribed, like those used for professional surveying monuments, can be obtained from companies that sell surveying and engineering supplies. These are buried in the ground with their tops exposed above the surface. A length of pipe (e.g., 2–3 feet) filled with concrete and buried with its top above the surface would be an adequate homemade datum. A large metal spike driven into the base of a conveniently located large tree also might serve as a datum. In urban or developed areas, a designated point on a building or other structure might be sufficient. If possible, position an elevation datum so that all other elevations to be recorded will either be above it (above datum), or below it (below datum). This will minimize potential confusion and error when reducing instrument or stadia readings to actual elevations.

The most straightforward procedure when using an instrument and stadia rod is to work with the scope of the instrument in a level position. A reading necessary to calculate elevation is taken off the stadia rod where the middle horizontal hair of the instrument's scope intersects the rod. Figure 6.54 shows the initial steps in this procedure. The instrument is set up over a selected point (A) and the height of the instrument above this surface is measured. Instrument height is typically measured to a marked central point on the scope. The instrument's scope is leveled and sighted to the stadia rod positioned over the elevation datum ("0" elevation in this example). Using the middle horizontal hair of the scope, take a reading from the stadia rod (5.2 feet). This reading tells you that the level of the instrument's scope is 5.2 feet above the datum. Since you also know the height of the instrument (4.5 feet), the elevation of point A relative to the datum can be calculated: 5.2 – 4.5 = 0.7 feet above datum. The elevation of additional

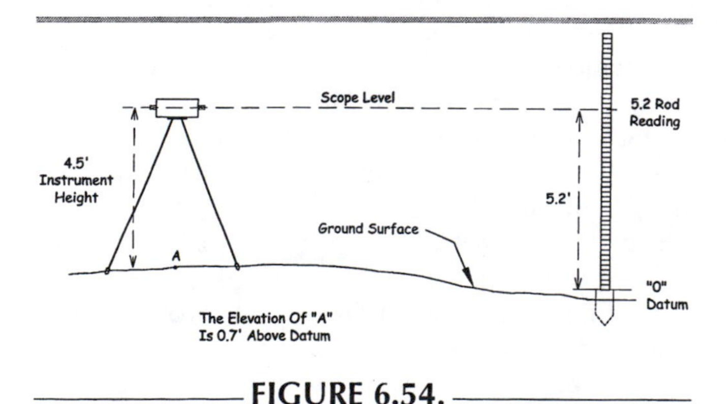

FIGURE 6.54.

Determining elevation with a stadia rod off an elevation datum or benchmark.

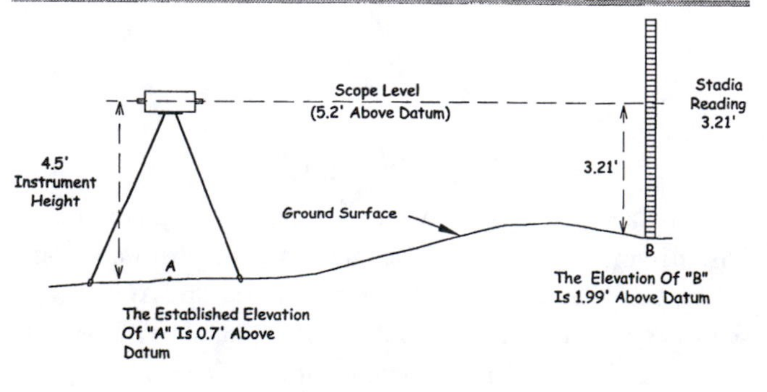

FIGURE 6.55.

Determining elevation with a stadia rod from a point of known elevation to a point of unknown elevation.

points can be determined from the instrument setup at A since you know the elevation of the level scope relative to the datum (Figure 6.55). If the position of A is marked in some way, it can be used in the future as a reference for calculating elevations because you have already determined its elevation relative to the datum. Always work from a known elevation to an unknown point, and every time you occupy a new position with your instrument, be sure to record the height of the instrument.

Elevations can also be determined using a stadia rod when an instrument's scope is inclined off the horizontal or level position. This is called ***trigonometric leveling.*** Data needed for the calculation include the height of the instrument, the elevation of the point where the instrument is set up, the vertical angle of the scope, the stadia reading (middle horizontal hair) and either the horizontal distance or slope distance from the instrument to the point of unknown elevation. Consult a text on surveying (e.g., Anderson and Mikhail 1998:169–173; Bettes 1992:68–70) for discussions of the formulas for making the calculation of elevation. Of course, total stations can generate required readings regardless of whether the scope of the instrument is leveled or inclined.

Collecting elevations sufficient to create a contour map of the topography is a relatively lengthy process. Topographic maps are useful or necessary when published maps are at a scale that masks or makes invisible the landform on which you are working. The maps are helpful in studies of geomorphology and site formation processes, regardless of whether the site is buried or exposed on the surface and are a must when the surface of the landscape is a cultural feature itself, e.g., earthworks, mounds, terraced gardens, or fields. Such maps are usually not created during the discovery or survey phase of fieldwork unless the landscape itself is an artifact, as I've just mentioned. Topographic maps are generally reserved for prolonged or intensive studies of localities.

The trick in planning the operations for a topographic map is deciding the level of detail needed. Do you need a map with a contour interval of 2 meters, 5 feet, 1 meter, or 1 foot to adequately characterize the landscape or surface features? For what purposes will the map be used? Is the purpose to document surface features that are artifacts of human behavior, make the correlation of stratigraphy between excavation units more understandable, or serve as a basis for unraveling site formation processes? Contour intervals need to be appropriate to the phenomena being mapped. In general, the smaller the contour interval that you intend to employ, the more shots you will have to take in the field. Remember, too, that you won't just be recording the elevation of specific points, but also the location of those points, since your ultimate goal is to depict changes in elevation over space.

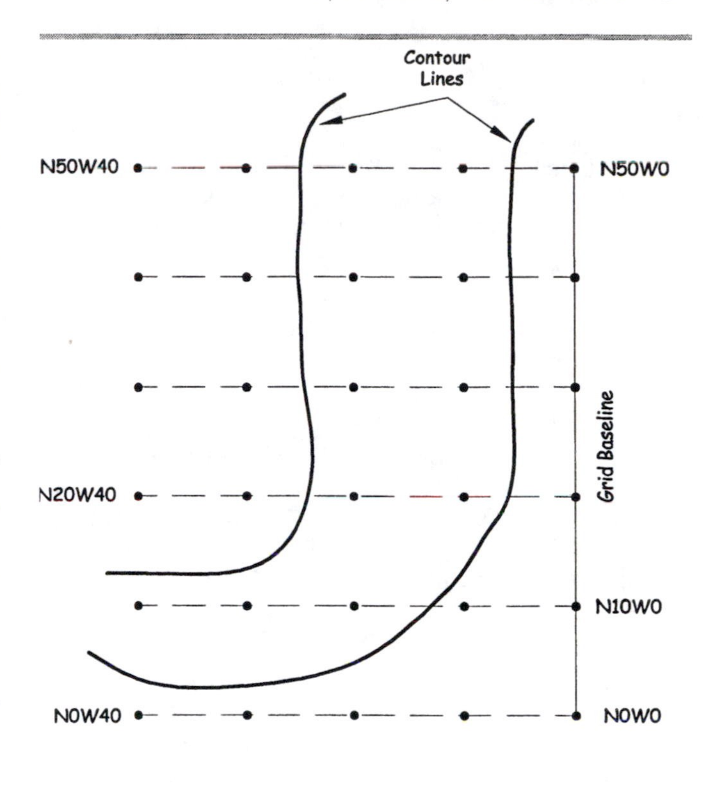

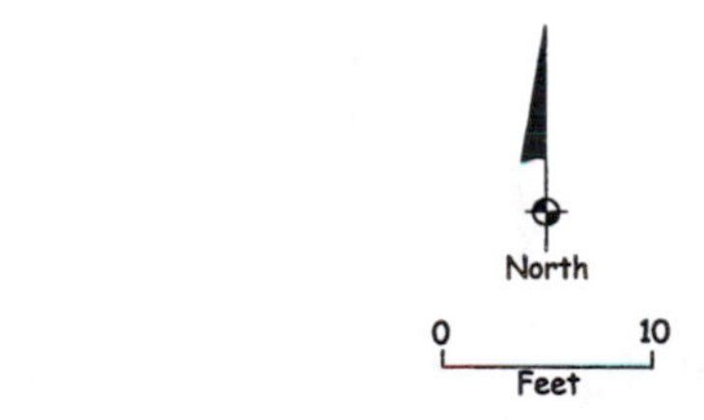

FIGURE 6.56.

A rectilinear grid system established for organizing the collection of elevations across an area. The intervals used in the grid should be appropriate for depicting the contours of the surface feature being mapped.

Rectilinear grids are often used to organize the collection of elevation data for contour mapping (e.g., Figure 6.56) and are constructed in the ways that we've already reviewed. The elevation of the points on the grid's baseline are determined with reference to an elevation datum or benchmark. Each point on the baseline is occupied with the surveying instrument and elevation readings are taken for the points on the line perpendicular to it. If large portions of the grid can be seen from just a few points on the baseline, it won't be necessary

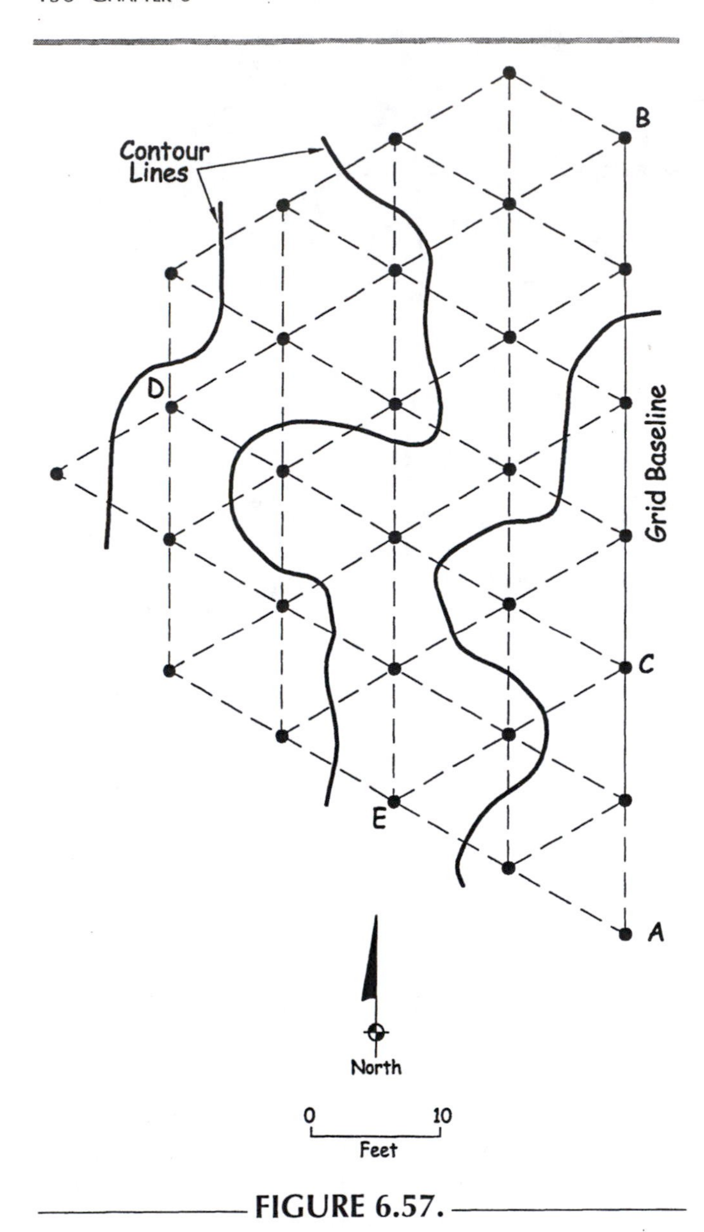

FIGURE 6.57.

A triangular grid system established for organizing the collection of elevations across an area. This type of system may make the interpolation of contour lines easier than a rectilinear one when topography is complex and varies over short distances.

to move the instrument as frequently. Make sure that you know the location of the grid points for which you are taking elevation readings.

A grid of equilateral triangles (all sides of same length, interior angles all 60 degrees) may be more useful when the topography of the surface to be mapped is complex and variable over short distances (Figure 6.57). The distribution of points on this grid system may make it easier to interpolate contour lines than what would be possible using a grid composed of squares (Bettes 1992:56). The procedure for laying out such a grid involves the initial creation of a baseline in the same way as for a rectilinear system, and using the ends of the baseline as points of reference in turning angles. The difference in creating a triangular grid is that rather than turning angles of 90 degrees off the baseline as you would for a rectilinear grid, the angles turned are 60 degrees. Plus, two grid lines can easily be laid out from the same position on the baseline of the triangular grid. For example, if your instrument is set up over point C, you can sight on point B with the horizontal circle/scale set to "0", turn 60 degrees to the left and set the points on line CD. While still occupying point C, you can sight on point A with the horizontal circle/scale set to "0", turn 60 degrees to the right and set the points on line CE. The distance between all points in the triangular grid is the same, just as in a rectilinear grid. The choice of what this distance or interval should be, of course, is yours.

Spot elevations can be taken for especially complex or noteworthy positions to enhance your map, even if you are working on a grid system. In fact, spot elevations alone can be used to generate a contour map if numerous shots are taken. But don't forget to determine the location of each point for which an elevation reading is taken!

A topographic map can also be created by "chasing contours" in the field. This means that you will use your instrument and rod, or total station and prisms, to deliberately seek out positions that have the same elevation relative to your benchmark or datum. Let's say that you want to chase an elevation of 3 feet above datum. The relative elevation of your level scope is 5 feet above datum. Therefore, any surface with a relative elevation of 3 feet above datum should have a stadia or electronically determined reading of 2 feet. Chasing contours requires constant and detailed communications between the person on the instrument and the person with the rod or prisms. As always, the position of each elevation reading must also be determined.

DRAFTING SURVEYING DATA

Actually drafting a map by hand with field data will make the utility of the various surveying operations that we've reviewed more apparent. You will literally be recreating on paper individual instrument setups, and from these locations turning angles and measuring distances, scaled to fit the desired format of the map. A protractor becomes the horizontal circle/scale of your

FIGURE 6.58.

Some tools useful in hand drafting maps: half- and full-circle protractors for measuring angles, a right triangle, drawing compass for triangulating distances to points, and an engineer's triangular scale for scaling distances for plotting.

instrument, a drawing compass the means for measuring distance in triangulation operations, and a triangular engineer's scale the means to scale down distance measurements (Figure 6.58). Any map that you draw, either in the field or the lab, should always have a title identifying its subject, date, north arrow, scale, and the name of the draftsperson. A key will be necessary to explain the use of symbols that are not conventional or self evident. Try using a protractor, drawing compass, and scales to recreate the surveying exercises discussed in this chapter.

The inspiring and enlightening experience of drafting a map by hand, however, is easily circumvented by the use of any number of mapping software programs on the market today. These programs allow you to manipulate and plot your data in ways that are either impractical or very difficult to do manually. Instrument- or GPS-derived data can be accommodated. In addition, rapidly advancing computer technologies and programs called ***geographic information systems*** (GIS) are being used by archaeologists and anyone who uses spatial data. GIS make use of computers to input, store, retrieve, analyze, and display any type of spatial data (Aldenderfer and Maschner 1996; Allen et al. 1990; Kvamme 1989b; Lock and Stanic 1995; Renfrew and Bahn 1996:83–84; Wescott and Brandon 2000). GIS are much more than systems capable of generating maps, and can play an important role in the interpretation of spatial data.

APPLYING YOUR KNOWLEDGE

1. Laboratory manuals that accompany many introductory textbooks on geology or physical geography contain exercises dealing with understanding and making topographic maps. One good example is:

   Strahler, Arthur N. and Alan H. Strahler. 1989. *Investigating Physical Geography: An Exercise Manual.* New York: John Wiley and Sons.

   Hone your skill at interpreting contour lines by drawing topographic profiles for a variety of landscapes, or visiting an area with the 7.5 minute quadrangle on which it is included. The DeLorme Mapping Company, Freeport, Maine, sells digitized 7.5 minute topographic maps by state on CD-ROM. The software included in this package is capable of generating a 3-D view of any chosen section of a topographic map. See suggestions in chapter 4 for sources of topographic maps.

2. Select an open area that gives you a line of site of at least 100 to 200 feet. Place a stake in the ground or otherwise mark a small spot over which you can stand as a point of reference. Pick something distinctive in the distance on which to sight, like a large tree, telephone pole, fence post, or building corner. Stand over your marked spot and take a compass reading to the designated target. Record your reading. Have others do the same. Try this with a variety of compasses if possible, especially ones that have built-in

Continued

level bubbles. Compare your readings and discuss what could account for the variation in them. Try the same exercise with a transit, theodolite, or total station.

3. This exercise works best with two or more teams of people. Its goal is to give you a sense of the replicability of compass-assisted mapping procedures. Each team develops a traverse for the other team to follow. Develop the traverse by picking a starting point and marking it with a stake or surveyor's pin, or anything that a person could use as a point of reference. Using a compass and tape designate a bearing and distance that will be traveled from the starting point. Execute this operation. From this second point, designate a second compass bearing and distance to be traveled. Execute this step. Do this two more times and mark the end of your traverse with a small object, nail, or pin that can be set in the ground and won't be visible at a distance. Give the instructions on how to follow the traverse to the opposing team, e.g., from the starting point go 50 feet on a bearing of North, 20 degrees East; from there proceed 100 feet on a bearing of North, 40 degrees East, and so on. The challenge is to see how closely the opposing team can come to the original endpoint. Try the traverse using a compass and measured pace, then using a compass and tapes. Compare the relative accuracy of the different methods.

4. Have your instructor select an area of approximately one acre on your campus. Construct a topographic map of the area employing an appropriately small contour interval (0.5, 1, or 2 feet). On your map also show trees or plantings, and any type of above-ground construction (e.g., fences, walkways, buildings, roads). As an alternative, construct a similar map of the yard or property surrounding the house or building in which you live.

5. Create a plan map of the trees or plantings, and above-ground construction in *Exercise 4* using fixed points and triangulation with tapes, rather than with a surveying instrument. Compare the accuracy of this map with the one generated using a surveying instrument.

DIG DEEPER

Construction and Use of Topographic Maps

- Thompson, Morris M.1988. *Maps for America: Cartographic Products of the U.S. Geological Survey and Others.* Reston, VA: U.S. Geological Survey. A thorough and clearly written introduction.

U.S. Public Land Survey System

- Bureau of Land Management. 1973. *Manual of Instructions for the Survey of the Public Lands of the United States.* Washington, D.C: U.S. Government Printing Office. Archaeologists working in states that employ the Public Land Survey System will want to check this out.

Using a Compass with Topographic Maps

- Jacobson, Cliff. 1999. *Basic Essentials of the Map and Compass.* Guilford, CT: Globe Pequot Press.
- Kjellstrom, Bjorn.1994. *Be Expert with Map and Compass: The Complete Orienteering Handbook.* New York: Maxwell Macmillan International.

Continued

❑ Many of the catalogs that offer compasses for sale also carry these and other publications that teach map and compass skills.

Introductions to Maps, Surveying, and Mapmaking for Archaeologists

- Bettes, F.1992. *Surveying for Archaeologists*. University of Durham, England: Department of Archaeology.
- Hogg, A.H.A.1980. *Surveying for Archaeologists and Other Fieldworkers*. New York: St. Martin's Press.
- Hranicky, William J. 1991. *Using USGS Topographic Maps*. Special Publication Number 20, Richmond: Archaeological Society of Virginia.
- Joukowsky, Martha. 1980. *A Complete Manual of Field Archaeology*. Englewood Cliffs, NJ: Prentice-Hall (see chapter 5, pp. 65–131).
- Napton, L. Kyle and Elizabeth Anne Greathouse. 1997. Archaeological mapping, site grids, and surveying, pp. 177–234. In Thomas Hester, Harry Shafer, and Kenneth Feder, *Field Methods in Archaeology.* Mountain View, CA: Mayfield Publishing Company.
- Spier, Robert F.G. 1970. *Surveying and Mapping: A Manual.* New York: Holt, Rinehart, and Winston. This book is interesting because of its emphasis on low-tech ways to do things in the field.
- Anderson, James and Edward Mikhail. 1998. *Surveying: Theory and Practice* (7th ed). Boston: WCB/McGraw-Hill. This is an extremely detailed and formal approach to surveying, a text used to teach those who will survey professionally.

GPS and Its Applications in Mapping

- Kennedy, Michael. 1996. *The Global Positioning System and GIS: An Introduction*. Chelsea, MI: Ann Arbor Press, Inc. A good basic introduction to the use of a GPS and its integration with GIS.

❑ Using GPS for Surveying

- Leick, Alfred. 1995. *GPS Satellite Surveying.* New York: John Wiley and Sons.

❑ Archaeological applications of GPS

- Brunswig, Robert H., Jr. 1999. *An Evaluation of Archaeological Applications of Mapping Grade Global Positioning Systems: Field Test in Northeastern Colorado's Plains and Mountains.* U.S. Department of the Interior, National Park Service, National Center for Preservation Technology and Training, Publication no. 1999-03.

GPS Tutorial

❑ The Trimble Corporation of Sunnyvale, CA, a supplier of GPS equipment, is focused on applying GPS to a wide variety of practical problems. They maintain a website with a tutorial in the nature and use of GPS at **http://www.trimble.com/gps/index.htm**> Remember that when using Internet resources, addresses can change. If this address doesn't work, use a search engine and the name Trimble or GPS to locate an appropriate website.

Continued

Equipment Suppliers

❑ Compasses, GPS receivers, and support equipment, and surveying instruments can be found in a variety of places. Check out the catalogs published by companies like Forestry Supplies (see chapter 5), or consult the yellow pages of your phone book. Search the Internet for companies that sell outdoors supplies, or engineering and surveying equipment, or search on the name of the piece of equipment or instrument in which you are interested.

Chapter 7

Sediments, Soils, Stratigraphy, and Geomorphology

Time is what keeps everything from happening at once.

Anonymous

INTRODUCTION

Archaeological evidence occurs on and within landscapes. It makes sense then to learn as much as possible about landscapes, what they are composed of, and the processes that shape them. This is part of understanding the context of archaeological evidence. It can be said that all archaeological fieldwork begins as a problem in understanding sediment, soils, stratigraphy, and geomorphology (Renfrew 1976:2). Few archaeologists become expert in all aspects of these topics, but everyone should be familiar with basic concepts and the language used by specialists like ***pedologists*** or soil scientists (those who study soils), ***sedimentary geologists*** (those who study sediments), and ***geomorphologists*** (those who study the shape of the earth's surface, its origins, and the processes that shape it). Archaeologists who develop expertise in one or more of these specialities are often grouped under the category of ***geoarchaeologists.*** In reality, geoarchaeology employs a variety of earth science techniques in the study of the archaeological record (Rapp and Hill 1998:1–17).

You need to be aware of the potential benefits of the analysis of sediments, soil, stratigraphy and geomorphology, and the types of issues and questions that guide analysis. If you are not, how will you formulate field strategies, collect pertinent data in the field, and use ongoing observations to adjust field strategies? At their most basic level, stratigraphy and the components of stratigraphy provide a means of organizing archaeological deposits and placing them in relative time. What artifacts and features are part of the same occupation or depositional episode; which are younger, which are older? Indeed, the use of stratigraphic excavations by archaeologists during the nineteenth century was one of the defining moments in the development of archaeology (cf., O'Brien and Lyman 1999; Trigger 1990; Willey and Sabloff 1993:97–108).

Beyond relative dating, the degree to which soils have been weathered, the number of individual soil sequences in a profile, and the landscape-forming processes that can be inferred from the characteristics of sediments and their layering all have implications for the age and dating of a specific stratigraphy. The organic component of buried strata can be subjected to radiocarbon dating or the procedure used to determine the oxidizable carbon ratio, which produces age estimates similar to those of radiocarbon. Some burned sediments and minerals can be dated with thermoluminescence, and sediments containing quartz grains that were exposed to sunlight provide dates through the technique of optically stimulated luminescence. Dates can be derived from the paleomagnetism of some sediments when calibrated with other chronometric techniques. The unique mineralogy and morphology of some sediments can be linked with specific events, like volcanic eruptions, whose age has been determined through other means (for summaries of dating techniques see Frink 1992, 1994; Renfrew and Bahn 1996:131–156; Taylor and Aitken 1997). In short, sediment, soils, stratigraphy, and geomorphology provide opportunities to tell both relative and absolute or chronometric time in relation to archaeological evidence.

The understanding of the natural world that comes from considerations of sediments, soils, stratigraphy, and geomorphology makes the recognition of some types of archaeological evidence easier. The identification of placed fill and humanly altered or created soils (***anthrosols*** or ***anthropogenic soils***), backfilled pits and excavations, and humanly altered landscapes is facilitated by this perspective. Understanding the sedimentary and stratigraphic signatures of the natural and cultural processes that have variably shaped and transformed the landscape is necessary in order to evaluate the integrity of archaeological deposits and contexts, and organize them for further analysis. The better our understanding of landscapes and their inherent variability, the better our ability to identify areas where the examination of surfaces might reveal traces of archaeological deposits, determine where on a landscape archaeological deposits are most likely to be buried and stratified, and determine the depth to which excavations must be taken to insure that all contexts that could contain artifacts and features have been examined. In other words, it impacts how we formulate field and excavation strategies. And finally, this understanding aids in the reconstruction of past environments, necessary background for understanding many of the choices and behaviors of the people that we study.

It's impossible to address any of these topics in detail in a single chapter in an introductory text. I'll focus on some of the basic vocabulary and procedures used by specialists. You need to be able to communicate with the specialists with whom you will work, ask appropriate questions, and be able to understand the answers given. If you can't perform the necessary studies and analyses yourself, what can you do to collect and describe data that will be useful to the people who will analyze them? An appreciation of how a knowledge of sediments, soils, stratigraphy, and geomorphology impacts field strategies will make the work that you are directed to do in the field more understandable, and thus more satisfying. The background learned from this chapter should also increase your ability to read and make use of published soil surveys and other environmental data. There are numerous texts dealing with the information that I will present. In particular I will be drawing from the work of Birkeland (1999), Brady and Weil (1998), Butzer (1976), Daniels and Hammer (1992), Foss et al. (1985), Hassan (1978), Hunt (1972), Stein (1987), and Waters (1992).

FUNDAMENTALS OF SEDIMENT AND SOIL

Uniformitarianism. Does the word conjure up spelling bee nightmares. The concept of ***uniformitarianism*** holds that the natural processes acting upon and affecting the earth today also operated in the past. Without it our understanding of the physical world of the past would be crippled. We observe stream behavior today, the variables that influence flooding, the creation of alluvial landscapes with their distinctive stratigraphies, and the nature of the sediments of which they are formed. We can use these observations to recognize ancient floodplains, make inferences about stream behavior, the frequency and energy of floods, and what all of this might imply about climate. So it is with any natural process that we can examine today or for which historical records are available. Uniformitarianism doesn't mean that the environment hasn't changed over time. It means that the rules governing the working of various natural systems are consistent over time. Our understanding of stratigraphy and geomorphology, and by extension their applications to archaeological research, hinge upon the concept.

SEDIMENT

Sediment has been defined in a number of fairly consistent ways:

- ☑ geologic materials formed by earth processes under ordinary surface conditions by the action of water, wind, ice, gravitation, chemical reactions, and biological organisms (Hassan 1978)
- ☑ solid inorganic or organic particles accumulated or precipitated by natural or human processes (Waters 1992)
- ☑ particulate matter that has been transported by some process from one location to another (Stein 1987)

The fragments of rock that detach from an outcropping of bedrock are sediments, the ash from a volcanic eruption is sediment, the silt left behind by flood waters is sediment, as is the sand washed up onto the beach by the ocean's waves and the salts that are left when the water evaporates. It also is useful to consider particulate matter that people create and move about, including artifacts, as sediments. They, too, are subjected to natural processes that can alter, transport, deposit, bury, and displace them. These processes in turn allow us to understand how an archaeological deposit and its stratigraphy have been formed and transformed.

Characteristics of Sediments

Sediments have a variety of characteristics or attributes that reflect their origin, mode of transportation, and deposition. Attributes that you reasonably will be able to deal with in the field include color, particle or grain size, texture, particle shape, particle orientation, and degree of sorting. Changes in these attributes over space (horizontal and vertical) make it possible to delineate stratigraphy and infer the geomorphic processes that were operative.

Color.

The color of a sediment reflects the parent materials from which it was derived, their composition and mineralogy. It can also reflect: staining or coatings of minerals and clay if the sediment was once part of a developing soil prior to being eroded, transported, and redeposited; and poor drainage conditions where color is a result of the degree to which oxidizing or reducing chemical reactions have taken place. For example, a sediment could be red in color because the original rock from which it was derived was red, or it could be stained with iron and clay minerals because it was once part of a developing soil.

Color is determined using a ***Munsell Color Chart*** (Kollmorgan Instruments Corporation 1994). A book of charts consists of hundreds of standard color chips that are systematically arranged by the three dimensions that are used to describe color: ***hue, value,*** and ***chroma.*** Hue is the dominant spectral color and indicates the relationship of a color to red, yellow, green, blue, and purple. It is used as the basis for organizing pages of color chips, their corresponding Munsell designations, and descriptive terms (Figure 7.1). Value refers to the lightness or darkness of the color, and chroma refers to the strength of this lightness/darkness relative to a neutral standard. A Munsell description of the color of a sediment or soil has two parts, a Munsell notation consisting of numbers and letters, and descriptive terms, e.g., 10YR5/8, yellowish brown.

Using the color charts is relatively simple. Begin with a small quantity of sediment on the end of a trowel or in a spoon. The sample should be relatively moist, or approximating what is called ***field capacity,*** the maximum amount of water that can be held in the pores of a sediment or soil without downward flow due to gravity. In other words, the sample shouldn't be so wet that water drips from it, or can be squeezed from it. Depending on field conditions, you may or may not have to deliberately moisten the samples that you are describing.

Find the basic hue page that you think will encompass the color of the sediment to be identified. Beneath each color chip is an open "window" or cutout in the page. In direct sunlight, with light coming over your shoulder, position your sediment sample behind the windows until you find a reasonable match. Rarely will you find an identical match. Choose the chip that most closely resembles your sample. Once you have chosen an appropriate color chip, consult the opposing page with the color name diagram to complete the process. The location of the color chip corresponds with the same location on the name diagram. For example, let's say that your sediment matches the color chip that is in the bottom row of chips at the far right shown in Figure 7.1. The chip is part of the hue category 10YR. Consulting the color name diagram we see that it has a value of 2, and a chroma of 2. The term "very dark brown" is used to describe it. The complete Munsell notation for the color of the sediment would therefore be 10YR2/2, very dark brown.

The overall color of a sedimentary deposit or layer may not be consistent and could include ***mottles*** or patches of contrasting colors. You must describe mottling as well as the dominant color of the sediment or soil. The color of mottles is described using the Munsell charts, but it is also necessary to record the abundance of mottles as a percentage of the exposed area deposit you are examining, and their size. Charts for assessing the proportions of mottles are included in Munsell books.

When using a Munsell color chart be careful not to dirty the color chips. This will alter their usefulness. In order to make color determinations more consistent, the same person should make the color comparisons whenever possible. If you're working in teams, designate someone as the color person. Although Munsell charts serve to standardize the way in which researchers talk about color, individuals perceive color in slightly different ways from one another. Munsell charts are also available for describing the color of rock and vegetation.

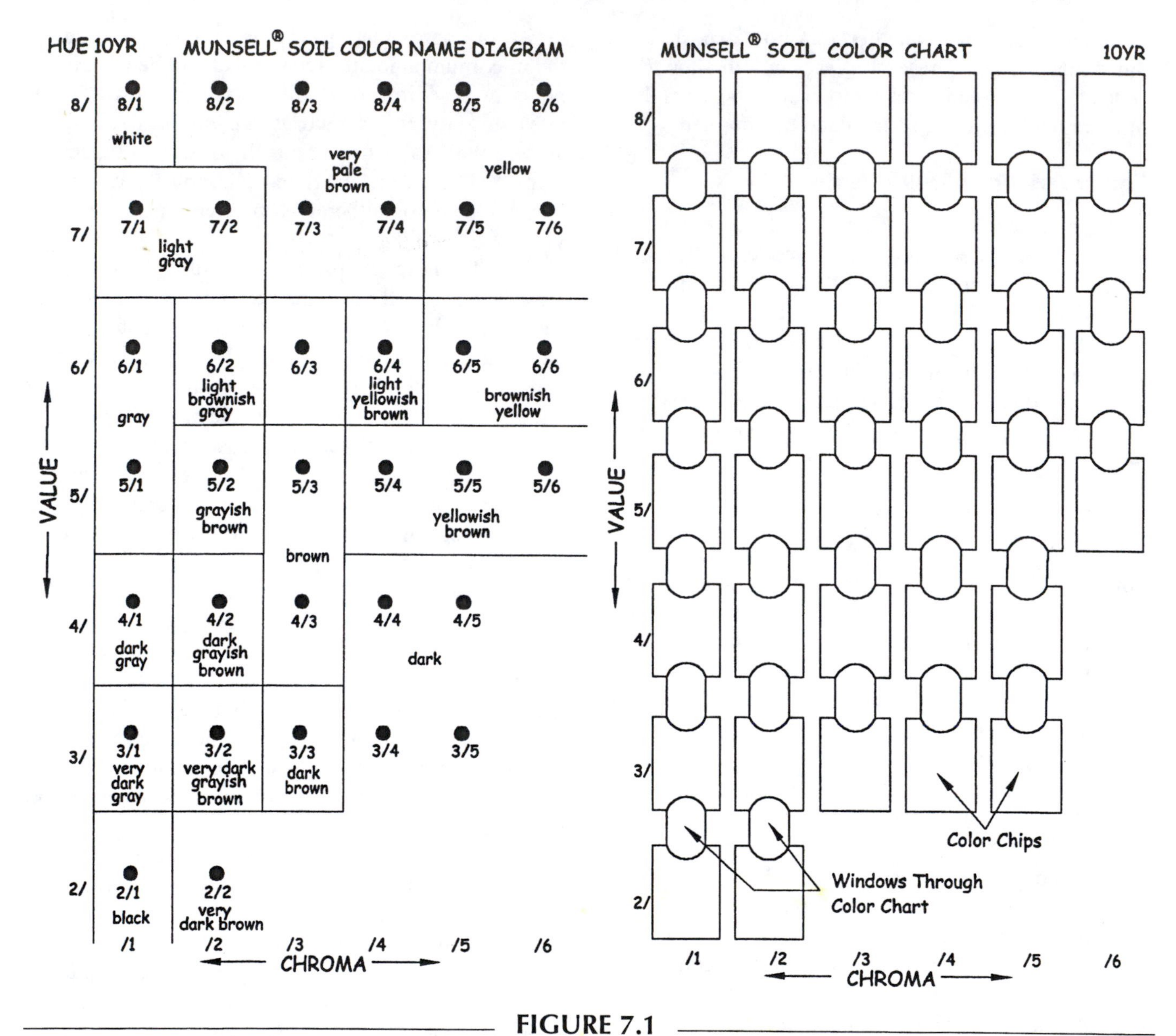

FIGURE 7.1

An example of a Munsell® Soil Color Chart and corresponding soil color name diagram (Macbeth Division of Kollmorgan Instruments Corporation, New Windsor, New York). The chips are colored in actual charts.

Particle or grain size.

The size of individual particles or grains of sediment is graded according to the ***Wentworth Grain Size Classification*** shown in Table 7.1. This classification applies to ***clastic*** sediments, those composed of mineral grains or derived from pre-existing rock, and most frequently encountered at archaeological sites. Notice that several terms used in general conversation, like gravel (Figure 7.2) or clay, have precise technical definitions. Particle size can reflect the energy of the process responsible for the deposition of sediments. The larger the particle size, the greater the relative amount of energy needed to transport it. The nature of the process itself also may be inferred, e.g., glaciers and high-energy landslides are capable of moving boulders, while weak floods are not; a weak flood may transport and deposit clay-sized particles whereas sand, silt, and clay may be associated with a high energy flood.

Texture.

The texture of a particular sediment (or soil) is determined by the mix of particle or grain sizes. The degree to which grains of a particular size are represented

TABLE 7.1

Wentworth Grain Size Classification for Sediments

Size Class	Measurement in Millimeters
Boulder	256–4096
Cobble	64–256
Pebble	4–64
Granule	2–4
(all of the above constitute gravel)	
Very coarse sand	1–2
Coarse sand	0.5–1
Medium sand	0.25–0.50
Fine sand	0.125–0.25
Very fine sand	0.0625–0.125
Coarse silt	0.0312–0.0625
Medium silt	0.0156–0.0312
Fine silt	0.0078–0.0156
Very fine silt	0.0039–0.0078
Clay	less than 0.0039

FIGURE 7.2.

Size classes of gravel: upper left, granules, 2–4 mm; lower left, pebbles, 4–64 mm; right, cobbles, 64–256 mm.

defines its textural or size class. Two different schemes of textural classification exist (Figures 7.3 and 7.4). The Folk classification is designed to describe all manner of sediments, from the finest particle or grain size to the coarsest. Descriptive terms are assigned based on the percentage of gravel present, and the ratio of the total amount of sand relative to silt and clay. Figure 7.3a depicts the scheme employed when a sediment contains gravel. The diagram shown in Figure 7.3b is used when gravel is not present in the sediment being described.

The USDA textural classification of soils can also be used to describe sediments. Unlike the Folk classification, however, it does not consider the amount of gravel that a sediment may contain. Instead it employs a range of adjectives to modify a textural class if it contains particles coarser than 2 millimeters, but only if the particles make up more than 15 percent of the sediment by volume. ***Gravelly*** is used to modify the textural class when particles between 2 millimeters and 7.6 centimeters are present, ***cobbly*** when they range between 7.6 and 25 centimeters, and ***stony*** or ***bouldery*** for larger-sized particles. A variety of additional modifying terms may be encountered in published soil surveys. Terms like channery, cherty, slaty, and shaly are used to describe the size and shape of gravel-sized particles that are not rounded and consist of specific materials like sandstone, slate, shale, limestone, and schist. Although the Folk classification scheme is more comprehensive, archaeologists tend to use USDA classifications more frequently, probably because of the relative ease with which assessments can be made in the field.

Texture, like particle size, can be indicative of the processes responsible for the deposition of sediments, and the relative energy of depositional events. The energy of depositional events have implications for the integrity of artifact deposits. For example, if an event is capable of transporting gravel-sized particles, it is also capable of transporting and redepositing smaller-sized particles, including artifacts. An event transporting and depositing clay-sized particles may have little detrimental effect on the surfaces or artifacts over which it passes.

The texture of a sediment or soil can be determined in a variety of ways. The most accurate is to take samples in the field and have particle size analysis performed in the laboratory. In the field you have two basic options. The first option is to collect and air dry a sample of known volume that you want to identify, then pass it through a series of graduated sieves (Figure 7.5). Ideally, this will enable you to separate the different size classes of gravel and sand, and to isolate silt and smaller-sized sediments from the rest of the sample. The method

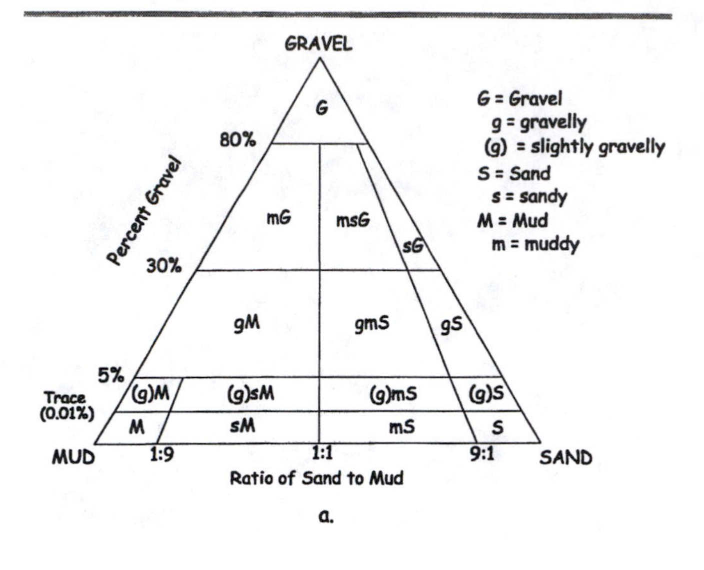

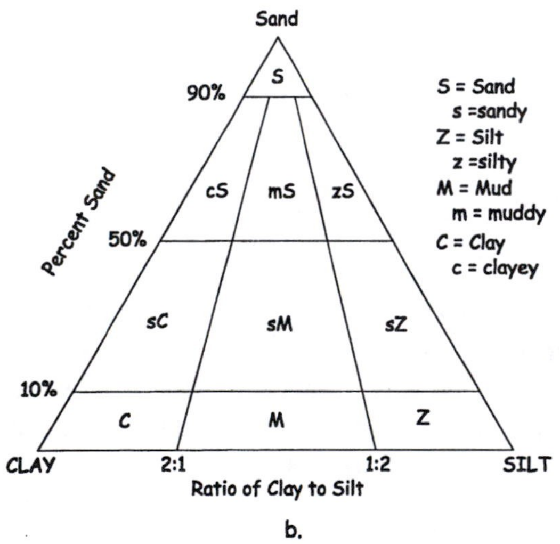

FIGURE 7.3.

The Folk Textural Classification of Sediments (Folk 1954, 1974). When gravel is present in a sample a. is used for the classification, b. for all other circumstances.

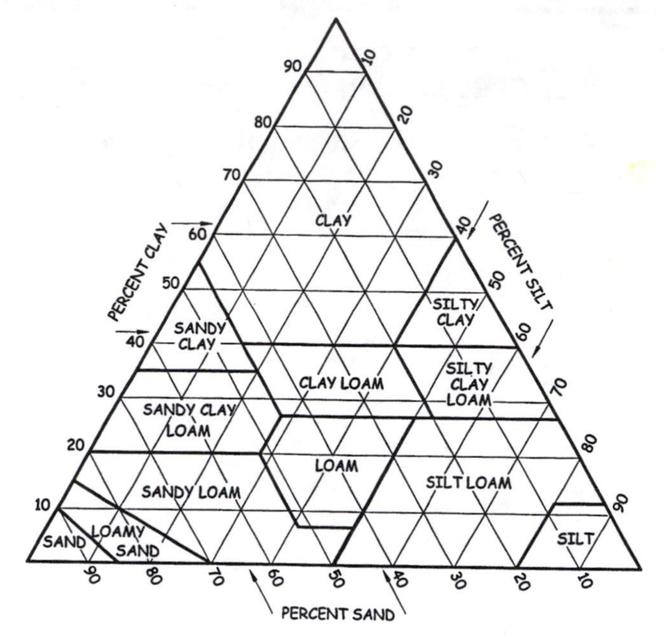

FIGURE 7.4.

The United States Department of Agriculture Soil Texture Triangle (Soil Survey Staff 1951).

is incapable of segregating grades of silt and clay, so some portions of a sample may still need to be examined in the lab. This method is also difficult to use when the sediments being examined are part of a weathered soil. Soil-forming processes can result in particles adhering to one another, forming masses that will be difficult to sieve unless broken up or milled in some way.

Field Characterization of Soil or Sediment Texture.

Table 7.2 outlines characteristics that can be used for estimating the textural class of sediments and soil in the field. However, local variations in organic matter content, clay mineralogy, parent material, and other factors may need to be considered and adjustments made (Foss et al. 1975).

Begin with a handful of the sediment or soil to be characterized (Figure 7.6). Slowly add water to it as you knead it in your hand to break down any chunks or aggregates. A squeeze bottle of water is helpful in this operation. Don't moisten the sample to the degree that water drips or can be squeezed from it. The sample should be moist enough to be moldable or plastic. If the sample becomes too wet, knead additional dry sediment into it. When the sample has reached the desired level of moistness, record your impression of the way that it feels. Do individual grains catch or drag across your fingers as you rub the sample (gritty)? After you apply pressure to the sample with your fingers does the sediment or soil adhere to them and stretch somewhat when your fingers are separated (sticky)? The feel of the sample will be tested again under wetter conditions following a few other observations.

Using the palms of your hands, roll the sample into a small ball. The ability to be molded into a ball and the stability of the ball reflect the percentage of silt, and especially clay in a sample. The higher the clay fraction in the sample the more moldable and stiffer it will be. Extreme percentages of clay in the sample will make the sample so stiff that it will be relatively difficult to mold. Of course, a sample that is composed primarily

TABLE 7.2

Field Criteria Used in Determining Textural Classes

Textural Class	Feel When Moist	Forms Stable Ball	Ribbons Out	Soils Hands	Plastic Properties	Sticky	Consistence Moist	Consistence Dry
Sand	very gritty	NO	NO	NO	NO	NO	loose	loose
Loamy Sand	very gritty	NO	NO	YES, slightly	NO	NO	loose	loose
Sandy Loam	gritty	YES, but easily deformed	YES, poorly formed with dull surface	YES	NO	NO	very friable	soft
Loam	gritty	YES	YES, poorly formed with dull surface	YES	YES, slight	YES, slightly	friable	soft
Silt Loam	velvety	YES	YES, poorly formed with dull surface	YES	YES, slight to moderate	YES, slight to moderate	friable	soft
Silty Clay Loam	velvety and sticky	YES, very stable	YES, well formed with shiny surface	YES	YES, moderate	YES	friable to firm	slightly hard
Clay Loam	gritty and sticky	YES, very stable	YES, well formed with shiny surface	YES	YES, moderate	YES	firm	slightly hard to hard
Sandy Clay Loam	very gritty and sticky	YES, very stable	YES, well formed with shiny surface	YES	YES, moderate	YES	friable to firm	slightly hard to hard
Silty Clay	extremely sticky and very smooth	YES, very resistant to molding	YES, well formed with very shiny surface	YES	YES, strong	YES, very sticky	firm to extremely firm	hard to very hard
Clay	extremely sticky	YES, very resistant to molding	YES, well formed with very shiny surface	YES	YES, strong	YES, very sticky	firm to extremely firm	hard to very hard

After Foss, Miller, and Segovia (1985).

FIGURE 7.5.

Examples of standardized geological sieves for separating particle sizes in a sedimentary deposit. Each is about the diameter of a small dinner plate.

FIGURE 7.6.

Quantity of sediment or soil used to begin field characterizations of texture.

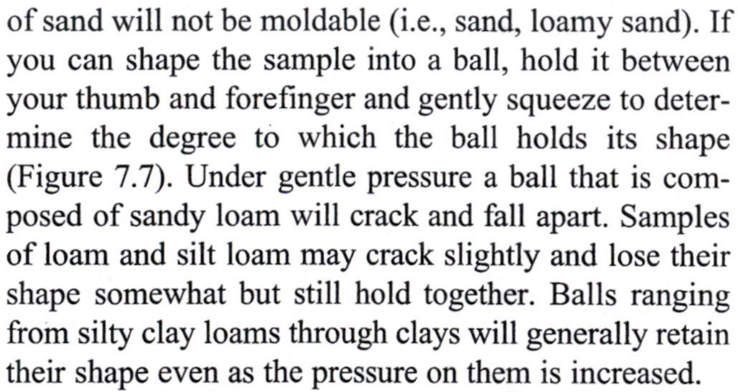

of sand will not be moldable (i.e., sand, loamy sand). If you can shape the sample into a ball, hold it between your thumb and forefinger and gently squeeze to determine the degree to which the ball holds its shape (Figure 7.7). Under gentle pressure a ball that is composed of sandy loam will crack and fall apart. Samples of loam and silt loam may crack slightly and lose their shape somewhat but still hold together. Balls ranging from silty clay loams through clays will generally retain their shape even as the pressure on them is increased.

Reform your sample ball. To ribbon out the sample, hold the ball between your thumb and forefinger. Using your thumb, push upwards to thin the ball into a flat sheet or ribbon (Figure 7.8). The ability of a sample to ribbon out is related to the percentage of clay-sized particles that it contains. The appearance of the surface of the ribbon, dull, shiny, or very shiny, also reflects clay content. A well-formed ribbon is one that holds together and has no surface cracks.

The ability of a sample to "soil" or dirty your hands relates to its organic content and/or the presence of silt- and clay-sized particles. The smaller-sized particles adhere to your skin more easily when moist than larger-sized particles of sand.

The feel and degree of cohesion or adhesion that particles exhibit is termed ***consistence.*** The higher the percentage of small-sized particles in a sample the greater the potential cohesion because of the tension created by moisture in the pore spaces between particles and the properties of clay minerals. This characteristic is typically used to refer to soils rather than sediments, as it is an indicator of structural development and soil-forming processes. Field observations of consistence are made under moist, wet, and dry conditions. Plastic properties and stickiness are aspects of consistence. Testing a sample for moist or dry consistence requires using sediments that you have not broken up or kneaded. For moist consistence the sample should be at or below the moisture level employed when doing the stable ball test. Use the following criteria to characterize ***moist consistence*** (Foss et al. 1985):

- ☑ loose or noncoherent—sample exhibits no cohesion, does not hold together in a mass
- ☑ very friable—sample crushes under very gentle pressure between thumb and forefinger
- ☑ friable—sample crushes easily under gentle to moderate pressure between thumb and forefinger
- ☑ firm—sample crushes under moderate pressure between thumb and forefinger but resistance is noticeable
- ☑ very firm—sample can barely be crushed between thumb and forefinger
- ☑ extremely firm—sample cannot be crushed between thumb and forefinger, greater pressure is required

Now knead and wet the sample to a greater degree than when you first started, or you can simply excessively wet the original sample with which you performed the stable ball test. Reassess your

(*a*)

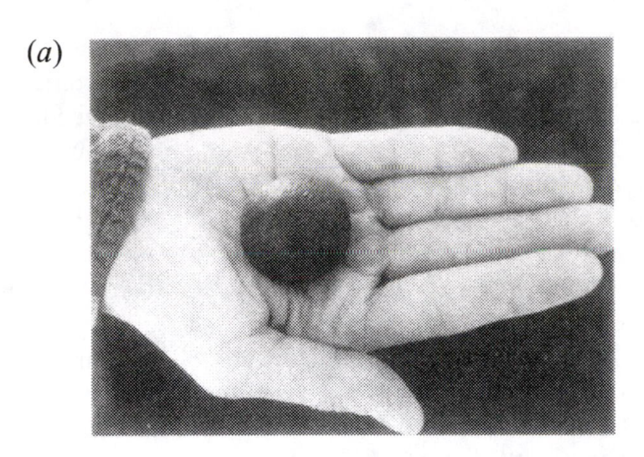

(*b*)

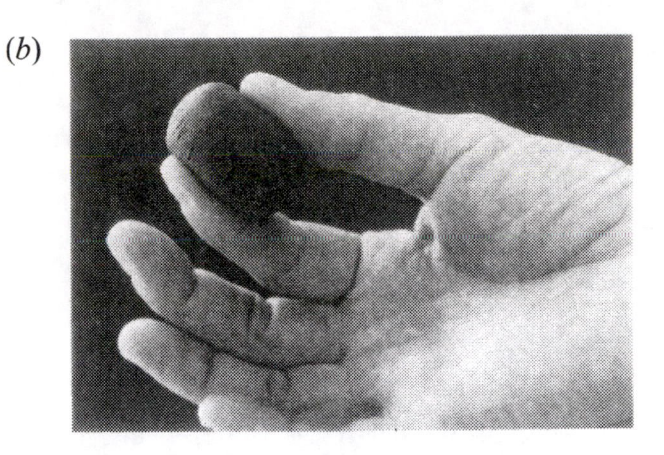

FIGURE 7.7.

Testing the ability of the sample to form a stable ball in the field characterization of texture. This sample formed a very stable ball (*a*) that deformed under moderate pressure (*b*).

characterization of its feel and its ***stickiness*** according to the following criteria (Foss et al. 1985:17):

- ☑ nonsticky
- ☑ slightly sticky—after pressure is applied to the sample, sediment adheres to both thumb and forefinger but comes off one cleanly
- ☑ sticky (moderate)—after pressure is applied to the sample, sediment adheres to both thumb and forefinger and stretches somewhat before pulling apart from either finger
- ☑ very sticky—after pressure is applied to the sample, sediment adheres strongly to both thumb and forefinger and is definitely stretched when your fingers are separated

The ***plastic properties*** of a sample refer to the degree to which it can be molded or deformed before breaking or rupturing. This property reflects the percentage of clay in the sample. You already tested for this characteristic to some degree when you attempted to form the sediment into a ball. With the sample still wet, roll it between your palms to form a fillet or wire (Figure 7.9). By holding the ends of the fillet/wire and bending it, you can characterize its plastic properties as follows (Foss et al. 1985):

- ☑ nonplastic—wire can't be formed
- ☑ slightly plastic—wire can be formed; it is easily bent and will crack at the bend
- ☑ plastic—wire can be formed, but it requires moderate pressure to bend or deform and there are no cracks at the bend
- ☑ very plastic—wire can be formed, a great deal of pressure is required to bend or deform it, and there are no cracks at the bend

Collect another sample and let it air dry. Test the sample for dry consistence using the following criteria (Foss et al. 1985:14):

- ☑ loose or noncoherent—sample exhibits no cohesion, does not hold together in a mass

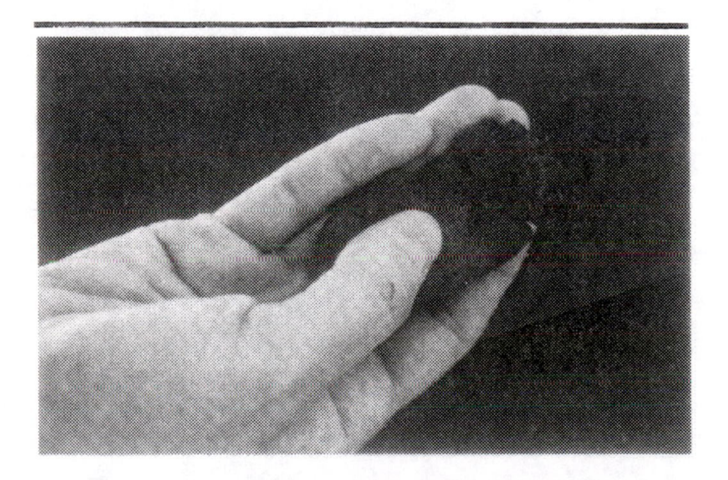

FIGURE 7.8.

Ribboning out a sample as part of the field characterization of its texture. This sample ribbons out moderately well and has a dull to slightly shiny surface.

- ☑ soft—sample is weakly coherent and fragile to the touch
- ☑ slightly hard—sample is easily broken between thumb and forefinger
- ☑ hard—sample can barely be broken between thumb and forefinger
- ☑ very hard—the sample cannot be broken between the fingers

The nature of the sand present in a sample can be determined using a 10 power hand lens and a sand gauge or sand-grain sizing chart. Sizing charts consist of actual samples of very coarse through very fine sand, and coarse silt, permanently attached to a card. They are often packaged in conjunction with visual guides for determining particle shape, roundness, and degree of sediment sorting. To decide what type of sand is in your sample, examine a small amount of sediment in the palm of your hand under a 10 power hand lens and compare it with the graded samples on a sizing chart. Hand lenses are also available that have incremental, measured circles imprinted on the reticule (comparators) that can also be used to identify the size of sand-sized particles. For sandy textural classes (sand, loamy sand, sandy loam) the predominant sand grain size is used as an adjective to modify the basic textural class. For example, if the sand in a sandy loam is mostly coarse, then the textural class would be described as coarse sandy loam. For other textural classes, if the fraction of sand in a sample is 25 to 50 percent then the adjective "sandy" gets added to the basic textural class, e.g., sandy clay loam.

The gravel content in a sample can be estimated by visual inspection or by passing a measured volume of sediment through graded sieves for a more accurate determination. When gravel makes up 15 to 35 percent of a sample by volume the adjective "gravelly" is added to the basic textural class, e.g., gravelly sandy loam. When percentages are between 35 and 60 percent, the adjective used is "very gravelly," and for higher percentages, "extremely gravelly" is added.

Assessing the field characteristics of a sediment or soil to determine texture will require practice and guidance from someone already familiar with the procedure. Deciding whether something is gritty versus very gritty, or whether the surface of a ribboned sample is shiny or very shiny is a relative judgement. It is best to practice assessing field characteristics with control samples, ones whose textural classes have been more objectively determined. You might want to include a laminated version of Table 7.2 in your field kit, along with an inexpensive sand-grain sizing chart.

Particle shape and roundness.

Descriptive terms for the shape of gravel-sized particles are organized by the quantitative relationship between the length of three axes of a particle (Figure 7.10): the ratio of the measurement of the intermediate axis (b) to the measurement of the longest axis (a), and the ratio of the shortest axis (c) to the intermediate axis (b). The four categories defined on the basis of this relationship are discoidal or oblate, spheroidal or equant, rod-shaped or prolate, and bladed. Visually, these categories conform to the shapes you would expect from the descriptive terms employed. Ideally,

(*a*)
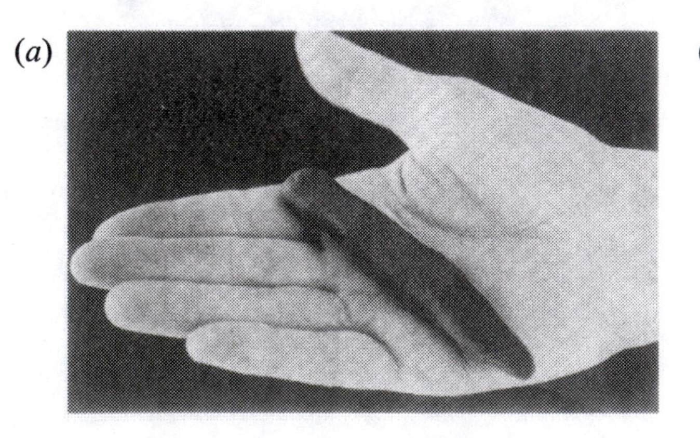

(*b*)

FIGURE 7.9.

Testing the plastic properties of a sample as part of the field characterization of its texture. The formed "wire" or fillet (*a*) is easily bent (*b*) and there are few cracks at the bend. It is slightly to moderately plastic.

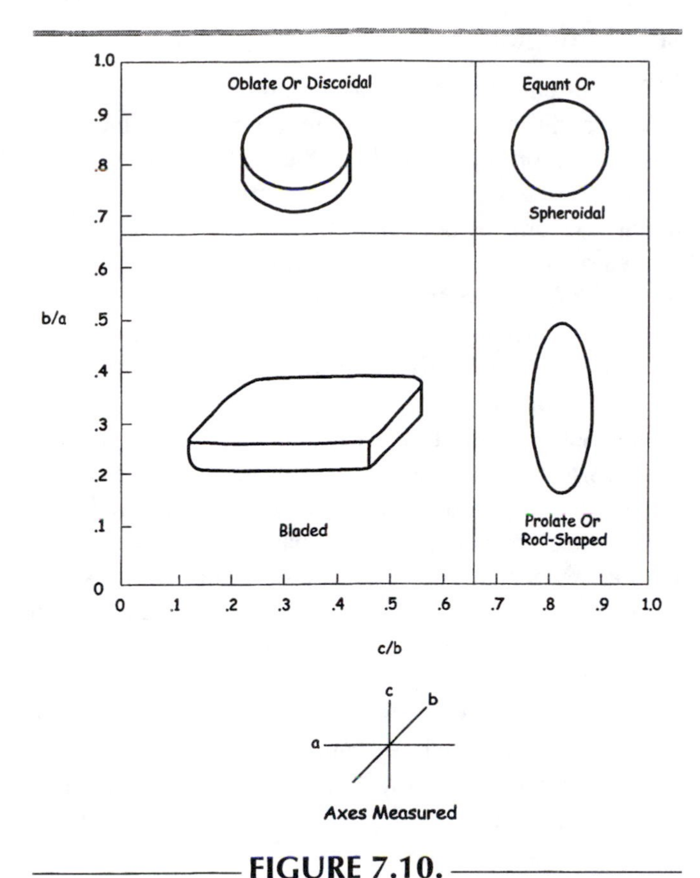

FIGURE 7.10.

The classification of the shape of gravel-sized particles (after Zingg 1935; also see Hassan 1978:Figure 6; Waters 1992:Figure 2.8).

the shape of gravel should be defined by measurements of samples, rather than simply by visual inspection. You won't be able to categorize the shape of particles smaller than gravel in the field. Roundness, or the degree to which the margins of a particle have been worn, is characterized by the visual guides shown in Figure 7.11. Field classification of this attribute will be limited to sand- and gravel-sized particles. The categories shown in Figure 7.11 can also be used to characterize the shape of sand-sized particles, as well as for describing their degree of roundness. Visual guides to particle roundness are often included with sand-grain sizing charts that can be obtained from companies that sell engineering, agricultural, and forestry supplies.

Particle shape and roundness are influenced by the nature of the parent material from which sediments have been derived, the nature and energy of the event that has transported the sediments, and whether they have been transported for short or long distances. For example, a rounded grain of sand could be the result of having weathered out of a sandstone, or extensive tumbling from being transported long distances by a stream.

Particle orientation.

The patterned orientation, or lack of patterning, of gravel-sized particles in a deposit determine their particle orientation. Particles may be aligned parallel to the direction of the stream flow which transported them, for example. Gravel transported and deposited by a landslide, in contrast, may have no single orientation. The patterned alignment of gravel-sized particles in some contexts could be the signature of human rather than natural activity. As should be obvious, particle orientation reflects the mode of transportation and deposition of sediments.

Sorting.

Sorting is characterized by the degree to which the sediments in a deposit are of the same size. Put another way, how many different size classes of particles account for the majority of the sediments (by volume) in a given deposit? Well-sorted deposits have particles that are mostly (68 percent or more) of the same size, or fall within a prescribed degree of variation or standard deviation (see Waters 1992:24–26). Particles of all sizes are evenly mixed in a poorly sorted deposit. Passing a measured volume of sediment through graded sieves is one way to assess the degree of sorting in the field. It can be estimated visually using a sorting guide that is often combined with grain sizing charts.

There are other characteristics of sediments that are worthy of attention for the information that they might provide on landscape and site formation processes, and stratigraphy. These include mineralogy, particle

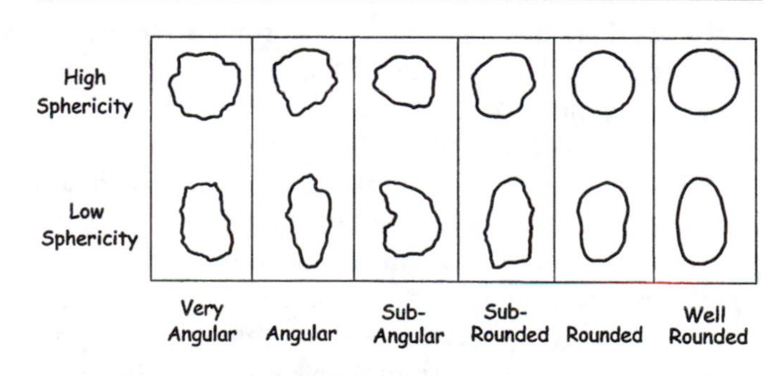

FIGURE 7.11.

The classification of the degree of roundness of sedimentary particles (after Powers 1953; also see Hassan 1978: Figure 7; Waters 1992: Figure 2.9).

fabric, pH and general chemistry, and sedimentary structures. Most of these characteristcs are best handled by a specialist.

Soil.

The terms sediment and soil are not interchangeable, as you can surmise from the following characterizations of ***soil.***

- ☑ weathered profiles developed by in-place physical and chemical alteration of pre-existing sediments; vertical sequences of distinctive horizons created by *in situ* weathering of a stable sedimentary deposit at the surface (Waters 1992)

- ☑ zones of sediment alteration at or near the surface of the earth influenced by biological organisms and climate (Hassan 1978)

- ☑ a natural body consisting of layers or horizons of mineral and/or organic constituents of variable thickness, which differ from the parent material in their morphological, physical, chemical, and mineralogical properties and their biological characteristics (Birkeland 1999)

- ☑ natural bodies in the upper portion of the earth's crust with unique morphologic characteristics resulting from past and present combinations of physical, chemical, and/or biological weathering processes (Brady and Weil 1998)

Soil is sediments that have undergone physical and chemical changes in place. This weathering or development on a relatively stable landscape results in the "unique morphologic characteristics" noted in the definitions above. All soils incorporate sediments, but not all sediments are soils. This relationship also means that soils share many of the same attributes ascribed to sediments, in addition to those that make them unique. A stratigraphic profile or sequence can contain both layers or strata of sediment, which are depositional phenomena, and layers or horizons of soil, which are developmental phenomena and result from *in situ* physical and chemical alteration. Their co-occurrence in a single profile reflects the dynamic nature of some landscapes, alternating between stability and instability or transformation. The terminology and classifications used by pedologists or soil scientists embraces sediments and depositional events that are not soils in the strictest sense of the word.

Soil Formation

The formation of soil is a function of four basic processes: additions of material to the ground surface, the transformation of materials at the surface and in other portions of a vertical section or profile, transfer of materials from one portion of the profile to another, and the removal of materials from the profile. The influence of each of these processes varies in importance from location to location.

Soil formation begins with one type of sedimentary deposit, or ***parent material.*** Parent material can be sediments newly transported into an area, or sediments created *in situ* as bedrock and gravel weather and are broken down into smaller-sized particles. Potential ***additions*** to the surface of this deposit of parent material include organic matter from vegetation and animals, ions and solid particles introduced with rainfall, and small amounts of particles transported and deposited by the wind. The additions of materials to an existing surface must not be too substantial in order for soil formation to proceed. If substantial deposits are added to an existing surface, they create a new surface and a different parent material on which soil-forming processes can act. In other words, for soil development to begin and leave its distinctive signs, a landscape must be relatively stable and not subjected to major erosion or deposition by water, wind, ice, volcanism, or the activities of humans.

As organic matter at a surface decays it breaks down into smaller-sized particles and organic compounds. Through mechanical, chemical, and biological weathering sediments near the top of the parent material, and those that may have been added to the surface, break down into smaller-sized particles. Primary minerals break down into secondary minerals and clays. Precipitation and surface water dissolve chemical constituents. These are the ***transformations*** that take place during the development of a soil.

Small-sized particles, especially clays, soluble organic material, and chemicals are moved downward into the ground by the action of water and gravity, and travel through the voids or spaces between pre-existing particles. They are deposited when the movement of water or the effects of gravity cease, when dehydration occurs, or when the chemical environment favors the precipitation of particles out of solution. The capillary action of water can also move materials upwards in the soil column in some instances. The action of plant roots, earthworms, burrowing insects and animals, and freezing and thawing cycles also serves to mix materials into different portions of the soil column. These processes can be grouped under the rubric of ***transfers.*** The movement and deposition of materials within the soil column is called ***translocation***. The portion of the soil column where translocated materials are deposited is the ***zone of illuvation*** or an ***illuvial horizon.*** Most illuvial horizons fall under the broader definition of B horizons in the classification of master soil horizons. ***Removals*** from the soil column consist of soluble material taken away by the action of groundwater.

Soil development is a near-surface phenomenon and is generally active to the depth that water penetrates a vertical section or column of sediments. The unique characteristics or properties of soil are a function of climate, organisms, topography, parent material, and time. The processes that I've summarized above form the soil while the factors of climate, organisms, topography, parent material, and time define the system within which the processes work.

Climate drives precipitation and temperature. Precipitation can add solid particles and ions to a surface, and as water supports the mechanical, chemical, and biological weathering of materials, the creation of solutions of organic compounds and chemicals, and the process of translocation. Climate has an influence on vegetation, which in turn contributes to the development of soil. Temperature has an influence on the mechanical and chemical weathering of materials and the behavior of substances in solution. Vary the climate and you vary the degree and rate at which some soil-forming processes occur.

Organisms (plants, animals, and humans) contribute particulate matter to the soil-forming process and through their activities can physically mix materials in a soil column or create conditions that alter the drainage characteristics of a landscape. Alter drainage characteristics and you influence all of the soil-forming processes that require water.

The topography of a landscape can have a dramatic impact on the amount of water moving through a soil system. Imagine a flat landscape with one margin adjacent to a slope. The portion of the landscape adjacent to the slope will have more water available for soil-forming processes simply because slopes funnel runoff from higher elevations downhill. Topography also influences the degree and frequency with which the wind, streams, seas, glaciers, and gravity can further sculpt the landscape. This is critical given the importance of landscape stability to soil-forming processes.

Don't forget that climate has an impact on the activities of wind, water, and ice! Parent material affects the initial color, chemistry, mineralogy, and texture of a soil, as well as the rate at which some soil-forming processes can occur. Some types of parent material are more resistant to weathering than others. Coarse-grained parent materials like sand and gravel have large voids or spaces between individual particles, which means that water, and everything that it may be transporting, can move through a vertical section quickly, favoring removals over translocations.

Finally, there is time. Soil-forming processes work over time. The longer a landscape remains stable, the more time for soil to form. The length of time that any factor influencing soil formation is held constant will affect the rate of soil formation.

I think that you can understand why it's called a soil system. Processes and factors are interrelated and create something that is more than the sum of its parts. Alter one part of the system and it affects other parts of the system. While the soil system may seem complex, it has some decided advantages for the practice of archaeology. Soil-forming processes alter the physical and chemical attributes of sediments. We can recognize and measure aspects of these attributes, and since they are time-sensitive phenomena we can use them in some cases to roughly estimate the age of portions of the soil column. The nature of soil development also reflects the relative stability of the landscape and the nature of the environmental setting.

Soil Characteristics

A review of some of the important characteristics or properties of soils follows. As with sediments, there are a number of significant characteristics of soils that could be discussed, but require laboratory analysis.

Soil color.

The color of soil can reflect the same things as the color of sediments, however, the color of soils can change as a result of the developmental process. Depending on the degree of soil development, the color of the original sediments can be obscured. The translocation of clay and iron minerals, and chemical reactions will alter the color of soils. The degree of organic matter in a soil will also have an influence on its color. Buried soils with high percentages of organic matter are useful for identifying former surfaces of the landscape. The color of soils is recorded in the same way as sediments, using a Munsell color chart.

Soil texture.

The examination of the texture of soils can provide many of the same benefits as those noted for the study of sediments. Like color, however, soil texture can change over time. The ongoing mechanical, chemical, and biological breakdown of materials contributes to this, as well as the translocation of materials through the soil column. The longer a soil weathers, the greater the chances of its texture being altered. The accumulation of clays in B horizons are helpful in estimating soil age. Soil texture is determined in the field using the procedures discussed above for sediments. Of course, submitting samples for the laboratory analysis of particle size is the most accurate way to determine texture.

Soil structure.

Aggregations of soil particles (those less than 2mm or gravel size) are called ***peds.*** The way in which peds or soil particles aggregate in masses or clumps, and the way that these masses separate from one another is called ***structure.*** Structure develops as a result of the natural cohesive properties of organic matter, silt, and clay-sized particles as they fill up the voids or pore spaces between individual grains. Particles can also be cemented by chemical or other precipitates. The more strongly developed the structure, the smaller the pore spaces between individual grains, and the more impermeable the soil. The field characteristic of soil consistence is an additional reflection of structural development.

Soil structure can be classified on the basis of the form/shape, size, and grade/stability of peds (Figures 7.12 and 7.13). Types of structure look like the terms used to describe them: ***granular, crumb, platy, blocky, subangular blocky, prismatic,*** and ***columnar.*** The difference between granular and crumb structures is porosity, or the relative abundance of pore spaces in the aggregate. Peds with blocky structure will not be perfectly square or cubelike. They are distinguished from subangular blocky peds by their sharp, angular margins. The primary difference between prismatic and columnar structure is the rounded nature of the tops of columnar peds. Structural types are bracketed by two structureless extremes, single grain and massive. ***Single grain*** means that there are no peds or cohesion of grains, like what you would see in a loose sand. ***Massive*** means that all of the grains of a deposit are cemented and constitute a cohesive mass that can't be broken down into peds. As seen in Figure 7.12, certain types of structures are often associated with distinctive soil horizons.

Evaluating soil structure.

There are two procedures that can be used in the field to evaluate the structure of a soil. Using the pointed end of a trowel or a knife, gouge portions of the exposed face of the soil that you want to evaluate. The shape of the casts visible in the gouged surfaces from which sediments have fallen can be compared to the diagrams of structural types. Second, remove about a palm-sized mass of soil, taking care not to break it up with your trowel or knife. With your hands, gently attempt to break the mass apart. With soils of granular or crumb structure you won't even get this far. What you will have in your hand is a pile of small sphere-like shapes (but not single grains) waiting to spill to the ground. The shapes of the other types of structure will become apparent as you pull the mass apart. Determine the grade or stability of the structure type by evaluating the gouged surfaces that you created and how firmly the peds in your hands hold together (adapted from Foss et al. 1985:12):

1. weak—poorly formed peds that are barely observable in place in the soil column and may not be evident in a mass of soil pulled apart in the hands

2. moderate—well-formed and distinct peds evident in place, and moderately durable in masses pulled apart by hand

3. strong—extremely well-formed peds easily observed and well defined in place, and very durable in masses pulled apart by hand

The size of peds can be assessed with reference to the scales shown in Figure 7.13. For granular and crumb structures the measurements refer to the diameter of the aggregate, for platy structures, the thickness of the ped. For blocky and prismatic structures, the peds must fit within one of the templates. Subangular blocky peds are measured using the blocky templates, and columnar peds using the prismatic templates. Munsell books also typically include size charts for different types of soil structure.

The surfaces of peds can become covered to various degrees with a coating of clay, referred to as ***clay films*** or ***clay skins.*** The translocation of clay and the coating of ped surfaces are time transgressive and, all other things being equal, reflect the amount of time that a soil has been developing. Continuous clay skins on ped surfaces represent the passage of more time than discontinuous clay skins. Once the clay skins have covered the entire surface of peds they will begin to thicken. The thickness of clay skins on peds is a criterion in assessing the degree of soil development and the passage of time. Given their typical association with A horizons and the effects of translocation, clay skins won't occur on peds with granular or crumb structures.

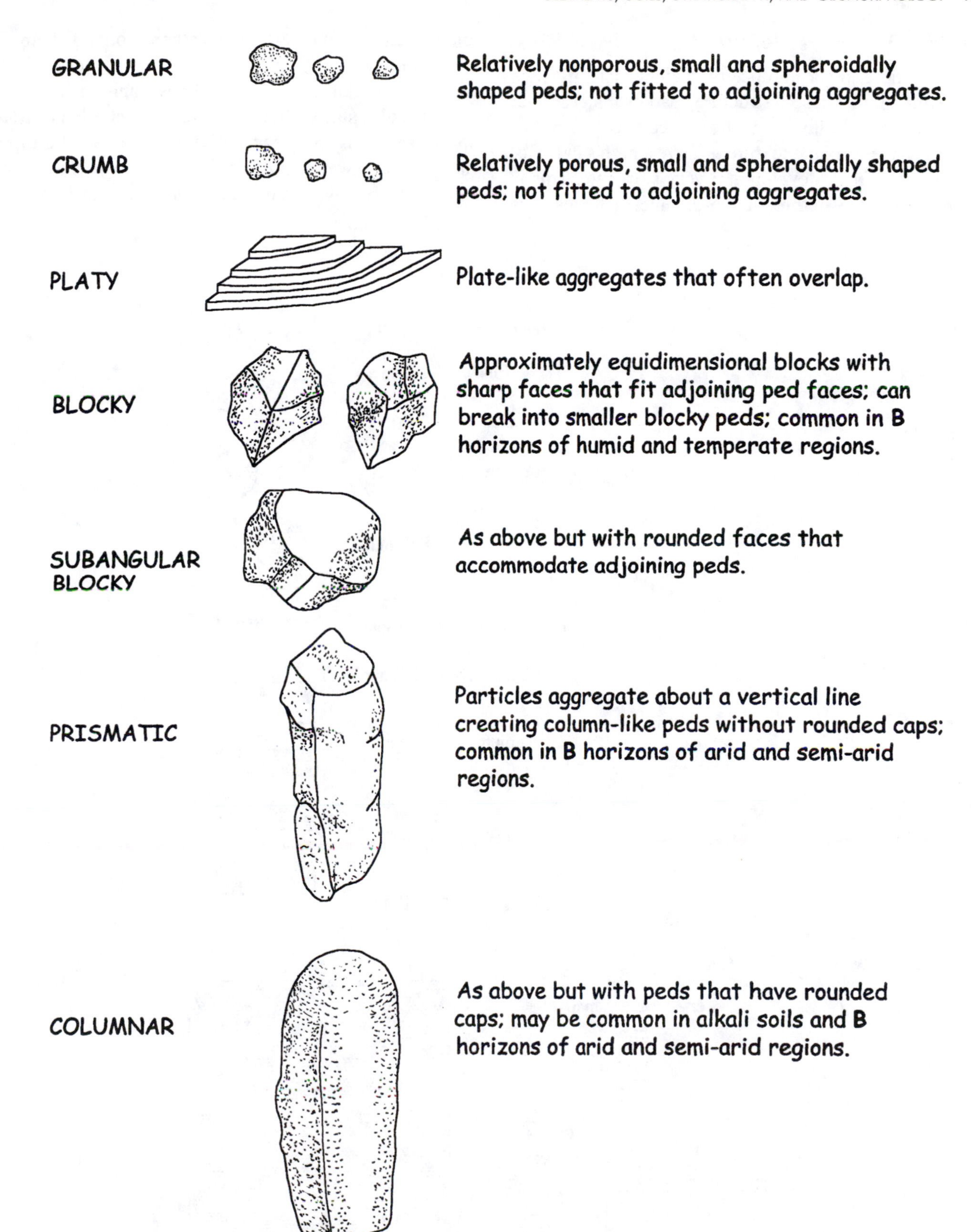

FIGURE 7.12.

Types of soil structure (adapted from Birkeland 1999; Foss et al. 1976; Soil Survey Staff 1951).

Formation of soil structure.

While there are a number of factors that influence the development of soil structure, parent material and the hydrology of a landscape bear special mention. For soils that have a coarse texture or have a high sand fraction, structure may develop at a very slow rate. This is due partly to the nature of the sediments themselves and what can and cannot be weathered out of them. If silt- or clay-sized particles can't be weathered from existing material, then structural development is inhibited. It is also slowed because of the large voids or pore spaces that exist between individual grains. The rapid movement of water through such a deposit may more frequently remove small-sized particles and materials

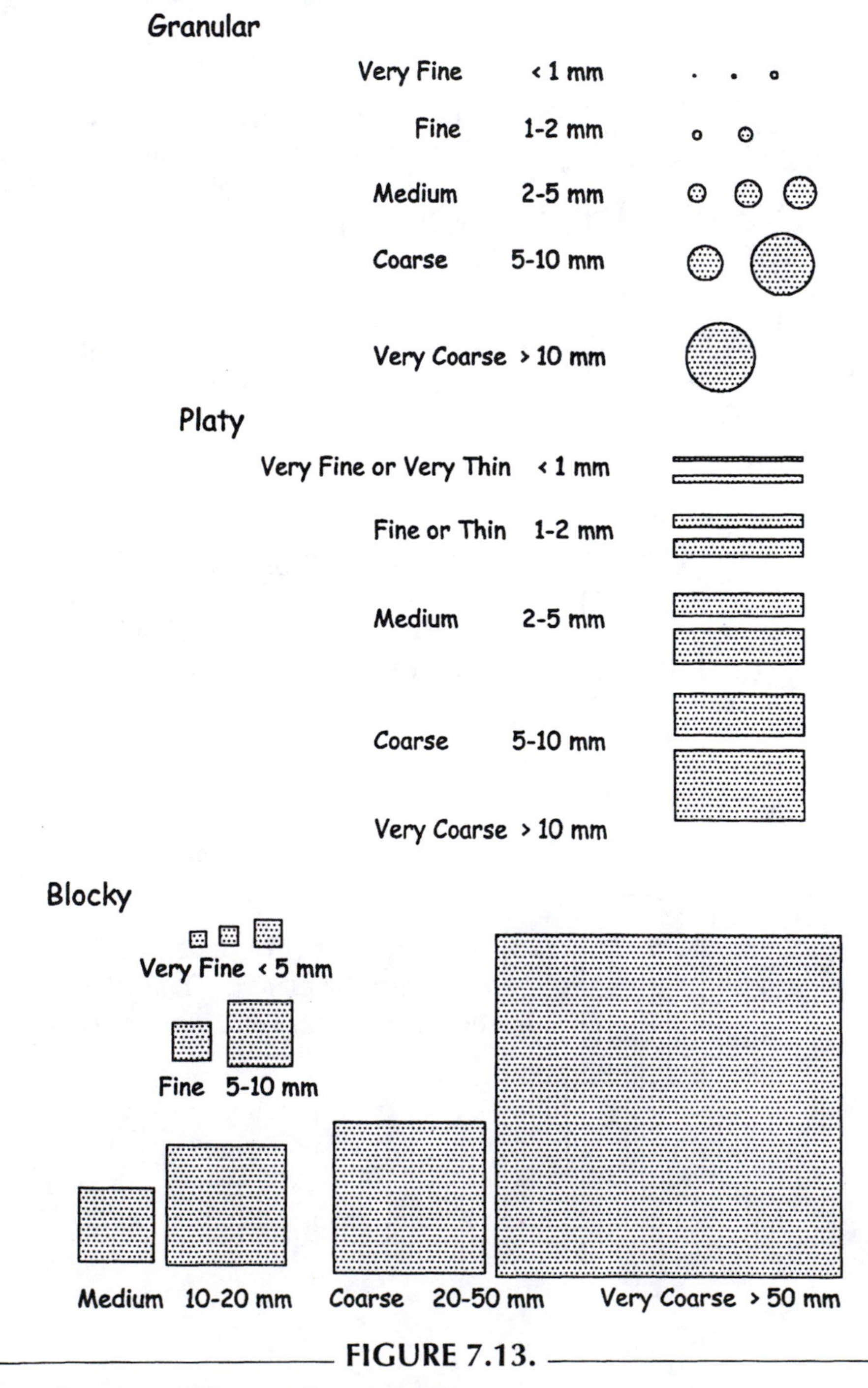

FIGURE 7.13.

Describing soil structure (from Foss, Miller, and Segovia 1985).

in solution than translocate them, plus the pore spaces that need to be filled are larger to begin with. Some sandy deposits may never exhibit structural development, even if they are quite ancient. In contrast, sediments that are already of a fine texture may express structure more rapidly because of the ready availability of small-sized particles that can be translocated and small pore spaces. As you might suspect, the hydrology, or water component, of the soil system also can dramatically influence the rate of structural development since it contributes to the weathering of sediments, the creation of solutions, and translocation. Dry climates or dry soil conditions inhibit the development of structure.

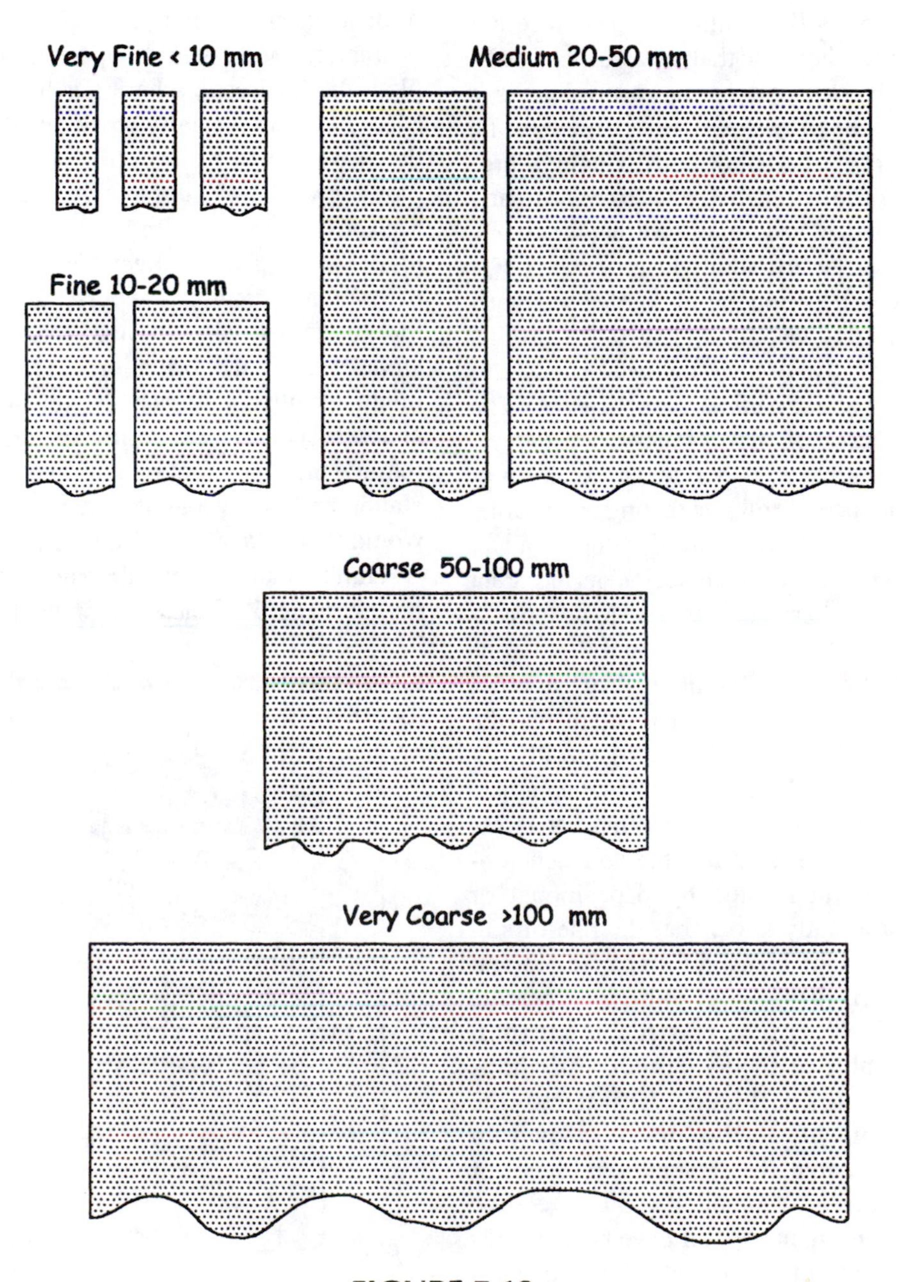

FIGURE 7.13.

Describing soil structure (Continued).

The degree of structural development can be an indicator of soil age and thus another source of information for evaluating the age of archaeological deposits. Since the size of voids or pore spaces is influenced by the degree of structural development, the structure of a particular layer or horizon of soil has implications for the integrity of archaeological deposits. Moderate to strongly developed structure can restrict the postdepositional movement of artifacts.

Soil Horizons

Soil horizons are natural, generally horizontal layers created by the processes that form soil. They are not depositional layers in the way that a deposit of sediments left behind by a flood or an earth slide is a depositional layer. ***Soil profile*** is used to refer to the layering or stratigraphy of soil horizons. The term profile is also used in a more generic sense to refer to any layering seen in a vertical section of sedimentary units, soil horizons, features, or artifact deposits.

Table 7.3 provides descriptions of master soil horizons, and Figure 7.14 illustrates an idealized profile. Organic A horizons form at or near the surface and comprise what you might already think of as "topsoil." The dark color of this mineral soil makes it easy to recognize. Of all of the master horizons, B horizons typically provide the greatest evidence of soil weathering in a profile. Depending on how strongly they are developed, all or much of the character of the original sediments or parent material is obliterated. B horizons also tend to be the reddest horizon encountered in a profile. C and R horizons are not true soils because by definition they have not weathered to any appreciable degree. The sediments of a C horizon can be brightly colored but will generally not match the redness of any B horizon developed upon it.

The development of a soil profile is time transgressive; it changes over time. Further, the soil development process can be interrupted by depositional or erosive events. You already know that the rate of soil formation is a function of many different factors and processes. What this means is that in a given profile, you may not encounter all of the master soil horizons, and you may not be able to subdivide the master horizons that are encountered. Figure 7.15 tracks the hypothetical development of the profile shown in Figure 7.14. In this example, the presence or absence of a master soil horizon is a function of the relative amount of time that sediments or soil have been weathering. Notice that through time, the absolute elevation of the existing surface changes, so any artifacts that might have been discarded on a surface gradually get buried as a result of in-place soil formation. At stage 2 of the development of our hypothetical profile (Figure 7.15), the top of the sediments on bedrock was a surface on which people could have lived, discarding artifacts and creating features. This "surface" eventually becomes covered, and new surfaces are created. By stage 5 in the development of our profile, any artifacts that were deposited on the top of stage 2 sediments are now part of what has become a B horizon, with neither the top nor bottom of this horizon representing a former surface. The detectable layering or stratigraphy in a soil cannot automatically be equated with former surfaces or depositional events.

Imagine what would happen if a farmer decided to plow the landscape where our hypothetical soil exists. Modern plows typically break up and mix materials to a depth of one foot below the surface. In our example, plowing would blend the O, A, and E horizons, creating a layer (plowzone) with a relatively homogenous color (dark from organic matter), texture, and at best, a crumb or granular structure. Of course, any artifacts in the original horizons get jumbled too. Continued plowing would prevent the formation of an O and E horizon, and might eventually impact the top of the B horizon. In another scenario, a series of erosive events could remove the upper foot of soil. The establishment of vegetation and the eventual formation of O and A horizons would then take place on top of the original B horizon.

Think about what would happen if flooding occurred during stage 5 of the development of our hypothetical

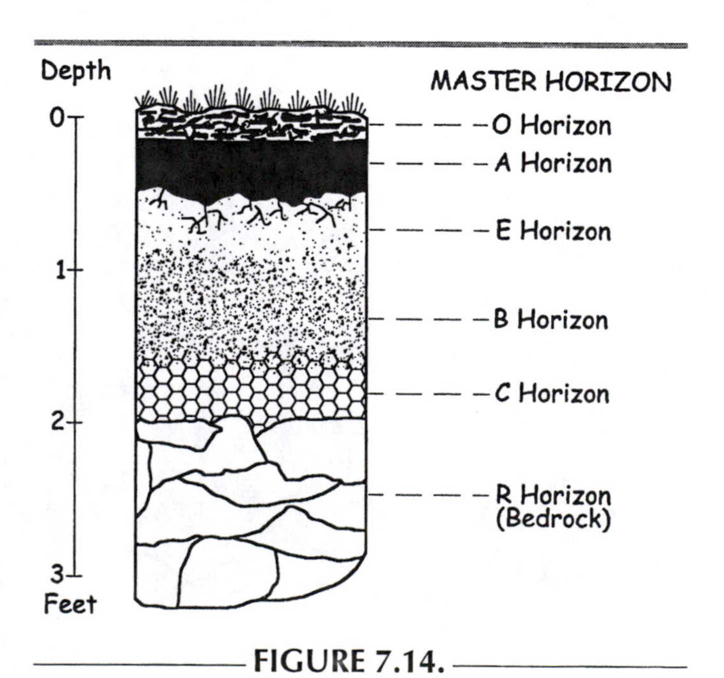

FIGURE 7.14.

An idealized soil profile.

TABLE 7.3

Master Soil Horizons and Subdivisions

Master Horizon	Possible Subdivisions	Description
O		Organic horizon dominated by fresh or decomposed organic material; over 25% organic matter by weight and dark in color; forms at the surface and on the top of a mineral soil.
	O1	Original form of the organic material is fresh and still recognizable or identifiable.
	O2	Decomposed organic matter; original material cannot be identified.
A		Predominantly inorganic or mineral soil horizon that accumulates well-decomposed organic matter (up to 25% by weight) and is dark in color; forms below an O horizon or at the surface; granular or crumb structures may occur.
	A1	Mineral horizon with dark colors indicative of the accumulation of organic matter (less than 25% by weight).
	A2	Older way to refer to E horizons that may still appear in soil surveys and older texts; lower portions of an A horizon where leaching of materials is intense.
	A3	Transitional horizon, but more like the A horizon than the underlying B horizon. Can also be designated as an A/B horizon.
E		Mineral horizon; zone of intense leaching or removal of well-decomposed organic matter, clay, iron, or aluminum, or combinations of these constituents; colors are usually gray or grayish brown and are lighter than the A horizon and the underlying B horizon; can exhibit platy structure; because of leaching this horizon is termed a zone of eluviation.
B		Mineral horizon; zone of illuviation or accumulation of clay, iron, aluminum, carbonates, gypsum, silica, illuviated organic matter (humus, generally less than 5–6% by weight), or combinations of these constituents; usually underlies an O, A, or E horizon; colors tend to be more yellowish brown to reddish brown or redder than overlying and underlying horizons; peds can exhibit a variety of structures, but this is where blocky, subangular blocky, prismatic, and columnar structures are typically found.
	B1	Transitional horizon, more like B than A horizon; just beginning to show the characteristics of a B horizon; could also be designated B/A horizon.
	B2	Zone of maximum accumulation of clay, iron, aluminum, illuviated organic matter, or combinations of these constituents; organic matter content is generally higher than what is seen in an A2/E horizon if present; strongest development of blocky, subangular blocky, prismatic, or columnar structure seen here.
	B3	Transitional horizon, more like B than C horizon; could also be designated a B/C horizon.
C		Relatively unweathered and unconsolidated material lacking the characteristics of an A or B horizon; sometimes called the parent material, although it is possible that it may not be the material from which the A or B horizons above it developed; any subdivisions of the C horizon are used to indicate differences in texture, mineralogy, etc.
R		Bedrock or consolidated rock underlying soil.

Sources: Brady and Weil (1998); Foss et al. (1976, 1985); Soil Survey Staff (1998, 1999).

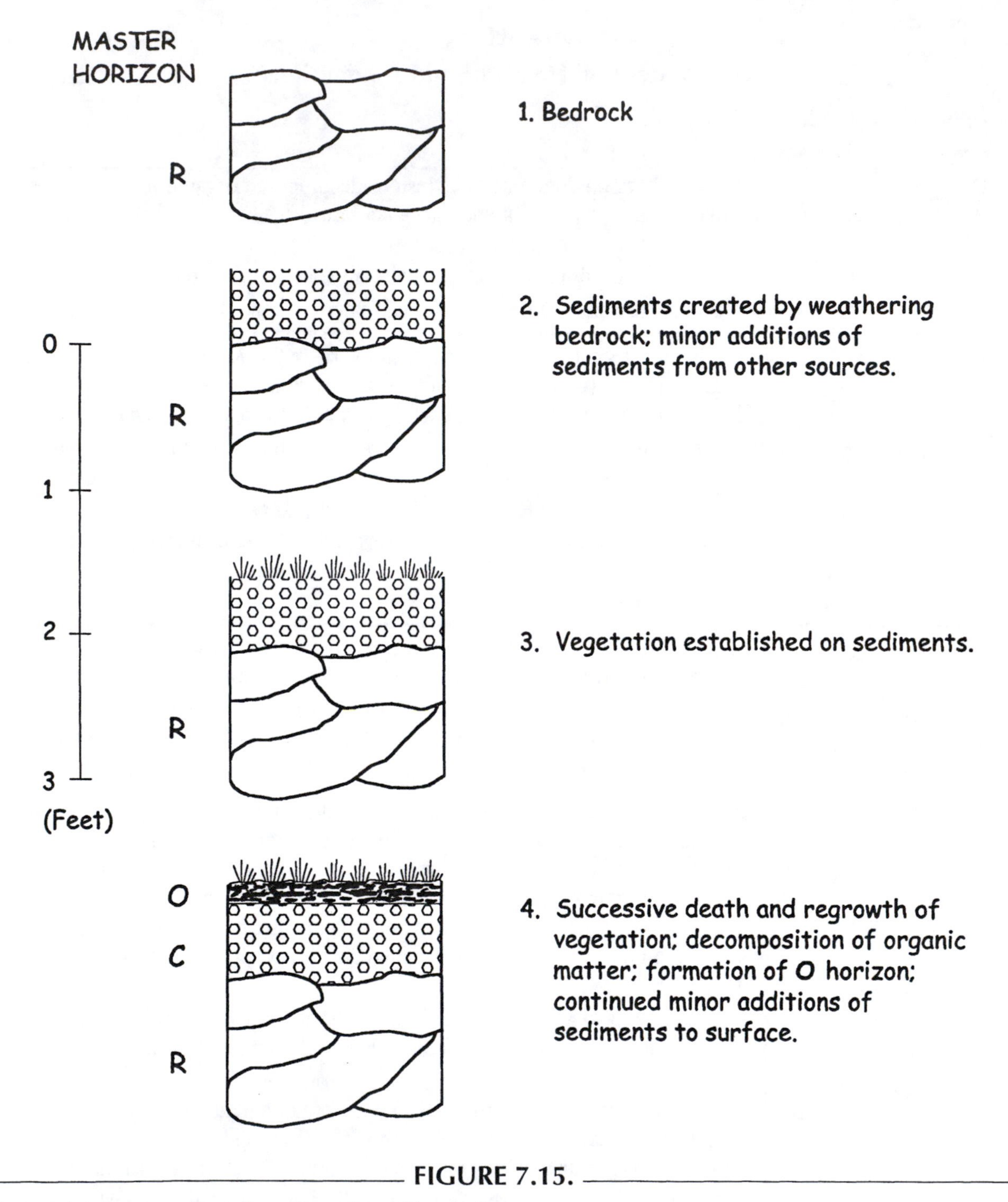

FIGURE 7.15.

Hypothetical development of a soil profile under stable conditions.

profile (Figure 7.16). The formation of the original soil is slowed or retarded, preserving its characteristics, while the most active formation of soil takes place in conjunction with the newly established surface and the sediments immediately beneath it. Any soil that is rapidly buried with an amount of sediment sufficient to remove it from the zone of active weathering will preserve many of its original characteristics into the future. Our example is a profile that contains both sediments and soils. There are two different soils in the profile resulting from distinct weathering histories: the O horizon at the top of the profile (stage 3), and the buried A horizon (stage 3). It is not that unusual for a profile to contain buried soils resulting from different developmental

MASTER
HORIZON

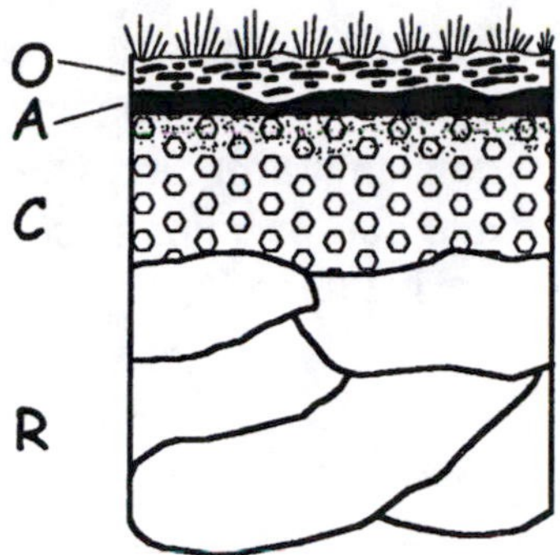

5. Vegetative cycle continues; some organic matter becomes thoroughly decomposed and mixes with sediments in top of **C** horizon ; **A** horizon established; sediments at top of **C** horizon continue to weather; translocation begins.

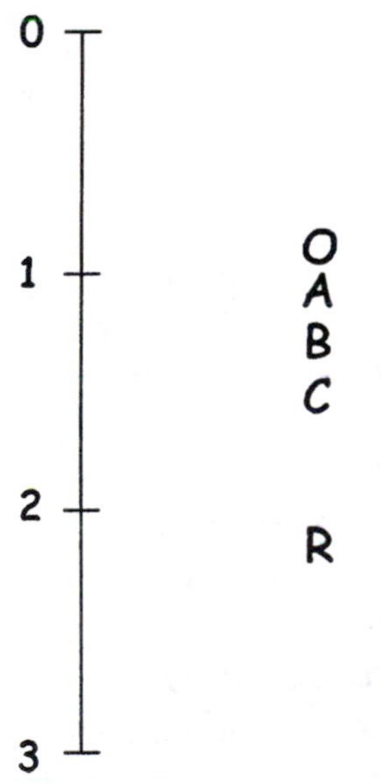

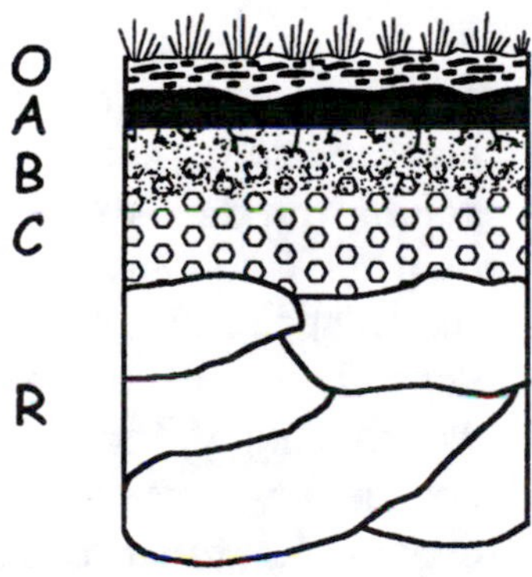

6. Thickening of the **A** horizon, weathering of its base; translocated materials create a **B** horizon in what was once the top of the original **C** horizon.

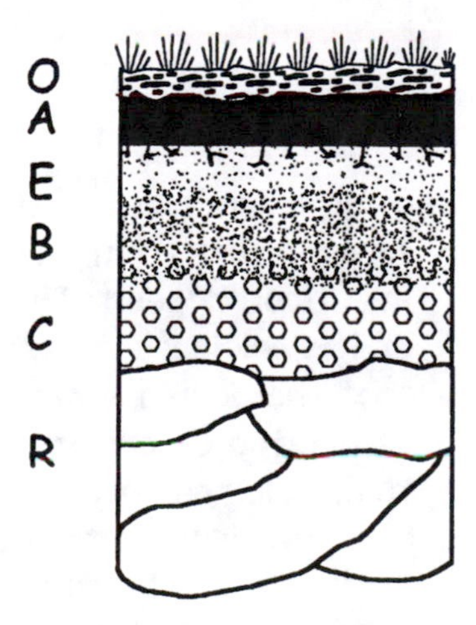

7. Continued vegetative cycle, decomposition of organic matter and minor additions of sediments to surface thickening the **A** horizon; leaching of base of **A** horizon, **E** horizon established; although near surface portions of profile gradually thicken, the weathering of the base of **A** horizon keeps pace with additions at surface; thickness of **A** horizon reaches equilibrium; translocation of materials supports the continued development of the **B** horizon, obliterating the character of the original sediments at this level of the profile.

FIGURE 7.15.

Continued

MASTER
HORIZON

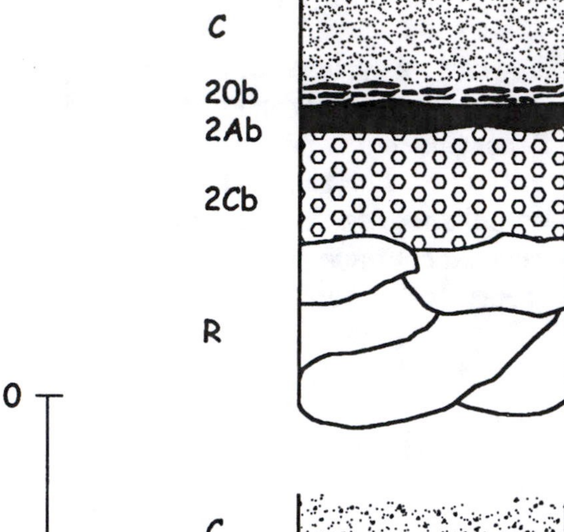

1. Flooding deposits fine sand on top of a pre-existing surface and soil column, retarding the weathering and soil forming processes once active upon it.

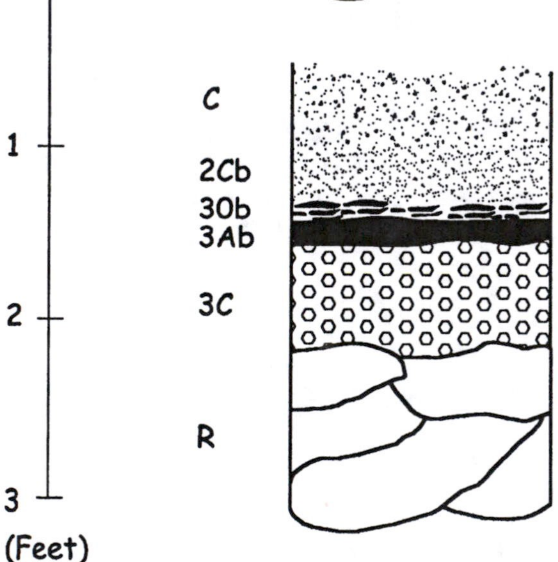

2. Before vegetation and organic matter can be established at the top of the deposit of fine sand, another flood impacts the landscape, depositing a coarse sand. This further retards the weathering of the original soil profile. Over the long term, the organic matter in the **O** horizon may fully decompose, blending with the **A** horizon.

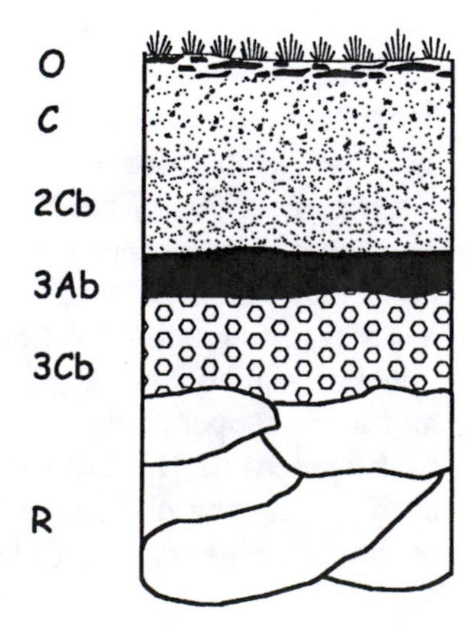

3. Vegetation is established and an **O** horizon formed at the top of the last flood deposit. The weathering of the original soil is so retarded that the buried **A** horizon and associated **C** horizon will retain their color and other characteristics into the future. Sediment weathering and soil formation will more actively impact the new surface and portions of the two upper **C** horizons.

FIGURE 7.16.

The influence of deposition on a developing soil profile. (Prefixes and suffixes are explained in the text.)

sequences, especially in dynamic environments like floodplains. Look again at Figure 1.8. The profile shows three buried A horizons with their associated subsoils.

There are a number of subscripts or suffixes that describe variations on master horizons, especially B horizons (for a complete list see Soil Survey Staff 1998:312–316; partial lists can be seen in Holliday 1992a:248–249; Waters 1992:Table 2.4). A few of them bear special mention here, and will aid your understanding of information contained in published soil surveys and the labeling of stratigraphic layers in the field. The suffix "p" denotes a soil that has been plowed, e.g., Ap (plowed A horizon or plowzone). Given that so much fieldwork takes place on land that has been farmed, archaeologists frequently encounter Ap horizons. B horizons that have developed to the point that they exhibit some color and/or structure are cambic, designated Bw, and those that have substantial accumulations of clay are argillic, Bt horizons. All other things being equal, Bw and Bt horizons represent progressively more well-developed, and therefore older soils, and are important for estimating the amount of time represented by a stratigraphic sequence.

The suffix "b" added to a horizon indicates that a horizon is part of a buried soil, e.g., Ab indicates a buried A horizon, Bb a buried B horizon, Cb a buried C horizon (see Figure 7.16). A buried soil and the use of the suffix "b" implies a break in one weathering sequence and deposition and weathering associated with another. This break is a ***discontinuity*** in the development of a soil profile. Arabic numerals are used as prefixes to master soil horizons to indicate discontinuities or discrete sequences of soil formation (Soil Survey Staff 1998:315–16). Numbering begins at the existing surface and proceeds downwards as necessary to label contrasting materials. The uppermost sequence is understood to be number 1, and is never used in the designation of master horizons (see master horizon designations used in Figure 7.16). There are instances where soil scientists forego numbering discontinuities, such as when buried horizons have developed in materials similar to those overlying it, and in the stratification of alluvial (flood) deposits where particle sizes are comparable between layers. However, distinguishing changes in depositional and soil-forming events with as much detail as possible is important to archaeologists who turn to stratigraphy to order and date archaeological deposits and understand the ever-changing environments in which they exist. I recommend numbering discontinuities whenever they are recognizable.

DEFINING AND INTERPRETING STRATIGRAPHY

From an archaeological perspective, ***stratigraphy*** can be defined as the vertical, horizontal, and chronological relations of sedimentary units, soils, and artifacts. In simple terms you can think of it as the layering of deposits. Strictly speaking, stratigraphy is how we interpret the stratification or layering that natural and cultural processes create. Recognizing and understanding this layering helps to achieve the goals outlined at the beginning of this chapter. Although not always interchangeable, the terms sedimentary unit, deposit, stratum, or horizon are used to describe the individual layers recognized in a vertical section or profile. Although the term deposit does tend to be used in a generic sense, it implies the act of deposition and would not be an appropriate way to refer to a soil horizon. Stratum is a fairly general term that can be used to refer to any distinguishable layer, or you can use the term layer itself. The term horizon is applied only to soils, and it would be inappropriate to refer to a soil horizon (excepting C and R horizons) as a sedimentary unit. As I have already noted, the term profile is used in a generic sense to refer to any layering seen in a vertical section of sedimentary units, soil horizons, features, or artifact deposits.

Understanding the vertical component of stratigraphy is dependent on the ***law of superposition*** and the ***law of ascendency and descendency*** borrowed from geology and earth science (e.g., Daniels and Hammer 1992:2). Taken together, these laws simply state that in undisturbed conditions, younger layers or deposits overlie older layers or deposits, and that the age of layers becomes progressively younger as you ascend through a vertical section, and progressively older as you descend through the section (Figure 7.17). In all cases, age refers to the act of deposition of a layer or sedimentary unit. The age of the contents (sediments, artifacts, features) of a layer is often associated with the age of the depositional event. So an archaeologist might assume that leaf-shaped spearheads found in stratum 6 of the hypothetical sequence shown in Figure 7.17 are older than triangular ones recovered from stratum 9. However, natural and cultural processes can complicate what seems like an otherwise straightforward relationship. Landfill is a classic example. Humans remove massive amounts of sediments and the artifacts they may contain from one location and redeposit them in another location. The fill used in the 18th century to

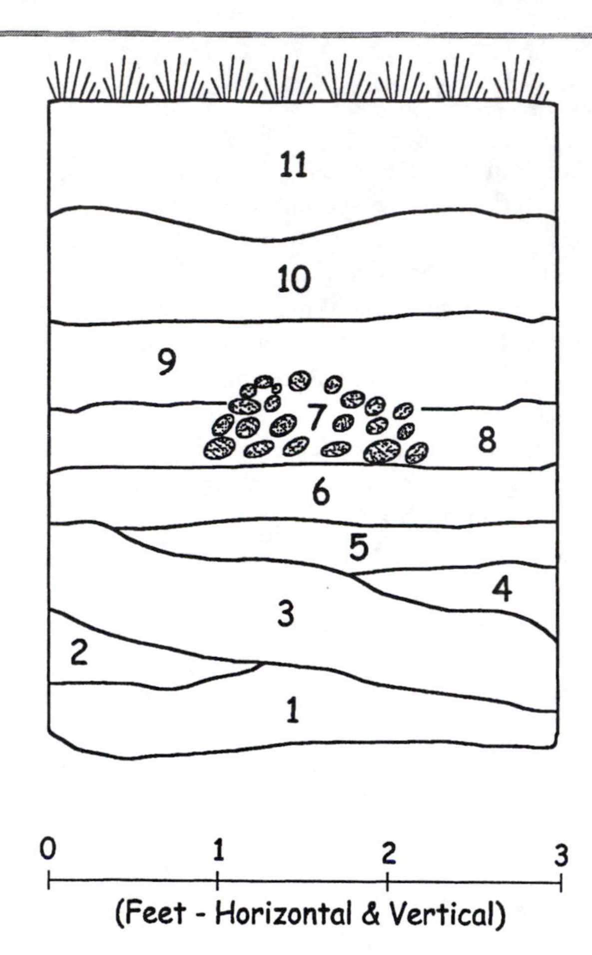

FIGURE 7.17.

A profile of deposits embodying the laws of superposition, ascendency. and descendency. Deposit 7 is a cached pile of rock (cultural). The order of deposition of the layers shown follows the sequence 1, 2, 3, etc. The relative age of these depositional events follows the same order, from oldest to youngest.

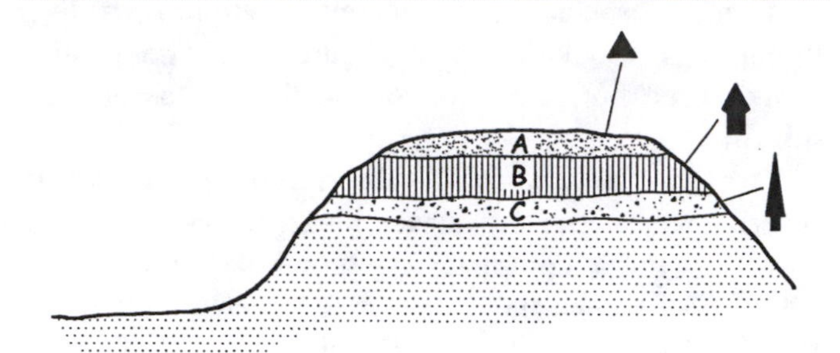

1. Undisturbed stratified deposits with associated chipped stone projectiles on hilltop position.

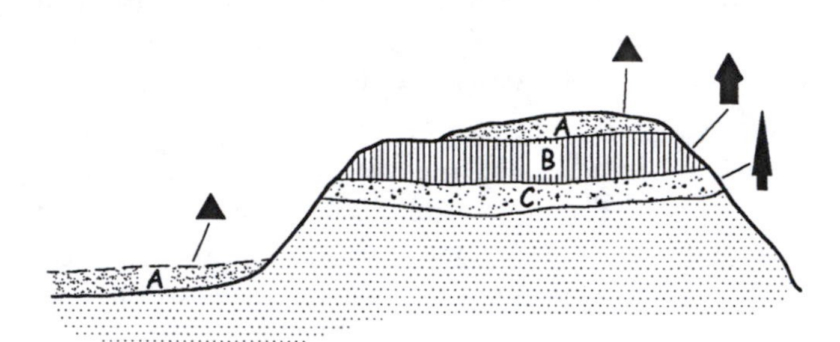

2. Erosion of hilltop deposits, down slope movement, and redeposition of sediments and artifacts.

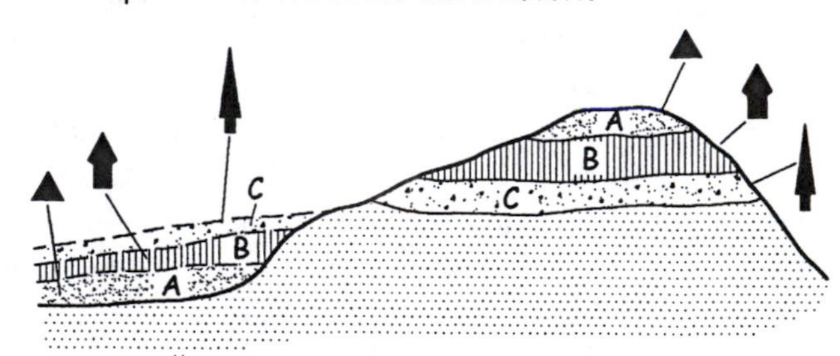

3. Progressive erosion and redeposition creates a reversed stratigraphic sequence of artifacts at the base of the slope.

FIGURE 7.18.

An example of natural processes leading to the creation of a reversed stratigraphic sequence of artifacts.

extend the shoreline along the Delaware River at Philadelphia included prehistoric Native American artifacts (LeeDecker et al. 1993). A flood erodes ancient sediments and some artifacts from an upstream location and redeposits them downstream thousands of years later. Natural processes can erode the upper portions of a profile, exposing or exhuming ancient sediments and soils. This newly exposed but ancient natural deposit is now a surface on which artifacts and features related to more recent times can be discarded and created. Relatively young artifacts eventually get worked down into the top of the older deposit.

A reversed stratigraphic sequence of artifacts created by erosion and redeposition is shown in Figure 7.18. The situation becomes understandable only by studying the broader landscape. Even though the color and texture of the sediments making up strata A, B, and C would be comparable between the hilltop and the base of the slope, soil structure would not. The presence of intact features in the hilltop position and their lack in the deposits at the base of the slope would be an additional clue that the hilltop deposits are intact and the down-slope deposits are derived from them. These cautionary tales are offered not to denigrate stratigraphic inferences, but to emphasize the importance of studying site formation processes and geomorphology. The stratigraphic world is much more complicated than what may be indicated by a single auger boring, bank exposure, or excavation.

When I first talked with artifact collectors about sites known to them along St. James Run, they were puzzled by the fact that they rarely if ever found artifacts on the surface adjacent to the stream. This bothered them because they typically found artifacts in these locations elsewhere. Instead, the sites associated with St. James Run are situated hundreds of feet away from the edge of the stream channel. We learned as a result of extensive surface survey, borings, excavations, and soil analysis that beginning no later than 6500 B.C., areas adjacent to the stream were occupied by a series of ponds. The ponds shrank and swelled through time in response to changes in climate and stream dynamics, gradually filling with sediments and finally disappearing sometime between 300–900 A.D. Native American activities in the area were focused around the margins of the ponds on well-drained ground. Today, both the areas of the ponds and the ancient well-drained ground with its archaeological deposits are part of the same flat to undulating surface along St. James Run (Shaw 1993; Stewart 1980:238–258; 1997).

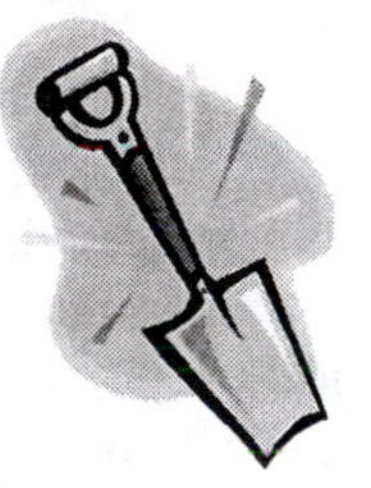

The Conococheague Site (18Wa54) struck me as a bit of an enigma when I first examined it. It was typical of other ancient Indian localities in the region in that it was situated in a relatively well-drained, flat stretch of floodplain adjacent to a large stream, Conococheague Creek (Figure 7.19). Artifacts diagnostic of the Late Archaic through Late Woodland periods (about 3000 B.C. to 1600 A.D.) occurred on the surface of the plowed field. But the site was atypical from the perspective of where it was located within the floodplain. Rather than being on the well-drained ground closest to the junction of Conococheague Creek with a small tributary, a pattern exhibited by the location of many other sites, the artifacts were clustered hundreds of feet away. Excavations and auger borings solved the mystery. The ancient surface of the floodplain was not flat, but sloped towards the stream junction. Successive floods gradually filled in and leveled out this lower-lying portion of the landscape, burying the Late Woodland surface (900–1600 A.D.) and artifacts in the process. Flooding more rarely affected the eastern or more highly elevated portions of the floodplain. Plowing brought artifacts to the surface in this eastern area but was unable to reach the more deeply buried deposits to the west. Today, as you walk east to west across the floodplain at 18Wa54, you move from an area where ancient deposits are on and near the present surface, onto deposits that are historic and modern in age (Stewart 1980:209–226).

We also cannot ignore the horizontal relations of sediments, soils, and artifact deposits. Horizontal or "flat" stratigraphy (Thomas 1998:224) means that the nature and age of deposits changes as you walk across what appears to be the common surface of a landscape. Any landscape differentially affected by erosion and/or deposition can possess time-transgressive surfaces; the age of the current surface varies across horizontal space. Horizontal stratigraphy, like vertical stratigraphy, is important for organizing artifact deposits in time, understanding site formation processes, and reconstructing the environments of the past. Two examples from my own research in western Maryland are described above.

The stratigraphy of sediments and soils does not automatically correlate with the way in which artifacts and features are layered. You have only to think of the postdepositional changes in color and texture that soil-forming processes promote in sediments to realize the validity of this statement. The challenge for the archaeologist is to understand how the cultural stratigraphy, the vertical and horizontal relations of cultural deposits, dovetails with what we could call the natural stratigraphy. In some respects distinguishing a natural and cultural stratigraphy is a false dichotomy because artifacts and features are part of some type of sedimentary or pedologic matrix, there are anthropogenic soils, and artifacts themselves can be thought of as sediments in the ways they are affected by natural processes. There are numerous types of sites or deposits where the stratigraphy is almost exclusively cultural in nature, e.g., towns and cities exhibiting repeated episodes of

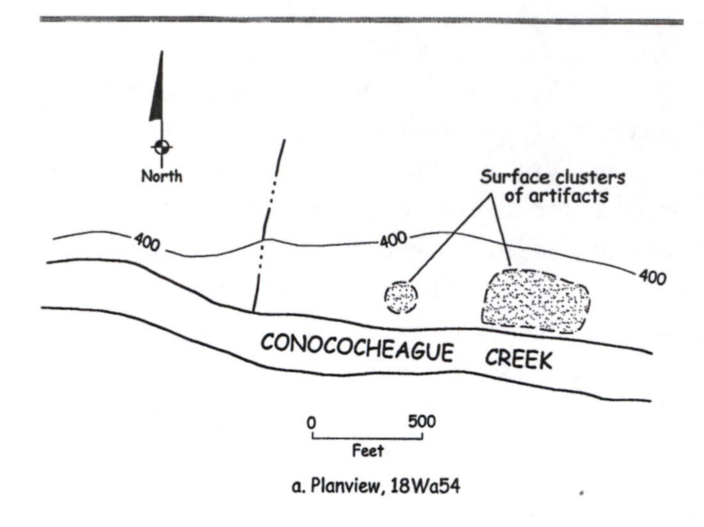

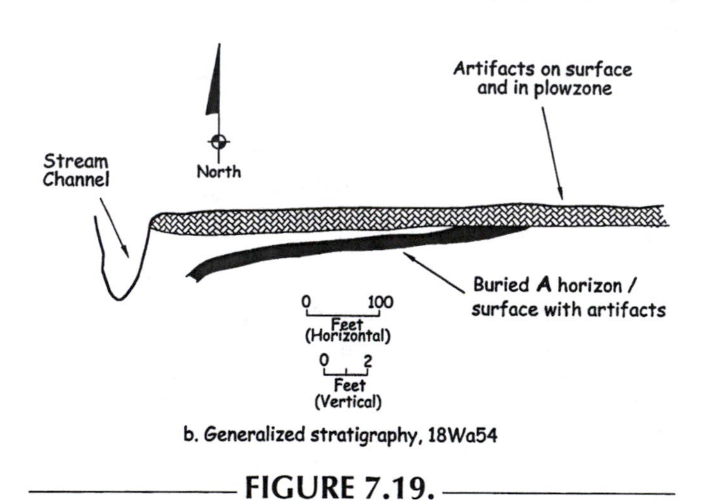

FIGURE 7.19.

Plan view and generalized profile of the Conococheague Site, 18Wa54.

architectural construction and destruction; temple or burial mounds. Yet not to be thinking explicitly about the impact of both natural and cultural processes on the stratification of a deposit is to miss interpretive opportunities. This is especially critical for archaeologists who study hunting and gathering peoples whose sites lack substantial architectural remains and whose artifacts and features are embedded in ancient sediments and soils.

We have already reviewed the essential characteristics of sediments and soils that contribute to the layers detectable in a stratigraphic sequence. The vertical boundaries of individual layers, strata, or soil horizons have implications for the natural and cultural processes influencing stratigraphy, and the identification of former surfaces upon which people could have lived and created archaeological deposits. A ***discontinuity,*** as defined in soil science or pedology, is the boundary between two periods of deposition, regardless of the amount of time that separates the depositional episodes, or a break in a soil-weathering sequence evidenced by erosion and/or deposition and a reinitiation of the soil-forming process resulting in the presence of multiple soil sequences in the profile. Discontinuities can be recognized as breaks in the patterned trends of color, texture, and structure that we typically associate with master horizons that are part of the same weathering sequence. A somewhat similar term, ***unconformity,*** comes from geology and the study of sediments. An unconformity is a break in stratification represented by erosion or the removal of pre-existing sediments or soils, and a prolonged hiatus or halt in sediment deposition. The weathering of sediments during this hiatus may result in some soil formation prior to the renewal of more active deposition.

Because of the nature of soil-forming processes, boundaries between soil horizons are often wavy or indistinct, i.e., they are hard to delineate and draw with precision. We can't assume that detectable boundaries between all soils have meaning in relationship to the layering of archaeological deposits. An exception is buried A horizons representing the remnants of former surfaces. However, since A horizons form under stable surface conditions, it is possible that artifacts and features representing periodic reuse of the landscape could occur on and into an A horizon. Boundaries between deposits resulting from depositional events are often sharp and distinct, and represent potential living surfaces for some period of time. The top of silt deposited by a flood functions as a surface until it is either buried by the deposition of additional sediments, eroded, or transformed by soil-forming processes. Relatively straight and well-defined boundaries between strata can also be the result of erosion prior to renewed deposition. Plowed soils (plowzones, Ap horizons) have a straight and well-defined lower boundary because of the consistent way in which the plow affects the materials it churns, and the frequency with which plowing occurs. Normally the action of roots and other biological activities make the lower boundary of an A horizon wavy and irregular. The areal extent or horizontal boundaries of strata are important for understanding the geomorphology or origins and evolution of the landscape.

GEOMORPHOLOGY

Geomorphology is the study of the shape of the landscape and the processes that form it. The surface of the earth can be sculpted by water, wind, ice, gravity, volcanism, and subsurface geological processes in a variety of ways. Landscapes whose development is dominated by a particular process or combination of processes are grouped accordingly: alluvial environments are those predominantly shaped by the behavior of streams or channelized flows of water; those where the wind is responsible for the erosion, transport, and deposition of sediments are eolian environments; colluvial environments encompass all manner of sloping landscapes where the action of gravity and the unchannelized flow of water are geomorphic agents; and sea level, waves, tides, and currents modify coastal environments. Volcanic eruptions resulting in ash, ejecta, and lava flows, and the glacial movement of sediments are other processes that can affect a variety of environments. Because climate and other conditions change over time, combinations of processes can be expressed in single landscapes, e.g., floodplains and stream terraces can be mantled with eolian deposits; hilltops, slopes, and stream valleys can be scoured by glacial activity, and some coastal areas were once upland flats and stream valleys prior to the rise of sea levels.

In addition to shaping the landscape, these geomorphic processes can leave behind distinctive sedimentary and stratigraphic signatures. Think of the erosion, transportation, and deposition of sediments as nature expending energy to do work. Certain types of work require more energy than others. The movement and deposition of gravel-sized sediments requires more energy than the transportation and deposition of silt-sized sediments. A natural event capable of moving and depositing gravel will also be able to move smaller-sized particles that may be available on the landscape being eroded. A high-energy flood is capable of manipulating gravel-sized sediments, the wind much less so and in a different way. Even during the course of a single event like a flood, landslide, or windstorm, the energy available to erode, transport and deposit sediments changes. The result can be variation in the types of sediment left behind by an event and the degree of its impact on the landscape as a whole; the energy of an event influences the degree to which it can impact a landscape. Intervals of time between erosive or depositional events contribute to the stable conditions under which soil may begin to form. These are key observations for archaeologists. In the field, our appreciation of this variation can mean the difference between finding a site and not finding a site, discovering nicely stratified deposits representing discrete intervals of time versus contexts in which artifacts from numerous occupations are mixed or conflated.

Even a summary review of the variety of geomorphic environments and operative processes that exist is beyond the scope of my effort here. To emphasize the type of variability that can occur within single landscapes and its potential impact on archaeological practice, we'll look briefly at two types of environments—floodplains and landscapes with slopes. Both are commonly encountered by archaeologists. Floodplains have always been appreciated for their potential to contain buried and stratified archaeological deposits. Landscapes with slopes, even very gentle ones, are contexts within which erosion and redeposition of sediments can be fairly common. Hopefully, this discussion will spur you on to future study and give you a basis for questioning and understanding the decisions made by those directing the fieldwork in which you are engaged. Books by Schiffer (1987, 1996) and Waters (1992) contain useful summaries and examples of how geomorphic processes affect archaeological sites.

Processes That Work on Floodplains

Floodplains or landscapes adjacent to streams can be very dynamic. Figure 7.20 depicts some of the features that can be associated with high-order streams, those that have numerous tributaries and drain a relatively large basin (for definitions of these features see Table 7.4). Floods will not affect all portions of a floodplain equally. When a stream in flood overflows its banks it is carrying with it sediments suspended in the water. The energy of the event influences the amount of work that it is capable of doing, i.e., the types of sediments that can be transported. Let's assume that a flood is transporting fine sand, silt, and clay as it flows over the bank. As the floodwater moves over the floodplain and begins to pool it loses energy. It continues to lose energy as the overbank process stops and the movement of the water in the floodplain slows. As the energy decreases,

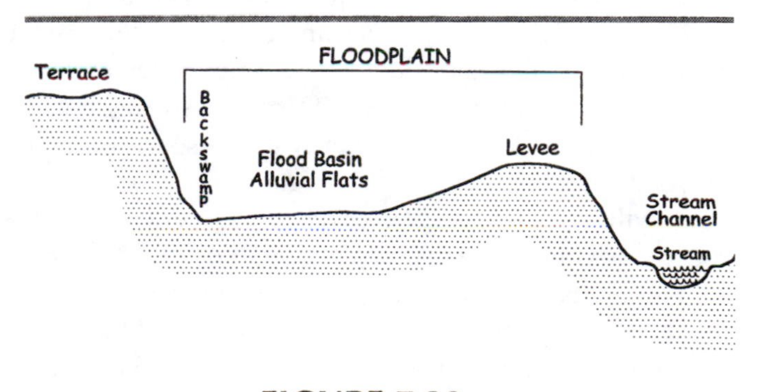

FIGURE 7.20.

A simplified diagram of a floodplain.

TABLE 7.4

Some Features of Streams and Stream Valleys

Term	Description
Alluvial deposition	deposition of sediments fostered by streams; deposited sediments referred to as alluvium.
Alluvial flats	general floodplain area behind the levee; also called the flood basin.
Backswamp	lowest ground in the floodplain, typically located on the outer margins or most interior section of the floodplain.
Floodplain	area characteristically affected by stream deposition.
Lateral deposits	alluvium deposited within the channel of the stream.
Levee	high, wedge-shaped banks adjacent to the most frequently inundated or flooded portion of the floodplain.
Overbank deposits	alluvium deposited as a result of water coming over the banks of the stream channel into the floodplain.
Stream	a channelized flow of water.
Stream order	hierarchical ranking of streams reflecting the relative size of the basin being drained; e.g., a first-order stream has no tributaries and drains a relatively small basin compared to a second-order stream, which has at least one tributary; the higher the stream order, the relatively larger the drainage basin.
Terrace	vertical escarpment or step separating obsolete floodplain from younger, more active floodplain.

Sources: Butzer (1976); Waters (1992).

the ability of floodwater to hold sediments in suspension also decreases. The larger-sized and heavier particles will fall out of suspension first, followed in time by progressively smaller-sized material as the energy of the event wanes. The relationship between decreasing energy and the progressive deposition of sediments results in two distinctive trends. First, when a vertical section of the material deposited by the flood is examined, particle size will "fine upwards." The basal portions of the deposit will be sandy in texture. Moving higher in the deposit the coarse fraction will decrease and finer particle sizes will dominate. Second, the overall texture of the deposit will become incrementally finer with increasing distance from the stream bank. This is because it takes less energy to transport silt- and clay-sized particles than sand, so more finer-sized particles will be found at a greater distance from the energy source driving overbank deposition, i.e., the flowing stream in its channel. The visibility of these fining trends depends upon the energy of a flood and the type of sediments available for erosion and transportation upstream. If a flood only has energy sufficient to transport silt- and clay-sized particles the fining trends in the alluvial deposit will be more difficult to detect.

Because of the relationship between energy, sediment transport, and deposition, the amount of material deposited in a given section of a floodplain can vary. In the overbank deposition of alluvium, more material typically is deposited closer to the stream channel than farther from it. The overbank deposit from a single flood would therefore have a wedge-like profile, thick near the stream channel and thinning with increasing distance from the channel. Depending on the size of the alluvial flats or flood basin, and the energy of a flood, not all sections of a floodplain might experience deposition, or deposition might be negligible. This explains why some floodplains can have the distinctive topographic profile depicted in Figure 7.20. Even the most minor floods can result in deposition near the stream channel while little or no deposition may occur in interior sections of the floodplain. Surfaces nearer the stream channel have the potential to build up faster than those at a greater distance. The area called the backswamp is thus the most low-lying portion of the floodplain because of its distance from the stream channel and the impact of this distance on deposition. As the topography of the floodplain develops, the backswamp receives surface runoff from the slope of the stream bank or levee, as well as from adjacent upland or terrace slopes. This supports its poorly drained character, at least on a seasonal basis. Individual floodplains may possess only a subset of the features noted here, or the many others that have not been mentioned. This variability results from factors such as the order of a stream, the stream system's relative age, surrounding topography, the width of the floodplain, and all of the variables that influence the nature and periodicity of flooding.

Because the rate and degree of alluvial deposition can vary across a floodplain, so too can the context of archaeological deposits. Levee or stream bank positions have the potential to contain buried deposits that will be more discretely stratified than those in more interior sections of the floodplain. Older surfaces or deposits can be deeply buried on the levee, but occur at or near the present surface in the interior. Figure 7.21 is a graphic example of these relationships. Moving into the interior of the floodplain, individual deposits thin and may become discontinuous, stratification decreases, and deposits of different ages are so closely superimposed that their separation becomes difficult. Add to this the consequences of agricultural plowing of near-surface deposits and we have a situation where the integrity and potential significance of any archaeological deposits changes dramatically depending on where you are on the landform.

The ways in which you would look for and excavate archaeological deposits in each of these areas would also differ. In a surface survey older artifacts would never be found on the levee, while a mixture of artifacts of different ages would be seen in the floodplain's interior. Excavations in search of cultural deposits would have to be deep on the levee, and because of this, the horizontal dimensions of excavation units would have to be large enough to permit working safely at depths. Shoring for excavation walls might be necessary, or the use of heavy machinery might be the most practical way to search for deeply buried materials. Relatively shallow excavations would suffice in the interior and wouldn't need to be of the same horizontal dimensions as those on the levee. The use of heavy machinery would be unnecessary. The stratigraphic separation of artifacts from different episodes of site use would be greater on the levee, less so with increasing distance from the levee, and in places so conflated or mixed as to be inseparable. Mapping the precise location of individual artifacts might be worthwhile in the levee excavations, while simple unit and level provenience would be adequate in excavations where artifacts from different occupations are mixed. Excavations across the floodplain would encounter sediments of different textures although they might be part of the same depositional event. Correlating the natural and cultural stratigraphy of the floodplain would require information from all portions of the floodplain. Clearly, you can see how an understanding of geomorphology and natural processes impacts archaeological strategies and methods.

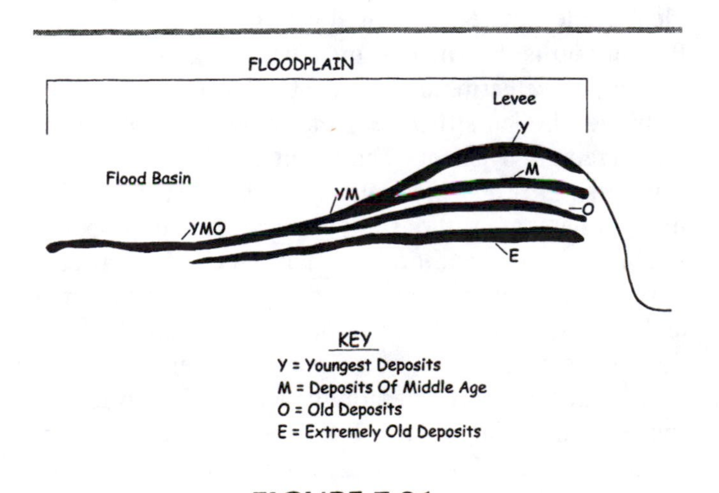

FIGURE 7.21.

Possible relations of deposits across a floodplain.

Processes that work on slopes.

Landscapes with slopes comprise more of the earth's surface than those affected by streams. Technically, a slope is any landform with a slope between 2 and 40 percent. Some slopes may not immediately strike you as such. A slope of 2 percent means that over a distance of 100 feet elevation has changed two feet. Native American habitation sites have been documented on slopes of up to 8 percent, and habitable rockshelters and caves can be positioned on much steeper slopes. Slopes greater than 40 percent are called scarps or cliffs. Between the summit, or top of a slope, and its base, or the toe of slope, straight, convex, and concave segments can be combined in many ways resulting in different types of slope profiles. A variety of processes that work at different rates can lead to the erosion, transportation and deposition of sediments along slopes (Table 7.5). All of these processes have the potential to bury artifacts and features that might exist on lower portions of a slope or adjacent landscape, or to disturb and displace them. Ozette, located on the Northwest Coast, is one of the country's most spectacular sites and known for the fantastic preservation of organic artifacts. It was buried by a mudslide (Ames and

TABLE 7.5

Processes on Slopes

Colluvium	sediments eroded, transported, and deposited along slopes; colluvial deposition.
Creep	grain-by-grain movement of sediments down slope; can be initiated by mechanical weathering of sediments and gravity, freezing and thawing cycles, or impact of rainfall upon the surface; a slow and incremental process; a form of mass wasting.
Flows	downslope movement of saturated materials that behave in a plastic manner; the rate of movement can be quick or carried out over a prolonged period; mudflow, earthflow, solifluction; a form of mass wasting.
Falls	mechanical weathering and free fall of rock; movement of coherent masses of sediment along a cliff or scarp (slopes greater than 40%); a form of mass wasting.
Mass wasting	downslope movement of materials under the primary influence of gravity.
Slides	rapid downslope movement of rock and unconsolidated sediments (i.e., material does not move as a block); a form of mass wasting that generally occurs on slopes of 20–34%.
Sheet wash	unconfined (unchannelized) overland flow of water and sediments; doesn't begin until the ground has lost its ability to absorb moisture at the rate at which it is encountering the surface; slope wash.
Slumping	sliding movement of blocky masses of sediment or soil resulting from a natural process aggravating some structural weakness in the deposits; downward movement of coherent units of material; generally occurs on slopes of 20–34%; a form of mass wasting.

Sources: Butzer (1976); Kauffman (1990); Ritter et al. (1995); Waters (1992).

Maschner 1999; Daugherty 1988). The more dramatic of these processes like slides, flows, and slumping result in distinctive topography, either because of the scar left behind when material was torn away from a portion of a slope, or the amount and nature of the material being transported and redeposited. Flows, slides, and slumping can redeposit large amounts of material in single episodes, sheet wash less so. Creep is an extremely slow, ongoing process. The degree to which redeposited materials are sorted also reflects the nature of the individual process. Slides create deposits that generally are poorly sorted and contain a wide range of particle sizes. Since slumping involves the movement of blocks or units of material, the new deposits can mimic the stratigraphy of the portion of the slope where they originated. Sheet wash can be well sorted. Some of the factors influencing processes active on slopes include the type of sediment or soils present, the degree of vegetation cover, climate, and rainfall.

One of the more mundane processes active on slopes is sheet wash, but it can have some surprising effects. As non-native colonists settled on the eastern seaboard of North America they cleared extensive tracts of land for farming. The clearing of vegetation sets up the conditions whereby rainfall and sheet wash can erode sediments from high points on the landscape, transporting and redepositing them at lower elevations. In some cases, the amount of sediment redeposited during a single episode can be substantial (Figure 7.22); in others, the buildup is incremental over time (Figure 7.23). At the Playwicki Farm, an 18th-century surface, remnants of burned Indian structures, and related artifacts were buried rapidly by sheet wash off an adjacent steep slope (Stewart 1999). The clearance of the landscape by colonial farmers after the site was abandoned probably established the conditions under which this process took place. The sediments burying the site are well sorted and thin with increasing distance from the base of the steep slope, reflecting the decreasing energy of the event and the ability of gravity and surface water to transport materials. The plowed agricultural field in which the site is located is gently sloping. Because of the wedge-like shape of the deposits that buried the 18th-century surface, later plowing disturbed the buried

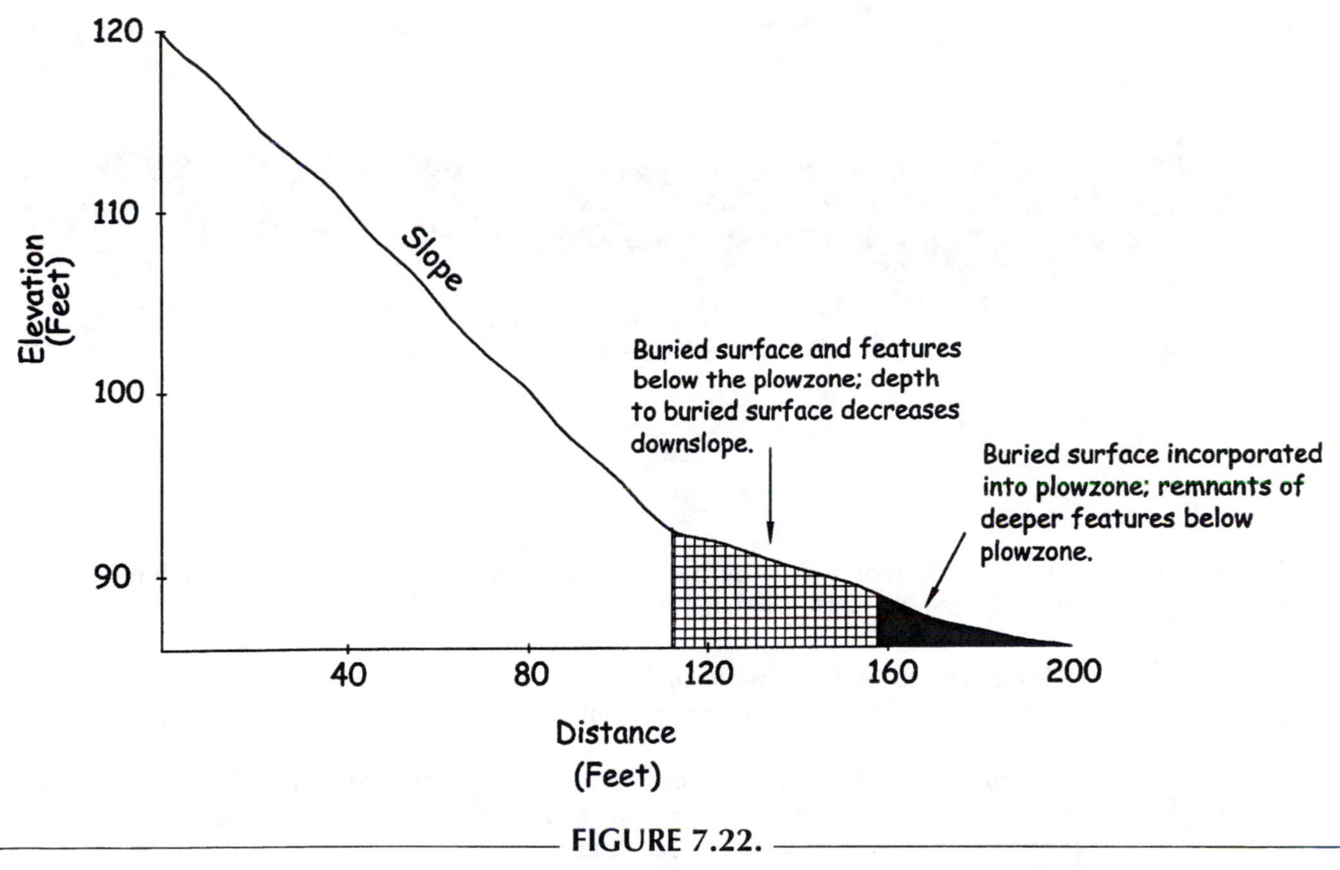

FIGURE 7.22.

The relationship of natural and cultural deposits with slope position at the Playwicki Farm Site (36Bu173). An 18th-century surface (A horizon) was buried by sheet wash off the adjacent steep slope. The thickness of the sediments burying the surface progressively thin with increasing distance from the base of the slope. Consequently, historic and modern plowing did not disturb portions of the buried site near the base of the slope but did impact site areas at greater distances from it.

deposits and brought artifacts to the surface in southern portions of the field, but left deposits to the north untouched. The Lister Site is on a landform with a 5 percent slope and yet incremental sheet wash was capable of burying an 18th-century surface (Stewart 1986). Again, historic land clearance seems to be critical in establishing the conditions for erosion and redeposition of sediments to occur. The burial of the surface also was aided by the sandy texture of *in situ* soils, developed on prehistoric eolian sediments. Sands can be more easily entrained (eroded and movement initiated) than silts and clays.

So much of the fieldwork done by archaeologists takes place in locations that are, or once were, agricultural fields. Although plowed sites are worthy of study, there is a limit to their value. This value is enhanced by the discovery of unplowed remnants of features and artifact deposits. The placement of excavations to search for unplowed remnants of deposits needs to take into consideration the processes, like sheet wash, active on even gentle slopes.

EXAMINING AND RECORDING A PROFILE

At some point during your field career you will examine and record stratigraphy in a variety of natural exposures, auger borings, hand-dug excavations, and trenches opened with heavy machinery. The creation of vertical sections for inspection, and the use of augers or coring devices are discussed in subsequent chapters. Procedures for examining a vertical section and drawing a profile are discussed here.

Depending on the type of deposits in the vertical section or profile, the tools that you will need for cleaning it and delimiting strata include a sharpened trowel, medium-sized paint brush or small whisk broom, root clippers or loppers, and a spray bottle of water. Make sure that the profile you are examining

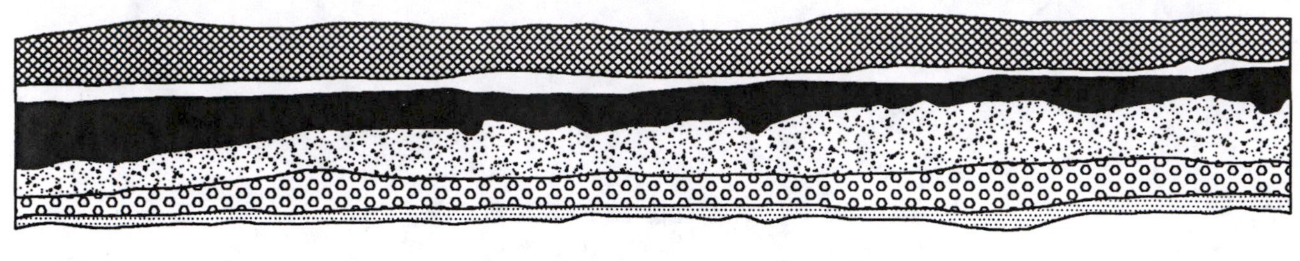

FIGURE 7.23.

Profile of deposits at the Lister Site (28Me1A). The buried A horizon dates from the mid- to late-18th-century and is capped by eroded and redeposited sediments from nearby gentle slopes.

does not represent a cut through slumped material if you are dealing with bank cuts, exposures along a stream, blowouts, cliff-faces, erosive cuts, or any type of ditch or hole in the ground other than an intentional excavation. You may already have some expectations about the natural stratigraphy of the area based on your review of a published soil survey or other background information.

If the profile has been exposed to the weather for too long and is overly dry or wet, consider recutting the face of the exposure. Before doing this, however, take note of any visible layers or boundaries. Extreme drying or wetting may reveal characteristics of strata not apparent under other conditions—

- ☑ strata with the greatest percentage of coarse sediments will dry out first since voids or pore spaces are larger and promote more rapid evaporation. Care should be taken when dealing with a stratigraphic sequence consisting predominantly of sands. Often the tension of water in pore spaces may be the only thing holding the profile together, and excessive drying could lead to the collapse or slumping of the vertical section or excavation wall.

- ☑ strata with the greatest percentage of fine sediments will retain moisture longer since pore spaces are smaller and retard evaporation.

- ☑ strata with a high clay content will develop well-defined cracks if the exposure is incredibly dry, or has been exposed to the sun for a long period of time. Special note should be made of soils that have the potential for such cracks to develop since they provide pathways along which artifacts can be displaced from their original plane of deposition. Areas with dramatic seasonal variations in rainfall and temperature will be the most susceptible to this phenomeon.

If a profile is too dry and recutting it is not an option, use a spray bottle or pressurized sprayer to gradually wet it. After making observations of a profile under normal circumstances, wet it to see if you have missed any interesting details.

In cleaning a profile, work from the top of the exposure downwards since your activities will create loose sediment and debris that will adhere to some degree to lower portions of the profile. Make sure that you use a sharpened trowel with a beveled edge during your examination (see chapter 5). "Cutting" through soil or sediments reveals the profile in the best way. Scraping or dragging soil/sediments with the trowel can obscure color and textural characteristics and the boundaries between strata. If you are faced with a profile where there are few discernible changes, try cutting the exposed face in different directions (e.g., left to right) since this will influence shading and the way your eye perceives color and textural differences. Regardless of the direction you choose, always cut the entire profile in the same direction so as not to blend or smear color distinctions. Cutting the profile with your trowel may also allow you to literally hear changes in the stratigraphy. The sound that your trowel produces will vary dramatically when you are cutting sediments or soils of different textures, e.g., a strata in which clay-sized particles predominate versus one that has a high percentage of sand. You should also be able to feel such textural differences—one will resist the cutting action of your trowel more than another.

Use root clippers or loppers to trim roots back to the face of the exposure. Don't attempt to remove roots, as this will damage the face of the profile. Use a brush or small whisk broom to clean sediments from rock or other large objects exposed in the profile (e.g., bone, architectural ruins, or materials). In some cases the brushing of sediments from objects may obscure the color of the object. While this may not be a concern when you are drawing a profile, it is something to be aware of if the profile will be photographed. Reclean these larger objects as necessary to enhance their color for photography. Avoid brushing sediments or soils in the profile since this will smear and obscure colors, an important characteristic in defining strata. Don't remove artifacts, rock, or large objects from the face of the exposure as their position should be noted in your drawing of the profile, and will be useful in interpreting the stratigraphy.

The profile may be photographed once it is clean, prior to defining stratigraphic changes, since the definition process may disturb portions of the profile. Not every profile may need to be photographed. Minimally, it is advisable to have photos of profiles that are representative of different portions of a site and landscape, unusual stratigraphy or phenomena, and those revealing any type of cultural feature. It will be impractical or impossible to photograph the profiles of shovel tests or auger borings. If it is feasible to photograph every profile, so much the better. Photographs are an invaluable complement to the paper records that are produced. The profile may be reshot after strata have been distinguished and visibly marked in some way, e.g., boundaries scribed with the sharp point of a trowel or defined with string or a sequence of pins or nails pressed into the face of the exposure. The before and after shots provide one means of evaluating the reliability of the way that stratigraphic layers and boundaries were defined.

Both color and black-and-white 35 mm shots are recommended. Color prints or slides will be useful as a record of the profile and will come in handy in any public presentations regarding the fieldwork. Black-and-white shots are taken as a matter of record, but also in anticipation of their use in any report or publication that might result from the work. It can be difficult to shoot profiles because of lighting conditions. A sign board listing the site or project, provenience of the profile, and date may be used. Sign boards are impractical if their use obscures significant features of the profile or if the scale of the shot is such that they are rendered unreadable. Take closeups of special aspects of the profile, like exposed portions of cultural features, or anything that might be relevant to the interpretation of the stratigraphy. Unless you are using top-of-the-line equipment, 35 mm film provides a better photographic record than digital photography. On the other hand, digital photos are handy for rapidly conveying information to colleagues and sponsors via the Internet, creating websites, graphics for presentations, and publications. Additional information on photography in the field is found in chapters 8 and 9. Video recording is beneficial but not as frequently employed as still photography. One special advantage of video footage is that it can include commentary about what is being seen and individuals pointing out significant features on camera.

DEFINING INDIVIDUAL STRATA

To define individual strata in the field you will rely on the characteristics of color, texture, gravel content, and in some cases the presence of artifacts or features. Full definition of cultural stratigraphy must rely on data from controlled excavations. But in profile, some things may be very obvious such as large artifacts, concentrations of artifacts, portions of features including architectural elements, construction or destruction rubble, and placed fill. On some types of sites, profiles dominated by features, architectural elements, construction and destruction rubble, or placed fill may be the rule. You should be supplied with guides for describing the basic characteristics of soils or sediments, sample bags or containers, hand lens, Munsell color chart, water bottle, rules/tapes, graph paper, and notebook. Answering the series of questions that follow will help you to define individual strata and the boundaries between them.

Where are obvious changes in color, texture, and soil structure in the profile? Color and textural differences will probably be the most visible characteristics to consider. I've already provided some suggestions about how to discern textural differences between strata. Another way to feel textural differences involves using the point of the trowel or knife to repeatedly stab the profile. Pick a starting point at the top of the profile and gently stab the sediments moving down the profile. You will feel the ease with which the tool penetrates the face of the exposure, which can reflect differences in the sand fraction versus finer particle sizes in a layer, and/or the degree of structural development in the soils contained in the profile. Gouging the profile with a knife or the pointed end of a trowel to retrieve hand samples of sediment and to see what type of scar is left on the face of the exposure is part of the procedure for evaluating the structure of any soils present.

Are there textural differences within any stratum defined on the basis of color? Are there structural differences within any stratum defined on the basis of color and texture? Pay special attention to overly thick strata (1 foot or more) to see if you can discern internal divisions. Internal differences in depositional strata have implications for the process promoting deposition, e.g., the fining upwards of texture in some alluvial deposits. With soils, internal textural and structural differences may allow you or a specialist to distinguish divisions or different types of B horizons.

Where are large artifacts or other objects, obvious clusters of artifacts, placed fill, features, architectural elements, or anomalies exposed in the profile? All of these details are worthy of being distinguished in your drawing of the profile, but some may be contained within a stratum, while others may constitute a stratum of their own. Things may be visible in the profile that aren't readily understandable (anomalies), but they should be recorded nonetheless. The points of origin of undisturbed features (their tops or where they are first recognized), and the vertical distribution of large and heavy artifacts exposed on the face of the profile are important for identifying former living surfaces whose original character has been obscured by soil development. For example, a pit is dug and backfilled from an existing surface, so the top of an undisturbed backfilled pit indicates the approximate location of the original surface. Large artifacts or objects are not as prone to postdepositional movement in a column of sediments as smaller ones. Their clustering at a common level in a profile could indicate the approximate location of a former surface.

To make further documentation of the profile easier, inscribe the boundaries of strata and features with the point of your trowel or knife into the face of the exposure. Other techniques that I have seen employed include pinning string along the boundaries, or using nails, tooth picks, or small wooden or bamboo skewers to delimit them. Your stratigraphic divisions should be approved by a field supervisor before you continue. With the strata delimited in some physical way, the profile can again be photographed. Prior to photography it will be necessary to clean portions of the profile that were disturbed in the process of testing for textural and structural changes.

All drawings are stylized to some degree. Decisions are made about what is important to show, how to depict it, and what can be safely ignored or documented in some other way. For example, precisely defining the vague lower boundary of some soils may be difficult if not impossible; the boundary must be generalized to some degree. Tree roots can disturb archaeological deposits, so it's important to show large ones that are exposed on the faces of a profile. But how big is big enough? Should every root be shown? How should the mottling of the color of a stratum be shown, or is it necessary to show it at all? To make reasonable decisions you must be aware of the possibilities stemming from the analysis of sediments, soils, geomorphology, site formation processes, and the impact of human activities on the environment. I've outlined some basic parameters to follow, but they are not exhaustive of the possibilities. Your interaction with your supervisors and specialists in the field will refine your appreciation of these issues.

DRAWING AND DESCRIBING A PROFILE

The most accurate drawings of a profile require the establishment of a level line from which measurements are taken (Figure 7.24). The use of a level line insures that the vertical relationships between phenomena in the profile are preserved with accuracy. A stake, surveying pin, metal spike, or anything that can be firmly anchored in the ground is placed at the surface on either side of the profile to be recorded. I prefer surveying pins or gutter spikes for this task. They provide firm support, can be placed in the ground by hand, and are

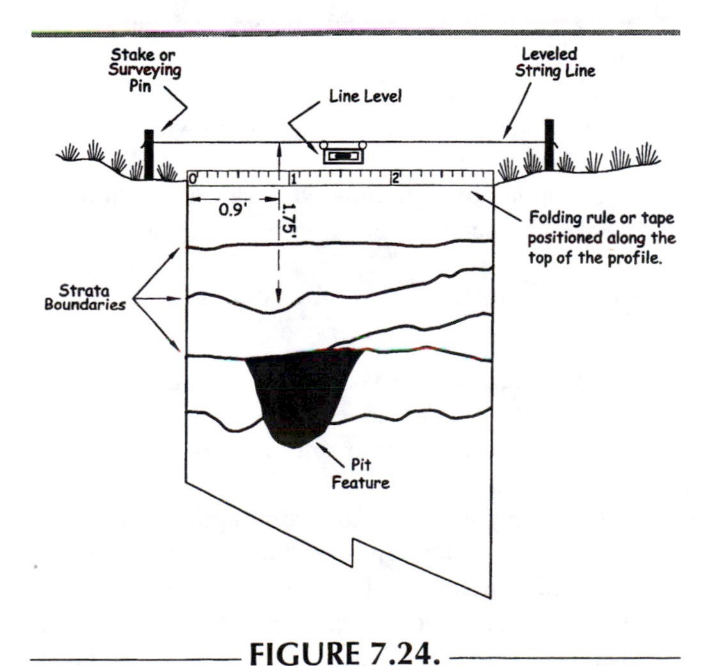

FIGURE 7.24.

The use of a level line in drawing a profile.

easily manipulated. The pins should be positioned so that a string tied between them would slightly overhang the face of the exposure. This will make measurements from the line to be strung between them down the face of the profile more convenient. Tie one end of a string to one of the stakes and thread a line level onto the string. A line level is a cylindrical bubble level about the length of a finger. With the loose end of the string, position yourself near the free pin and move the line level to the center of the string. Pull the string taut, holding it next to the pin. Gradually move the free end of the string up or down until the bubble on the line level is centered. Carefully wrap the string around the pin and tie it off. If the bubble on the line level needs to be recentered, one of the pins on either end of the strung line can be pushed slightly deeper into the ground. If the bubble is off center to the right, push the right pin slighter deeper, if off center to left depress the left pin.

Before taking any measurements, set up a piece of graph paper for the drawing. If you are working in engineer-scaled feet or meters, use graph paper that is organized in divisions of 10 between the bolder guidelines. Choose a scale for the drawing that will accommodate the length and width of the profile, and that will allow the intricacies of strata boundaries, features, etc. to be shown. Drawings of overly long, overly wide, or complex profiles may have to be drawn in sections or employ larger sheets of graph paper. It may also be useful to create a series of drawings of the same profile that focus on particular aspects of it, e.g., natural stratification, cultural features (Buccellati 1992). If practical, use the same graphic scale for drawing all profiles for a particular area or site. This will make comparisons and the correlation of stratigraphy between exposures or excavations easier. Pick a horizontal line on the graph paper to represent the level line that you have established.

Lay a folding rule or a metal tape along the surface at the top of the profile. The "0" end of the tape should be oriented with one edge of the cleaned profile. This will be one of your guides in recording the measurements necessary to make the drawing. If the surface of the ground at the top of the profile is sloping, you will have to establish control points on the ground. Have two people pull a tape between the pins supporting the level line, holding the tape parallel to the level line. With a plumb bob, drop measured points onto the ground at a regular interval, e.g., every 0.5 foot. Mark the points on the ground with nails.

With your control points established, take measurements down from the level line to points along the boundaries of strata, features, etc. that you have delineated on the face of the profile. The points to be measured include where an individual boundary line meets the edge of the cleaned profile or excavation wall, and sufficient points between these edges so that you can accurately draw boundary lines by connecting the dots representing your individual measurements. The more complex the boundary to be drawn, the greater the number of points that will have to be measured. The surface at the top of the profile is also a line to be drawn. Each point to be located requires two measurements: a horizontal one indicating its distance from the "0" end of the tape at the top of the profile; and a vertical one indicating its distance or depth from the level line to the point. Figure 7.24 shows an example of recording a single point (A) along the boundary of a stratum. Point A is 0.9 foot from the horizontal starting point and 1.75 feet below the level line. Similar types of measurements along other portions of the same boundary will allow you to recreate it on the graph paper. It is best to work with one boundary at a time, from the top of the profile to the bottom, beginning with the surface.

With the basic drawing complete, label the strata or other phenomena that you have shown. There is no one system that archaeologists use for labeling strata. Some assign numbers to individual strata, from top to bottom beginning with the number 1. Others use letters. I recommend that the use of letters be avoided since this can be confusing if soil horizons are subsequently identified in the profile. Later, specialists defining lithostratigraphic units may use numbers or letters and will assign them from the bottom up in a sequence (Waters 1992:62–68). If you are profiling the wall of a completed excavation you will undoubtedly be using the system already established for labeling strata (see chapter 9). Any cultural features or anomalies portrayed will be given a designation by your field supervisor.

Describing each stratum that you have defined, as well as the features, anomalies, and other details included within them is the next task. Procedures for determining color, texture, gravel content, and structure, and the conventions for describing them have already been presented. The methods for using field characteristics of sediments and soils to determine texture may not always be adequate for revealing differences that you believe exist in the profile. Any such differences should be discussed in detail in notes prepared to accompany a profile drawing.

If you are in a situation where you don't have the necessary guides to describe the characteristics of sediments and soils, or you are uncertain about your ability to accurately characterize color, texture, and structure, employ a relational strategy until you can get guidance from others. Rank strata in terms of criteria that make some sense to you, or that you are comfortable observing, for example

☑ ***Color***—which is the darkest stratum, next darkest; which is the brightest colored stratum, next brightest

☑ ***Texture***—which stratum has the greatest percentage of clay, the next greatest; which stratum has the greatest percentage of sand, next greatest

☑ ***Soil Structure***—which stratum has the most well-developed structure or is the most compact, which stratum has the next greatest ranking for these attributes; which strata have no discernible structure

If nothing else, this method will allow you, or anyone, to tell whether there is patterning in the characteristics of the soils or sediments you have observed, and whether these patterns conform with particular landscape-shaping processes or soil horizons. Remember that if you are working in teams let one person do all of the color descriptions. Every profile drawing should possess the following minimal characteristics—

☑ a title block noting the project and site, what is being drawn, the provenience or spatial location of the profile, the date, the name of the person or persons creating the drawing, and a catalog number for logistical tracking and linking it with other relevant documents, artifacts, or features (see chapters 7 and 8)

☑ a scale for the horizontal and vertical dimensions depicted

☑ a key to any symbols used in the drawing

☑ descriptions of individual strata or a notation that such are attached or exist on separate sheets

☑ some indication that the drawing has been approved by a field supervisor

You may be able to make other important observations about the profile to accompany your drawing. These observations relate to somewhat higher levels of interpretation, some of which may stretch your personal skills. Do any of the defined strata possess the characteristics of a master soil horizon? Do trends in color, texture, and structure with depth correspond with the expectations of in-place development of soil? Are there any reversals in color, textural, and structural trends in the profile? Where are they located? In other words, can discontinuities or unconformities be identified? Which strata are depositional? Which strata or portions of strata might be former surfaces?

COLLECTING SAMPLES

There are many types of samples that could be collected from a profile. Sediment and soil samples from individual strata can be collected from the face of a cleaned exposure and submitted for laboratory analysis of particle size, chemistry, and trace elements. Samples for determining whether pollen or plant phytoliths are present, and the nature of the environments they represent, can be collected from more interior sections of the exposed profile to avoid contamination from materials adhering to the face of the exposure. Samples of sediment, rock, or burned clay can also be collected for dating by optically stimulated luminescence, thermoluminescence, or paleomagnetism; sediments or soils with a high organic content can be collected for radiocarbon or oxidizable-carbon-ratio dating. These samples are best collected under controlled conditions where their context is not in doubt; deriving them from the walls of a profile may not always be appropriate. In any case, the collection of samples for specialized analysis may often be accomplished by the specialist themselves, or under their guidance.

Column samples are samples of constant volume taken in small, continuous vertical increments down the wall of an excavation or profile. Nothing is removed from the samples as they are taken, including artifacts. Because they are taken in standardized continuous increments, some samples may crosscut the defined boundaries of strata. If the column is taken down the wall of a completed excavation, it is likely that samples also will crosscut the levels used in the excavation of the unit (see chapter 9). This crosscutting of strata and levels is acceptable. The vertical increment employed in taking columns samples should be smaller than the vertical dimension of any excavation level. Data from column samples can be used as an independent test of inferences based on data from controlled excavations, so you don't want to use the same sized increment in your column sample as that employed in excavation levels. Think of the increment as a way of controlling the context of whatever is included within the sample. The smaller or more fine grained your control over context, the greater the flexibility you have in manipulating and using the data from that context. I regularly use a vertical increment of 0.2 foot (about 6 centimeters) and one square foot (about 30 centimeters on a side) as the horizontal dimension of an individual sample. Once collected, samples can be passed through graduated geological sieves as an initial means of sorting contents, and then flotated (see chapter 9).

Column samples make available fine-grained information on the vertical distribution of gravel-sized sediments, artifacts, and botanical and faunal remains. Vertical distributions of materials can be examined by type, frequency, size, or weight and compared with data from general excavations to evaluate cultural stratigraphy, the postdepositional movement of artifacts, and the integrity of archaeological deposits. Processed samples provide information on the presence/absence of micro artifacts that can be missed during normal excavations where fine-meshed screens or flotation are not employed. An example of the utility of column samples is the research conducted at the St. Anthony Site, located in the floodplain of the Susquehanna River in Pennsylvania (Stewart 1994). The variable stratigraphy across the site (Figure 7.25) has resulted from the impact of flooding on the ancient undulating to gently sloping topography, and the plowing of the landscape during historic times. Controlled excavations into these deposits revealed two major concentrations of artifacts, one in the plowzone, and one in a buried surface and A horizon dating from approximately 800 A.D. to 1290 A.D. The analysis of column samples (Figures 7.26 and 7.27) confirmed the basic interpretation of the cultural stratigraphy. It also clearly showed the vertical distribution of artifacts within defined strata, and the degree to which artifacts had been displaced both above and below major planes of deposition. Analysis of the micro botanical remains in the column samples allowed probable historic and recent contaminants to be identified and factored out of the botanical assemblage from the buried surface.

Should all exposed profiles be treated in the same way? No. Some types of excavations like shovel tests (see chapter 9) don't provide an exposure sufficient to support all of the procedures that I've described. The stratification that they reveal is important, but can be documented by taking depth measurements to the generalized boundaries of strata and describing the characteristics of the sediments encountered. Some profiles may be redundant, i.e., they provide the same information, so a detailed drawing, description, and sampling of a representative example from the group could be completed, with others recorded in a more generalized fashion. For example, all four walls of an excavation unit may reveal the same stratigraphy, so a detailed description of one could stand for the others. Redundancy aside, detailed profiles from all areas of a site or landscape are necessary for correlating stratigraphy and understanding geomorphology.

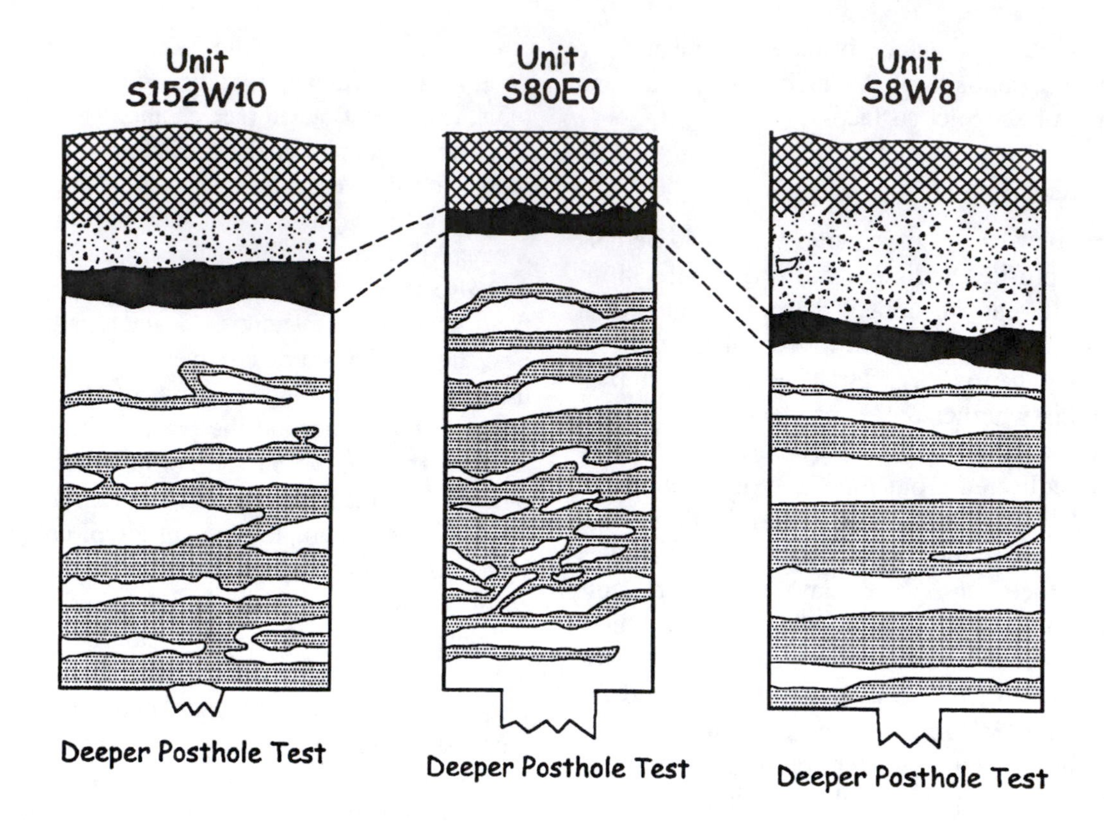

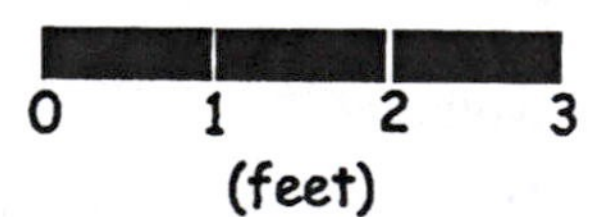

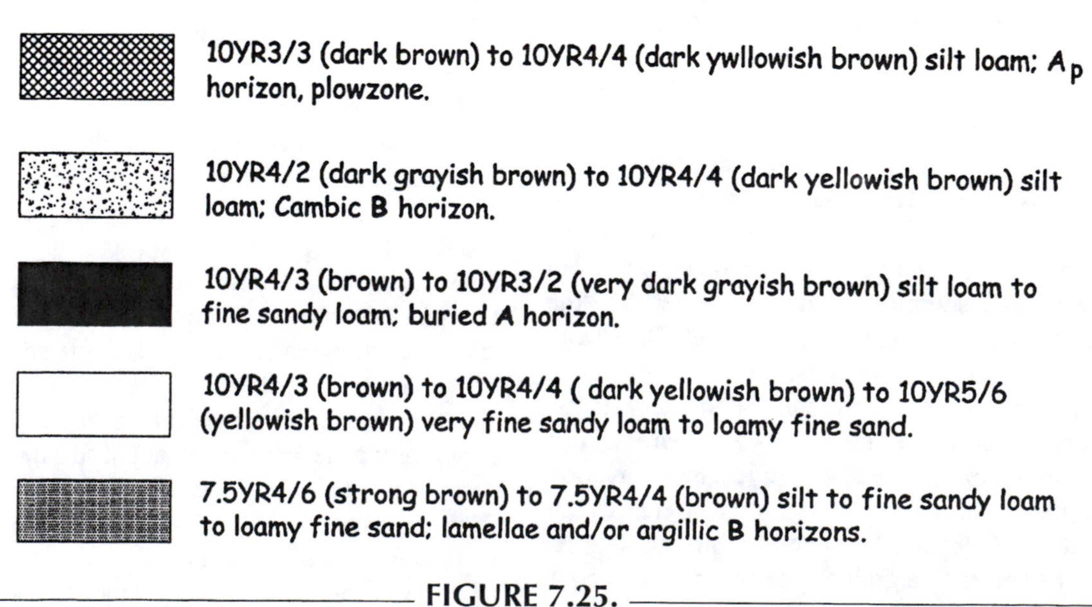

FIGURE 7.25.

Representative soil profiles from the St. Anthony Site, Pennsylvania.

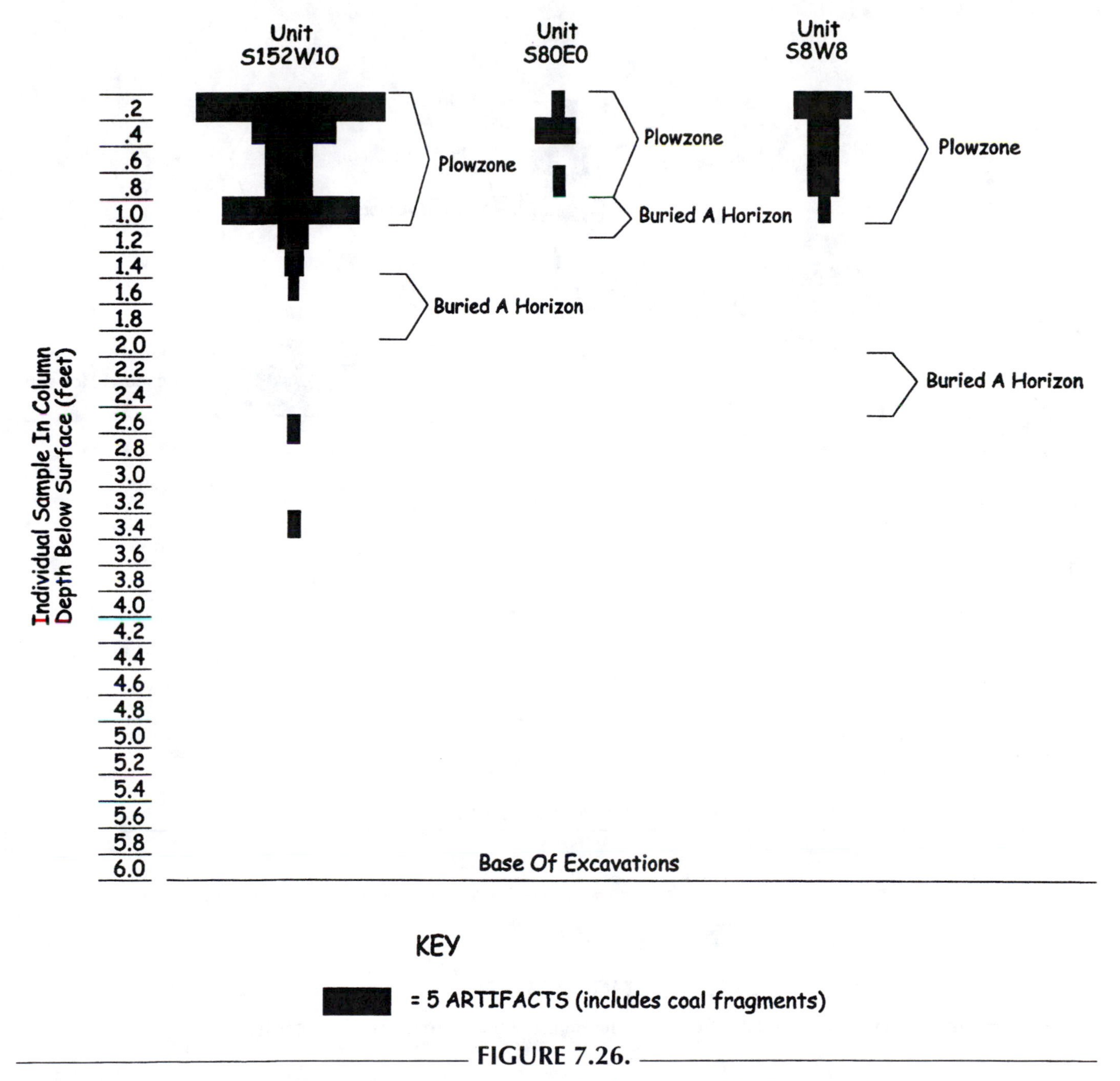

FIGURE 7.26.

Vertical distribution of historic artifacts from column samples at the St. Anthony Site, Pennsylvania.

DATING AND CORRELATING STRATA

Whenever possible, the dating of a deposit or stratum should employ multiple lines of evidence. Any good introductory text on archaeology will contain summaries of the variety of relative and chronometric techniques available to date deposits (e.g., Renfrew and Bahn 1996; Sharer and Ashmore 1993; Thomas 1998). Stratigraphy obviously reflects relative time following the law of superposition. Analysis of sediments and soils in their stratigraphic contexts can also be the basis for more precise estimates of the amount of time represented by a stratum, soil horizon, or stratigraphic sequence.

The formation of soil is time dependent. Remember, however, that to understand the time factor in soil formation, the differential effects of the other factors of the soil system (climate, organisms, topography, parent material) must be considered. This means that no single model for estimating the age of soil will work

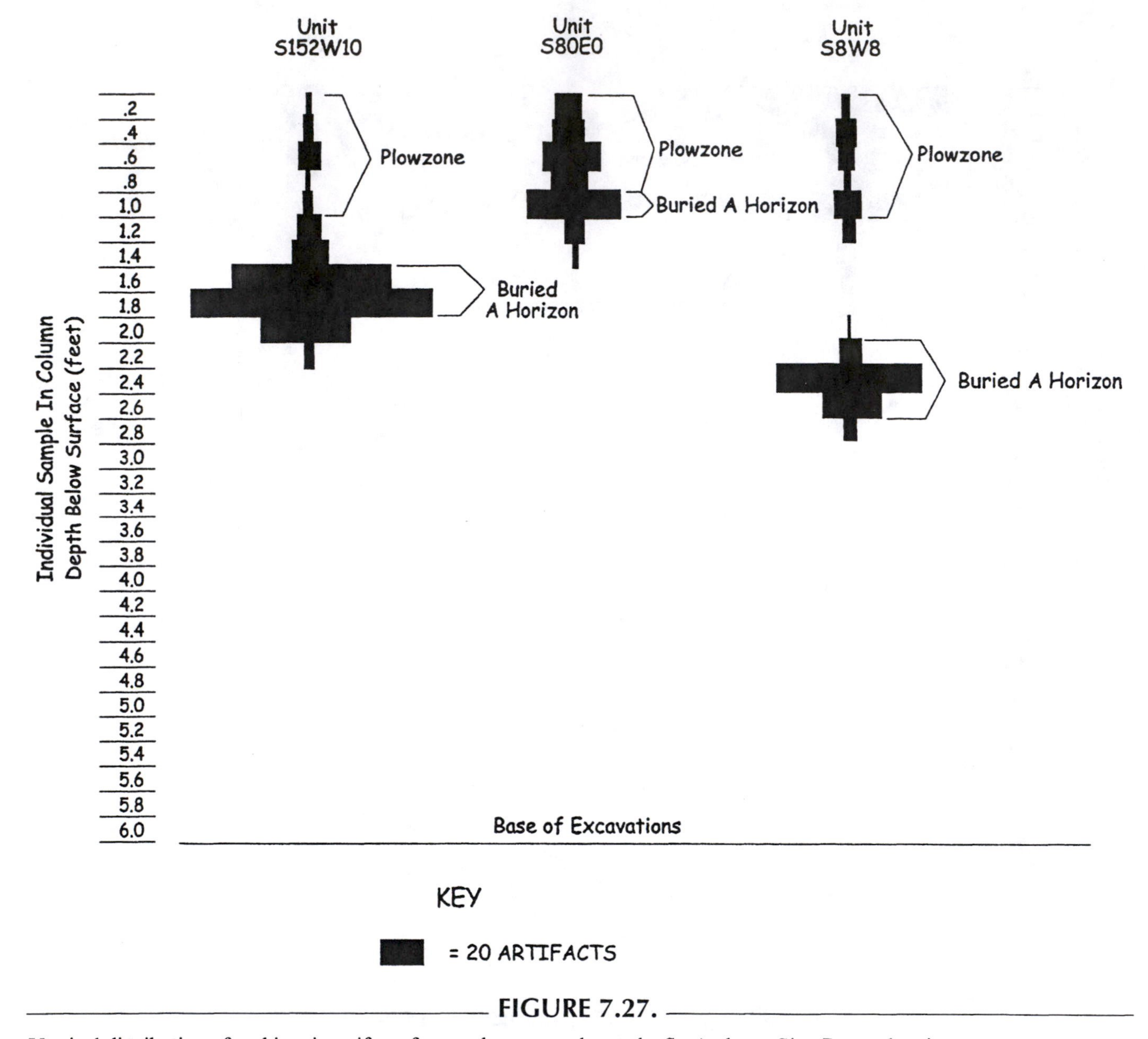

FIGURE 7.27.

Vertical distribution of prehistoric artifacts from column samples at the St. Anthony Site, Pennsylvania.

everywhere, even though all may rely on the evaluation and quantification of the same soil properties (cf. Bettis 1992; Bilzi and Ciolkosz 1977; Birkeland 1999; Daniels and Hammer 1992; Foss and Segovia 1984; Harden 1982, 1986; Holliday 1992b; Meixner and Singer 1981; Vento et al. 1990). Models for estimating age must take into account local environmental and geomorphic factors.

A and B horizons are the most obvious expressions of the passage of time represented by the weathering of soils. For an A horizon to reach a steady state, i.e., under stable surface conditions its thickness remains constant because additions of decomposed organic matter to its surface are balanced by the weathering and leaching of materials from its base, may take from a few centuries to over 1,000 years. Climate and vegetation are major factors influencing the rate of formation.

The color, clay content, structure, presence and thickness of clay skins on peds are properties that reflect the amount of time that B horizons have been developing. Color may be the first property to appear in forming B horizons. Cambic B horizons (Bw) are relatively young and just beginning to show red coloration and/or the development of structure, but little accumulation of clay. Progressively more well-developed B horizons show brighter colors, stronger and larger/more massive expressions of structure, and higher accumulations of clay (B2 horizons). Significant accumulations of clay are used to define argillic B horizons (Bt). In the southeastern United States, cambic B horizons have minimum ages of 2,000 years, argillic horizons 4,000 years. Continuous clay coatings on the peds of an argillic horizon are an indication of from 8,000 to 10,000 years of soil weathering (cf. Foss et al. 1992; Foss and Segovia 1984). Parent material and hydrology are two factors that can have a disproportionate impact on the rate that these properties appear and develop. For example, the accumulation of iron oxides (red colors) and clay in sands, or sediments with sandy textures, is slower than in sediments with finer textures. The same relationship holds for the development of structure. The amount of water moving through the soil column affects rates of translocation, and therefore the rate at which color, clay, and structure are expressed in B horizons.

Multiple soil sequences and sedimentary deposits complicate the process of estimating the age of a profile. Individual sequences of soil weathering should first be evaluated on their own. The presence of multiple weathering sequences means that discontinuities or unconformities exist and that erosion and/or deposition has interrupted soil formation. Estimating elapsed time between episodes of deposition is difficult, if not impossible, if based strictly on the characteristics of sediment or stratigraphy. When erosion is indicated, e.g., a discontinuity where an argillic B horizon overlays a cambic B horizon, we can't be certain about what has been removed by the erosive event and the amount of time associated with the accumulation and weathering of this material.

Correlating stratigraphy between exposures or excavations at a site or on a given landscape is a procedure that bridges fieldwork and laboratory analysis. Correlation involves answering some basic questions. What are the horizontal boundaries of the strata visible in profiles? Which strata or vertical boundaries of strata represent the same event, and which strata represent the same points or ranges of time? Correlation serves a variety of purposes. It is part of reconstructing a history of how the landscape developed. The effect that geomorphic processes have had on site formation and the integrity of archaeological deposits helps to determine the significance of the deposits and the need for, and nature of, ongoing or future fieldwork. It is a way of organizing archaeological deposits into meaningful units for description and analysis.

The spacing of excavations and the use of augers in intervening areas is a direct way of tracking the horizontal extent of individual strata, whether cultural or natural. Following similar-looking deposits in similar stratigraphic relationships over space is just one part of this procedure. Because of geomorphic processes, sediments deposited as a result of the same event can exhibit different characteristics over space, and exist at different depths below the surface in different stratigraphic sequences (recall the discussion of alluvial processes in a floodplain). The erosion of a landscape can sweep away sediments and soils of different ages and leave exposed at a common surface sediments and soils of different ages. Thus an understanding of geomorphic processes and their sedimentary and stratigraphic fingerprints is essential to correlating strata that represent the same event, or a boundary between strata that represents an erosive event. Horizontally extensive anthropogenic soils, architectural or large cultural features (like roadbeds) may link deposits across a landscape. In turn, the differential removal, exhumation, deposition, or removal of materials on a landscape by humans can create stratigraphic relationships over space as complex as those resulting from natural processes.

Correlations of strata and boundaries representing deposition, erosion, and soil formation do not automatically equate with correlations of the age of deposits. The range of time or age that can be assigned to a deposit from the same event can be different, depending upon how subsequent geomorphic processes affect the landscape. For example, sediments deposited by a flood in interior sections of a floodplain may remain exposed at the surface for a much longer period of time than those deposited nearer the stream channel (see Figure 7.21). The interior floodplain sediments could therefore contain artifacts from numerous occupations, while those nearer the channel contain only those of a single component. As emphasized above, the dating of deposits must employ multiple lines of evidence.

The ***Harris Matrix*** is a means of organizing the units of stratification into a sequence of deposition using a schematic diagram (Harris 1975, 1979). It involves the correlation of levels, layers, or strata (referred to as units of stratification) across a site or excavation area and provides a visual synthesis of these relationships that is in essence a depositional history. In this case, history means a sequence of events, not the absolute age of the events or the range of time assigned to a deposit. The Harris Matrix enjoys its greatest use in the United States among historic archaeologists who generally deal with sites that present more complex stratigraphic situations than those explored by prehistorians. There are some pitfalls associated with the way this system has been used in the field. The importance of ongoing stratigraphic correlation while in the field and its impact on the day-to-day evolution of excavation strategies is unquestionable. However, some archaeologists attempt to label strata and excavation levels as they are encountered with designations implying correlation. Ideally Stratum 1 in excavation unit S10E40 is equivalent, in terms of depositional history, to Stratum 1 in unit N20E40. Attempting to embed correlations in the labeling of strata while in the process of discovering and understanding stratigraphic variability seems doomed to constant revisions. And revisions confound the paperwork documenting the work already completed.

IMPACT ON FIELD STRATEGIES

With an understanding of sediments, soils, and geomorphic processes you will look at landscapes with different eyes and will be able to make greater use of background data like that found in soil surveys for planning and executing fieldwork. We've come full circle from the consideration of prospects at the beginning of the chapter. This understanding bears on four different aspects of fieldwork, the utility of surface inspections, the placement (spatial distribution) of excavations and subsurface tests, excavation procedure, and distinguishing natural anomalies from cultural features.

Given the topographic setting of the project area, what kind of geomorphic processes might have affected the landscape? Which portions of the landscape are likely to be dominated by deposition, erosion, or stability? Where would sediment be expected to accumulate the fastest? The slowest? Where could archaeological deposits be expected to be deeply buried, stratified, shallow and possibly mixed? Answers to these questions can guide decisions about the utility of surface inspections in locating sites since portions of an existing surface may be of different ages. We want to distribute excavations across a landscape not only to help us find archaeological deposits of all ages, but to find them in contexts where they are most well preserved. To get the most complete picture of the geomorphology of a landscape and its stratigraphic variability, excavations and other subsurface tests may have to be placed in areas where we might not expect to encounter archaeological deposits.

Deciding how deep to take an excavation or subsurface test relies on estimates of the age of the deposits being encountered. In turn, the depth of an excavation has a bearing on its horizontal dimensions and safety precautions taken. In some cases, it may not be practical or possible to have an exposure of sufficient depth because of safety concerns or logistical constraints. Augers may be employed at the base of an exposure to extend your observations on stratigraphy to greater depths. During the site discovery phase of fieldwork the extreme depth of deposits of potential interest may make the use of earth moving machinery for excavation advisable, bringing with it a new batch of safety concerns. Subsequent excavations of deeply buried deposits would require opening a large, stepped excavation block (see chapter 9).

How an individual excavation proceeds, how the ground is taken apart, follows from assumptions or knowledge about how it was formed and resulting stratification. In early stages of fieldwork, excavations generally progress by levels that conform to changes in the color, texture, or some obvious property of the sediments (including artifacts), or are embedded within grossly defined strata. Overly thick strata are dismantled in small arbitrary levels until their nature is understood. Subsequent adjustments of levels will depend on your understanding of the depositional processes at work, the impact of soil formation on the stratigraphy, and how disturbed or undisturbed any given deposit may be, and, of course, accumulating information about the vertical distribution of artifacts and features. The three-dimensional location of artifacts may be plotted during the excavation of an undisturbed deposit or one that represents a single component or narrow slice of time. Unit and level provenience may suffice for artifacts from less well-preserved or more broadly defined contexts (see chapter 9).

The more background you have the better your ability to distinguish natural variations or anomalies in sediments, soils, or stratigraphy from those that are cultural in origin. This harks back to one of the points about the recognition of archaeological evidence in chapter 2. Understanding the phenomena that nature is capable of producing makes it easier to detect the residues of human activities, many of which may have no correlates in your personal experience.

Field strategies must be adaptable and in fact do change over the course of fieldwork. Fieldwork constantly generates information. Your comprehension of sediments, soils, and geomorphology, and the variability that is part of any landscape will be cumulative. Take advantage of your growing understanding of the landscape.

APPLYING YOUR KNOWLEDGE

1. Determining sediment or soil texture and color is a group exercise and requires a room with lots of table space that can get dirty, and sinks if possible. Each participant must collect and bring in three different sediment or soil samples. They can be from three different strata or horizons of the same profile, or they can be from different profiles or contexts. Each sample should fill at least half of a quart-sized plastic storage bag. Record the type of landscape from which your samples were collected, e.g., floodplain, non-floodplain, stream terrace, hilltop, etc., and label the sample bags with this information, a sample number, and your name. Distribute the sediment samples around the room in which you will be working. You will need copies of the table *Field Characteristics of the Major Textural Classes,* one or more books of Munsell Soil Color Charts, squeeze bottles of water, paper towels, and materials for taking notes. Work in teams of two. Characterize the texture and color of as many samples as you can in the time allotted for the exercise. Compare your results with those of other teams. How do they compare? What might account for the variability in results? What might be done to reduce this variability?

2. Get the published soil survey for your county, or one in which you are interested. Review it and try the following: 1) Pick the soil series and phase that exhibits the most complicated or dynamic depositional history and explain your choice. 2) List the soil series and phases that exhibit an unconformity. Do they occur in the same type of environment as the soil noted in #1? 3) Pick a series/phase that represents a stable landscape for the longest period of time and explain your choice. 4) Pick the soil series/phase that you think represents the youngest landscape, or the smallest amount of time. Explain your choice.

3. Create a scale drawing of a hypothetical profile that contains both sedimentary deposits and soils that embodies the evolution of a landscape. On your drawing describe the color, texture, gravel content, structure, and nature of the boundaries for each layer or stratum in your profile. **Don't** label any master soil horizons. Exchange your profile with a classmate and challenge them to correctly label any soil horizons present, interpret the profile and describe the sequence of events and processes that led to its formation, and identify any boundaries that might have been surfaces upon which people could have lived and discarded artifacts, and explain your choices and or the lack of any such boundaries in the profile.

4. Find an archaeological site report in your library or obtain a cultural resources management report from your SHPO. Prepare a review of the report with the following questions in mind. Did a consideration of soils and geomorphology figure into the planning stages of the project? How? How did a consideration of soils and geomorphology influence the placement or location of excavations or subsurface tests, and the excavation of individual units? Summarize the stratigraphy of the site and the methods used to describe it. Were basic descriptive data for sediments and soils recorded? How were strata dated and correlated? What methods were used to define the cultural stratigraphy and organize excavated artifacts and data for additional analyses? Given your understanding of sediments, soils, stratigraphy, and geomorphology, did the investigators take full advantage of the concepts and techniques available? What would you have done in their position?

DIG DEEPER

❑ There are numerous introductory texts on sediments, soils, and geomorphology written by specialists in these fields, and finding one in a college library or buying one online should be an easy task. There is also a substantial number of works written by archaeologists showing the application of these earth sciences to archaeology.

Continued

- Hassan, Fekri A. 1978. Sediments in archaeology: Methods and implications for paleoenvironmental and cultural analysis. *Journal of Field Archaeology* 5:197–212.
  This is a great, article-length introduction to the potentials and usefulness of sediment and stratigraphic analysis in archaeology.
- Waters, Michael R. 1992. *Principles of Geoarchaeology: A North American Perspective.* Tucson: University of Arizona Press.
  My favorite introductory text for archaeologists. I use it in the classroom for teaching an introduction to sediments, soils, and geomorphology for archaeologists.
- Holliday, Vance T. (ed.). 1992. *Soils in Archaeology: Landscape Evolution and Human Occupation.* Washington, D.C.: Smithsonian Institution Press.
  A collection of case studies documenting the different ways in which soils data are used in archaeological analysis and interpretation.

❑ For a detailed introduction to soils try:

- Birkeland, Peter W. 1999. *Soils and Geomorphology* New York: Oxford University Press.
- Brady, Nyle C. and Raymond R. Weil. 1998. *The Nature and Property of Soils.* Upper Saddle River, NJ: Prentice Hall.
- Davidson, Donald A. 1985. Geomorphology and archaeology, pp.25–56. In G. Rapp and J. Gifford (eds), *Archaeological Geology.* New Haven, CT: Yale University Press.

❑ This and the following reference provide a succinct look at the integration of geomorphology and archaeology.

- Stafford, C. Russell. 1995. Geoarchaeological perspectives on paleolandscapes and regional subsurface archaeology. *Journal of Archaeological Method and Theory* 2(1):69–104.

❑ If you want to pursue geomorphology in more detail start with introductory texts like those below, or look for books dealing with specific geomorphic environments, e.g., alluvial/fluvial geomorphology, aeolian (eolian) geomorphology, desert geomorphology, hillslope geomorphology, coastal geomorphology, glacial geomorphology.

- Butzer, Karl W. 1976. *Geomorphology From the Earth.* New York: Harper and Row.
- Easterbrook, Don J. 1999. *Surface Processes and Landforms.* Upper Saddle River, NJ: Prentice Hall.
- Ritter, Dale F., R. Craig Kochel and Jerry R. Miller. 1995. *Process Geomorphology.* Dubuque, IA: W.C. Brown.

❑ Relevant articles are typically found in the following journals or series: *Journal of Archaeological Science, Journal of Field Archaeology, Geoarchaeology, Quaternary Research, Geomorphology, Soil Science.*

CHAPTER 8

Working on the Surface

Supposing is good, but finding out is better.

Mark Twain, Eruption

DISCOVERING ARCHAEOLOGICAL DEPOSITS

You have arrived in the field. Landowners have been contacted and permission secured to trespass. Background research has been completed and your head is bursting (or soon will be) with what is already known about the environment of your project area, previously recorded cultural resources, and land use history. There is a framework or research design for how fieldwork generally will proceed. A sampling strategy, if necessary, has been formulated. Expectations have been generated about what types of archaeological deposits might be found and where. You and your crew mates are appropriately dressed and well equipped. Individual responsibilities have been made clear and a chain of command established. Useful local services, especially hospitals or emergency squads, have been identified and there are travel plans for how to reach them if an accident should occur. Everyone has been shown the topographic map for the area and is oriented.

Your primary goal is to find archaeological deposits and begin to learn something about their nature, i.e., characterizations of size and general content. Background research and predictive modeling will have generated expectations about what might exist in a project area. Deposits discovered and registered in the past may also be relocated. In the context of an archaeological survey, it may not be possible to take full advantage of the information embodied by a surface site or deposit, although some projects may be designed to move into more intensive work once a discovery is made. It may also be practical to thoroughly study low-density archaeological deposits when they are encountered rather than schedule additional stages of fieldwork to do so. For the purposes of this discussion, I'll distinguish between what is involved in surface surveys to discover archaeological sites or deposits and the more intensive study of these archaeological resources.

Factors affecting discovery are outlined in Table 8.1 and apply to both surface surveys and those that employ excavations. Site type and size can vary dramatically so even if deposits are abundant they may not be easily discovered. Generally, the more abundant an archaeological resource the greater the chance that examples of it will be discovered. The more clustered a resource, the better the chance that it could be missed, especially when an area is only sampled. How effective a particular technique is for discovering sites or artifacts depends on the size and other physical properties of artifacts and features, and the size of a site and the density of artifact deposits within it (McManamon 1984). Visibility is also affected by these same things, as well as the nature of the environment in which artifacts and sites are being sought. Artifacts along a shoreline are more easily seen during low tide than at high tide. More artifacts will be visible on the surface of a newly plowed field that has received rain than one that has not received any rain. Irregularities in topography representing human activities (e.g., house mounds, earthworks) may be more visible when vegetation has died back, or is just starting to grow, than when in full growth. And of course, sites or artifacts in areas that are physically difficult to reach are less likely to be discovered than those that are convenient to roads and developed areas.

A ***surface survey, surface inspection,*** or ***pedestrian survey*** (terms used interchangeably) is the first step in an archaeological survey or the site discovery phase of most projects. Even when you plan for the systematic

TABLE 8.1

Factors Affecting the Discovery of Artifacts and Archaeological Deposits

Factor	Description
Abundance:	the frequency or prevalence of a site or type of artifact within a given area; density of archaeological deposits per unit area (e.g., habitation sites are more abundant than cemeteries; tools and implements are more abundant than ornaments or ritual objects).
Clustering	the degree to which sites or artifacts are spatially aggregated (e.g., sites may cluster in a single portion of a project area rather than being widely dispersed; artifacts within a site are evenly distributed rather than organized in discrete activity areas).
Obtrusiveness	the ease with which a site or artifacts can be detected using a particular discovery technique (e.g., different sizes of excavation units, surface survey versus excavation; flotation versus dry screening).
Visibility	the ease with which something can be seen (e.g., large artifacts are more easily seen than small ones; artifacts are more easily seen on the surface of a plowed field than on the surface in a wooded area).
Accessibility	the effort required to reach an area or location.

Source: Schiffer et al. (1978).

coverage of an area with subsurface testing, examining exposed surfaces and natural "windows" into the ground is still the first line of attack and may help you streamline your approach to site survey in a given area. Any surface survey usually entails some type of subsurface investigation (see below and chapter 9). A prefield analysis of project area geomorphology and land use may have indicated that some surfaces are too young to have artifacts exposed on them, precluding a surface inspection. Such assessments need to be confirmed in the field by examining the stratigraphy associated with relevant landscapes.

Why should artifacts or archaeological features be exposed at the surface? Any number of natural and cultural processes could be responsible for erosion and the exposure of once-buried deposits (Schiffer 1987:121–140). Some environments are dominated by erosional processes. In others, the deposition of sediments and the development of soils can be extremely slow and balanced out by periodic erosion (Figures 8.1 and 8.2). Some natural processes may expose and rebury artifacts on a daily or seasonal basis. The influence of waves and tides on shoreline sites is a good example of this activity (Figure 8.3). One high tide may erode and expose artifacts on a shoreline while the next reburies them. Seasonal variation in tides and winds can amplify either the erosion or reburial process. Similar fluctuations are possible in arid environments where eolian processes are active, or in areas with little vegetation cover and dramatic seasonal fluctuations in precipitation. Plowing is probably the most familiar cultural process that leads to the exposure of artifacts on the surface.

Any surface exposure presents a biased view of an archaeological deposit (cf., Dunnell 1988; Dunnell and Simek 1995; Frink 1984; Shott 1995). Depending on the geomorphology of a landscape and the processes responsible for exposure, only artifacts of a certain age or size range may be visible. What and how much is visible on the surface can change through the year or by season, again depending on the operative processes. Plowed agricultural fields are a good illustration of this variability. The deep and rough plowing used at the beginning of the farming season produces large clods and deep furrows (Figures 8.4 and 8.5). Until the field has been washed with rain only large or brightly colored artifacts will be visible. Artifacts exposed in deep plow furrows may be reburied quickly as surface runoff and sheet wash move sediments from high points to low points in the field. Artifacts in a wide

FIGURE 8.1.

The Badlands of Nebraska, an environment dominated by erosional processes where exposed surfaces are of various ages.

FIGURE 8.2.

A surface scatter of artifacts in an arid environment in southeastern New Mexico. The scatter includes fragments of ground stone implements, fire cracked rock, and chipping waste from stone tool manufacturing and maintenance.

FIGURE 8.3.

A shoreline along the eastern shore of the Chesapeake Bay in Maryland. Shells and artifacts have been cast up on the beach by the action of waves and tides. They are from an inundated prehistoric Indian site just offshore and below the level of the current beach. Subsequent tides and wave action may remove or bury the shells and artifacts.

FIGURE 8.4.

The rough plowing that can characterize the initial working up of an agricultural field. Note the large clods of earth that this activity produces. Fewer artifacts will be visible on such surfaces in comparison with more finely worked surfaces.

FIGURE 8.5.

Initial plowing of a field can result in a surface distinguished by deep, parallel furrows. Any artifacts exposed within the lower portions of the furrows will become reburied as the field continues to be worked. Rainfall and the very localized erosion and redeposition of sediments that it promotes will also help to rebury artifacts.

FIGURE 8.6.

An agricultural field that has been more finely worked (disk harrowed) and planted. Large clods of earth have been broken up and deep furrows eliminated. The nature of the artifacts that might be exposed on the surface here will be different than those visible when the field was roughly plowed.

range of size classes may be exposed in a more finely worked field (Figure 8.6). The numbers of artifacts visible on the surface will also change through the plowing and weathering cycle of a field. By the end of the growing season (Figure 8.7), the surface may be so well washed and smoothed by the localized erosion and redeposition of sediments that many artifacts are reburied, and a once-visible archaeological deposit has become invisible.

Plowing not only vertically displaces artifacts but it shifts their position from side to side over time, distorting the size and shape of initial clusters of material (cf. Amerman 1985; Boismier 1997; Lewarch and O'Brien 1981a; Redman and Watson 1970; Roper 1976). While this lateral distortion may not be great, the patterned spatial distributions seen at the surface are not a precise reflection of distributions in the deposit prior to plowing. Recognizing and documenting patterning in surface distributions of artifacts remains important, but its analysis and interpretation must take into account the current understanding of the effects of plowing.

The effects of the weather and moisture can also change the way in which the color of surface sediments and patterns in color are perceived (Figure 8.8). In turn,

FIGURE 8.7.

The extremely well-washed surface of a plowed field at the end of the growing season. Very localized erosion and redeposition of sediments have filled in even the shallowest of furrows, giving the surface a flat and smooth appearance.

FIGURE 8.8.

The dryness of deposits can affect the recognition of colors and patterns of color at the surface that might be indicative of cultural features, or provide clues about subsurface stratigraphy. The darker colored surface in the foreground has been recently plowed. This has churned moister subsurface sediments to the surface and mixed them with drier sediments. The lighter colored surfaces in the background are dried out from having been exposed to the weather for a prolonged period.

the utility of aerial photography and surface inspections to detect cultural features and make inferences about subsurface stratigraphy is impacted.

The troubling conclusion—the inspection of surfaces, especially those in plowed fields, can have very different results depending on the time of the year and the conditions under which they are examined. This variation influences the interpretation of individual localities and in turn, analyses of regional collections of sites. The bias inherent in surface inspections can be reduced by repeated surveys of exposures under different conditions or times of the year. Repeated inspections almost never happen during Phase I CRM or site-discovery projects. However, the surface bias in these instances is often offset (or can be) by the use of excavations to sample the same areas. Repeated surveys should definitely be a part of more intensive studies of surface sites.

A surface survey can be ***systematic*** in that the area to be covered is repeatedly traversed by crew members spaced at regular intervals, noting the location of artifacts or making observations about other phenomena that might relate to archaeological deposits (Figures 8.9 and 8.10). The spacing between members of the team is such that their view of the surface between them should slightly overlap. The spacing can change, depending on field conditions, e.g., open space versus dense undergrowth. Compasses may be needed to keep crew walking parallel lines over long distances, or when working in wooded areas. It's okay to step off the precise transect line that you may be walking during a surface survey to get a closer look at something that you may have spotted, to collect it, or flag its position. When resuming forward progress through the survey area return to the line that you were following. In the woods it may be necessary to clear brush with machetes or brush axes following predetermined compass bearings that represent parallel transects through the survey area.

A systematic surface survey need not yield ***systematic coverage*** of an area. Exposed surfaces may be small and unevenly distributed. A survey area may include newly cultivated fields, fallow fields, and woodland. The age of surfaces in the survey may vary

FIGURE 8.9.

Beginning a systematic surface survey in the Pecos River Valley of New Mexico. A range pole is used to mark the starting point for the first series of transects to be walked. Crew are spaced at even intervals and a compass is used to take a bearing that will keep the crew on-line through this portion of the survey area.

FIGURE 8.10.

Systematic surface survey in the Delaware Valley of Pennsylvania. Spaced at even intervals, crew are traversing a plowed field in search of artifacts and other evidence of past human activity. The relatively small size of the field makes it easier to maintain the lines or transects walked by individuals without the use of a compass or other surveying equipment.

dramatically. Systematic coverage means that the search for archaeological deposits has been equally thorough in all portions of the survey area, regardless of the specific methods employed. In some portions of the area this might mean a combination of surface inspections and excavations, in others only excavations.

Even in wooded areas with few exposed surfaces a systematic walkover is necessary. Remember that you're not just looking for artifacts but suggestive patterns or anomalies in topography, vegetation, and drainage that might relate to past human activity. Things to watch for are summarized in Table 8.2, and some examples are shown in Figures 8.11 through 8.19. As I've said before, the greater your understanding of the natural world and what it is capable of producing, the easier it will be to recognize things as cultural in origin. Even steep slopes and rock outcroppings can't be ignored since these are locations where inhabited caves or rockshelters, quarry works, or rock art might exist. Sediment or soil colors apparent at the surface can be clues as to the existence of below-ground cultural features, but they can also furnish information about vertical and horizontal stratigraphy. For example, in a plowed field with slopes you might notice that the surface colors of high points are redder or brighter than surface colors of lower elevations. These colors reveal the different ways in which plowing and incremental soil erosion and redeposition have affected the field. Erosion occurs more frequently on the exposed slopes and high points of a field, redepositing sediments at lower elevations. Because of this, plowing will cut into weathered sediments or B horizons and churn them to the surface more frequently than at lower elevations. Since weathered sediments or B horizon soils are generally brighter than those of the topsoil or A horizon, there is a contrast in surface colors depending on slope and elevation. This means that the oldest artifacts that may be buried in the field have a greater chance of being exposed on the high points rather than the low points, even though the same archaeological deposit may extend throughout the field. It also means that the impact of plowing on archaeological deposits may not be as great in the low points in contrast to the high points. Stratigraphy or cultural features may have been completely destroyed by the plowing of the high points while remnants may survive elsewhere in the field. Agricultural fields are probably the easiest places to make these types of observations.

Recognizing standing structures as historic requires background in architectural history and/or comparing the location of existing structures with those derived from historic maps and documents. The nature and orientation of above-ground construction can provide clues about associated features and artifact deposits. For example, the ruins of a historic barn imply the existence of a nearby farmhouse or dwelling. The yard area associated with the public facade of a standing structure is less likely to produce buried artifacts than the yard area associated with the rear of the building.

Of course during a surface survey you also are looking for artifacts wherever exposed surfaces exist (Table 8.3; Figures 8.20–8.25). Some types of exposures actually provide a view of deposits that exist at or near the surface, such as in agricultural fields or places where vegetation cover is lacking. Some provide windows into what may lie even deeper below the surface, like stream banks or tree throws. Surface exposures may be examined as is, as in the case of agricultural fields, or may require some gentle troweling to see what may be masked by a thin veneer of sheet wash, like at the base of large trees or other random exposures that can be encountered in wooded areas. Some archaeologists may systematically rake small areas along transects through the woods to improve surface visibility. If surface conditions are poor in an unplanted plowed field, arrangements may be made to have the field retilled. Fallow agricultural fields likewise can be mowed and replowed in order to enhance surface visibility. If areas are extremely large, strips might be plowed through them rather than working up the entire area. The extra expense and scheduling that this involves should be considered during the planning stages of the project. Once plowed, a field needs to be washed with rain to heighten the visibility of any artifacts at the surface. Never have an area plowed that hasn't been cultivated in the past, and never plow deeper than the existing plowzone. You don't want to disturb deposits that haven't already been disturbed. Confirming that a landscape has been plowed can be accomplished quickly using an auger or coring device (see Chapter 9).

TABLE 8.2

Surface Observations Checklist

	Description	Examples
Topography	unusual trends in the way elevations rise or fall, unnatural topographic patterns unusually straight stream courses, stream courses with abrupt sharp angles mounded areas depressions caves rockshelters modified rock exposures or cliff faces	abandoned roadbeds, lanes, trails, railroad lines, terraced gardens, edges of old farm fields, placed fill and made land, quarry pits, dumps, structures, graves, petroglyphs (carvings on rock exposures) and pictographs (painted rock art), mill races.
Vegetation	introduced species species out of context patterned growth (or lack of growth)	linear arrangements of trees along old roads, field edges, house yards; ornamental plants or shrubs in the woods, lush growth over pit features backfilled with organic sediments.
Drainage	patterning in the shape and size of extremely wet or extremely dry areas (also may be expressed in patterned vegetation growth)	backfilled wells or any substantial subterranean feature.
Sediment/Soil Color	unnatural colors (for the context) patterns of color (also may relate to drainage)	middens, tops of backfilled cellar holes, tops of pit features in general.
Above-Ground Construction		standing structures, ruins, grave markers, patterned alignments of rocks or petroforms (e.g., medicine wheels).
Artifacts		

FIGURE 8.11.

A part of the extensive rock overhang at the Meadowcroft Rockshelter in Pennsylvania. The excavated and covered portion of the site is in the background. This type of natural feature was often inhabited by Native Americans.

FIGURE 8.12.

The depression that this archaeologist is standing in is a quarry pit in the Blue Ridge Mountains of Maryland. Here ancient Native Americans procured metarhyolite for the manufacturing of chipped stone tools.

FIGURE 8.13.

The circular depressions in the foreground and center of the photo represent the former location of kivas, or semi-subterranean ceremonial structures, constructed by ancient Puebloan people at the Village of the Great Kivas site on the Zuni reservation in New Mexico.

FIGURE 8.14.

The "V" in this portion of the Potomac River at the West Virginia state line is a wall of stone constructed to trap fish. The movement of fish downstream is blocked by the wall, making it easier to spear or net them. Wicker or basket traps may be set in the opening at the apex of the "V." It is still debated whether the fish trap was constructed by colonists or Native Americans.

FIGURE 8.15.

The isolated low mound covered with brush and mature trees in the midst of this fallow agricultural field is the location of a historic farmstead. The "mound" is a result of years of plowing the fields surrounding the former house and barn complex and subsequent erosion. The house and barn yards were not affected by similar rates of erosion. The mature trees once stood in the yard area of the house. Church Creek, Maryland.

FIGURE 8.16.

The low vegetation to the left of the trail through the woods is vinca, an ornamental ground cover. It is a clue to the location of the buried ruins of a 19th-century house. Trenton, New Jersey.

FIGURE 8.17.

The dark line cutting across the surface of this plowed field is a soil anomaly representing a cultural feature, albeit a recent one. The darker colored soil is fill in a utility pipe trench that was dug through the field. The tops of historic and ancient features can also appear as sediment and color anomalies on existing surfaces.

FIGURE 8.18.

An obvious above-ground cultural feature built by the prehistoric Anazasi Indians, Cliff Palace at Mesa Verde National Park, Colorado.

FIGURE 8.19.

Another obvious above-ground cultural feature. These figures (petroglyphs) were carved by Native Americans on rocks along the Susquehanna River in Pennsylvania. Little Indian Rock (36LA185) has over 150 designs. Petroglyphs were created by Native Americans throughout many parts of the country and over vast periods of time. They indicate special use landscapes that might not otherwise be indicated by the presence of archaeological deposits.

FIGURE 8.20.

A view of stratigraphy provided by the bank of an arroyo. Any bank cut or vertical exposure can provide information for interpreting the geomorphology and age of landscapes. They also can be scraped or vertically peeled to search for artifacts and cultural features.

TABLE 8.3

Types of Exposures or Disturbances to Look For and Examine*

Exposed surfaces in general	Animal burrows
Environments that naturally have little vegetation, arid and eolian environments	Agricultural fields
Banks or vertical sections of streams, ditches, arroyos, erosion gullies, blowouts, roadgrades, railroad grades	Fence lines and tree lines bordering agricultural fields
Freshly slumped areas, scars from earth slides	Orchards and tree nurseries
Margins of slopes	Activity areas of animals on the range or in pastures
Margins of ponds, lakes, reservoirs, shorelines or beaches of streams and tidal areas	Trails, dirt or undeveloped roads
Base of large trees	Utility right-of-ways
Tree throws	Areas of new and ongoing construction
	Logged areas
	Burned-over areas

*See Schiffer (1987 or 1996), Wood and Johnson (1978) for discussions of disturbance processes and their effect on the formation of the archaeological record.

FIGURE 8.21.

Vertical exposures afforded by a drainage ditch alongside of a road.

FIGURE 8.22.

Wooded areas may offer few exposures to examine during a surface survey. Erosion along trails gives patchy but useful views of near-surface deposits.

FIGURE 8.23.

A tree throw is a window into deeper subsurface deposits. The rootball of the tree can be examined for artifacts, as can the hole created by the toppled tree.

FIGURE 8.24.

The activities of animals can provide views of near-surface and subsurface deposits. Pictured are eroded sections of a pasture in which horses are kept.

FIGURE 8.25.

The entrance to an animal burrow with sediments excavated from subsurface deposits.

Exposures of vertical sections serve many purposes. When cleaned, their stratigraphy can be examined and recorded, generating useful information about the geomorphology of the landscape and the potential age of near-surface and subsurface deposits. Their faces can be checked for artifacts and features, which, when found, will be in a stratigraphic context. The faces of cleaned exposures can also be vertically peeled, a layer at a time, to provide an excavation of sorts through any stratified deposits (see chapter 9).

Stream banks and shorelines are also useful for getting a quick look at the geology of the local region and what would have been available for use by past inhabitants. Break open samples of gravel with a rock hammer and identify the range of lithic materials represented. Do the same at rock outcroppings.

The integrity of surface contexts varies dramatically, reflecting the different ages of surface and near-surface deposits, and the processes responsible for exhuming artifacts. For example, the fluctuation of water levels in any body of water fosters erosion that can expose and redeposit artifacts on shorelines and beaches. Artifacts, especially large ones, may be gleaned from the surface of agricultural fields in the way that any rock or obstruction would be by a farmer, and tossed into the adjacent tree line or placed along the margins of the field. In both of these cases, the artifact finds are out of context; they are not in direct association with the deposit from which they were derived. But the location of the finds are vital clues as to where more intact archaeological deposits may exist. In contrast, I have worked on surface sites on farms where plowing had just begun to tear into the top of buried cultural features and artifact deposits. I assumed this because of finding bone and other artifacts in very tight clusters on the surface, sometimes associated with soil discolorations (Figure 8.26). This kind of thing is to be expected given the incremental erosion that is connected with agricultural land and the variety of methods that farmers can use to cultivate their fields. Some archaeological deposits are shallowly buried and immediately disturbed by plowing. Others are more deeply buried and may only be disturbed after generations of plowing and erosion. Signs that once-buried deposits are relatively newly exhumed consist of:

- ☑ tight spatial clustering of artifacts
- ☑ artifacts fashioned from perishable materials
- ☑ "fresh" looking artifacts made from materials that generally weather rapidly when exposed at the surface
- ☑ patterned soil discolorations
- ☑ human remains (Dunnell and Simek 1995).

Any find is worthy of attention regardless of whether it seems to be part of something that might be defined as a site. The degree of attention accorded a find is another matter, and will depend on an understanding of its context and associations, and the research design of a project. Artifacts and any other surface observations of note typically are flagged or marked in some way prior to mapping, collection, or further study. The easiest way to mark finds and locations is with wire or PVC stake flags available from forestry, surveying, and engineering suppliers (Figures 8.27 and 8.28). Flagging artifact locations and relevant surface features gives a visual impression of the

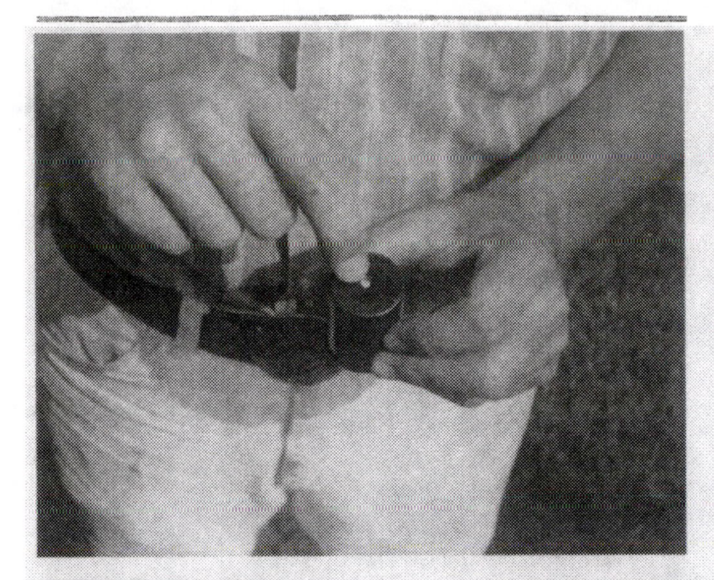

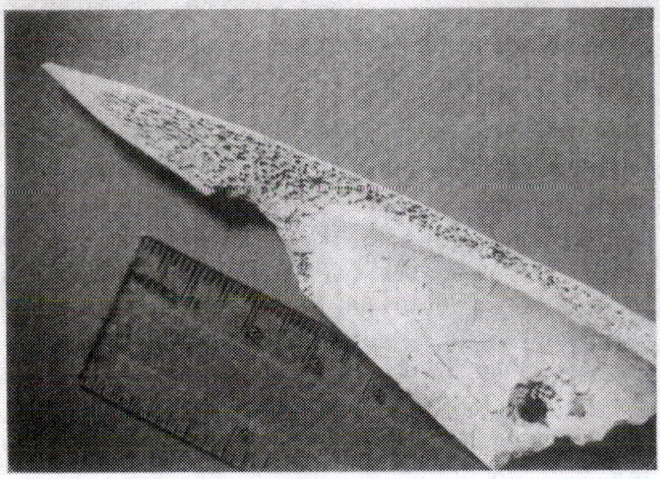

FIGURE 8.26.

A Late Woodland (circa 900–1600 A.D.) shell bead (top) and bone tool (bottom) from the surface of a plowed field at 18WA23 along Antietam Creek in western Maryland. Surface finds of artifacts fashioned from organic materials may be indicative of recently disturbed features. The locations of such finds can be important for positioning excavations to examine subsurface archaeological deposits.

FIGURE 8.27.

Stake flags useful in marking the location of artifacts and features during a surface survey. These are brightly colored vinyl flags mounted on wire shafts measuring over one and a half feet long. They can be obtained in greater lengths, with larger flags in a variety of colors, and PVC shafts.

potential size and internal organization of an archaeological deposit. It is also useful for making decisions about how to map or record the provenience of finds. The flagging of individual artifacts may become impractical with high-density deposits involving several hundred or thousands of artifacts. Flagging in these cases ceases to be useful as a tool for visually discerning artifact clustering and the internal organization of a surface deposit. The collection of artifacts within an established grid system and the tabulation of results is necessary to detect patterning in these cases (controlled surface collection). Selective flagging of finds in a high-density deposit will be useful for defining its potential boundaries. The variety in the density and distribution of artifacts, field conditions, and a project's research design are reflected in the techniques used for mapping and recording provenience.

FIGURE 8.28.

The small, white-colored flags are marking the location of artifacts found on the surface along an eroding slope and dirt road. Flagging artifact locations gives a visual impression of the potential size and internal organization of an archaeological deposit. It is also useful for making decisions about how to map or record the provenience of finds.

FIELD NOTES: WHAT TO RECORD AND HOW

GENERAL ISSUES

Field conditions, methods, observations and results need to be documented. This can be accomplished through the use of journal entries, standardized forms, maps, photographs and video recording. Field notes are generated throughout the day and are organized on a daily basis. The type of field notes taken by an individual will vary depending on their position of responsibility. Crew chiefs or field supervisors must take different types of notes than what is expected of a field technician or crew member. These will include descriptions of field conditions, methods, and a general narrative of the day's activities. The recording of finds, their provenience, and observations about the contexts in which they are found may be delegated to a larger segment of the field crew. I'll describe note taking and record keeping from a comprehensive perspective. Never forget that the records and notes that you and your coworkers create are documents for the future. They are part of the realization of the worth of nonrenewable archaeological resources and may outlive the resource itself. They will be the basis of interpretations, syntheses, and reports, and will be vital to planning additional fieldwork and making preservation decisions.

Daily entries in a field book should lead off with the date, location, weather, and a list of field crew members. Narrative entries can be organized around the following topics:

- ☑ activities performed
- ☑ rationalization for specific activities
- ☑ field conditions under which activities are performed
- ☑ results of activities
- ☑ interpretation of results
- ☑ relevance to research design
- ☑ implications and/or recommendations for further action

Don't just describe, think on paper. Don't take things for granted assuming that they aren't worth mentioning, or that you will remember everything significant. Your memory of events may be fresh now but what about four months from now,when you are in the laboratory working on the report for the project? Or four years from now when you want to use your field notes to address other research topics? Or 40 years from now when a scholar refers to them to better understand the method and theory of the time? It may be the case that the person responsible for the day-to-day management of fieldwork is not the person who writes the technical report for a project. This seems to be increasingly true in CRM where companies have to balance the need to keep crews in the field on numerous projects, with report deadlines for individual projects. Detailed field notes, organized and written in a way that can be understood by professional colleagues, are critical. The best way to proceed is to assume that other people will be using the data that you collect and the records that you create.

If you are serving in a supervisory position it is always good to make note of any accidents or injuries that crew members experienced and the action that was taken to assist them. These notes may figure in subsequent treatments for individuals and any insurance or legal actions arising from an incident. Make note of any visitors or contacts with people made in the field. The landowner on whose property you are working may require this information, and sponsors of fieldwork may want it to document public interest and outreach.

Describing the ***field conditions*** under which a surface survey takes place will enable you and others to evaluate the reliability of survey results and their impact on additional phases of fieldwork. For wooded areas and open fields, comment on:

- ☑ the relative maturity or age of the forest
- ☑ the type of ground cover that exists in the area (e.g., grass, dense bushes, and green briars)
- ☑ the types of surface exposures encountered
- ☑ the distribution of surface exposure over the survey area

When working in cultivated or fallow agricultural fields, make note of how long ago they were plowed or worked. If possible, determine how they were worked. There are a variety of methods used by farmers to remove existing vegetation or crop cover and prepare a field for planting. Not all of these involve the traditional plowing and disking that results in the best surface exposures (and the most disruption to archaeological deposits). Describe the nature of the surface. Has it been well washed by rain? To what degree do planted crops, crop stubble, or other ground cover obscure views of the surface? What crop or ground cover is in the field and how mature is it? Always record any evidence of previous artifact collecting or pothunting. This evidence can range from obvious signs of digging to numerous footprints in a plowed field.

Whether you are dealing with an isolated find, low density cluster of artifacts, surface feature, or anything that fits the definition of an archaeological site, basic types of information need to be recorded in the field. Data categories are summarized in Table 8.4 and reflect what appear on the inventory forms used by states to officially register sites (see chapter 4). Defining the horizontal boundaries of a site begs the question of what constitutes an archaeological site (see chapter 1). Attention must be paid to the spatial distribution of finds across a landscape, regardless of their density or the activities that they represent. For the purposes of state registration, "site" may be thought of as a device for organizing spatial distributions of artifacts and surface features on discrete landscapes, even if distributions are not continuous and clusters are spatially isolated from one another. I see this as a reasonable compromise of the concerns of "siteless" archaeology and those who advocate precise definitions of sites for management purposes. It requires, however, that spatial distributions and the internal variability of any locality be adequately documented.

For rockshelters it is important to note the linear extent of the rock face or overhang and the size of the area beneath it. The size and shape of cave openings and habitable floor space should be recorded. The

TABLE 8.4

Basic Field Observations for Surface Finds, Artifact Clusters, or Sites

type of site or feature (e.g., open air, rockshelter, cave, ruin, rock art, mound, isolated find, etc.)

landowner name and address

surface conditions

topographic setting

exposure/slope direction (compass bearing) and degree/percent

soil type

bedrock

immediate vegetation

type of previous disturbances (e.g., erosion, plowed field, logging, etc.)

evidence of vandalism, possibility of destruction

nature of and distance to nearest source of water

related primary drainage

size and shape of cluster, feature, site, rockshelter or cave

patterning or clustering of artifacts within a site

evidence of horizontal stratigraphy

degree of integrity of context

number and types of artifacts or samples collected

number and type of artifacts observed but not collected

number and type of photographs taken

complete documentation of rock art will require expertise that may be beyond the skills of the average field technician. Basic observations that can be made consist of:

- ☑ the type and size of rock exposure involved
- ☑ the area covered by designs
- ☑ the type and number of designs
- ☑ production technique (e.g., painting, pecking, incising, grinding)
- ☑ Munsell readings of any colors used
- ☑ any association with archaeological deposits
- ☑ the standard locational data gathered for any archaeological site

There are a variety of survey-level and region-specific publications that deal with the scope of the phenomena and provide guidance on data collection, drawing, and photography (e.g., Grant 1983; Lee 1991; Sanger and Meighan 1990; Schaafsma 1985; Wellmann and Arndt 1979; Whitley 2000).

Any artifacts that are collected need to be accompanied by a tag or label that describes their provenience:

- ☑ project or site name or designation
- ☑ surface find
- ☑ locational data (reference to some type of coordinates or collection unit)
- ☑ field specimen catalog number (see below)
- ☑ date
- ☑ collector's name

When the exact location of an artifact is mapped this means using either mapping coordinates or a unique artifact number that relates it to the records of the person operating the GPS or surveying instrument. For example, the person running the total station sights the first artifact to be located, calls it artifact #1 in her notes, and records its angle relative to an established point of reference and its distance from the instrument's position. The person collecting the artifact creates a label that lists the site/deposit designation, artifact #1, the date, and his/her name. Each additional artifact is treated in the same way with a unique number. The only importance of the artifact number is that it relates the object to the mapping notes. Similarly, if artifact clusters are serving as collection units, the mapper will assign each cluster a unique number or letter that is associated with her notes. The tag accompanying any artifact from a cluster employs only the cluster designation and not an artifact number, e.g., 36BU173 (site), cluster 1, 10/12/98 (date), J. Doe (collector). Tags for artifacts collected by grid unit will employ the grid coordinates of the unit rather than an artifact number or cluster designation. This information may be written on the bag in which an artifact is placed or on a tag or piece of paper placed in the bag with the artifact. Labeling the exterior of the bag and placing a provenience tag inside of a bag insures that this information will not be lost if one or the other is damaged or obscured by moisture, dirt, or wear.

Collected artifacts and samples are generally listed in a ***field specimen catalog*** (Figure 8.29). These catalogs are used for materials derived from both surface and subsurface investigations. Not only do they provide a rough inventory of artifacts and samples collected, but they are important for tracking materials from the field to the laboratory. Catalog numbers are assigned to individual or unique contexts—one context or provenience = one catalog number. If you were using a grid of units to collect artifacts from a surface site, the artifacts from collection unit N10E20 would have a different catalog number than those from collection unit N10E25. The order in which catalog numbers are assigned is meaningless. Collection unit N10E20 could have catalog number 2 and collection unit N10E25 catalog number 22. Catalog numbers are assigned sequentially (1, 2, 3,....n) on an as-needed basis as materials are collected and logged. The date that materials are collected provides a link with dated field notes. Later, the catalog is used to check that everything collected in the field has made it into the laboratory for processing and analysis. Field specimen catalog numbers may also be employed in marking/labeling artifacts in the laboratory. Since the catalog numbers are linked to specific contexts, they can be used as a shorthand to denote the context. Rather than write 18WA23 (site), N10E20 (collection unit), and surface (level) on a relevant artifact, 18WA23 (site), 2 (catalog number) will suffice. The catalog number 2 encodes the information about the intrasite provenience of the find. For multiple finds from the same context/catalog number, secondary numbers can be assigned in the laboratory, e.g., specimen catalog number 2-1, 2-2, 2-3.

FIELD SPECIMEN CATALOG

Site or Area Designation______________________________________

Catalog #	Surface provenience Surface collection unit Excavation unit	Excavation level Feature # and level	Summary of Finds & Diagnostic Artifacts	Date Field	Date Lab

FIGURE 8.29.

An example of a field specimen catalog used for tracking materials collected from both surface surveys and excavations.

I make it a habit of drawing diagnostic or significant artifacts in the field. Sometimes the drawings are nothing more than the traced outline of the piece with a few surface characteristics added, some descriptive notes, and its provenience. At other times they are elaborately stippled and shaded. For me the drawings serve as a reminder of finds as I review my notes, think about interpretations of field data, and consult reference materials. The attention that you focus on an object while making a drawing often leads to observations about style, manufacturing, use, and wear that might be glossed over during a casual inspection. The drawings also function as a backup if artifacts become misplaced or lost, something that can happen between the field and laboratory. Appropriate drawing methods vary for the type of object being illustrated. Adkins and Adkins (1989) and Dillon (1992) provide a good introduction to the subject in all of its complexities.

When all artifacts discovered during a surface survey are not collected, decisions have to be made about what to record other than their provenience. What to record depends on the project's research design and the issues to be addressed with the analysis of artifacts, and whether additional phases of fieldwork are planned. Minimally, the range of artifact classes present (e.g., flakes, pottery, bifaces) should be listed and an estimation of their representation in the greater assemblage made. The large size and high density of some deposits precludes more exacting inventories during the site discovery phase of fieldwork. With other deposits it may be possible to construct more precise inventories. In situations that require accurate inventories and in-field, literally *in situ,* artifact analysis, the use of standardized data sheets is a must. For each class of artifact, a standardized form can be created that lists attributes or combinations of attributes to be observed and checked off where appropriate. The need for written comments is minimized as much as possible. Forms are customized to meet the specific inventory and analysis needs of a project.

Whether you are using standardized paper forms or keeping notes in a journal, neatness counts. Completed record sheets invariably will be checked by a supervisor and you may be asked to redo them if they are sloppy and hard to read. Small, hand-held tape recorders can be effective for recording impressions and reminders about things to do when there isn't time to stop and make notes in a field book. Transcribe tape recorded notes on a daily basis. Small, hand-held electronic data recorders, notebook or tablet computers are experiencing some use in the field (e.g., Schneiderman-Fox and Pappalardo 1996; Zeidler 1997). Notes, standardized data sheets, drawings, and mapping data can be accommodated by these systems. They are downloaded into a database on a PC at the end of each day in the field. This allows field data to be manipulated quickly in the field, and circumvents the need to transcribe paper-based records into computerized databases back in the office or laboratory. The systems are battery powered, which means that care must be taken in the field to always be equipped with power. It's essential that downloading occur at frequent intervals to insure that any mistakes in data recording or failures in hardware or software can be addressed quickly, and while still in the field. Only time will tell if electronic data recorders can stand up to the abuse that any equipment experiences in the field.

The quality of notes that you take will change as you learn more about archaeology, gain experience working in the field, and fill positions with more responsibilities. But from the beginning your note taking can be structured in a way to keep you thinking critically about what you are doing, just like the folks who are supervising your work. Make it a habit to make daily entries in your field book, regardless of the other types of documentation that you may prepare in your official role as a crew member. Begin by noting the date, location, and project on which you are working. What were the day's activities and what was their purpose? What techniques or procedures were followed? Are there other ways that the work could have been accomplished? How do the day's accomplishments affect plans for ongoing fieldwork? On the next page is a sample entry that would be appropriate for both the person guiding fieldwork and the field technician giving serious thought to the day's activities.

PROVENIENCE AND MAPPING

The location and spatial distribution of artifacts and surface features are important for a variety of analyses and interpretations. In the planning of ongoing or future fieldwork, they are key for designing sampling strategies and the positioning of excavations. There are three different approaches to mapping or recording the provenience of surface finds of artifacts found during a survey:

1. general site or area provenience;

2. site/area provenience and reference to some type of baseline, grid, or mapping unit; and

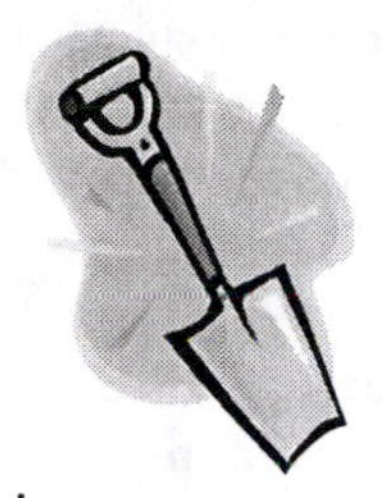

May 19, 2000
Midwest Pipeline Survey
Allerton, Iowa

We spent the entire day doing a surface survey of wooded project areas on the Smith Farm. There weren't many exposed surfaces to examine, just random areas around the bases of large trees, places where trees had been blown down, some eroding banks along a small stream, and places where trails went up and or down slopes. There's no way that this represents systematic coverage of the project area. This seemed like a pretty haphazard way to determine if there were any archaeological deposits in the area. Surface exposures might not be in the same areas where you might predict sites to be. What if the sites are really small? And there's still the chance that very ancient deposits occur nowhere near the existing surface. Transects of small excavations will have to be used to get a better idea of what might be out here.

On the other hand, looking for things on the surface is easier and less time consuming than digging dozens of test pits and has the potential to turn up useful information. We did actually discover a few things. A portion of the stream, 200 feet long, is very straight and precise and lined with rock. This doesn't seem to be natural and may have something to do with a 19th-century water-powered mill that is mentioned in a history of the area but we can't find on any historic map. There are some suspicious looking mounds of grown-over rock near the stream too. We also found a few chert flakes along a trail where kids ride their ATVs. There may be an Indian site there. Really didn't expect to see anything in that area given the predictive models that we're working with. It wasn't even included in the sample of probability areas where we were going to do transects of shovel test excavations. On Monday we'll do some mapping and open some test pits to figure out what's going on in the area with the mounds and the flakes. We may have to rethink the excavation sampling strategy when we're through. I have to remind myself that probability sampling is not a technique for discovering sites, it's a way to organize fieldwork so that results are representative of the whole area.

3. site/area provenience and exact location within it.

Each requires a different level of effort. Deciding which approach to take depends on the nature of the context in which finds are made, the size and density of the archaeological deposit, whether additional work or repetitive surface surveys are planned, and whether subsequent excavations will be keyed to specific artifact locations.

General site or area provenience means that the location of an artifact, cluster of artifacts, or surface feature is simply recorded as existing within the boundaries of a defined site/area or cluster. This is the most generalized form of recording location or provenience and requires the least amount of effort once the overarching boundaries have been documented. Site/area or cluster boundaries can be documented by reference to topographic features or landmarks that appear on the USGS 7.5 minute quadrangle or other map being used in the field. Compass bearings, pacing and/or tapes can be used to determine the details of these boundaries and translate them onto available maps. If convenient, a transit, theodolite, or total station can be used to perform the operations. In situations where it is difficult to relate site/area or cluster boundaries to local topography or landmarks appearing on available maps, it will be necessary to establish a mapping datum and run traverses to tie it into a visible or easily relocated landmark. The use of a GPS receiver is reasonably accurate and probably the easiest means of documenting boundaries. Simply walk the boundaries to be recorded with the GPS receiver taking readings at appropriate intervals. Once downloaded into a computer and using suitable software, the boundaries can be drawn and related to UTM or latitude and longitude coordinates. The sole use of general site or area provenience is reasonable when

- ☑ additional fieldwork is planned that will examine and record artifact and feature locations in more detail

- ☑ non-clustered, low-density finds occur over a relatively small area and artifact locations won't be used to position subsequent excavations or subsurface tests

- ☑ the context of finds is questionable, e.g., finds made along a beach or eroding shoreline that are related in some way to intact deposits inland or offshore

Reference to an established baseline, grid, or mapping unit may be used in conjunction with general site/area provenience to more precisely document artifact and feature locations. Grids are useful for attempting to distinguish patterning or changes in the density of artifacts over space where visual perceptions of patterning are inadequate. This would be the case in

- ☑ open areas where surface exposures and the occurrence of artifacts are patchy, but too numerous to have individual patches, clusters, or artifacts mapped individually

- ☑ open areas with extensive surface exposures and low-density artifact deposits

Within a grid unit, observations are made about the frequency and nature of artifacts. The grid system may also be used in a controlled surface collection. In open areas where clusters of artifacts are clearly discernible, the boundaries of individual clusters may be mapped using any of a number of techniques already reviewed. Referencing finds to increments along an established baseline is helpful for areas with very linear exposures and occurrences of artifacts like those along slopes, eroding banks, and shorelines. As with the grid approach, the baseline has to be tied into a mapping datum or easily relocated landmark.

There are a variety of situations when recording exact locations or "exact" or "point" provenience is appropriate. When the time is taken to map the location of individual artifacts then they usually are also collected. To do so during a later stage of fieldwork is an unnecessary replication of effort. Remember that the discovery of an archaeological deposit and a more thorough examination of it may be combined in a single field effort, rather than staged. The location on the surface of certain types of artifacts, or clusters of artifacts, may closely reflect the location of remnants of undisturbed deposits or features. The spatial relationships they reveal are important in themselves and also because of their potential use in positioning excavations to explore subsurface deposits. The processes that expose artifact deposits on the surface, like plowing or periodic erosion, are progressive, their effects on a deposit changes over time. Thus there are instances when portions of features or deposits are first exhumed. Perishable artifacts once preserved within them are exposed on the surface. The original clustering and associations of artifacts may also be mimicked to a close degree. In other words, the contexts of the finds and their spatial relationship to subsurface and less disturbed deposits is still very close, and thus more informative than those that have been subjected to generations of disturbances and exposure. I've noted some ways to recognize such deposits above. Even when the context may not be exceptional, the location of artifacts typically associated with burials, caches, or ritual features is noteworthy in providing information about the general areas in which remnants of these significant features may be sought. Mapping the exact location of individual artifacts may be the expedient or practical thing to do when finds are widely scattered or isolated, as they may be in wooded areas, or when small, low-density clusters are encountered. One should have some confidence that the surface contexts in these situations are somewhat reliable. In other words, you don't want to map the exact location of artifacts found on a landscape that was created with placed fill, or on a beach or shoreline where artifact locations are more a reflection of the natural processes that have disturbed intact deposits than patterning resulting from human behavior.

The exact location of surface features, their size and shape should be precisely recorded. Documenting the topography of some surface features (like mounds) may require an additional effort separate from the archaeological survey. Surface features may be the focus of excavations used to confirm their identity as cultural phenomena or employed to take advantage of the information they may contain. It is therefore important to be able to precisely relocate them. Exposed vertical sections that were cleaned and profiled, and especially those where artifacts were recovered, should be precisely located.

The recording of locations in widely scattered or isolated situations can be accomplished quickly and accurately with a high-end GPS receiver. If these locations will not be important for the subsequent placement of excavations, than utilizing GPS equipment with a more moderate degree of accuracy will suffice. In open areas, total stations can be used to map numerous artifact and feature locations more quickly than a transit or theodolite. The efficiency of these techniques declines in wooded areas where lines of sight are limited. Here it may be necessary to run traverse lines or establish baselines to which finds can be related. If artifact locations will not be used to specifically position later excavations, then compass bearings and tapes or pacing may be sufficient to link find spots with an established baseline or traverse line. As usual, any baseline or traverse line needs to be tied into a landmark or datum that can be relocated and used as a base for reconstructing mapping operations. Depending on the quality of available maps for a project area, sufficient topographic landmarks may exist to which artifact finds can be referenced with a compass bearing, triangulation, and a measured distance.

To make the most of dense and large surface deposits it is necessary to perform tasks that go beyond those associated with an archaeological survey. Deposit boundaries and the configuration of discernible internal clusters should be mapped, as well as the location of surface features and the range of artifacts noted above. In addition, the provenience of artifacts and samples that are collected must be documented. At some point the quantity of things to be mapped becomes more typical of a controlled surface collection. A controlled surface collection may reasonably employ a grid system as a frame for artifact collection and spatial observations, rather than the exact provenience of individual pieces. Decisions about how much to collect and map during the discovery phase of fieldwork must be based upon a project's research design and budget, whether additional work is planned or possible, and the experience of the investigator.

Sketched plans showing surveyed areas, find spots, and artifact distributions are a useful supplement to the more formal maps that are created (Figures 8.30 and 8.31). They serve as a cross-reference if mistakes are suspected in any of the measurements taken to create

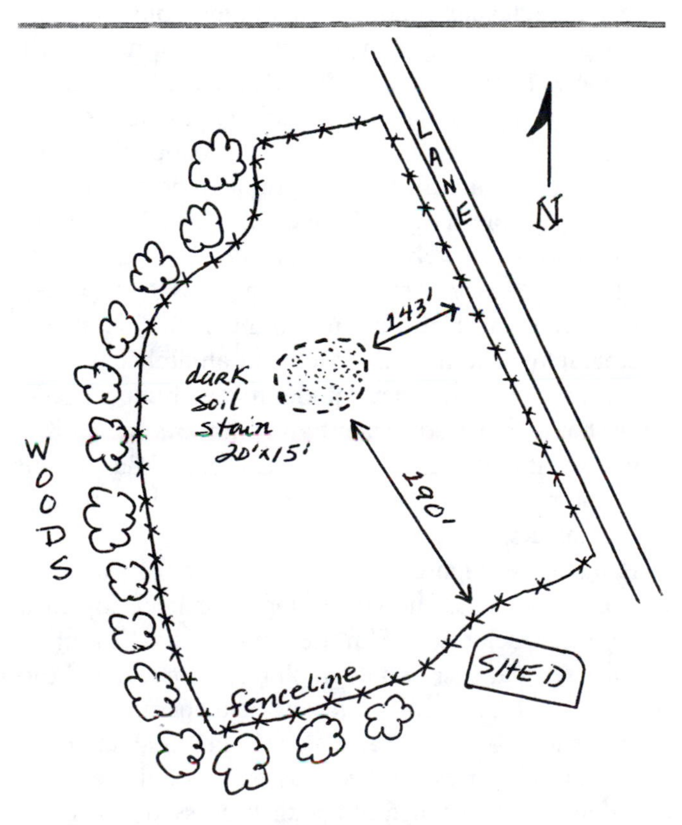

FIGURE 8.30.

A sketch map in a field book showing the location of a potential archaeological feature noted during a surface survey. The juxtaposition of the shed, lane, fence line, and soil stain shown in the sketch make it easier to relocate the feature in the future.

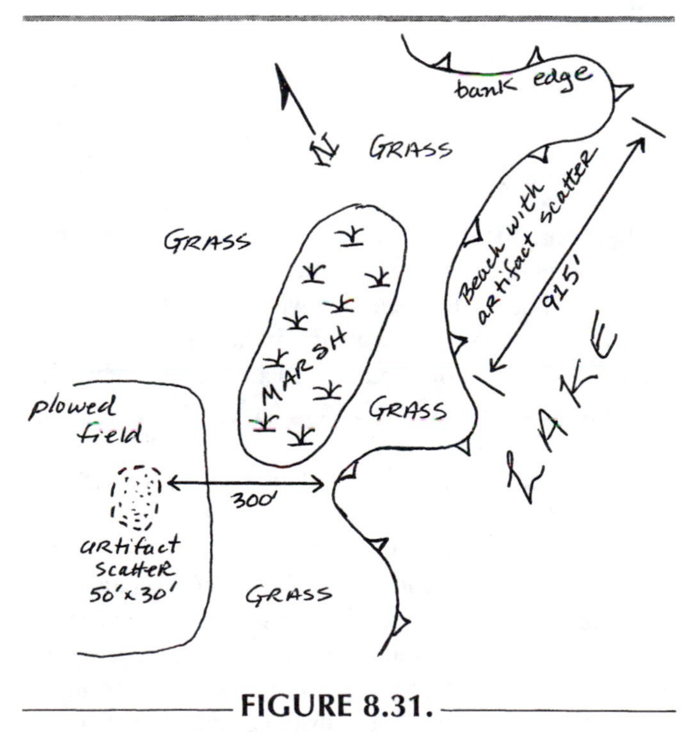

FIGURE 8.31.

A field book sketch depicting a survey area and surface finds. Here the distinctive shape of the shoreline enables the area to be identified on published maps and easily recognized on the ground during a relocation effort.

the formal maps. They also can be an important component of site registration forms, making it easier for others in the future to relocate what you have documented. Include in your sketches obvious natural features or cultural landmarks and distinctive trends in topography that would enable someone on the ground to recognize the area that you are depicting. The cultural traits (land, shed, fence line) sketched in Figure 8.30 are easily distinguished on the ground and are a spatial reference for relocating the potential archaeological feature shown, as is the shape of the shoreline and location of the marsh in Figure 8.31. The drawings need not be done to scale but should contain paced or measured distances relating significant features to one another. The drawings should, however, fairly represent the general spatial relationships between the things being depicted. Some of the things incorporated in your drawings also may be visible on published maps, such as the lane and shed in Figure 8.30 and the shape of the shoreline in Figure 8.31.

PHOTOGRAPHY

Photography is a visual complement to the other records that are generated to document all aspects of fieldwork. Photographic images are also an important means of illustrating fieldwork and its results in public presentations and publications. Be aware of these different purposes for photos when planning what to shoot in the field. If you are not familiar with the basics of photography there are a number of good introductions that can be consulted (e.g., Dorrell 1989; Howell and Blanc 1995).

Photos should be taken of all aspects of field procedures to create a visual record of methodology. Take action shots of people working, shots of survey areas and the environments in which the work takes place. For area shots, positioning people at judicious points on the landscape or around a feature of interest provides a sense of scale and depth. Photograph the areas occupied by archaeological deposits and sites from a variety of different directions, as well as any discovered surface features and notable artifact clusters. It may be necessary to photograph some things throughout the day as their visibility is enhanced because of changes in lighting. Sign boards listing the subject of a photo, its provenience, and the date should be used when the scale of the shot is small enough that they are visible and easily read. Smaller scale photos should also include a measured scale and a north arrow (a trowel is a traditional substitute).

It may be necessary to photograph individual artifacts in the field either because of their fragile condition, or some constraint that prevents their collection and removal from the field. Lighting conditions will be the critical variable in getting reasonable field photos of artifacts. Because of the need for mobility during a surface survey, it is unlikely that portable photo stands and lighting equipment will be part of a team's field gear. Artifact photos should include a scale or a background consisting of graph paper.

Ideally, documentary shots should be done in both color and black and white, necessitating the use of two cameras in the field. Color transparencies/slides are typically used in presentations. Whenever lighting is questionable it is always wise to bracket the f stop (graduated openings of the camera's aperture or diaphragm) used for the ideal exposure, one stop above and one stop below. This increases the chances of getting a useful shot. Although digital photography has yet to match the resolution of 35mm photography, its ability to generate images in a variety of formats is an advantage. Images can be downloaded into a laptop computer with a modem and transmitted to the office or laboratory if field staff have access to a phone line. This may make the identification of problematic artifacts or features possible. Digital images can also be used to maintain a project website for education and public outreach, or keep project sponsors up to date. Video recording also is useful but not to the exclusion of still photography. The mobility factor in an archaeological survey may preclude its use under many circumstances. Any photographic equipment and film are susceptible to the harmful effects of heat, moisture, and dust in the field and should be transported and stored in appropriate containers.

Photo logs (Figure 8.32) are as necessary as field specimen catalogs. Individual logs are kept for each camera and each roll of film used in the field. This practice insures that every photograph has a context and can be linked with a specific subject or field operation. Large numbers of photographs are taken in the field and crew members' memories can't be trusted to provide an adequate inventory when faced with processed negatives, contact sheets, prints and transparencies. The log can also be used later as a way to organize processed negatives, contact sheets, prints and transparencies for storage. Use an indelible marker to label the film roll with its number or designation when removed from the camera.

PHOTO LOG

Project______________________________
Camera Designation______________________
Camera Type___________________________
Film Type________________**ASA/ISO**______
Roll Designation________________________

Frame #	Subject	Provenience	Direction Faced	Lens or Filter Used	Shutter Speed	f stop	Photographer	Date & Time

FIGURE 8.32.

A example of a Photo Log used to link photography with field operations and organize the negatives, contact sheets, and images produced.

WHAT TO COLLECT

Do you collect every artifact that is found during a surface survey? The answer to this question is not always yes (e.g., Butler 1979). Deciding what to collect means thinking about a project's research design, the context and quantity of finds, ethical responsibilities, logistical and economic constraints, and what the future may hold for a given locality. Not all of these factors are given equal weight in the decision-making process, and the importance of a factor can change from case to case. Any artifact that is collected must always be accompanied by provenience data.

Let's start with a consideration of what you would like to collect from a research perspective, all other things being equal and understanding that we can't anticipate the specifics of every problem that might be addressed with data from a survey. Any artifacts that could aid in determining the age and cultural affiliation of a surface deposit should be collected. In most cases this means artifacts whose style, form, or way of being made can be associated with a particular period of time. Regardless of their style, ceramics and heat-treated lithic artifacts can be chronometrically dated by thermoluminescence (Dunnell and Feathers 1994). The degree of weathering of artifacts fashioned from certain types of rocks also can be of use for dating purposes and estimating the number of cultural components in a deposit (e.g., Beck 1994; Stewart 1984).

Samples of artifacts representing the range of artifact classes present in the deposit as a whole, and within recognizable clusters, should be gathered. What constitutes an adequate sample will change from case to case. Artifacts derived from the examination of vertical exposures or profiles are always kept, as they are associated with a stratigraphic context and have the potential to provide more information than artifacts found on the surface. The research issues that a project hopes to address could heighten the relative importance of certain classes of artifacts. For example, artifacts made from exotic materials whose sources are known could be used to address trade or settlement movements. Some objects, as well as human remains, may consistently be associated with burials, or spiritual or ritual features. These artifacts and human remains should be gathered and consultation with the SHPO/THPO and representatives from appropriate groups initiated. All isolated finds should be collected, as this may be the only chance that anyone will have to deal with them. Samples of locally available lithic material and clay may be useful for comparative purposes and replicative experiments.

Perishable types of artifacts or those requiring conservation in order to survive should be collected (see chapter 9 for a discussion of artifact conservation in the field). Artifacts from sites or portions of deposits threatened by ongoing erosion should be collected, for example buried deposits being exhumed by rising and falling water levels along a shoreline. This may be the only record that is created concerning the deposit before it is completely destroyed. If the threatened deposit is large, full collection may not be feasible; a representative sample should be taken. Looting may be the greatest threat to some sites, in which case it would be wise to remove any chronologically diagnostic artifacts, recognizable tools, or ornaments, as these are typically sought by collectors.

The collection of any artifact carries with it a professional obligation to adequately document the location and context of the find, provide for its processing in the laboratory (which minimally should involve cleaning, stabilization or conservation if necessary, and labelling with provenience data), and curation. These obligations obviously have practical and economic implications. Is the budget, logistical support, or volunteer labor available for the project sufficient to meet these obligations? Who owns the property where the artifacts were collected and who retains legal ownership of the artifacts? State and federal agencies who sponsor research on public property must adhere to guidelines for research and site evaluation, and for artifact and data curation (e.g., National Park Service 1991). Unfortunately a real crisis exists in the curation of artifacts and archaeological data, driven in large part by the tremendous amount of work promoted by federal and state laws, the relatively small number of facilities that exist to deal with the products of this work, and the standards that need to be upheld (Marquardt et al. 1982; National Park Service 1996). Private landowners do not have to abide by the same rules, and depending on the relationship with a property owner, an investigator may decide that she has only one chance at examining a newly discovered deposit. The project's field strategy would be adjusted accordingly with the hope that any problems arising from the greater volume of artifact and data processing and curation can be solved. The "one chance" scenario also might apply to situations where a site or deposit is located in an area frequented by looters. The archaeologist's choices boil down to study it now, or risk never seeing it again.

The quantity and type of material collected can be influenced by the expectation of additional fieldwork, especially in the case of CRM projects. In-field evaluations of a deposit may convince an investigator that its value cannot be realized simply by collecting the basic survey-level data. Or the quantity of material and data that need/could be gathered are so great as to be beyond the scope of the discovery phase of fieldwork. The investigator is convinced of the need for additional work and is prepared to make the case with appropriate analyses and presentations in the report to the sponsor. Further, it is not unusual for additional work to be negotiated with a client and regulatory staff while a crew is still in the field, streamlining the reporting, evaluation, and proposal process and resulting in a lessening of the long-term costs of a project. A research design can also be structured as a multiyear or multiphase effort that includes returning to any discovered surface deposits of potential significance.

No-collection policies have been advocated by some agencies and institutions for sites on public lands that are not threatened by development (see Butler 1979; Schiffer et al. 1978 for review and comment). The rationale given for no-collection policy is that all necessary observations can be accomplished in the field, the deposit is preserved in-place, and the trouble and expense of artifact processing and curation are avoided. This policy ignores the reality of what can happen to archaeological sites on a day-to-day basis, the nonrenewable nature of archaeological deposits, problems with defining adequate types of field analysis and the shifting nature of research issues, and professional ethics. Dealing with artifacts and archaeological data certainly has its costs, but in the words of William Butler (1979:798), "what is the value of an irreplaceable resource?" There may be situations where a private property owner allows work to proceed with the condition that artifacts can't be removed, or must be turned over whenever the crew leaves the field. Mapping and data collection would have to be adapted to meet these circumstances.

REMOTE SENSING

Remote sensing is an umbrella term for describing noninvasive techniques used to learn about the surface and subsurface of the earth. These techniques are another tool for discovering archaeological sites. The use of photographic and radar reflection imagery from aircraft and satellite platforms fits the classic definition of remote sensing (Heimmer 1992). Various land-based or geophysical techniques pass sound, radio, and sonar waves, or some type of electromagnetic energy into the ground and detect how the medium responds to it. Patterns and anomalies in characteristics of vegetation, topography, drainage, subsurface deposits, and the earth's magnetic field can be recognized with remote sensing. Recognizing a pattern or anomaly is one thing, figuring out what it means is another. It is up to the archaeologist to determine which are cultural in origin, a process called ground truthing. So although remote sensing is a "surface" technique for helping to locate archaeological deposits and sites, in many cases it requires excavations to realize its potential benefit.

In chapter 4 we talked about the potential benefits of aerial photographs for various aspects of fieldwork. Aerial photos may directly reveal components of sites like earthworks, mounds, architectural ruins, or shell middens, or they may hint at potential subsurface deposits based on crop marks, patterned vegetation growth, and drainage characteristics. The use of special film like ***false color infrared*** can enhance the search for patterns or anomalies. The false colors in infrared photos show the varying degrees to which ground surfaces retain and reflect heat. The buried ruin of a house foundation may absorb, retain, and radiate more heat during the day than the sediments surrounding it. An aerial infrared photograph of the field in which the buried ruin lies would reveal a rectilinear pattern of heat reflection that would stand in contrast to the reflections of the surrounding area. Heat radiation can also be detected from the air and imaged using thermal sensors rather than cameras and infrared film. Radar imaging of topography produced by aircraft-mounted equipment (***side-looking radar***) also reveals aspects of the surface that are related to human constructions, like roads, earthworks, and mounds, and can work in forested areas. Landsat, or Earth Resources Technology Satellites, records the intensity of reflected light and infrared radiation from the earth's surface. Collected data can be converted to photographic images and used in ways similar to aerial photographs, looking for subtle patterns or differences in vegetation, drainage characteristics, and radiated heat to provide clues about human alterations of the surface and the existence of subsurface features.

LAND-BASED TECHNIQUES

Land-based techniques have a number of things in common. They all require some type of device that passes sound or energy into the ground, and a means of recording and measuring the behavior of the sound or energy in the ground. Testing is organized along mapped transects or grids so that readings and interpretations can be related back to specific locations. This method is essential for ground truthing. It may be necessary to clear brush and vegetation along transects or at grid points to facilitate the use of equipment. The nature of topography, sediments, and stratigraphy can influence the effectiveness of some techniques, like ground-penetrating radar, making the development of regional approaches to their use necessary (e.g., Conyers and Cameron 1998).

The ***standing wave technique,*** like other acoustic techniques, employs a device that passes sound waves into the ground. Sensors placed at distant points record the speed at which the sound travels through the ground. Comparable speeds between the sensor points means that there is consistency in the density of the below-ground materials through which the sound travels. Inconsistent speeds indicate a change in subsurface material, some of which might relate to architectural or other cultural features.

Magnetometers measure local magnetic distortions against the background of the earth's magnetic field. Distortions can be caused by ferrous metals or materials, rock, and sediments with magnetic properties, like volcanic stone and fired clay. So concentrations of metal artifacts or features like rock- or clay-lined hearths, walls, ceramic kilns, and burned sediments may be detected. Anomalous readings can be caused by natural deposits of igneous and metamorphic rocks and sediments that contain magnetite, or highly organic soils with maghematite, the magnetic form of the mineral hematite. Nearby power lines, buried pipes or electrical lines, metal fences, signs, and vehicles can also distort readings. Some archaeological phenomena may provide too small of a contrast in magnetic distortion to be clearly recognized using this method.

Passing an electrical current through the ground between control points and measuring the resistance, or loss of voltage that it encounters, is called ***electrical resistivity.*** Four electrodes must be placed in a line for each test of resistivity (Figure 8.33). Resistivity readings

FIGURE 8.33.

Using electrical resistivity in the field. Each of the circular spools spaced at regular intervals are attached to rods that are the electrodes that record the loss of voltage over distance. The electrodes are attached by wires to the small boxlike device left of center in the photo. This device is the source of the current that is passed between the electrodes and the meter that records the electrical resistivity or loss of voltage.

between points where sediments or materials are relatively uniform will contrast with those between points lacking uniformity, possibly because of the presence of cultural features. The damper the sediment or soil, the more easily electricity is conducted and the less resistance a current faces. The drainage characteristics of backfilled pit features or ditches therefore might contrast with those of adjacent undisturbed sediments, influence electrical resistivity, and result in their detection. Having some knowledge of subsurface sediments and soils prior to conducting tests is very helpful, if not mandatory. The small size of some archaeological features relative to natural anomalies that influence resistivity can also hamper detection.

Electromagnetic conductivity (EM) is the reciprocal of electrical resistivity. A self-contained device induces an electromagnetic field into the earth above a test location and a voltage meter records conductivity. It is somewhat easier to use than resistivity equipment because it does not require the sequential placement and repositioning of electrodes into the ground.

With ***ground penetrating radar*** (GPR) pulses of electromagnetic energy are passed into the ground from a sled pulled along the ground (Figure 8.34) or a truck mounted antenna. The speed with which the radar waves are reflected back to the surface are recorded. By comparing and contrasting subsurface velocities of the radar waves sent into the ground at controlled points, potential cultural features and stratification can be identified—changes in subsurface materials result in changes in subsurface velocity. Like other electromagnetic techniques, unusual variation in subsurface materials is interpreted as a potential cultural feature like a pit or architectural ruins. The larger the vertical and horizontal extent of the anomaly, the easier it is to recognize in GPR readings. Additionally, the strength of the reflected signal provides information about subsurface materials. Metal objects produce a strong reflected signal. Some materials attenuate the strength of the signal. Clayey sediments or soils, and those that are saturated with water, are highly conductive, causing the radar pulse to be dissipated within them, thus limiting reflections to the surface and making the detection of potential cultural features extremely difficult.

A ***metal detector*** is a hand-held electromagnetic device probably known to most readers for its ability in locating metallic objects relatively near the surface. Controlled sweeps along mapped transects or grids can be used to find and plot the spatial distributions of metal objects. Such data are useful by themselves but also can provide clues to the location of cultural features like former wooden structures.

USING REMOTE SENSING

Remote sensing techniques have the ability to quickly locate potential archaeological sites and features in a nondestructive manner. Aerial photography and Landsat imagery may be particularly useful when project areas are very large, or when cultural features are of such a large scale as to be hard to recognize on the ground (e.g., ancient roads or extensive earthworks and mounds). Even though ground truthing is a necessity, in the end it involves comparatively less digging than if subsurface features were being sought using traditional

FIGURE 8.34.

A ground penetrating radar sled in use. The sled is being pulled along a measured transect line (the string to the right of the sled). It emits a radar pulse into the ground and records the timing and strength of the signal's reflection. A cable connects the sled to a data recording device.

excavation strategies. The foreknowledge that remote sensing can provide about the nature and distribution of potential features within a site can alter research designs for the better. It allows for the more judicious placement of excavation units, doesn't require as much excavation to generate basic data about a deposit, and thus preserves more of the undisturbed portions of the site.

On the downside, the use of remote sensing can be expensive (excepting metal detectors). Equipment for land-based techniques costs in the thousands of dollars, making the use of rented equipment and consulting specialists a more cost-effective tactic. The need for readings to be taken on mapped transects and the extensive clearing of brush needed to accommodate the operation of some devices are investments of time and labor that might not otherwise be made.

When deciding whether or not to use remote sensing, you must take into account not only time and budgetary constraints, but the nature of the site or sites being studied, the geophysical contexts in which you will be working, and the limitations of a particular technique. Just about anyone can take advantage of aerial photography because coverage of many areas already exists. The use of a metal detector on historic sites is easily accommodated in most budgets, or finding someone in the local area who has one and would be willing to volunteer some time to a project shouldn't be too difficult. If you're working on or anticipating complex sites with numerous features, then one or more land-based techniques may be worth the expense and easily justified in a proposal. Cemeteries, Native American villages and towns, and historic sites in general are obvious candidates for remote sensing. Urban or industrial sites present difficulties for electromagnetic techniques because of the ubiquitous presence of metal and power and utility lines, and the havoc they wreak on readings.

LEARNING MORE ABOUT SURFACE AND ABOVE-GROUND ARCHAEOLOGICAL RESOURCES

Controlled surface collections involve the intensive study of archaeological deposits in areas where surface exposures are extensive, as in agricultural fields, arid or other environments with little vegetation cover. Field workers document the provenience and collect all artifacts they observe. Surface or plowzone deposits probably comprise the bulk of the archaeological record in this country and their study is worthwhile in itself. Intensive study of these deposits also contributes to an understanding of the subsurface or sub-plowzone deposits to which they may be related. The nature and patterning of surface deposits are frequently used in planning how and where to locate excavations.

An initial surface survey will have provided information about the area occupied by a deposit, its density, and discernible spatial clustering of artifacts. Appropriate field methods for a controlled surface collection are those that can reveal patterning and meaning in the spatial distribution of archaeological evidence given the geomorphology and land use history of a locality, i.e., the relative integrity of the context. They also should reflect an understanding of how surface data can be manipulated and interpreted. The most important decision to be made is choosing a suitable way of documenting the provenience of artifacts.

Former agricultural fields should have existing vegetation mowed. The field then should be replowed and disked or harrowed, taking care not to go deeper than the existing plowzone or disturb unplowed deposits. The field should be rained on one or two times to make artifacts more visible. Too much rain will promote the reburial of artifacts once exposed. The best conditions for collection in an active agricultural field are the same, after tilling and a few episodes of rainfall. Scheduling a controlled surface collection in an active farm field can be difficult. Some farmers may not allow archaeological fieldwork once a crop has been planted, while others will permit walking over a field until plants have sprouted. Controlled surface collections outside of agricultural fields must be timed according to the season of the year and how the weather and other natural processes either promote or detract from the visibility of objects on the surface.

Ideally, the controlled surface collection of a locality should be repeated over different seasons to reduce the bias inherent in surface deposits. This can be tough to accommodate in projects that have a limited time frame for completion, as is often the case in CRM. When repetitive collections are planned it is essential that the same mapping and artifact collection procedures be employed so that results can easily be compared and contrasted.

In all cases, information regarding the nature of subsurface soil or sedimentary deposits is gathered through the examination of any exposed vertical sections, auger borings or cores, or small excavations that expose stratigraphy (see chapter 9). This information becomes the basis for evaluating the geomorphology of the locality, providing estimates about the age of surface deposits, and the existence of any horizontal stratigraphy. Exposed vertical sections should be profiled, described, and their location precisely mapped. Similarly detailed records should be kept for auger borings or cores used to examine stratigraphy. The positioning of the borings should reflect an understanding of the type of natural processes typically associated with a given environment/ landscape and the spatial variability in stratigraphy that can result. Hand excavations to expose stratigraphy may be necessary in areas where the stratigraphy is very complex or where it is not practical to use augers or coring devices (e.g., deposits with a high percentage of gravel).

LOW-DENSITY ARTIFACT DEPOSITS

Artifact positions are flagged. Mapping the exact location or provenience of individual artifacts will probably be just as time consuming as establishing any type of grid system, large or small, or mapping the boundaries of evident clusters and using individual clusters as provenience and artifact collection units. This is especially true if a total station or high-end GPS unit is available for use. The exact provenience of artifacts will be especially useful if repeated surface collections are carried out. It will allow for greater analytic flexibility than mapping and collecting within a grid system, and may in fact reveal that there is spatial clustering of materials where none had been evident. Grid systems employing units measuring from 100 feet on a side and larger have been used to collect artifacts in low-density deposits.

MODERATE- TO HIGH-DENSITY ARTIFACT DEPOSITS IN DISCRETE CLUSTERS

The quantity of artifacts involved makes it impractical to record the exact provenience of every artifact, yet control over the spatial distribution of individual objects is necessary for analysis and interpretation. The boundaries of individual clusters are flagged and mapped. An approximate center point of each cluster is chosen, marked with a stake or surveying pin, and mapped. Within a cluster, the provenience of artifacts will be based upon their distance from the cluster's center point using what has been called the ***"dog chain method"***(Figure 8.35). Using a tape, circles with regularly increasing radii are configured around the cluster's center point. All of the artifacts within a specific radius of the center point are collected as a group and given the same provenience. The regular interval for the radii should be chosen to reflect the overall size of the cluster, the quantity of artifacts present, and any internal variability perceived by the investigator. In the example shown in Figure 8.35 the interval for each radius is 10 feet. One crew member holds the "0" end of a tape at the cluster's center point. Another crew member holds the tape taut at 10 feet and slowly circles the center point while a third crew member collects all of the artifacts that fall between the center point and the radius of 10 feet. All of these artifacts would be given the same provenience, e.g., 45WR50 (site), Cluster #1, 0–10 feet. The tape is then extended to 20 feet and the operation repeated, the new provenience being Cluster #1, 10–20 feet. The operation is repeated until all artifacts within the cluster have been collected.

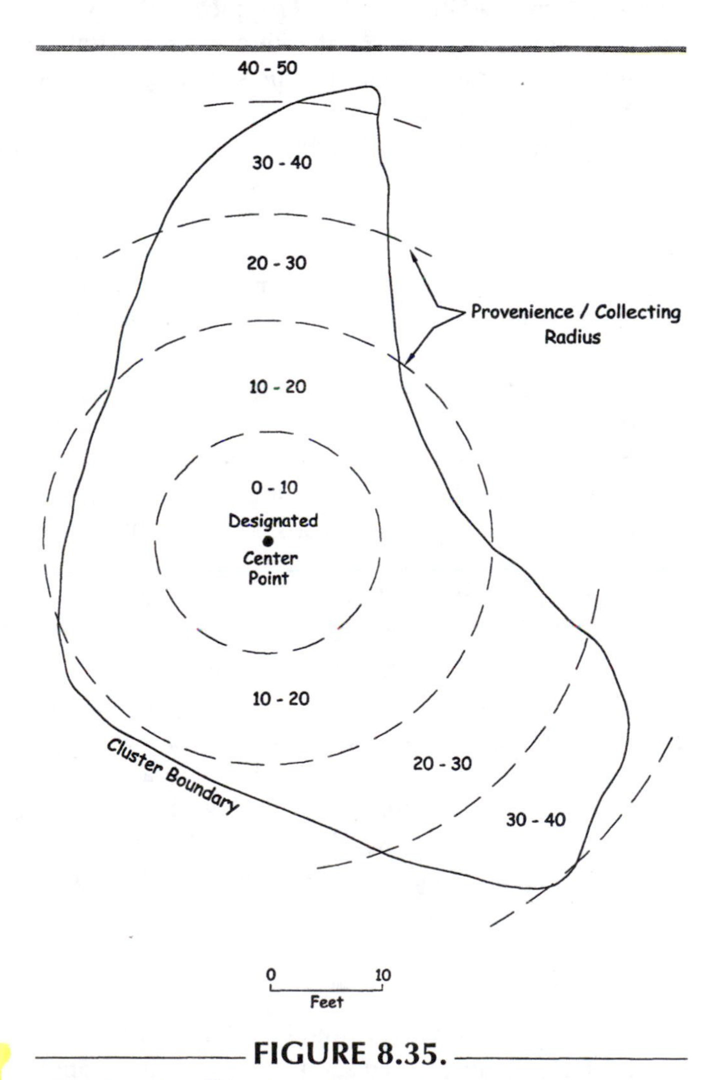

FIGURE 8.35.

The "dog chain" method of recording provenience and collecting artifacts within a defined cluster. Using a tape, circles with systematically increasing radii are configured around the cluster's center point. All of the artifacts within a specific radius of the center point are collected as a group and given the same provenience.

MODERATE- TO HIGH-DENSITY ARTIFACT DEPOSITS WITH NO EASILY DEFINED CLUSTERS

Again, the quantity of artifacts involved makes it impractical to record the exact provenience of every artifact, but control over the spatial distribution of individual objects has to be documented in a way that enables general artifact associations to be examined and patterning to be sought in the frequency distributions of artifact classes or other data sets. This entails establishing a grid system over the area to be collected. The major decision to be made is how big to make the individual units within the grid. The impact of plowing and geomorphic processes on the surface distribution of artifacts is variable enough that one size of grid is not suitable for every controlled surface collection. In plowed fields, grid units measuring 10 feet on a side and larger are typically employed, and are rationalized as a compromise between the acknowledged effects of plowing on artifact clustering and the need to document and understand the distorted remnants of that patterning. Smaller units are used to reflect the investigators attempt to adjust the size of the grid system to their perception of surface patterning and the integrity of the deposit (Figure 8.36). The size of grid units used on unplowed surface sites is equally variable; an investigator balances his/her understanding of site formation processes with the need to collect spatial and contextual data at a scale suitable to get the most information from a deposit. All artifacts collected within a single grid unit are given the same provenience, e.g., site/deposit designation, and the coordinates of the grid unit.

Suggestions for record keeping and photography during a controlled surface collection parallel those given for the discovery phase of fieldwork. If a program of excavation is to immediately follow a controlled surface collection, separate plots of the distribution of artifacts and artifact classes will be valuable for developing a sampling strategy.

FIGURE 8.36.

A gridded portion of the Summer Bay Site located in the Aleutian Islands of Alaska. Positioned on a slope, erosion had removed sediments that once buried the archaeological deposits, revealing artifacts and features. The alignment of rocks in the foreground is a portion of what was once a semi-subterranean native structure. The grid was constructed to map artifacts and features, and serve as collection units and a guide for the placement of excavations. Each grid unit measures two meters on a side. The small size of the grid units reflects the relative integrity of the deposit and the scale of the features and artifact distributions.

ABOVE-GROUND FEATURES

Following their discovery and initial documentation, more can clearly be learned about rock art, ruins, standing structures, and other above-ground features. Guidelines for the rigorous documentation of rock art are found in the references that I've already mentioned in this chapter. Minimally, architectural ruins should be documented by measured drawings and photographs using a camera that produces large-format negatives. Low-altitude photos from camera rigs mounted on captured balloons or small-scale aircraft (e.g., Noli 1985; Walker 1993), overhead photo mosaics utilizing tripod

FIGURE 8.37.

A boom or hydraulic lift, sometimes referred to as a "cherry picker." A railed platform capable of holding 2–3 people is situated at the end of a mechanical arm that can be extended, raised, and lowered to different heights (30 feet and greater) and at different angles. It is excellent for taking overhead photographs of surface features as well as excavations.

or bipod mounted cameras (Sterud and Pratt 1976), or shots taken from a "cherry picker" (Figure 8.37) can enhance documentation of ruins and other above-ground features. Standing structures are evaluated by architectural historians and can be subjected to the formal documentation standards (Russell 1990) of the Historic American Building Survey (HABS) and the Historic American Engineering Record (HAER).

INTEGRATION WITH EXCAVATION AND SUBSURFACE TESTING STRATEGIES

It is hard to imagine a surface survey that couldn't benefit from some type of subsurface testing. Just because nothing was found on the surface doesn't always mean that there isn't anything out there to be found. Stratigraphic variability across a landscape and patchy surface exposures require subsurface testing to achieve systematic coverage of an area. Since artifacts exposed on the surface of plowed fields are biased in terms of size and weight, excavated samples of plowzone are necessary to realize a balanced view of assemblages. Assessments of the potential age of a surface and the materials found upon it are partially dependent on a study of stratigraphy and geomorphology. Some of these data are furnished by the examination of vertical exposures during a surface survey.

How an understanding of a surface deposit contributes to an excavation strategy is as variable as surface deposits themselves. While it can be said that surface deposits are not the mirror image of what exists beneath the ground or the level of plowed soil, some deposits are less distorted than others and can be used as guides for sampling the subsurface component of artifact clusters and activity areas seen at the surface. I've already made a number of suggestions about how certain types of artifacts found on the surface may relate to subsurface features in the process of being exhumed. Dunnell and Simek (1995) propose that with a sufficiently large and controlled collection of surface materials, sub-plowzone deposits can be located and characterized. This is done by looking at modalities in the size distribution of degradable types of artifacts in surface collections (e.g., low-fired pottery, bone), assuming that over time plowing and exposure reduces once-large artifacts into smaller fragments. On the other hand, some surface deposits have no substantial subsurface component or are an extremely distorted version of what remains below ground. Further, the location and spatial patterning of older and more deeply buried archaeological deposits need not have any relation to later occupations represented by artifacts found on the surface or in a plowzone.

In devising sampling strategies and determining where to place excavations, archaeologists cannot rely solely on the surface distributions of materials. The next chapter looks at the variety of subsurface testing and excavation techniques employed in discovering and studying archaeological deposits.

APPLYING YOUR KNOWLEDGE

1. For class discussion: Very limited funds are available for a site survey (surface) of a section of a large river valley. One-third of the valley consists of farms, one-third is woodland (some privately held and some public), and the remainder is occupied by a series of small towns. Nothing is known about the Native America archaeology of the valley. Histories and a variety of documents exist that deal with the non-native settlement and development of the area. What type of survey would you design? How would it be organized? Would some areas get more attention than others? Would you use a sampling scheme? Why or why not? What sort of additional information would you want to design a better survey? Think about the practical realities of site survey as well as theoretical and interpretative issues in framing your responses. Try this exercise using a square area measuring 7 miles on a side and centered on the town where you live.

2. Here's an interesting experiment to try if you're patient and have a garden or access to one. It will allow you to see firsthand the vagaries associated with finding artifacts on the surface. You're going to "plant" some artifacts in the garden and watch how they change position, appear, and disappear over the seasons. Select 10–20 modern objects that are fairly durable, easily recognized, and of different sizes (but nothing bigger than the palm of your hand). You might use marbles, the tiles from a Scrabble game, bottle caps, or even different sizes of road gravel. Be creative. Coat each of the objects that you select with some outrageously visible dayglo paint. Make a scaled map of the boundaries of your garden, then arrange your painted "artifacts" on the surface of the garden in any way that you like. They can be widely dispersed or in a cluster. Next, map the position of each artifact on your plan of the garden. Now ignore them. Treat the garden as you normally would. Every two weeks check to see how many and which of the artifacts are still visible, and whether or not their positions have changed. Do this throughout the growing season. Periodically remap the position of visible artifacts. Were there any noticeable cycles in the appearance and disappearance of artifacts, or in their horizontal movements? Can trends be related to the size or weight of individual artifacts? What other factors influenced the visibility of artifacts?

DIG DEEPER

Theory and Method of Surface Studies.

❑ There is a large body of literature dealing with the theory and method of surface studies of archaeological deposits. The following provide good examples of how surface explorations fit into the planning, execution, and interpretation of fieldwork.

- Dunnell, Robert C. and William S. Dancey. 1983. The siteless survey: A regional scale data collection strategy. *Advances in Archaeological Method and Theory* 6:267–287.
- Lewarch, Dennis E. and Michael J. O'Brien. 1981b. The expanding role of surface assemblages in archaeological research. *Advances in Archaeological Method and Theory* 4:297–342.
- O'Brien, Michael J. and Dennis E. Lewarch (eds.). 1981. *Plowzone Archaeology: Contributions to Theory and Technique.* Vanderbilt University Publications in Anthropology No. 27, Nashville, TN: Vanderbilt University.
- Redman, Charles L. 1987. Surface collection, sampling, and research design: A retrospective. *American Antiquity* 52(2):249–265.
- Sullivan, Alan P. (ed.). 1998. *Surface Archaeology: Method, Theory, and Practice.* Albuquerque: University of New Mexico Press.

Continued

- Wandsnider, LuAnn and Eileen L. Camilli. 1992. The character of surface archaeological deposits and its influence on survey accuracy. *Journal of Field Archaeology* 19:169–188.
- Wandsnider, LuAnn and James Ebert (eds.). 1988. *Issues in Archaeological Surface Survey: Meshing Method and Theory. American Archaeology* 7(1) (A thematic issue).

Remote Sensing Techniques.

❑ Most introductory texts on archaeology contain sections summarizing the different types of remote sensing techniques that can be used to discover archaeological deposits and features. To begin your exploration of techniques in more detail I recommend a number of books and the journal, *Archaeological Prospection.*

- Clark, Anthony. 1996. *Seeing Beneath the Soil: Prospecting Methods in Archaeology*. London: B.T. Batsford Ltd.
- Conyers, Lawrence B. and Dean Goodman. 1997. *Ground-Penetrating Radar: An Introduction for Archaeologists.* Walnut Creek, CA: AltaMira Press.
- Heimmer, Don H. 1992. *Near-Surface, High Resolution Geophysical Methods for Cultural Resource Management and Archaeological Investigations.* Manual prepared for Interagency Archaeological Services, Rocky Mountain Regional Office, National Park Service, Denver, CO. Golden, CO: Geo-Recovery Systems, Inc.
- Jensen, John R. 2000. *Remote Sensing of the Environment: An Earth Resource Perspective.* Saddle River, NJ: Prentice Hall.

Photography.

❑ Bungling photography in the field can be critical, since photos are important records of what has been done and you may not be able to retake your shots. The following are good introductions to the subject.

- Dorrell, Peter. 1989. *Photography in Archaeology and Conservation.* New York: Cambridge University Press.
- Howell, Carol L. and W. Blanc. 1995. *A Practical Guide to Archaeological Photography.* Archaeological Research Tools, Volume 6. Los Angeles: Institute of Archaeology, University of California.

Digital Photography.

❑ The following short article looks at the pluses and minuses of digital photography.

- Rick, John. 1999. Digital still cameras and archaeology. *Society for American Archaeology Bulletin* 17(3):37–41.

CHAPTER 9

Subsurface Investigations

But what be bones that lie in a hole?

J.R.R. Tolkien, Lord of the Rings

The goal of any type of excavation is to discover and document the stratigraphic and spatial relationships between artifacts and features, i.e., their context. As with working on the surface, excavation or subsurface testing strategies differ depending on whether you are engaged in survey-level work, testing a discovered locality, or intensively excavating a significant deposit. This chapter presents basic excavation procedures and the elaborations associated with different types of field investigations.

THE BASICS

EXCAVATION EQUIPMENT

Digging Tools

A number of the hand tools and implements used in the field have been discussed in previous chapters. The most critical tool is, of course, you the archaeologist! Archaeologists can be as passionate about the types of shovels they use as they can about their trowels. Some prefer ***long-handled, round-pointed shovels.*** The shovel's point makes it easier to break ground and deal with dense and gravelly deposits. Others, myself included, prefer long-handled, square-pointed or ***flat shovels.*** The straight cutting margin of the flat shovel can be used as a guide to create and maintain the horizontal lines and walls of an excavation. The shovel's margin can also be used as a "ruler" in establishing the horizontal margins of shovel tests that are not laid out in the same precise way as an excavation unit. The cutting margin of a flat shovel, or any shovel, should be periodically sharpened with a milled, common bastard file. Sharpen the edge on a bevel from the top side, in the same way you sharpen your trowel. Shovels with shorter handles become necessary in deeper narrow excavations where you don't have room to wield a long-handle shovel. Traditional ***short-handled spades*** can be useful for breaking up especially hard or rocky sediments and soils. Some models are fashioned completely from heavy-duty steel to enhance their performance. Picks and mattocks are also used in these tough situations.

Caution: The force that is required to wield some tools can severely damage artifacts. Their use in certain contexts should be considered carefully.

Screening Devices

Excavated matrix is sifted through hardware screen in order to retrieve artifacts that were not seen *in situ* or noticed during the process of excavation. Excavated matrix may be shaken through a screen or pushed through the screen by hand or with a wooden block. Wearing gloves will prevent cuts and scrapes that can result from this process. A wooden implement for breaking up chunks of matrix and getting them through the screen is favored over a metal one. Metal implements, like trowels, have a greater potential to damage artifacts in the screen. A trowel can be used to move the matrix back and forth across the screen to coax it through without fear of causing damage. Using a trowel to mash or cut up masses of sediment or soil should be done with caution. Sediments or soils with a high clay content, and soils with developed structure or a firm consistence make screening a more laborious and problematic operation. In these cases more effort is needed to break the matrix up into fragments small enough to pass through the mesh of the screen. In some cases the force required to do this may be substantial, as when dry clayey sediments are involved, and could damage any artifacts embedded in the chunks of material tossed into the screen. Wet clayey sediments are equally problematic since they are plastic and easily adhere to materials. They literally have to be rubbed or pushed through a screen, and this action can also damage artifacts to various degrees. The excessive abrasion of lithic artifacts against the metal of the screen can obscure any original edge damage or use wear that they may possess. Likewise, the surfaces of sherds of low-fired earthenware or terra-cotta pottery can be damaged, masking important surface treatments or decorations, or the sherds may simply break.

The most frequently used mesh size employed in ***dry screening*** is 1/4 inch. Screens with 1/8-inch mesh are also used in dry screening in the field, but less frequently. As the mesh size is decreased, the time it takes to sieve and examine a screenload of excavated matrix increases. Artifacts small enough to pass through the mesh of a screen—small flakes, micro debitage, and small fragments of plant and animal remains—can be lost during the screening operation. The lack of these materials, or their underrepresentation, impacts interpretations of an archaeological deposit.

> *For example, in an experiment utilizing the flake waste from the re-creation of 10 small-stemmed projectiles, only 2% of the total number of flakes, representing 25% of the total weight of flakes, was recovered in a screen with 1/4-inch mesh (Kalin 1981).*

Using a very small mesh increases the recovery of small artifacts but also dramatically increases the amount of time needed to screen excavated matrix. The bias of using 1/4-inch mesh is offset partially by the dry screening or water screening of samples of matrix through 1/8-inch or smaller mesh screen, and collecting samples for flotation. The choice of mesh size is a compromise between

- ☑ artifact recovery rates
- ☑ the time and cost devoted to screening
- ☑ the type of fieldwork being performed (e.g., survey, test excavations, intensive excavations)
- ☑ a deposit's potential for organic preservation
- ☑ the type of deposit or feature being dug

Fine-grained recovery techniques (use of screen mesh smaller than 1/4 inch, wet screening, and flotation) typically are not used during surveys or site discovery. Their use normally begins, and progressively increases, through the testing and intensive excavation phases of fieldwork. Deposits with the potential for organic preservation, like those associated with caves, rockshelters, or pit features in general, warrant the concerted use of fine-grained recovery techniques from early on in an investigation. The thorough investigation of stone tool manufacturing, the use and economic importance of plants and animals, and burial features also necessitate their application.

Screening operations can be set up in a number of ways, all involving an open type of box with solid sides, usually made of wood, and a screened bottom. A small box screen, 2′ × 2′ or less in dimension and without a stand or attached legs, is extremely portable and therefore serviceable on surveys. In the field the screen is either hand held and shaken or propped on an improvised stand of logs or rock. The small volume of individual loads of matrix that it can handle slows screening relative to what is possible with a larger piece of equipment. However, the smaller screenloads improve the chances of even spotting very small artifacts before and during sifting. Shaker screens, a larger box mounted on a pivotal pair of legs (Figure 9.1; also see Figure 5.13), are fairly lightweight and can be carried by hand, or outfitted with backpack or carrying straps. When not in use the screens legs fold up around the frame of the screen. Shaker screens are very effective when sediments or soils have a relatively low clay content, crumb or granular structure, and loose or friable consistence. The effort required to deal with dense, clay rich sediments and soils with well-developed structure can be awkward with a shaker screen, since the rig must be balanced upright when not being shaken. Screens suspended from tripods (Figure 9.2) or any type of framework offer the same sifting advantages as a shaker screen since they can be manipulated in a back-and-forth motion. Screen frames constructed to roll back and forth on metal wheels achieve the same purpose (Figure 9.3). However, both of these lack the portability necessary for use in surveys or in any situation where equipment has to be transported over appreciable distances by hand. Screens propped on sawhorse legs or fixed to a rigid stand (Figures 9.4 and 9.5) are impractical for use during surveys. Because they are heavily constructed without portability in mind, they can better withstand the wear and tear associated with processing problem sediments and soils. They also afford greater mobility to the person doing the screening. Screen boxes may be positioned on top of wheelbarrows (Figure 9.6), which makes it easy to move backdirt around an excavation.

FIGURE 9.1

A portable shaker screen in use. A box screen is mounted on a centrally positioned pair of legs, which allows the screen to be pivoted in a back-and-forth motion promoting rapid sifting when sediments are relatively loose, have a crumb or granular structure, and a relatively low clay content. In this scene sediments from a bucket auger are being emptied into the screen.

FIGURE 9.2.

Screens suspended from tripods provide some of the advantages of shaker screens but are not as portable.

FIGURE 9.3.

A screen box is mounted on metal wheels that allow it to be rolled back and forth along the top of the screening stand to enhance sifting.

FIGURE 9.4.

A heavier-duty screen setup useful for dealing with problem sediments and soils. The sturdy box screen is propped on sawhorse legs. Note the skill of the excavator in the background. The shovelful of matrix that she has tossed toward the screen is traveling as a clustered mass rather than as a trailing spray.

FIGURE 9.5.

A box screen mounted on a rigid frame that is assembled on-site. The smaller suspended screen is 1/4 inch mesh and can be used like a shaker screen. The lower screen has a smaller mesh. When fine-grained recovery techniques are employed, nested screen setups like this provide for the rapid separation of larger-sized artifacts and materials from the matrix, speeding up the screening operation and reducing the potential damage to some artifacts.

FIGURE 9.6.

Screen boxes set up on top of wheelbarrows. This technique makes it easy to move backdirt around an excavation and stockpile it in a common location if necessary.

Water or ***wet screening*** (Figure 9.7) can be useful for breaking down problem sediments and soils. Using screen meshes smaller than 1/4 inch is less time consuming and increases the visibility of any artifacts in the screen. Wet screening may be accomplished by repeatedly dipping hand-held screens into a source of water (e.g., a large tub or barrel, stream), or by using a low-pressure hose in conjunction with any dry screening setup. Obviously, a water source must be convenient. When studying Native American archaeology, water screening may not be appropriate because of its effect on the recovery and analysis of residues, pollen, phytolith, and starch grains adhering to some types of artifacts. The shrinking and swelling of organic materials as they become wet then dry can lead to their degradation if the process is not handled with care. Low-fired pottery can also suffer from this procedure. Constant volume samples of excavated matrix may be collected from excavations and transported to the lab or to a convenient location for wet screening or flotation.

FIGURE 9.7.

A simple water screening operation using a hose and a screen on a rigid stand. Water screening can be useful for breaking down problem sediments and soils and making the use of screen mesh smaller than 1/4 inch less time consuming. It also increases the visibility of any artifacts in the screen.

Deciding where to locate a screening operation and backdirt piles can be tricky (Figures 9.8 and 9.9). They must be located a sufficient distance from an open excavation so that screened sediments won't fall back into it or collapse the edge of an excavation. These are potential threats both to excavators and the integrity of the excavation unit. The location should also ensure that rainfall won't be able to wash sediments back into the excavation. On the other hand, it is convenient if a screening operation can be situated within a shovel throw of an excavation. If excavations are to be backfilled by hand, you want backdirt piles from screening to be as close to them as safely possible. Another factor to consider is whether or not an excavation will be expanded and in what direction. You want to minimize the number of times that a backdirt pile has to be moved. An old joke in archaeology is that the best features and most outrageous artifacts in a deposit are buried under backdirt piles. There are OSHA standards applicable to these issues. While someone is working in an excavation, others must remain at a safe distance in order to prevent the possibility of cave-ins. In general this distance is twice the height of the vertical side wall of the open excavation (OSHA 1991). Materials should not be piled closer than 2 feet from the edge of an open excavation (OSHA 2000). The type of soil being excavated and whether it is loose or coherent should be considered in deciding whether these guidelines are adequate.

FIGURE 9.8.

The appropriate juxtaposition of screened backdirt and the excavation from which the sediments were derived. Screening operations and backdirt need to be located so as not to pose any threat to excavators or open excavations. This backdirt pile is far enough from the excavation so that rain won't wash sediments back into the unit, while still being close enough to the unit to make backfilling easily accomplished with shovels. The unit is a meter square.

Augers and Coring Devices

Augers and coring devices are indispensable for all levels of field investigations. A discussion of their role in archaeology is provided by Stein (1986) and Schuldenrein (1991). Auger and coring devices generate information on subsurface sediments, soils, and stratigraphy with minimal impact on deposits, and with substantially less effort than that invested in a shovel test or excavation unit. They are indispensable for gathering data on stratigraphic changes over space, so necessary to the study of the geomorphology of a landscape. Smaller artifacts may be retrieved in some types of these devices. Samples of pollen, phytoliths, and other botanical remains can be retrieved from cores. Augers and coring devices can be used to determine where best to place other types of subsurface tests. They can be used to delimit the horizontal and vertical extent of pit features prior to initiating a controlled excavation. A variety of devices are pictured in Figure 9.10.

FIGURE 9.9.

This screening operation is too close to the open excavation, allowing sediments to fall back into the unit. In deeper units, this can pose a real threat to the safety of the excavator.

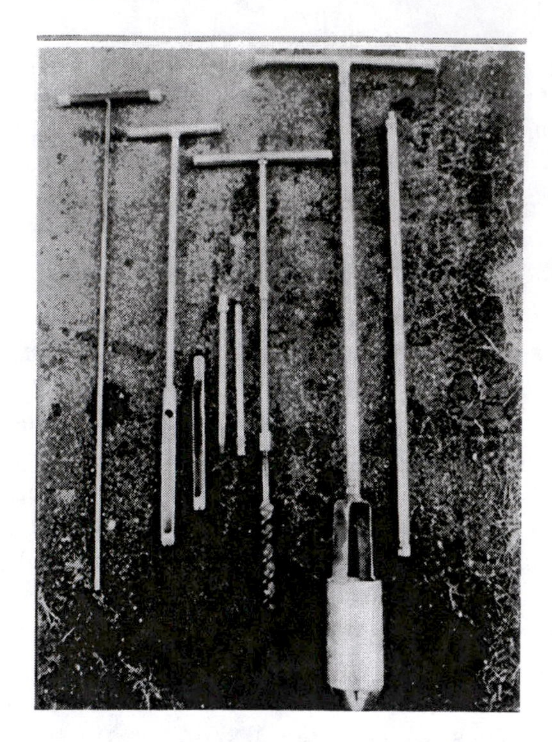

FIGURE 9.10.

Probes, augers, and coring devices useful in the examination of subsurface deposits. From left to right: a pointed metal probe or "tile probe"; a coring probe with an open or split sampling tube; a sampling tube attachment for the coring probe; extensions for the shaft of the coring probe; a screw auger; a bucket auger; extension for the shaft of the bucket auger.

Coring devices produce relatively continuous vertical samples of subsurface deposits. Many archaeologists prefer split-spoon or ***split tube samplers.*** A partially open sampling tube is attached to the end of a metal shaft with a "T" handle. Sampling tubes can be up to 21 inches long on single-probe devices, but are typically 12 to 15 inches long on those used to take deeper and longer cores. The diameter of the hole created by the tube sampler is less than one inch. To begin, the device is pushed into the ground the length of the split tube sampler, then withdrawn. A continuous core of sediments is exposed in the open part of the spoon. The face of the core must be cut with a trowel or knife because the act of pushing the device through the ground smears the core's exterior. A better view of sediment color, texture, and stratigraphic breaks is obtained after cutting. After examination, the core is extracted from the tube and the device is reinserted into the initial hole in the ground. It is pushed into the ground another length of the sampling tube; if the first core from the hole went to one foot below surface, the second probe and core will extend the hole to roughly two feet below surface. Again, the face of the core in the tube sampler is cut, examined, and removed. This procedure is repeated to the desired depth below surface. A folding ruler or stiff tape is periodically inserted into the coring hole to keep track of depth and relate this to individual core sections. Extensions are available to lengthen the shaft of the coring device.

The ease with which this type of core/probe device is used depends on the nature of the deposits to be tested. Under favorable conditions the split tube sampler functions well to depths between 3 and 6 feet. Dense, clayey sediments and deposits with large amounts of gravel may prevent its use. Special tips can be obtained for the end of the tube sampler to help overcome difficulties arising from excessively wet and excessively dry and loose sediments.

Artifacts are rarely recovered from a split tube sampler because of its small diameter. The small diameter of the core enables the texture of sediments to be evaluated but does not retrieve a large enough sample to accurately detect anything other than crumb or granular soil structure. The primary advantage of the split-spoon or split tube sampler is the relational view of sediments and stratigraphic breaks they provide, and the extremely small area they disturb.

Hand-operated devices exist for taking larger diameter (2–3 inches) cores when sediments are loose, soft, or wet, as is the case when coring ponds or marshes to search for and recover pollen and botanical remains. *Mechanical coring* devices are available for extracting large and deep cores on dry land and in poorly drained conditions (cf. Canti and Meddens 1998; Stein 1986). These larger core samples make it possible to evaluate more fully the characteristics of sediments and soils and search for artifacts, floral, and faunal remains. Mechanical devices can also be used in dense, rocky, or hard-to-penetrate sediments where it would not be possible to use a hand-operated tool. Mechanical devices are not as portable as hand operated ones and in some cases are mounted on trucks. The location and accessibility of locations to be tested must be considered when deciding to use a mechanical coring device and choosing among the ones available.

Augers, both hand-operated and mechanical, drill into sediments to provide samples and information on stratigraphy. A small-diameter (1–2 inches) ***screw auger*** attachment can be fitted on to the same "T"-handled metal shafts used with split tube samplers (see Figure 9.10). A screw auger can be used in dense and rocky deposits where a hand-operated coring device cannot. While it derives samples of sediment from progressively deeper levels, distinguishing the thickness of a layer and stratigraphic breaks cannot be done as accurately as with a coring device. The device is screwed or twisted into the ground the length of the screw head, then removed for examination. Sediments are collected in the screw's threads and are visible when the device is pulled from the ground. The auger is reinserted into the original hole, screwed an additional length, and removed. This incremental process continues until the desired depth is achieved. A folding ruler or metal tape should be inserted into the bore hole after every sample is removed. This type of screw auger can be effective to the same general depths as a split tube sampler. However, because it is twisted into the ground and is harder to remove than a tube sampler, there is a greater potential for damage to the device's shaft as depth increases. Like a tube sampler, this screw auger does not produce a large enough sample for soil structures other than crumb and granular to be adequately evaluated.

Bucket augers also are screwed into the ground but produce a much larger and more useful sample of deposits than screw augers. Samples are large enough to evaluate soil structure and to be used in a variety of laboratory analyses. Bucket augers can retrieve many types of artifacts. Mounted on a "T"-handled metal shaft (see Figure 9.10), the bucket portion of the auger ranges from 2 to four inches in diameter. On regular models the sides of the bucket are fully enclosed. The top and bottom (auger bit portion) of the bucket are open. The vertical location of stratigraphic breaks can only be approximated with this type of device since sediments are being drilled into the bucket and there is no side view of the sediments inside of the bucket. The implement is twisted into the ground the length of the bucket portion of the auger, then withdrawn. The depth of the hole should be noted. Examine the color and texture of the sediments in the top and bottom of the auger and determine whether they indicate any stratigraphic changes. The removal and handling of sediments can be done by a single experienced individual, but is more easily accomplished with two persons. Turn the entire auger upside down, resting the "T" handle on the ground. Repeatedly tap the handle on the ground to loosen the sediments in the bucket and get them moving (Figure 9.11). With your free hand, or with the help of another, cradle the sediments sliding out of the bucket. Place them on a section of tarp, plastic, or cleared ground for further examination and for comparison with other samples taken from the same bore hole (Figure 9.12). Take special note of any stratigraphic changes discernible in the sample. Reinsert the auger in the original hole and derive another bucket-length sample. Always measure the depth of the hole after each sample is removed. Specially designed bucket augers deal with clay and mud. Portions of the sides of the "buckets" on these augers are open, allowing samples to be examined before removal. Using shaft extensions, and depending on the type of deposit, bucket augers can examine deposits at depths well beyond six feet. The more gravelly a deposit, the more difficult it will be to use a bucket auger.

FIGURE 9.11.

The process of removing a sample from a bucket auger. Once removed from the bore hole, the auger is turned upside down. Tapping the handle on the ground will loosen the sample and cause it to slide out of the bucket, where it can be collected and examined.

A ***posthole digger*** or ***postholer*** (Figure 9.13) is capable of providing the same types of information as a bucket auger, although it is only effective to approximately three feet below surface. It minimally creates a hole from 9–10 inches in diameter. With the wooden handles pressed together, the postholer is rammed into the ground. Spreading the handles causes the jaws of the device to close around whatever it has penetrated. The material can then be lifted from the hole, dumped and examined. Like an auger it generates samples large

FIGURE 9.12.

A bucket auger in use. Extracted samples are laid out in stratigraphic position on the plastic tarp on the right.

FIGURE 9.13.

A posthole digger can serve many of the same purposes as a bucket auger in shallow deposits, and is useful for recovering artifacts. The force involved in the use of a posthole digger, however, can severely damage any artifacts that it strikes.

enough for field and laboratory analysis of sediments and soils. Also like an auger, it doesn't produce a continuous vertical section of material, so stratigraphy must be documented by repeated measurements of the depth of the "posthole" and examination of the samples retrieved between measurements. The force involved in the use of a posthole digger can severely damage any artifacts that it strikes.

A simple ***probe*** is a slender shaft of metal or fiberglass with a "T" handle (see Figure 9.10). The typical shaft ranges from 3–4 feet in length. Longer models are available and some are designed so that extensions can be added. They are sold commercially as "tile probes." Archaeologists have long used homemade versions of this implement. Probes are great for locating underground obstructions or dramatic changes in soil texture or structure by "feel." Depending on what the probe is being pushed through, the user feels different degrees of resistance. The size or horizontal extent of a solid obstruction or a distinctive stratigraphic break or layer can be tracked by offsetting probes in the vicinity of an obstruction or conducting probes on a measured transect or grid. Probes are most effective if you already have some knowledge of the nature of subsurface deposits. For example, if you are working on an alluvial landscape where little or no gravel is expected to occur, then solid obstructions encountered with a probe become even more meaningful. It is impractical to use probes where deposits are dense, clay rich, or gravelly. Probes are helpful in locating and mapping out the spatial extent of buried structural ruins, shell middens, any type of feature containing numerous or large solid objects (e.g., rock-lined hearths), or very large artifacts (e.g., stone bowls, grinding platforms). Probes are not meant to retrieve samples, but their tips may become smeared with material that may be suggestive, such as a smear of white powdery material from probing the top of a buried shell midden.

EARTH-MOVING MACHINERY

The use of heavy, earth-moving machinery can be appropriate in a variety of field situations, from site discovery to intensive excavations (cf. Van Horn 1988). Backhoes, bulldozers, and gradalls are the most popular (Figures 9.14–9.16). For most of the operations involving heavy machinery, it is important that the cutting edge of a bucket or front-end loader be smooth or straight, not toothed. Smooth edges will create relatively smooth cuts in the trenches being dug or on the surfaces being stripped that will require less effort to clean for closer examinations. Request that a bar or small plate be welded across the teeth of any machine you plan to use (Figure 9.17).

FIGURE 9.14.

A small bulldozer with a straight-bladed front-end loader. This type of heavy machinery is effective in stripping large areas of fill prior to excavation, exposing extensive surfaces in the search for subsurface features, or backfilling extensive excavations.

FIGURE 9.15.

The best of both worlds. A backhoe (left side of machine) for cutting deep trenches and removing fill from small areas coupled with a straight-bladed front-end loader (right side of machine).

FIGURE 9.16.

A gradall. Its telescoping boom allows the machine to sit to the side of deposits while excavating. Deposits are never compressed or deformed by the weight of the machine passing over them.

FIGURE 9.17.

A backhoe in operation with a plate welded over the toothed portion of its bucket. This plate results in smoother cuts of trench walls and stripped surfaces, making them easier to clean and examine.

Backhoes are relatively lightweight from the perspective of heavy machinery. They will not deform or compress the deposits over which they pass in the way that other heavy machinery will. The backhoe, with its digging bucket attached to a movable arm or boom, is useful for cutting trenches 10–12 feet deep while positioned on the surface. It is fairly maneuverable and can be used to excavate in small, hard-to-reach areas. The backhoe may be used to strip small areas without riding over them. Machines with front-end loaders are more efficient for stripping large areas but must ride over them to do so.

Bulldozers are tracked vehicles used most frequently to strip large areas or to excavate large blocks to depths where hand excavations will begin. Their weight can compress and deform near-surface deposits. The degree of disturbance depends on the type of sediments or soils that are present. For example, coarse-grained sediments are more easily affected than those with moderate to high clay content, wet sediments are more easily deformed than ones that are not saturated. Bulldozers or any machine with a front-end loader makes the backfilling of extensive excavations a breeze. Care must be taken that unexcavated areas are not damaged during backfilling. ***Gradalls*** are excellent for the highly controlled stripping of relatively small areas. A gradall's telescoping arm or boom allows the machine to sit to the side of deposits while excavating. Deposits are never compressed or deformed by the weight of the machine passing over them. Note: The use of heavy machinery in specific field operations is detailed in later sections of this chapter.

Working with Heavy Equipment Operators

Develop a rapport with the machine operator, explaining exactly what you want to do and how you would like it to be accomplished. Emphasize that speed in completing an operation is not a requirement and that the safety of the crew comes first. Only one person will give directions to the operator when the machine is in use. Make this clear to the operator. Work out a series of hand signals with the operator since you won't be able to hear one another over the noise of the equipment. These signals should encompass things like: stop, shut down the machine, continue, dig deeper, thickness of cut to take, dump, move left, move right. When a machine is in use, stay in sight of the operator. Don't stand beside or behind the machine, or within the arc of movement of the boom arm of a backhoe or gradall. When you want to move closer to an excavation to examine something, signal the operator to stop before moving within the range of the backhoe or gradall boom. Don't be shy or hesitant in directing an operator, assuming that they know what you want. You're the boss. Remember that it is necessary to wear a hard hat when working around heavy machinery. There are OSHA standards for shoring, stepping, or sloping excavations that are more than five feet deep, whether machine cut or dug by hand.

TYPES OF EXCAVATIONS

Cut banks (also referred to as bank cuts or ***strata cuts***) or any cleaned vertical exposure can be turned into an excavation of sorts. Once the profile has been recorded, the face of an individual stratum can be sliced/peeled with a trowel in search of artifacts and features. ***Shovel tests*** or ***shovel probes*** are small types of excavations, typically 1.5 feet or less on a side. Their small horizontal dimension restricts the depth to which they can be excavated, generally three feet or less. The term "probes" may also be used to refer to auger or posthole borings used to look for subsurface deposits of artifacts. The term ***unit*** refers to larger excavations that are generally excavated in a more controlled fashion than shovel tests. Units may be square or rectangular. They generally are distinguished from ***trenches*** and ***area*** or ***block excavations*** by their smaller size and because they may not have to be horizontally subdivided to provide minimum standards of provenience or context control. Some typical dimensions of units are 1 × 1 meter, 5 × 5 feet, and 2 × 2 meters. Trenches are linear excavations, at least two times as long as they are wide. Area or block excavations are horizontally extensive and may take on a variety of geometric shapes. Because of their larger size, trenches and block excavations must be subdivided into a series of smaller units in order to maintain proper control of artifact and feature provenience. One could say that small units are the building blocks of any extensive excavation. Discussions to follow regarding the appropriate use of different types of excavations are organized in terms of discovering archaeological deposits, testing them, and mounting intensive investigations.

Through time, natural and cultural processes create the discernible layers and features that are encountered below ground. In an excavation, an archaeologist removes layers and features in the reverse order in which they were created, latest to first. The most discrete chronological and stratigraphic contexts that can be defined in a deposit are used as excavation levels for the removal of artifacts and features. Deciding how to remove a layer or feature demands that we make assumptions about them, or have some foreknowledge of them. Many of these assumptions are quite basic and derive from a fundamental knowledge of sediments, soils, geomorphic and cultural processes that transform landscapes, and the criteria used in defining stratigraphy (see chapter 7). Although making assumptions is necessary, excavation procedures need to be structured to allow for the recognition of the unexpected. They also need to be flexible to accommodate new information as digging progresses. Excavation strategies evolve as the understanding of a deposit evolves.

The simplest way to proceed with an excavation when beginning work in a new area is by discernible stratigraphic layers. A core from a split tube sampler or an auger boring easily provides advance notice of the types of stratigraphic changes or even the features that can be expected in the immediate area of an excavation. There may be bankcuts or other subsurface exposures of which to take advantage. It makes sense to excavate strata recognizable as single events or depositional episodes as single excavation levels. This minimally includes plowzones and placed fill that usually can quickly be identified at the outset of fieldwork. The confident identification of other single-event deposits is normally only possible after some degree of excavation, data collection, and analysis have been completed. Any highly disturbed strata or placed fill may be removed as individual excavation levels. A sample of the matrix may be screened if artifacts are sought for the purpose of determining the age of the event or understanding the fill's origin. In most cases, fill is not screened, although any potentially informative artifacts spotted during excavation may be collected. Extremely thick deposits of placed fill, such as those encountered in urban areas, may be removed in a series of gross levels simply as a matter of convenience or as a means of verifying its depositional history. Natural deposits resulting from relatively modern events can also be dug as single levels. Again, your understanding of the age and nature of such deposits may or may not require screening a sample of the matrix or collecting any artifacts.

Removing each identifiable stratum as a single excavation level, regardless of its nature or thickness, is called ***excavating by natural strata.*** This practice doesn't always guarantee that you are working in the most discrete chronological or stratigraphic segments into which a deposit might be divided. For example, a plowzone that is one foot thick can be dug as a single layer without losing any of the information it contains, while a reddish brown B horizon of equal thickness may not. The plowzone by definition is a churned mixture of materials and digging it in smaller incremental levels can do nothing to change this fact. The color and texture that are associated with the B horizon have developed over time and don't represent a single depositional episode of sediments, artifacts, or features. Therefore, in order to see if different depositional episodes or contexts are encompassed by the B horizon, it needs to be dissected in pieces.

Excavating by arbitrary levels within natural strata is a way of maintaining contextual control and deriving useful data in situations where strata are not fully understood, are overly thick, or represent well-developed soil horizons. ***Excavating solely by arbitrary levels*** is obligatory when meaningful stratigraphy cannot be defined in a deposit. Arbitrary levels are just what their name implies, levels whose thickness has been decided subjectively, but with a purpose in mind. Since they are used to maintain tighter control on the provenience of any finds or data collected during an excavation, an arbitrary level is generally thin. A scan of archaeological site reports dating from the past twenty years reveals levels that can be 5–10 centimeters, 3–6 inches, 02.–03. foot, and sometimes greater in thickness. The thickness of an arbitrary level should be chosen to reflect an excavator's current understanding of a deposit and the things that he/she wishes to learn from it. Given the inherent variability in archaeological deposits from place to place, it shouldn't be surprising that the definition and use of arbitrary levels also varies. For example, the choice of arbitrary levels can reflect the type of sediments being excavated. An excavator may decide to use small arbitrary levels in the initial exploration of a sandy stratum knowing that the postdepositional movement of artifacts in such sediments can be moderate, and reasoning that smaller levels will provide better contexts in which to recognize and understand this movement than larger ones. Once this process is understood and other information has been gathered to define the cultural stratigraphy of the deposit, the size of arbitrary levels can be adjusted accordingly and used in future excavations. As I have already noted, excavation strategies evolve over the course of fieldwork as relevant information accumulates.

Working within a definable stratum with arbitrary levels may not result in levels that are all the same thickness, e.g., using 0.3-foot levels in a stratum 0.7 foot thick. Because of the sloping nature of the base of a stratum, a level may be 0.3 foot thick over a portion of the floor of the excavation and only 0.1 foot thick in other areas. This is simply the nature of the game. Although levels of equal size and volume make some postexcavation analyses easier, the extent of manipulation required in the field to achieve them is unrealistic and probably unachievable. As you might surmise, excavating by natural strata, employing arbitrary levels with natural strata, or digging solely by arbitrary levels are not mutually exclusive approaches. Within a single excavation some natural strata may be removed as a single level, others as a series of arbitrary levels.

Any type of level may be excavated by ***quadrants*** or other equally sized subdivisions to enhance the contextual control over the location of artifacts. In addition, the precise location of select artifacts may be recorded before they are removed from the excavation floor. This highest level of contextual control has been called ***exact provenience, point provenience,*** and sometimes ***piece plotting.*** These same terms also are used to describe excavation technique, e.g., excavating a level using point provenience.

Strata and Level Labeling

Strata and levels are labeled progressively beginning at the top of an excavation and moving to its base using numbers or combinations of letters and numbers. The uppermost stratum in an excavation could be labeled "A" and the first level excavated from it designated as "1." All artifacts or other materials found or collected from this context would have the provenience Stratum A, Level 1. Subsequent levels excavated within this stratum receive a sequential number, e.g., Stratum A, Level 2; Stratum A, Level 3, until the stratum has been completely removed. Even if Stratum A is completely removed as a single excavation level it should still be considered as Stratum A, Level 1. The use of letters for designating strata and numbers for specifying levels within strata is viewed by many as a way to avoid confusion and errors in record keeping. They argue that if numbers were used to signify both strata and levels it would be too easy to juxtapose them on artifact tags, record sheets, and other field notes. In turn, other archaeologists eschew the use of letters for designating strata since these are used in the labeling of soil horizons that may exist in a deposit. They prefer the use of numbers for both strata and levels. As stratigraphic changes are encountered in an excavation, stratum designations change in a regular fashion, i.e., A, B, C, or 1, 2, 3.

There are different schools of thought regarding the labeling of excavation levels associated with different strata—one system uses a single sequence of numbers for all of the levels removed from an individual excavation, e.g., Stratum A, Level 1; Stratum A, Level 2; Stratum B, Level 3; Stratum C, Level 4. Using this system, a person examining artifact labels or field notes always has some sense of the relative depth of a context in an excavation because excavation levels are numbered 1 to *n* from top to bottom. They needn't always refer back to depth measurements on data sheets or an excavation's profile.

Others employ a new sequence of numbers every time excavation into a different stratum begins, e.g., Stratum A, Level 1; Stratum A, Level 2; Stratum B, Level 1; Stratum C, Level 1. In this system one immediately gets a sense of the thickness of individual strata, but their relative depth below surface is not as apparent. Still another system uses the actual vertical position of a level within a stratum, e.g., Stratum A, 0–10 centimeters; Stratum A, 10–20 centimeters; Stratum B, 0–10 centimeters; Stratum C, 0–10 centimeters. Features are numbered sequentially as they are discovered within a site or deposit although their provenience will include the strata and levels in which they occur, e.g., site 44WR50, Stratum A, Level 1, Feature 1.

Features are frequently dissected by natural or arbitrary levels independent of those used to take the overall excavation down. Feature levels may be distinguished by sequential numbers or by listing the vertical position of a level within the feature: Stratum A, Level 1, Feature 1, Level 1; Stratum A, Level 1, Feature 1, 0–10 centimeters, respectively.

Each of these systems is designed to document the excavation sequence and distinguish different contexts from one another. Ultimately, some strata and/or levels from a single excavation may be combined in defining the cultural and natural stratigraphy of a deposit. Correlating the natural and cultural stratigraphy between excavations and across a landscape may result in a variety of recombinations of strata and levels. Because the correlation of stratigraphy requires information derived from a series of excavations, and because there is no way of knowing the range of variability that will be encountered prior to excavation, strata and level designations don't have to mean the same thing from excavation to excavation. In other words, Strata A, Level 1 in Unit 1 does not have to physically equate in all of its details with Strata A, Level 1 in Unit 2. Attempting to assign strata and level designations so that correlations are implied is problematic except on the simplest of sites, and will require constant revision.

Provenience Recording Systems

Stratum and level are the most basic ways of recording the provenience of artifacts, features, or other material evidence in an excavation; e.g., the provenience of this projectile point is site 44WR50, Unit 1, Stratum A, Level 2. An excavation level may be horizontally divided into quadrants or smaller subdivisions to heighten contextual controls; e.g., the provenience of this projectile point is site 44WR50, Unit 1, Stratum A, Level 2, NW quadrant. Any artifact mapped *in situ* (given point provenience) within a specific stratum and level is given a map number that relates it to a plan view of an excavation floor, or a series of coordinates relative to an excavation datum; e.g., the provenience of this projectile point is site 44WR50, Unit 1, Stratum A, Level 2, Artifact #7. Of course, when there is no definable stratigraphy and a deposit is being excavated with strict arbitrary levels, there will be no stratum designation. Provenience recording systems are often used in tandem. You may be excavating with the intention of mapping every artifact found *in situ,* but may still find some artifacts in the screen. So for a single excavation level it is possible to have some artifacts or materials with a basic stratum and level provenience, and others with stratum, level, and artifact map number. Provenience recording systems vary depending on the type of subsurface test or excavation, the stage of fieldwork involved, and the integrity of the context or deposit being excavated.

EXCAVATING A SIMPLE UNIT

As a prelude to seeing how excavation strategies can vary with the stage of fieldwork involved, let's examine how a basic unit is excavated when a deposit is first being examined. Let's assume for the purpose of this discussion that the unit will be five feet or less in depth. Deeper excavations must cope with a variety of standards for ensuring the safety of the field crew, which in turn necessitate elaborations of some archaeological techniques. We'll walk through the steps for excavating by natural strata and arbitrary levels first, then see how the process differs when a level is dug in quadrants, or when the intent is to see and map all artifacts *in situ.* We'll review the procedures for defining and excavating a simple feature. Data collection and preservation, record forms, and field notes are addressed in a subsequent section.

Initial Procedures

The unit's location has been selected and it has been laid out using the triangulation technique discussed in Chapter 6. We'll assume that we are working with a unit 5′ × 5′ in horizontal dimension. The corners of the unit are marked with surveyors pins, spikes (galvanized gutter stakes are great), or some type of stakes. If wooden stakes or hubs (a short, thick stake with a square cross section) are used, they are pounded in close to the ground and a small nail or tack is placed in their top to represent the actual corner point of the unit (Figure 9.18). Using stakes and nails to establish the corners of an excavation requires a two-staged triangulation effort. After the first point or corner is set the position of each remaining point is triangulated, first to locate the position where the stake will be driven, and again after the stake has been driven to locate where to place the nail in the top of the stake. This method is more laborious than using surveying pins or metal spikes to mark corners, but staked corners are less likely to be disturbed than ones that are simply pinned. A compromise is to use a wooden stake and nail for the initial corner of the unit and employ surveying pins or spikes for the remaining corners. The sturdy wooden stake and its nail can then be used as the elevation datum for the unit. No matter what is used to fix the corners of the unit, care must be taken not to dislodge or disturb the corner markers. These corner markers, and the line that will be strung between them, are important for maintaining straight side walls during an excavation. In turn, straight side walls are critical to some mapping procedures performed within an excavation unit. A small balk, or unexcavated area within the unit, is created adjacent to each corner to insure that it will be stable throughout the duration of the excavation. Stakes require a larger balk than when corners are marked with surveying pins or spikes. In fact, the balk needed to protect corners with surveying pins or spikes is fairly negligible. Wrapping corner stakes, pins, or spikes with brightly colored vinyl flagging will make them more visible, aiding in their protection from accidental disturbance.

A datum for recording elevations within a unit can consist of a firmly placed wooden hub and nail marking one of the corners of the unit. It also can be a separate hub and nail in the vicinity of the unit that might serve as elevation control for a cluster of excavations. When choosing the location of a shared datum stake consider the possibility that initial excavations may need to be expanded to fully define any features or interesting phenomena discovered. It is essential that the datum stake be firmly embedded and protected from disturbance during the course of fieldwork. When the hub is set, leave enough of it above ground so that it is higher in elevation than surrounding surfaces or anything to be measured. In this way all elevation readings will be below datum (BD). The arbitrary "0" point of the datum is the top of the hub. Place a nail in the top of the hub to serve as an anchor for the string that will be used in conjunction with a line level to record unit-specific elevations. The top of the hub should be related to the primary elevation datum for the site or a fixed benchmark using a surveying instrument. In this way the arbitrary "0" elevation of the unit datum is translated into

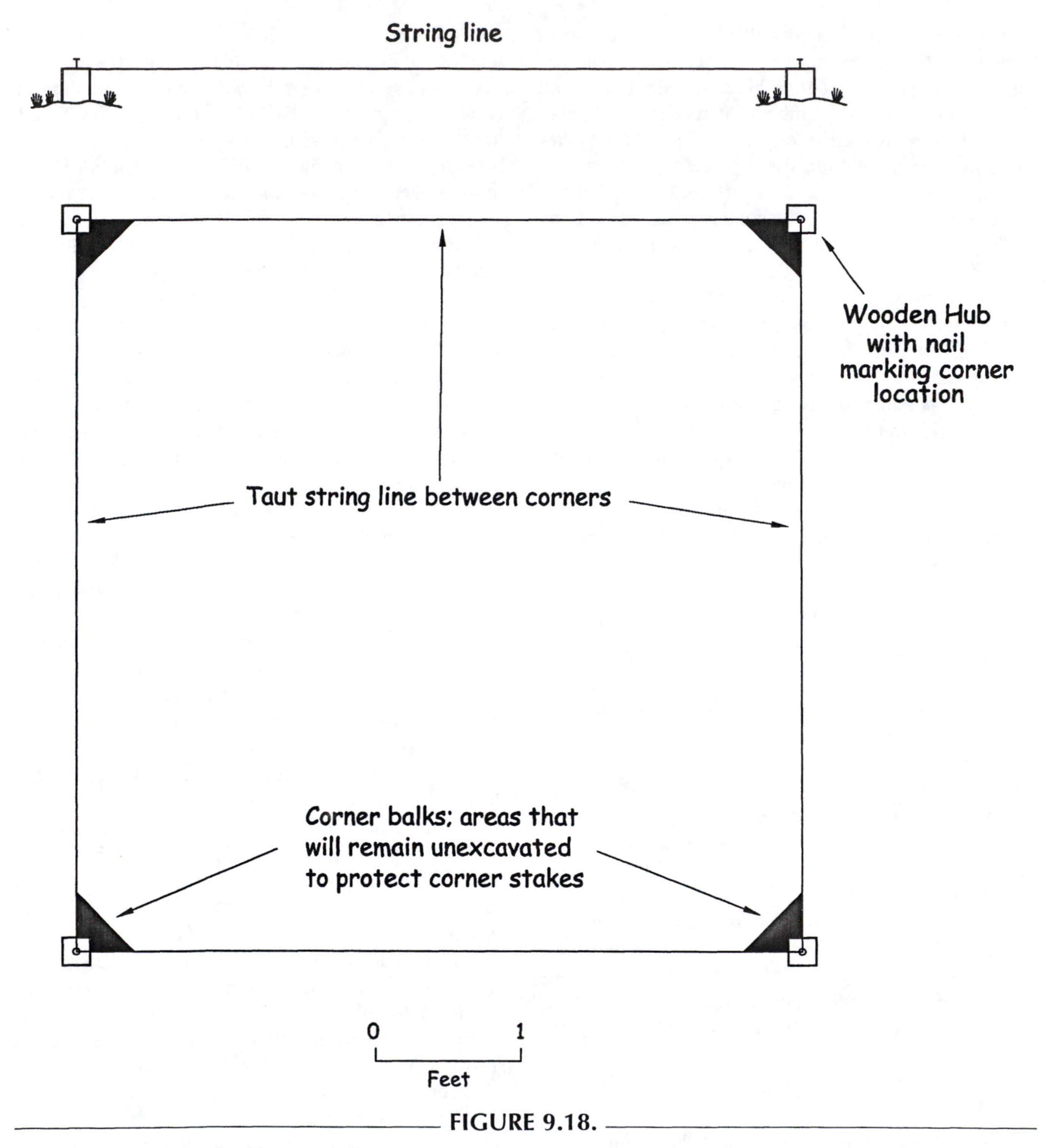

FIGURE 9.18.

Plan view of a staked unit with corner balks.

an actual value relative to all other elevations recorded on-site. Doing this before the unit is opened saves the step of having to convert excavation-specific readings at a later date.

Determine the opening or ground elevations for the perimeter of the unit using a line level suspended on a string pulled off the datum hub (Figure 9.19). Make sure that the string is anchored at the base of the nail on the datum. The string should be long enough to reach the farthest margins of the unit. You will also need a folding rule or a rigid tape. Position yourself adjacent to the point to be measured. In one hand, hold the extended folding rule. Rest the "0" end of the rule on the point to be measured, holding it in a vertical position. With your other hand, pull the string with the suspended line level taut. The line level should be about midway between the point to be measured and the datum stake. Position the taut string against the face of the vertical rule. Move the string up or down until the bubble on the line level is centered. When this is achieved, see where the taut string intersects the face of the rule. This will be your elevation reading, indicating how far beneath the datum your point on the ground is. Take elevation readings for all corners of the excavation and its center point.

Secure a length of string to the nail on the stake, pin, or spike at one of the excavation's corners, then take the string and move to all of the remaining corners, pulling the string taut and wrapping it around each of the other corners. Tie it off back where you started, making sure that the whole network remains taut. You have now defined the shape of the unit with this strung line (see Figure 9.18).

With a split tube sampler or an auger, examine the stratigraphy adjacent to the unit. If you are using a split tube sampler you can take the core from within the boundary of the unit since this device creates such a tiny disturbance. Plan how the excavation will proceed, which strata can be removed as single levels, and which will require the use of arbitrary levels. For illustrative purposes let's assume that our hypothetical unit contains no layers of placed fill or highly disturbed contexts, so all excavated matrix will be screened.

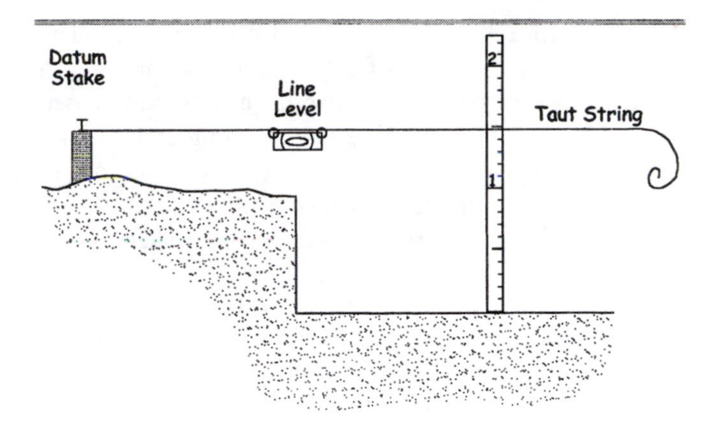

FIGURE 9.19.

Determining an elevation off a datum stake using a line level and rigid ruler or tape. The string is moved upwards or downwards until the bubble on the line level is centered. The ruler is read where the level line intersects it. In this example, the surface beneath the ruler is approximately 1.49 feet BD or below datum.

Basic Excavating Rules

There are basic rules to follow in proceeding with any excavation—

- ☑ **Don't just mechanically dig a level, whether it is natural or arbitrary.** "The boss says that this stratum is supposed to be 0.5′ thick. I'm digging until I'm 0.5′ below our starting elevation."

- ☑ **Don't hesitate to terminate a level if you encounter significant changes in what's being dug.** Remember that you are opening an excavation in a new area to discover and learn. Are there cultural deposits? What is the natural and cultural stratigraphy and how do they relate to one another?

- ☑ **Always look for changes in the color, texture, or structure of the matrix that you are excavating, regardless of what assumptions have been made based on the exploratory core or auger boring.** Sediment color and texture will be your most visible clues for defining stratigraphic boundaries and spotting anomalies that might be cultural features. There are two places where changes will be seen, the floor of the excavation, and the sidewalls. It is vitally important to keep the side walls of an excavation straight and clean, both to track stratigraphic changes and to retain the horizontal dimensions of a unit from its surface to its base. To make sure that you have a clear view of the excavation floor, avoid creating or stockpiling a lot of loose matrix. In other words, don't let the digging get too far ahead of the screening. Remove matrix from the floor as it is created.

- ☑ **Work from surfaces or stratigraphic changes that you have exposed and understand into those that you don't understand or have questions about.** If you have defined the base of an excavation level adjacent to a wall or in a corner of the unit, work away from this exposure, don't just turn your attention to another part of the unit. The team member doing the screening will also have an important perspective on how the excavation is progressing. They are staring at the matrix as much as the excavator and will notice changes in the frequencies of artifacts as well as variations in the matrix itself.

- ☑ **Never mix materials from different levels or contexts in the screen.** This rule should be obvious.

- ☑ **Don't forget that cultural stratigraphy doesn't have to correspond exactly with the natural stratigraphy but may be embedded within it.** A buried occupational surface or former depositional plane need not be associated with a visible change in sediment color or texture, but may simply be represented by the tops of features or the relative clustering of artifacts and other materials at a particular elevation. As you dig through a level ask yourself if the artifacts that you are finding tend to cluster in a particular segment of the level (top, middle, bottom) or whether they seem to be distributed throughout.

- ☑ **Pay special attention to the depth at which large and heavy artifacts occur within a level since these may still lay near their original depositional plane.** Take special note of any chronologically diagnostic artifacts observed *in situ.* It is always a good idea to take spot elevations and map the location of significant artifacts because of their potential importance in defining cultural stratigraphy, even when the intention is not to excavate entire levels using exact or point provenience techniques.

- ☑ **Begin any excavation with the cutting of the side walls using the taut string defining the unit's perimeter as a guide.** The shape of a flat shovel has an advantage here. You're going to use it to drop a vertical line from the string onto the surface of the ground. Without putting pressure on the string, orient the flat back of the shovel against it with the blade of the shovel touching the ground. The alignment of the flat portion of the shovel should be vertical. Holding the shovel in place, strike the butt of the shovel with the heel of your boot so that the shovel blade penetrates the ground. You should end up with a relatively straight cut in the ground that parallels the strung line and is situated beneath it. Strike the shovel repeatedly to get a deeper cut but don't go beyond the presumed depth of your first stratum. Repeat this procedure around the edges of the unit. When a corner is approached, cut a slanting line linking adjacent sides of the unit to create a corner balk (see Figure 9.18). The blade of the flat shovel is a handy measure for creating corner balks. Orient the shovel blade so that each end is touching the line of one of the side walls. The position of the shovel blade and the adjoining side walls should form a rough isosceles triangle with the shovel blade as its base.

From the inside of the unit you can now make shallow diagonal cuts with the shovel towards the side walls and begin the removal of the level. Work along each wall to bring a small area down to the base of the first natural stratum before moving into central portions of the unit. Stop your vertical progress when you notice a change in the color or texture of the matrix that you are excavating on the floor of the excavation or in the wall. This is one of the reasons for beginning your work along a wall; it gives you a view of stratigraphic changes. Don't expect excavation level floors to be flat when digging by natural strata, and don't try to make them so. Follow the trends defined by the color and texture of the matrix. You also should halt your vertical progress if you encounter anything that might be construed as a feature or evidence of an occupational surface within the stratum. With the side walls defined and the base of the level exposed in a strip along each wall, you have visual control points that will make digging the rest of the level easier. Since the exact provenience or position of every artifact within the level is not being plotted, a good deal of the level can be removed with a shovel. Repeatedly skimming (not chunking) the areas to be excavated with a flat shovel is relatively quick, lets you see many artifacts *in situ,* and gives a good impression of the level from top to the bottom (Figure 9.20). Shovel skimming across a level's floor also allows you to see anomalies/features that might appear within a level. Chunking the level out with a shovel may be quicker but you will lose this perspective. However, taking diagonal or vertical cuts with the shovel may be necessary in dense or gravelly deposits. Use your trowel to create clean cuts for examination on the floor or side walls, or to more fully expose objects that you have decided to inspect *in situ.* An excavation level also can be removed by skimming with a trowel, but is a much slower process.

FIGURE 9.20.

Shovel skimming an excavation floor. The contours of a flat shovel are especially suited to this. Shovel skimming is relatively quick, lets you see many artifacts *in situ,* and gives a good impression of the level from top to the bottom. It also allows you to see anomalies/features that might appear within a level before making any dramatic cuts into them.

Arbitrary Levels

If your first level is arbitrary, the base of the level will be determined strictly by depth measurements. The depth of the level is adjusted to the surface elevations of the unit. How this adjustment is made will determine whether the level's floor is flat or contoured. Let's say that your opening surface elevations for the four corners of the unit are as follows: northeast = 0.3′ BD, northwest = 0.35′ BD, southwest 0.4′ BD, southeast 0.4′ BD. Excavating an arbitrary depth below each of these points will result in an excavation floor that mimics the topography of the existing surface and will have a slight slope to it. If a flat level floor is desired, then excavations must proceed off a single standard elevation. This could be an average of the surface elevations, its highest point, or its lowest point. Choices will vary depending on how complex or sloping the topography of the existing surface is. In general, the use of arbitrary levels with flat floors makes fewer assumptions about the nature of previous surfaces or depositional planes than do levels whose floors are contoured to match the existing surface. With any type of arbitrary level there is always the danger of crosscutting what may have once been depositional planes or surfaces that are no longer distinctive from a sedimentary or pedologic perspective. Remember that arbitrary levels are for control in learning more about the natural strata which are the primary unit of analysis in the initial exploration of a deposit. Therefore, how an arbitrary level is used will vary depending on the current understanding of the deposit. If you knew ahead of time which strata were depositional and which were the result of soil-forming processes, then the use of arbitrary levels, and excavation strategies in general, could be modified accordingly.

To open a unit with an arbitrary level requires excavating control points to the base of the level at the corners of the unit and the center point of each wall. Let's assume that your intention is to excavate arbitrary levels that are 0.3′ thick. The surface elevations of your unit are those noted above. In order to achieve a flat level floor you've decided to use the average of the surface elevations as a reference point, 0.36′ BD. This means that when the excavation of your first arbitrary level is complete, the level's floor should measure 0.66′ BD all across the unit. Not all portions of the excavated level will be 0.3′ thick, but this is the compromise that must be made when the existing surface is uneven and a flat level floor is desired.

Using a trowel or working very carefully with a shovel, begin removing matrix from a small area (a square foot or less) in a corner of the unit. Repeatedly check the below datum reading for the surface that you are creating with a line level and ruler until you have reached the desired 0.66′ BD. Do the same for the other corners of the unit. If you are comfortable with the visual reference that these control points provide for the base of the level, begin shovel skimming the areas between them. If you feel that more control points are necessary, establish additional ones at the approximate center points of the side walls, then shovel skim the remainder of the level. This same procedure can be used anytime you want to dig an arbitrary level beneath any sloping or uneven excavation floor.

The procedure is a bit different if instead you are contouring the base of your arbitrary level to match the existing surface. The base of the excavated level will be an arbitrary measure below each of the unique readings for the opening of the level, in this case the surface. You still need to bring control areas in the unit's corners down to level, but rather than constantly refer to below datum readings taken with the line level, you can simply measure down from the existing surface in each corner with a folding ruler or tape.

All artifacts derived from the same level are assigned the provenience for that level and may be bagged together. However, some artifacts may be fragile or subject to damage if mixed with other materials, e.g., pottery or bone, and may be placed in individual bags. Avoid placing stone tools that will be examined microscopically for edge damage and use wear in with other lithic artifacts, since this may inadvertently alter their margins.

Spatial Coordinates

Record spot elevations and locations (exact or point provenience) for artifacts that you think are important within your excavation level. Do the same for any samples that might be collected, such as charcoal for radiocarbon dating. This can be accomplished in a number of ways. The spatial coordinates of an artifact may simply be listed along with an elevation, or an actual map of the excavation floor showing the artifact's position can be drawn. Coordinates are measured to the center of the artifact relative to the southwest corner of the unit (Figure 9.21). The *X* coordinate is the distance west to east. The south to north measurement is the *Y* coordinate. Even though the southwest corner is the theoretical datum for these coordinates, the actual measurements are taken off the south and west walls of the unit where a line perpendicular to each wall intersects the artifact's position. The elevation of the find (*Z* coordinate) below datum is also recorded. Assign a number to every artifact that has its coordinates recorded. Any artifact whose exact position or elevation has been recorded and assigned a map number should be tagged and bagged separately from the other artifacts in the excavation level. The method used to record coordinates assumes that the walls of the unit are straight and that corners have right angles. It also requires "eyeballing" perpendicular lines from the unit's side walls to the artifact. If two folding rules or metal tapes are used to take the measurements, a visual impression of perpendicular lines and right angles is easily obtained. This entire procedure can be completed by one person.

Triangulation

An alternative method for determining an artifact's position is triangulation using measurements from two corners of the unit (see Figure 9.21). One person holds the "0" end of a tape on the corner nail/pin/spike and the tape is pulled by a second person to the artifact's position. This person must insure that the tape is oriented in a horizontal position. A plumb bob suspended (by hand) above the center of the artifact is used to extrapolate the artifact's position to its intersection with the horizontally oriented tape. This procedure is repeated from a second corner of the unit. Be sure to note the corner from which each measurement is taken as well as the

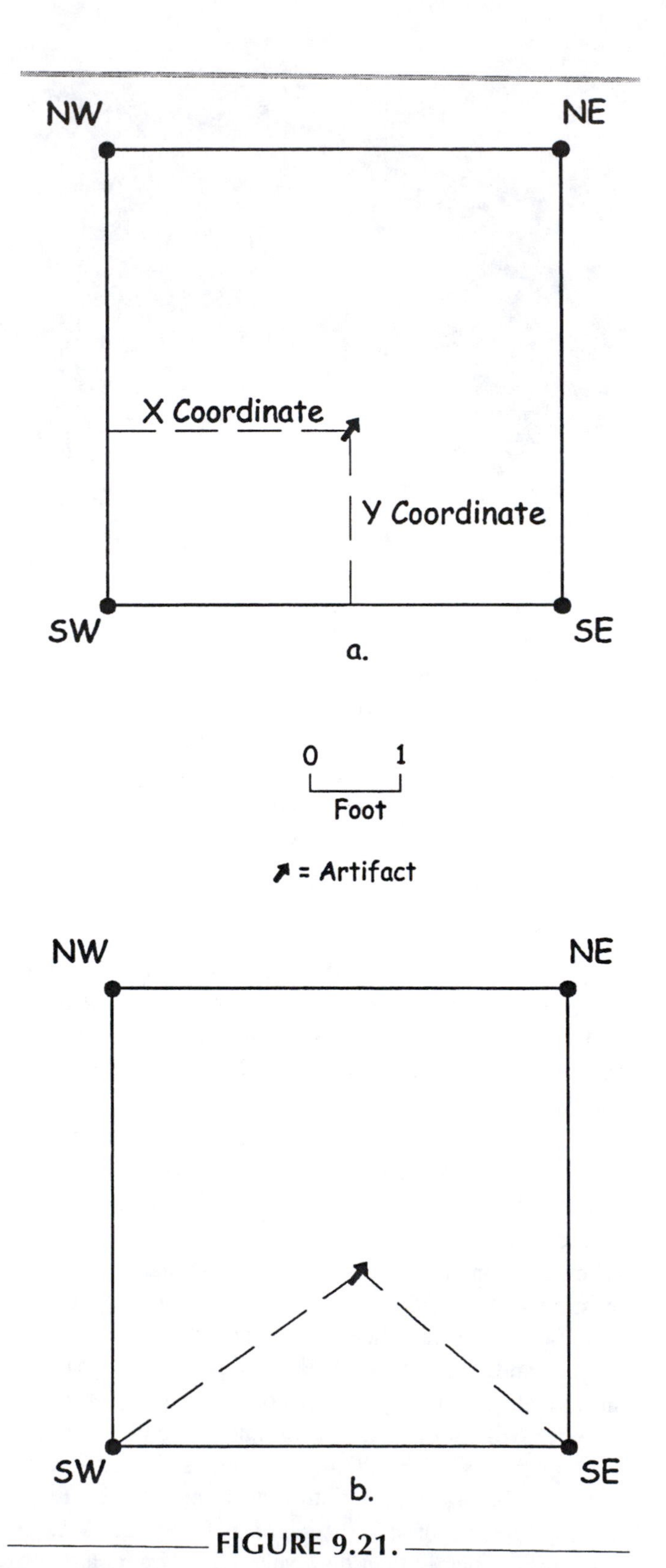

FIGURE 9.21.

Determining the spatial coordinates of an artifact on an excavation floor. a) Distances to the center of the artifact are measured from south to north (*Y* coordinate) and from west to east (*X* coordinate). b) A measurement from two corners that allows the artifacts position to be triangulated.

measurement itself. Two people are needed to perform this operation. A length of string with an attached line level may be substituted for the tape in the initial part of this operation. The string is anchored on the corner nail/pin/spike and extended to the artifact's position. The line level on the string insures that the string is held in a perfectly horizontal position, increasing the accuracy of the operation. Determine the intersection of the plumb bob line above the artifact's position with the horizontal string. The length of the string used to reach the artifact's position can then be measured with a tape.

The deeper an excavation the more awkward determining an artifact's location can become. Straight and plumb excavation side walls must be maintained from the opening of the unit to its conclusion or the measurements taken off the walls in the coordinate method will be skewed. The triangulation method becomes awkward when an excavation approaches chest height or more in depth. Drafting floor maps of artifact distributions is easier with coordinate measurements rather than triangulation ones. An engineer's scale and graph paper make the plotting of coordinates easy. A drawing compass must be used to translate triangulation measurements to paper. First the compass must be adjusted to the measurement from one corner, then positioned on a mapped corner and a small arc drawn in the interior of the unit map. The compass is then adjusted to the second measurement, positioned on the second corner, and an additional arc drawn. The artifact's position is where the two arcs intersect.

Scaled Map

Rather than just recording coordinates or triangulation measurements, a scaled map showing the location of select artifacts may be drawn in the field. You will still determine an artifact's position using the methods described above. To construct this type of map, first establish the dimensions of the unit on a piece of graph paper. Label the map with its provenience and subject matter, e.g., 18WA23 (site), Unit 4, Stratum 1, Level 1, Mapped Artifacts. Include a north arrow and a scale for the drawing. Provide a key if you use symbols to designate different types of artifacts on the map, e.g., pottery, flake, biface. Using either an engineer's scale or a drawing compass, translate the spatial measurements onto the map. Larger artifacts may be drawn to scale. This will require some additional measurements and observations. Once the center point of the artifact has been fixed on the map, measure the length and width of the piece. Observe the orientation of the artifact's long axis. With an engineer's scale, work off the designated center point of the artifact on your map to recreate its shape with the new measurements. For extremely large artifacts, measuring the spatial coordinates to various points along its margin will be more expedient. Next to each artifact's location on the map place its appropriate map number. This number links the position shown on the map to specific spatial coordinates and the artifact itself, since it will have been tagged and bagged separately.

Closing Out a Level

Before taking the closing elevation for any level, trowel the floor clean to inspect for any anomalies that might indicate the tops of features. Using a sharp trowel cut the excavation floor in the same direction, beginning on one end of the unit and proceeding to the other. Don't walk over or otherwise muddle sections that have already been cleaned. Before cleaning the floor for this final inspection make sure that the side walls of the unit are straight. This process will generate matrix that will have to be screened and may result in additional artifact finds. Of course it will also dirty the floor of the excavation. When an excavation is still relatively shallow, the taut string between corners can be used as a visual guide for cutting vertical walls (Figure 9.22). Position yourself over the string so that one foot rests in the unit while you kneel on your other leg on the existing surface. Looking straight down the string you can tell whether or not the excavation's side wall is plumb. If it bulges beyond the string, use your trowel to gradually shave the wall until it looks plumb. The back of a flat shovel can also be used while sighting down the taut string to straighten the walls (Figure 9.23). If you want to be absolutely fanatical about having straight walls hold a torpedo level against them to see if they are plumb. A torpedo level is long and rectangular, and contains a series of bubble levels. Some are mounted parallel to the length of the level while others are perpendicular to it. This enables the torpedo level to be used in both horizontal and vertical positions, i.e., it can tell if something is perfectly flat or straight up and down. Torpedo levels may be a necessity in maintaining straight walls in excavations that are relatively deep.

Closing elevations for an excavation level are taken in the same way as opening elevations. These and other data are recorded on standardized record sheets. A practice that I find useful is to place nails marked with a piece of vinyl flagging into the walls of an excavation at the base of each excavation level. This provides a visual record of the levels that have been removed as the excavation gets deeper and is a good comparison with the natural stratigraphy as it is exposed in greater detail. The nails can also be used as control points for measuring the depth of subsequent arbitrary levels. Place flagged nails at every corner and the center point of every wall.

FIGURE 9.22

Straightening the side wall of an excavation using the strung line that defines the unit's edge as a visual guide. Looking down the taut string the excavator can see whether or not the wall is plumb, or straight up and down. A trowel is used to shave portions of the wall until it is straight.

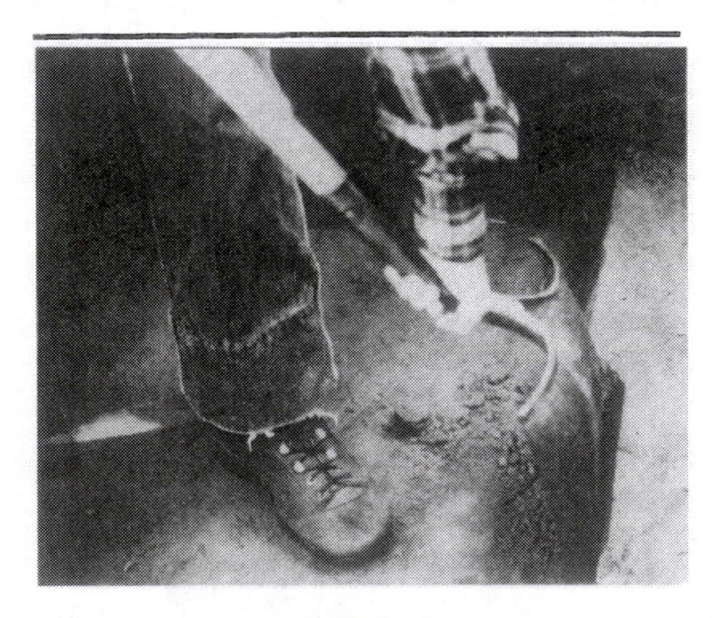

FIGURE 9.23.

The orientation of a flat shovel when it is used to make vertical cuts, such as those necessary to straighten the side wall of an excavation.

Cataloging

The final step before starting a new level is to assign a field specimen catalog number for the artifacts and record sheets related to the context or level that has been completed. This cataloging system works in the same way as the one described in chapter 8. Every unique context, i.e., excavation level or feature, is assigned a unique catalog number. For example, all of the artifacts derived from Unit 1, Stratum 1, Level 1 might be assigned catalog number 3, if that is the next available number in the site's catalog. Subnumbers are used when artifacts within the level were given map numbers, e.g., Catalog #3-1, 3-2. The range of subnumbers employed should be listed in the field specimen catalog under the summary column (see Figure 8.29). Additional subnumbers will be assigned in the laboratory as part of the more detailed cataloging of all of the artifacts derived from the same contexts. It is convenient to check the field specimen catalog to see what the last assigned subnumber was, rather than having to sort through all of the artifact bags and labels in the laboratory. Once a catalog number has been assigned in the field, it should be placed on all relevant artifact tags, bags, and record sheets and maps.

DIGGING DEEPER

The second and subsequent levels of an excavation may be a combination of natural strata and arbitrary levels, or strictly arbitrary levels. The procedures involved in their excavation parallel those I've described. When digging a natural stratum you follow the color, textural, or structural attributes that define the stratum, and halt your vertical progress whenever and wherever the matrix changes. When a continuous series (two or more) of arbitrary levels will be dug there are easier ways to control for the base of the level than constantly taking below datum readings with a line level and tape. One is to measure down from flagged nails placed into the sidewalls around the perimeter of the unit at the base of the previous level. You can quickly establish your control points for the base of the level around the edges of the excavation floor and shovel skim the remainder of the level. Another method involves creating a shelf of unexcavated matrix around the unit's perimeter that is the desired thickness of the arbitrary level (Figure 9.24). Rather than start the excavation of a new level against the side walls of the unit, excavations are offset slightly from the walls. Small linear areas paralleling the unit walls are troweled down to the desired base of the level with measurements from the top of the "shelf" indicating when the appropriate thickness of matrix has been excavated. There is no need to constantly check the depth of the ongoing excavation with below datum readings. This works because the top of the shelf is the base of the previous excavation level. Once fully defined, the shelf also provides a visual reference for shoveling or troweling out the remainder of the level floor. The shelf itself is excavated once the remainder of the unit is down to level. As with any level, always make sure that straight side walls are maintained.

Subdividing a Level

Excavating a level by quadrants or other spatial subdivisions provides more control over artifact provenience than a general level excavation but is more time consuming. This approach is rarely used until an archaeological deposit has been discovered and tested to some degree. However, because cultural features may be encountered at any stage of fieldwork and the quadrant technique can be useful for their exploration, I'll discuss it here. Subdividing a level means that you are creating new contexts that must be excavated, screened, and documented individually. The first task is to physically divide the level floor to be excavated into quadrants. This is easily done by identifying the center point of the unit and each of the side walls and playing connect-a-dot. Pull a tape between two adjacent corners of the unit and have someone hold it there. Find the midpoint of the distance between the corners on the tape. Hold the string of a suspended plumb bob adjacent to the midpoint with the point of the plumb bob not quite touching the excavation floor. When the string and plumb bob have steadied, allow the point of the plumb bob to stick into the ground. This is the center point of your side wall. Remove the plumb bob and place a nail, surveying pin, or small spike in its place. Do this for the remaining walls of the unit. The unit's center point can be found by pulling a tape diagonally across the unit between two opposing corners. Find the midpoint of this diagonal measurement and use a plumb bob to superimpose this point onto the excavation floor in the same way you did for the side walls. There are now five points on the floor of the excavation. Anchor one end of a string to the nail at the center point of the north wall. Pulling it taut, wrap it first around the nail at the unit's center then around the nail marking the center of the south wall. The unit is now divided into halves. Do the same with another string, moving from west to east across the unit. The unit is now divided into quadrants. The string lines define two of the four boundaries of each quadrant.

The same general principles apply when digging a quadrant as when excavating any other type of level. Establish the base of the level in small areas around the perimeter of the quadrant to guide your major excavation effort. Use the string lines on the excavation floor as guides for cutting the additional side walls of the quadrant. In terms of measurements and data collection, each quadrant is treated as if it were a level within a unit. The matrix excavated from a quadrant is screened, tagged, bagged, and cataloged separately from the other quadrants in the level. In small units, excavating by quadrants is often used as a substitute for exact or point provenience excavations. Removing a level by quadrants can also be useful when artifact densities are too high to make point provenience excavations practical.

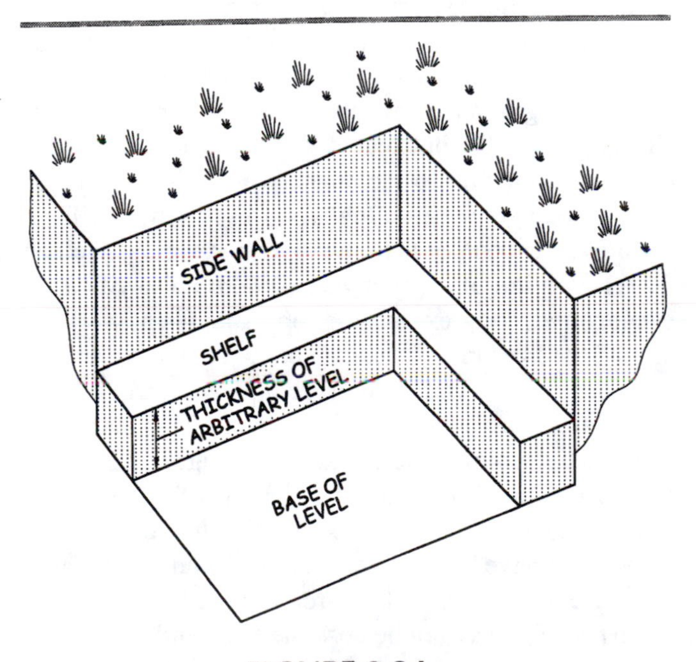

FIGURE 9.24.

One method of providing a visual reference for the depth of an arbitrary excavation level. Rather than start the excavation of a new level against the side walls of the unit, excavations are offset slightly from the walls. An unexcavated shelf is left along an excavation's side wall for control during the removal of an arbitrary level. The top of the shelf is the base of the previous excavation level. Measurements from the top of the shelf aid excavators in determining when the base of the arbitrary level has been reached rather than having to constantly take below datum readings.

POINT PROVENIENCE EXCAVATION

Excavating using point provenience for all artifacts in a level represents the extreme in contextual control, which is one reason why it is employed with many types of cultural features. It is rarely used during discovery or initial testing phases of fieldwork, or before the integrity and potential value of an archaeological deposit is understood. Both the digging and record keeping are very time consuming. First decide whether artifacts will be mapped and pulled as they are found, or pedestaled and left in place until the entire level has been completed before being mapped and pulled. Leaving exposed artifacts *in situ* until the level's floor is complete is the most painstaking and time consuming of the two procedures. Its only advantage is that it provides a visual impression of artifact distributions and clustering (which can be photographed). Both the map-and-pull and *in situ* strategies require an excavator to work cleanly, only opening small areas of a level at a time. Mapping artifact locations may follow any of the methods previously described. A total station may be used instead for the mapping procedure if available, and will speed up the recording process. In this case the total station would have to be set up over a datum or fixed point whose spatial relationship with the unit under excavation has been determined and mapped. The angle turned off a point of reference, distance from the instrument position to the artifact, and artifact elevation would be recorded for every artifact. Remember that any artifact afforded point provenience must be assigned a unique artifact/map number, and tagged and bagged individually.

Map-and-Pull

For the map-and-pull procedure, begin against one wall and with a trowel attempt to cut/skim a small area (nothing bigger than a square foot) down to the base of the level. As an artifact is encountered you may stop and map it, or wait until a number of artifacts are exposed to record provenience and remove the artifacts. When mapping is done by hand, some excavators may decide to only record a sample of artifact elevations, relying on their impressions for determining whether materials tended to cluster in a specific vertical segment of the level or were evenly distributed. Constantly scoop up and remove loose matrix for screening. Keep a bucket and scoop or dustpan handy for this purpose. Do not let excavated matrix accumulate on the level's floor—it will only get in your way. The exact provenience of artifacts seen *in situ* will of course be recorded, but those found in the screen or seen in loose matrix will be assigned a general level provenience. Work your way along the unit's side wall. Since this type of excavation is relatively slow it is often possible for two people to work side by side along the same wall during the initial stages of the excavation. Once a linear area along one wall has been cleared, back up slightly and clear another contiguous area paralleling the first. Repeat this process until you have worked your way across the floor of the unit. The entire floor is then lightly troweled as a final check for anomalies, and closing depths are taken as you would for any excavation level.

In Situ

The key to successfully exposing and pedestaling artifacts *in situ* is to work the smallest area possible until it is as complete as possible before expanding into new areas. This work is done with a trowel and in many cases will require the use of thinner or smaller implements. Begin by excavating a very narrow (0.5′ or less) trench down to the base of the level all along one side wall. Thinly slice the vertical face created by this initial excavation with a trowel all along its margin to reveal the location of artifacts. Rather than removing a level by skimming it in a horizontal plane, you are vertically slicing/peeling the unexcavated face of the level from one side of the unit to the other. Watching the vertical face as you peel it will allow you to see some artifacts before you encounter them with your trowel, as well as detect stratigraphic changes. When an artifact is encountered, fully expose the top of it. Then, working around the margins of the artifact, make gentle vertical cuts into the excavation floor, removing matrix from a small area surrounding the artifact. This forms a pedestal upon which the artifact rests (Figure 9.25). Try to extend your vertical cuts and clearing of matrix to the base of the level. Artifacts exposed at the very base of a level will have little or no pedestal. As additional artifacts are encountered the procedure is repeated. In some places it may not be possible to extend your excavation to the base of the level because of the density of exposed artifacts. In these places you must be even more fanatical about instantly clearing away any loose matrix from the work area. Every time you trowel through an area you run the risk of dislodging artifacts that have already been exposed. When you have exposed as much of the excavation floor as possible begin mapping and photography.

FIGURE 9.25.

Artifacts exposed and pedestaled in the Paleo-Indian level of the Thunderbird Site, Front Royal, Virginia. Leaving artifacts *in situ* during the excavation of a level provides a visual impression of clustering and spatial associations. Both hand-drawn maps and photography are used to document the completed level floor.

FIGURE 9.26.

A mapping frame consisting of a sturdy wood and metal frame with perforations spaced at regular measured intervals. String is passed back and forth through the perforations to create a grid of squares. The mapping frame is oriented over the corners of a unit, superimposing a grid on the excavation floor. This device makes it easier to map complex distributions of artifacts on the level floor.

Mapping Frame

When large numbers of artifacts are involved, recording artifact locations is made more efficient with the use of a mapping frame, or by superimposing a grid over the unit floor (Figures 9.26 and 9.27). A mapping frame breaks down a large, complex mapping procedure into a series of smaller and less complicated operations. The mapping grid is easily translated to graph paper. Each grid square becomes a frame of reference for mapping the position of artifacts falling within it, rather than taking measurements off the unit's corners or side walls (Figure 9.28; compare with Figure 9.21a). Mapping frames are preconstructed to particular horizontal dimensions, e.g., 1 meter on a side or 5 feet on a side. Some are fabricated so that the size of the frame is adjustable. Once the grid lines of the frame have been strung, it is oriented over the corners of a unit. If the excavation is shallow enough there is no need to superimpose grid points onto the level's floor. If the mapping frame is at some distance from the base of the excavation a plumb bob can be used to translate the points of the grid onto the level's floor. Do this by holding the string of a suspended plumb bob against the intersection of each of the frame's grid lines and letting the plumb bob drop to the excavation floor. The points on the floor may then be marked with nails and a bit of vinyl flagging to heighten their visibility. Preconstructed mapping frames are not very portable and are not a regular part of the gear carried into the field during an archaeological survey. They should, however, be included in the equipment amassed for more intensive types of excavations.

Mapping frames are easily improvised in the field with nails/spikes and string (Figure 9.29). Follow the procedure already described for locating the center point of a unit and its center walls and superimposing these on the floor of the excavation. String the lines between these points dividing the unit into quadrants. Using the quadrant string lines and unit walls as guides, offset measurements from each of these initial points to establish the new points to be used for the final grid. Set nails or spikes at each of these new points. Remove the quadrant lines and the five points/nails used to construct them. Finally, string lines between the new points to create the grid of squares for mapping. On both improvised and preconstructed mapping frames the size of the grid squares can be altered depending on which grid lines are actually strung.

FIGURE 9.27.

An improvised mapping frame. Nails have been placed at one foot intervals around the perimeter of this unit and connected with string to create a grid of squares over the rock hearth exposed on the excavation floor.

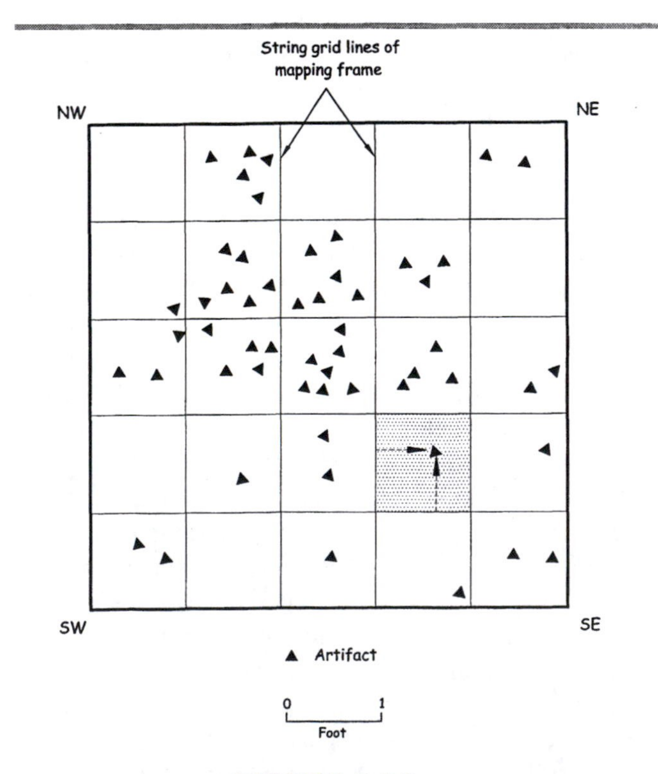

FIGURE 9.28.

A mapping frame breaks down a large, complex mapping operation into a series of smaller and less complicated operations. The mapping grid is easily translated to graph paper. Each grid square then becomes a frame of reference for mapping the position of artifacts falling within it, rather than taking measurements off the unit's corners or side walls.

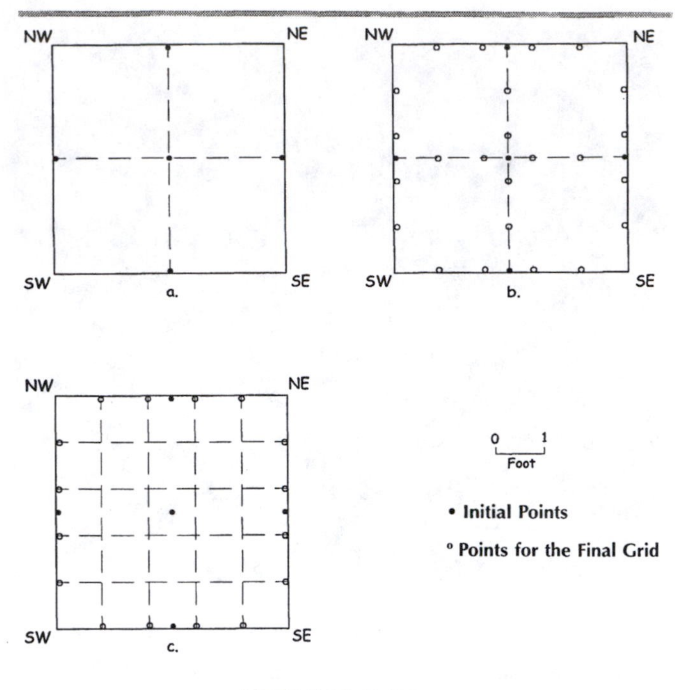

FIGURE 9.29.

The steps for improvising a mapping grid for excavation floors: a) Establish the center point of the unit and each of the side walls and string lines between them. b) Using each of the initial five points as a base, offset measurements to establish grid points at desired intervals. In this case, measurements offset from each starting point are 0.5′ and 1.0′, resulting in a grid of squares one foot on a side. c) Remove the quadrant string lines and the initial five points used to construct them. String the new grid lines.

After the mapped artifacts have been pulled, tagged, and bagged, remove the mapping grid. It will be necessary to re-excavate the floor of the unit, breaking down artifact pedestals and bringing all portions of the floor down to the appropriate level. Additional artifacts may be discovered in the process and should be carefully exposed and added to the existing map. If the initial map is too crowded for additions, a new map will have to be drawn.

FEATURE EXCAVATION

Cultural features can be encountered in any type of subsurface test, but working with them is easiest in the context of units or larger types of excavations. Many types of features are first recognizable as some type of color or textural anomaly in the matrix being excavated. These features would include all manner of back-filled pits, below-ground excavations or disturbances, including burials and some architectural features (e.g., pit houses, cellar holes, wells, wall trenches, postmolds). It also includes materials that have decomposed in place and are only recognizable as a sedimentary or soil anomaly (e.g., wooden artifacts or architectural elements, organic refuse, skeletal material), as well as anthropogenic soils, placed fill, or sediments. Clusters or concentrations of artifacts representing specific activities may also be considered as features, for example, a cluster of flakes representing the manufacturing of a stone tool, or a cache of stone axes. Architectural features consisting of patterned alignments or associations of building materials (e.g., foundations, enclosures, posts, tent rings, mounds, earthworks) are another major category.

How you approach the study of a feature depends on the size of the excavation in which it is first encountered, the depth of the excavation when the feature is discovered, the known or anticipated size and type of the feature itself, and the purpose of fieldwork, i.e., site discovery, testing, or intensive study. Experience with the archaeology of a particular culture, time period, or region is invaluable in the decision-making process. Be assured that if a feature is discovered, your handling of it will be closely supervised. All endeavors have two things in common:

1. determining the horizontal extent and characteristics of the feature in the open excavation
2. determining the vertical extent and characteristics of the feature in the open excavation

Horizontal Definition

Defining the horizontal extent of a feature is a fairly straightforward process. Working no deeper than the elevation at which it was first noticed, follow the indications of the feature across the unit's floor until its boundaries are defined or the walls of the unit reached. Artifact clusters or concentrations attributed feature status will be defined through point provenience excavation techniques. Any artifacts or samples taken from the area of the feature during the definition process should be given a feature provenience and tagged and bagged separately from other artifacts found in nonfeature portions of the excavation level. Assign a feature a unique number that will become part of its provenience and that of any artifacts or samples associated with it, e.g., Unit 1, Stratum 2, Level 2, Feature 1. Feature numbers are assigned sequentially within a site or the deposit being dug, with no numbers being reused.

Map the exposed horizontal portion of the feature. When clusters of artifacts are involved, the techniques for recording point provenience described above may be applied, or a mapping frame or grid employed. Sediment/soil anomalies, the tops of pits or any backfilled excavation, or architectural features may be mapped using the same basic principals (Figure 9.30). In these cases, the primary goal is to define and draw the shape of an area. Consider the shape of the feature and visually break it down into a series of straight lines that are connected at various angles. Once the spatial coordinates of the end points of these lines has been determined and placed on a map, drawing in the feature's outline is relatively easy. Large artifacts or objects within the bounds of a feature will require locating additional points in order to accurately portray them, like the rocks in the feature shown in Figure 9.30b. Measurements taken off unit corners and side walls may be sufficient for outlining many features. A mapping frame or superimposed grid should be used for those features with more complex shapes or internal components of interest.

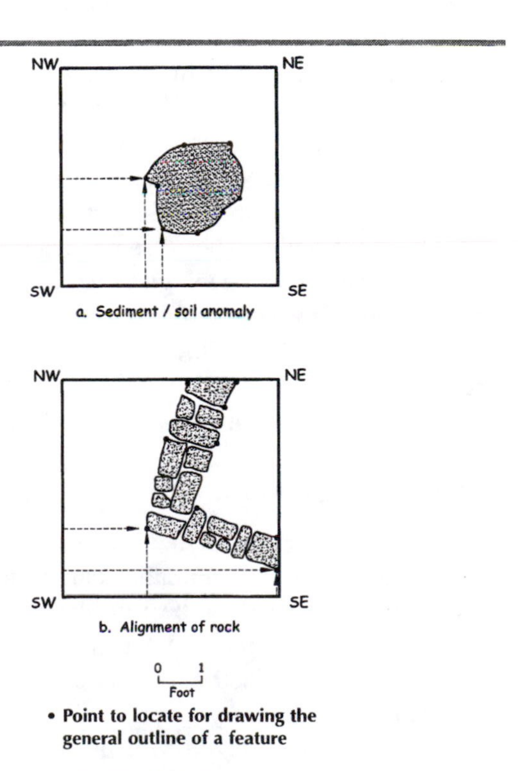

FIGURE 9.30.

Mapping the outlines or boundaries of a feature. Visualize the shape of a feature as a series of straight lines that are connected. Once the spatial coordinates of the endpoints of these lines have been determined and placed on a map, drawing in the feature's outline is relatively easy.

Before exploring the vertical component of a feature it is ideal to know its full horizontal dimensions, which may indicate what the feature is and, in turn, imply things about its vertical dimensions, what it might contain, or what may be associated with it. Knowing the dimensions will also impact excavation strategy, like deciding how to control provenience within a feature or position cross sections. For some types of artifact clusters (e.g., a chipping cluster) the initial process of defining the feature is by necessity an exploration of its vertical component. In contrast, architectural features like stone or masonry walls, or patterned alignments of rock, will rarely be dismantled until they have been seen in their entirety and associated deposits have been dug. The excavation of adjacent deposits in these cases reveals how deeply buried the construction may be. Once a feature has been recognized as a burial it may not be dug at all, depending on the outcome of interactions with all of the appropriate authorities that must be consulted in such cases.

If a feature extends beyond the unit in which you are working, exposing the rest of it means opening a new excavation or series of excavations appended to the existing one, and digging all of the levels to the elevation where the feature should first appear. In other cases, a feature may occupy all, or a large portion of the excavation floor, making further work impractical unless the excavation is expanded. Of course the deeper the level at which the feature occurs, the longer it will take to reach it in other units. Fully exposing some features may not be possible during an archaeological survey, but is the meat and potatoes of more intensive excavation programs. On the other hand, if only the very edge of a feature is encountered, its excavation may reveal little in the absence of knowing what lies beyond the open unit.

Vertical Definition

The vertical aspects of features that are first recognized as sediment/soil anomalies can be tested prior to excavation with a split tube sampler and information about the total depth of the feature and any internal stratification amassed. Whether the dissection of the feature will advance by arbitrary levels or detectable stratification, and how many levels will need to be excavated, can be estimated on the basis of these data. It is essential that the natural stratigraphy in the vicinity of the feature be known in order to evaluate any core taken from a feature. Rarely does the excavation of this type of feature proceed in toto. Rather, one or more ***cross sections,*** or test slices, are removed from the feature.

Cross Sections

A cross section allows you to investigate feature attributes and internal variation, recognizing that whatever is learned can be turned to good advantage in the removal of the remaining portions of the feature (Figure 9.31). Cross sections provide a view of the subfloor profile and dimensions of the feature, the sediments and artifacts that fill it, and what they may imply about how quickly or slowly the feature was created or filled. Does the feature represent a single, short-term event or a series of events? What do its contents suggest about the function or meaning of the feature? Information of this type will undoubtedly influence the strategy for excavating the remainder of the feature.

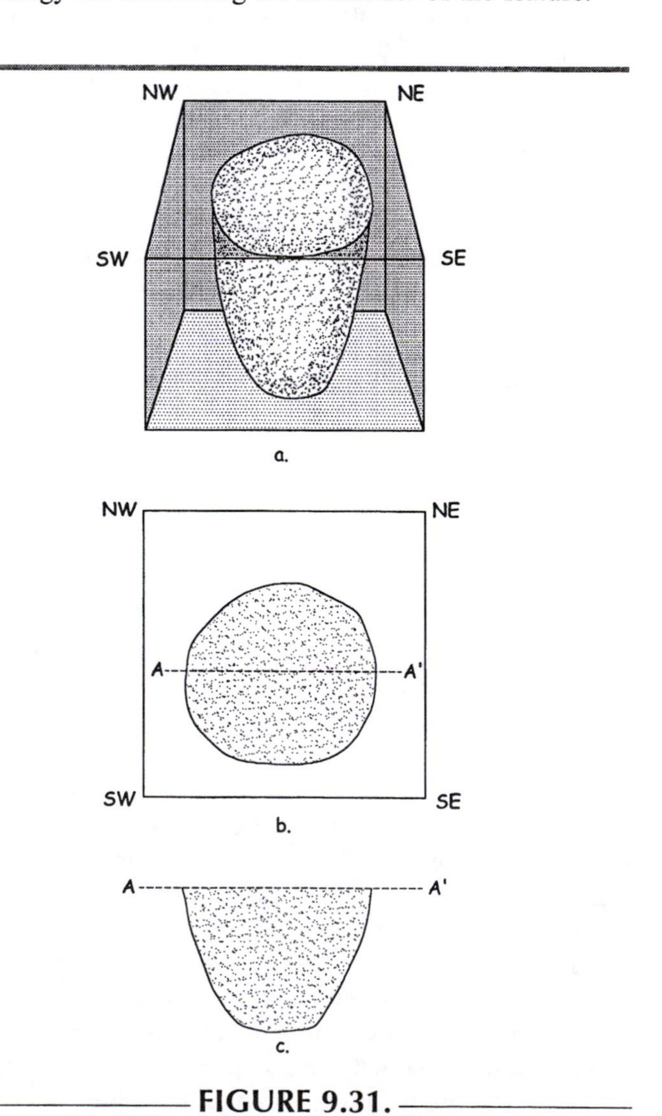

FIGURE 9.31.

The three-dimensional aspects of a pit feature (a) are investigated incrementally through plan views exposed on a unit's floor (b) and the excavation of cross sections revealing its subfloor profile (c).

To sample internal variation, cross sections are typically oriented along the length (***longitudinal sections***), the width (***transverse sections***), or diameter (***bisections***) of a feature as it appears in plan view (Figure 9.32). When a cross section is positioned to segment a feature into relatively equal halves the term bisection may be applied. Features with complex horizontal shapes make bisection in the strictest sense of the word problematic. The more complex a feature is in horizontal plan the greater the need for ***quarter sections*** or smaller subdivisions to investigate potential internal variation. The need is the same the larger a feature is in plan view. Never dig a cross section that is larger than the minimal spatial unit being used for control in general excavations. If anything, contextual controls should be greater in dealing with features given their importance relative to data garnered from general excavation levels. So while the bisection of a large feature may adequately reveal internal characteristics and stratification, control over the spatial context of things inside of the feature may be too generalized if the entire section is dug at once. Proceeding by quarter sections or smaller subdivisions enhances control over artifact and material provenience. The ways in which larger features were filled can also be more complex than smaller features and possess complicated stratification best revealed in a series of variably oriented cross sections. When the top of a feature is fully exposed, cross sections may be oriented in the most efficient and potentially informative ways possible. This is less likely when sectioning only partially exposed features. When a feature is not fully exposed this means that some portion of it extends into one or more of a unit's side walls. Because of this, a subfloor profile of the feature will be preserved in the side wall no matter how an exploratory cross section is oriented, so the cross section should be positioned to complement this guaranteed view (Figure 9.33). If the feature is large enough, quarter or smaller sections may be appropriate. In some cases the exposed segment may be small enough to excavate in toto without sectioning.

Features that are deep relative to their horizontal dimension (i.e., about 1.5 to 2 times as deep as their greatest horizontal dimension) can be physically difficult to completely section in one operation. Imagine trying to bisect a pit feature that measures two feet in diameter but is three feet deep. You will be laying on your stomach, hanging headfirst into the open feature, trying to remove the deepest level with your trowel. It may be necessary to take the initial cross section down to the greatest depth possible, then remove the opposing section to the same depth before going deeper with the first section. With small and deep features, like postmolds, it may not be possible to excavate a cross section without also removing some of the surrounding, nonfeature matrix (Figure 9.34). Keep the excavation of surrounding matrix to a minimum in such cases. Care must be taken to associate artifacts found in the surrounding matrix with the appropriate general excavation level, rather than the feature being sectioned.

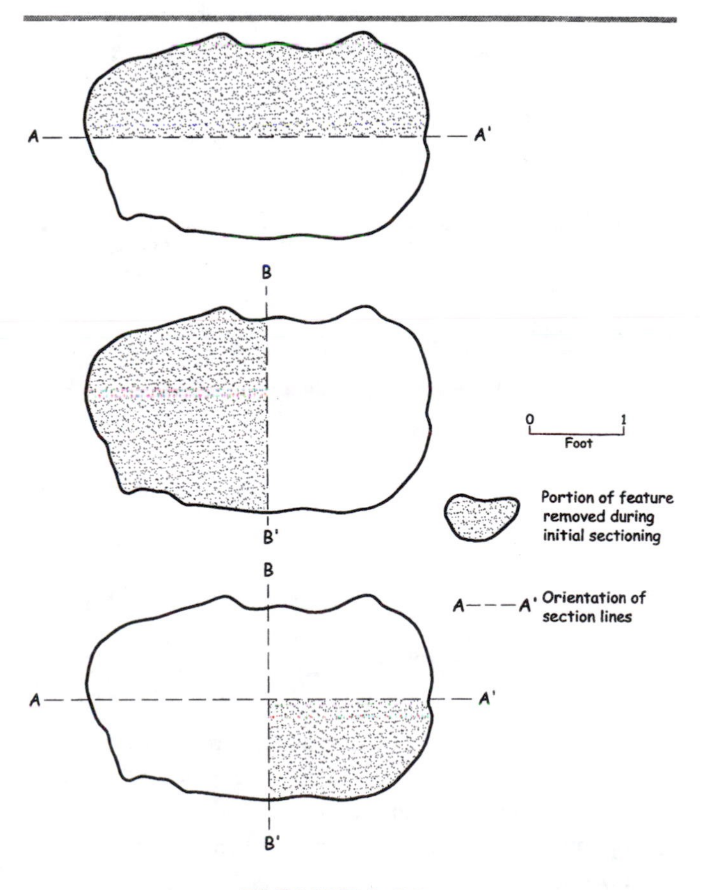

FIGURE 9.32.

Plan view of a sediment/soil anomaly feature showing the different ways that it might be cross sectioned. From top to bottom are a longitudinal section, a transverse section, and a quarter section. The longitudinal and transverse sections are also technically bisections.

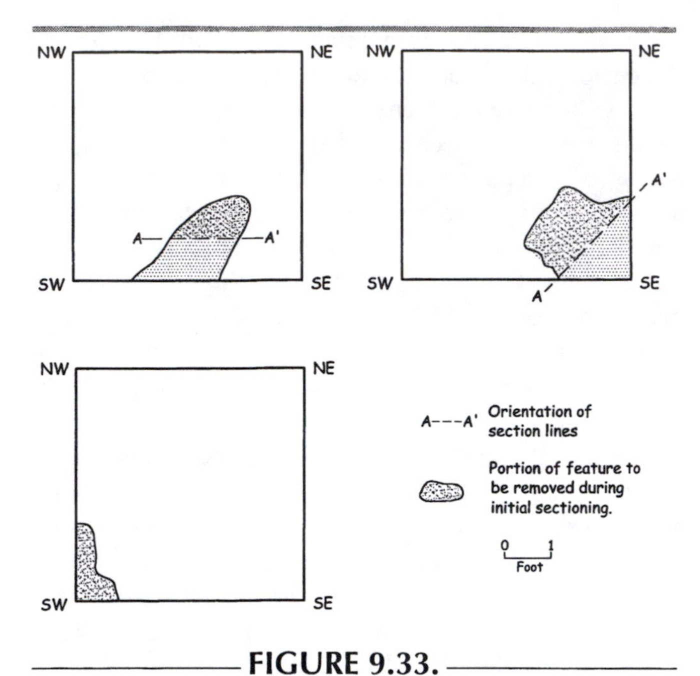

FIGURE 9.33.

Cross sections (top and right) of partially exposed features should be oriented to complement feature profiles that will be preserved in the side walls of the unit following excavation. The example on the bottom is small enough that it can be excavated in toto without sectioning.

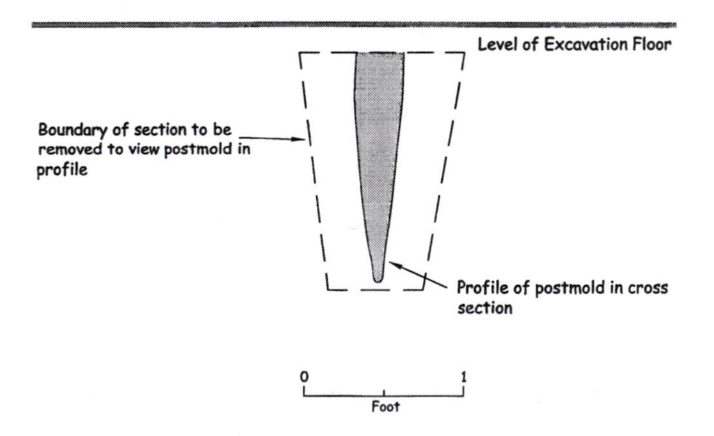

FIGURE 9.34.

Cross sectioning small and deep features, like postmolds, may require removing nonfeature matrix from the area immediately adjacent to the feature. Care must be taken to associate artifacts found in the surrounding matrix with the appropriate general excavation level, rather than the feature being sectioned.

Once the orientation of a cross section has been chosen, physically create the cross section line. Place nails in the excavation floor on opposite ends of the section line and connect them with a taut string set flush to the floor. With your trowel make vertical cuts along the string line to initiate the cross section. Work from the section line back to the outer boundaries of the feature. Natural or arbitrary levels are removed in a feature cross section the same way they are in any excavation, with attention to the same details (Figure 9.35). However, feature fill can be highly variable and it can be difficult at times to determine whether you are encountering meaningful stratification or simply mixed loads of fill. Of course, that is one of the purposes of a cross section, to allow you to better understand the feature prior to its full excavation. Only remove matrix that can be identified as feature fill. Stop your excavation when you encounter natural, undisturbed soil. In this way you will recreate the subfloor shape of the feature. I have seen feature excavations done so carefully that marks from the digging tools used by ancient people can be discerned on the outer walls of a pit. How the matrix removed from the first cross section is treated varies depending on the type of feature. Some may be dry screened in the manner of matrix removed from general unit excavations. All or portions of it may be set aside for flotation or wet screening through fine hardware mesh.

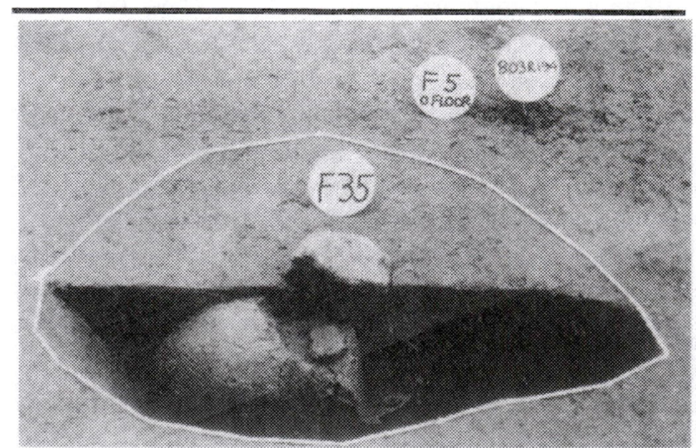

FIGURE 9.35.

A longitudinal bisection of a pit feature (#35) at the prehistoric Shermans Creek Site, Pennsylvania. The boundary of the feature as seen in plan view is outlined with white string. A large cobble has been left *in situ* at the top of the feature and most of a Clemsons Island pottery vessel is exposed in the cross section.

Upon completion, the cross sectioned feature should be photographed and a profile drawn of the basal contours and any stratification evident in the wall of the section (Figure 9.36). The procedure for doing this is the same as for drawing any profile (see chapter 7). A level line is established for depth measurements. It should be positioned immediately above the cross section line so that a tape can easily be used to measure up from the base of the cross section to the level line. A folding rule or tape is laid out on the excavation floor across the top of the feature for horizontal measurements. Set up a piece of graph paper for the drawing, showing the level line and the "0" point from which horizontal measurements will be taken. The points to be recorded in order to draw the feature's profile can be at regular small intervals, or selected to reflect how the profile's overall shape might be broken down into a series of discrete lines. For example, to record point *c* in Figure 9.36, measure over 0.73' from the "0" end of the horizontal tape at the feature's margin, then measure down from the level line. In this same way, a line representing the level of the excavation floor and any strata within the feature should also be recorded.

Excavate remaining portions of the feature using insights gained from the removal of the cross section. Adjustments could range from reorganizing the way that levels are removed from the feature to deciding whether or not to map artifacts within the feature by point provenience, to decisions about screening and taking flotation or other samples. The feature should again be photographed when it has been completely excavated and one or more profiles drawn to document the contours of its sides and base. The position of the feature should be indicated on the floor plan of each subsequent excavation level through which it has cut.

Careful cross sectioning becomes even more vital when features intrude or cut into one another. In some circumstances, the intrusions are obvious, as can be seen in Figure 9.37 where a series of oval to circular anomalies overlap one another. Superimposed and intrusive features may be recognized in plan view as patterned variations of sediment color or texture within a larger anomaly (Figure 9.38). The anomalies visible on the excavation floor shown in Figure 9.38 could reflect only deposition, digging and subsequent filling of excavations (deposition), or combinations of all of these processes. Think about how the plan view may reflect the superposition of deposits, and therefore a sequence of events. If we rule out the possibility that the anomalies were created all at once (e.g., someone laying out different types of sediments to create a mosaic on a surface), then some suggestions about a sequence of events can be made that could guide the placement of cross sections. Anomaly #4 seems to be superimposed on, or may cut through anomalies #1–#3, making it the latest event. In turn, anomaly #3 may be superimposed on, or cut through #2 and possibly #1. It therefore would have to be a later event than #2, but occur before #4. Anomaly #2 seems to be superimposed on, or may cut through #1, making #1 the earliest of the events in the sequence. Of course, an understanding of superposition requires information from the vertical plane. The cautious use of a split tube sampler may confirm the reality of interpretations prior to opening a cross section. In any case, it is

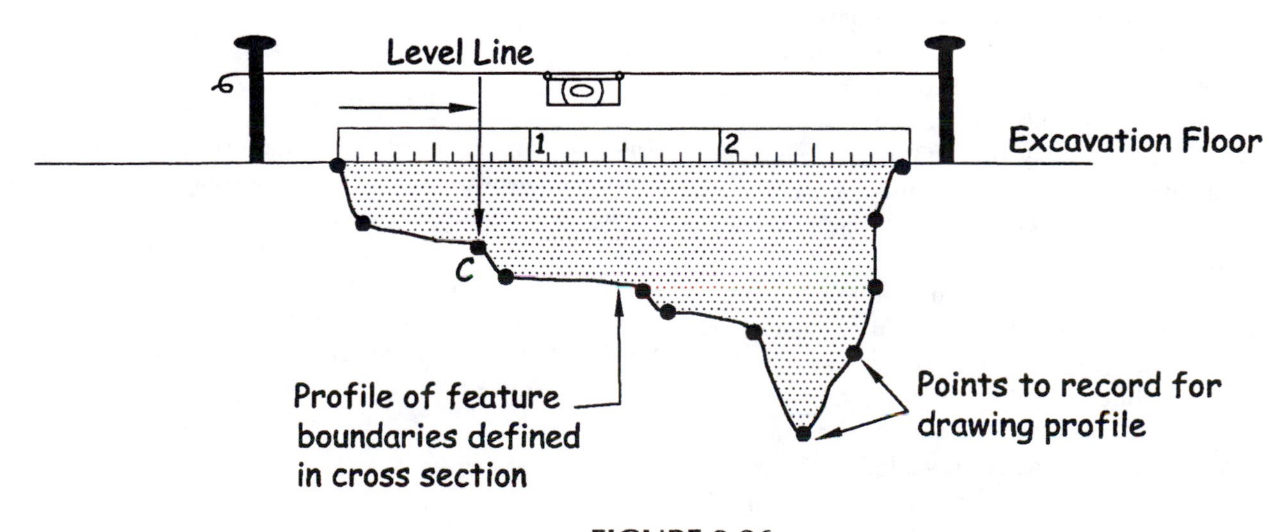

FIGURE 9.36

Recording the profile of a cross section through a feature.

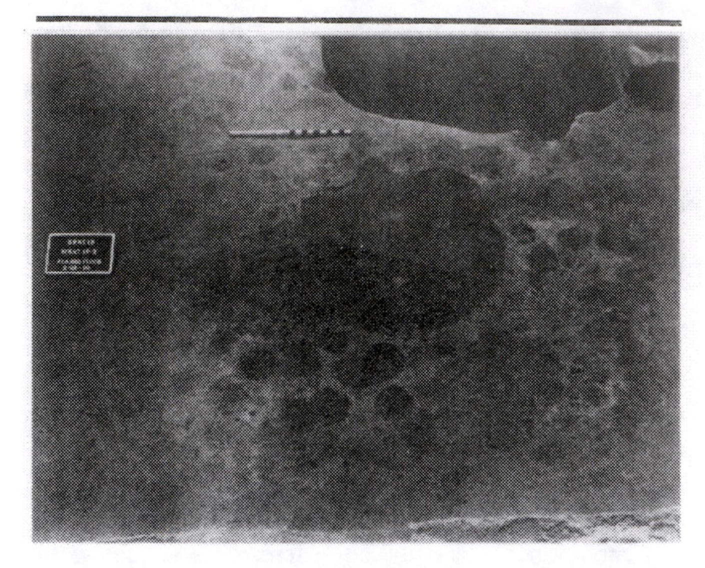

FIGURE 9.37.

An example of features that intrude into one another from the Middle Archaic deposits at Area D in the Abbott Farm National Landmark, New Jersey. The large, dark-colored circle in the picture is a reused hearth basin filled with a mixture of organic and burned sediments. The irregular borders on its left side are a result of the shape of the basin being slightly modified over the period of its use. The hearth feature is surrounded by a series of postmolds represented by smaller ovals and circles, many of which overlap one another, again suggesting the reuse of this activity area over time.

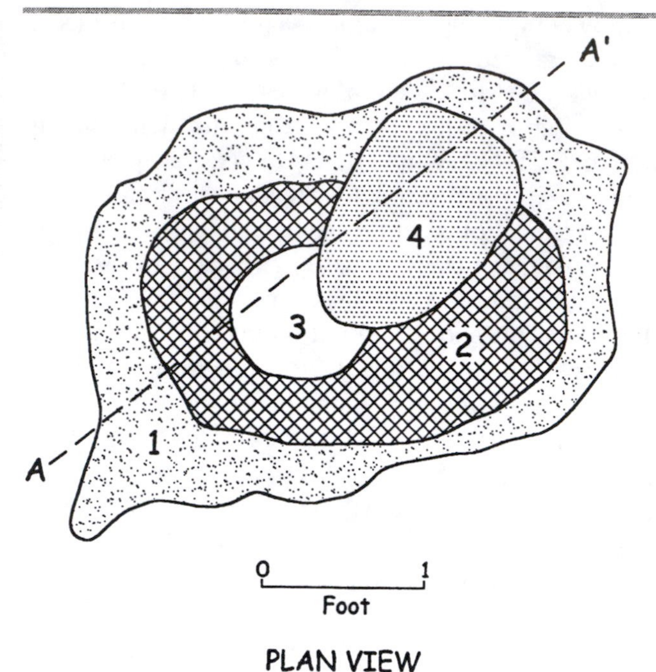

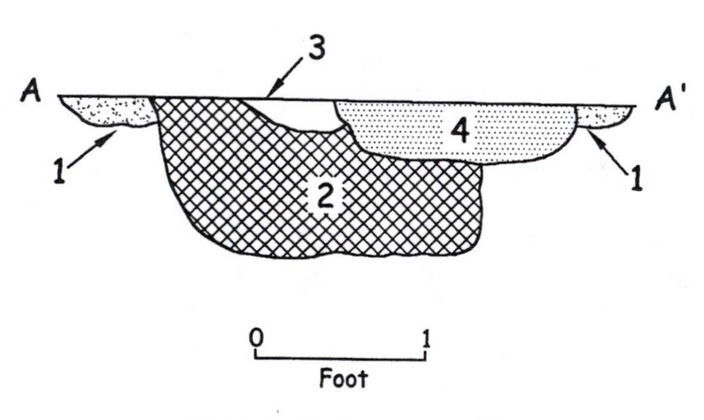

FIGURE 9.38.

The plan view of a variety of sediment/soil anomalies suggests that a number of features are superimposed on or cut through one another. The cross section confirms this and reveals the sequence in which the distinctive deposits were emplaced: #1 is emplaced and subsequently cut by the digging and filling of #2; #3 cuts into the top of #2 and is filled; and finally, #4 is dug cutting through #1, #2, and #3 before being filled.

imperative that the full horizontal boundaries of any anomaly be defined prior to excavating what may be a series of superimposed and intrusive features. A cross section of the anomalies shown in Figure 9.38 fleshes out the sequence. Even though there is one common cross sectional line, anomaly #1 was excavated separately from #2, #2 separately from #3, etc. In completing the excavation of the features, the logical focus should be on removing them in the reverse order of their creation, i.e., the last first.

In some cases the presence of multiple features may not be suspected until during or after an anomaly has been cross sectioned (Figure 9.39). In the example pictured, a large but thin anomaly of mottled organic and burned sediments (***a***) masks the tops of two distinctive oval shaped features (***b***) that are revealed as the cross section is removed. With the complete excavation of the larger anomaly, the two smaller features are fully defined and can be independently sectioned (***c***).

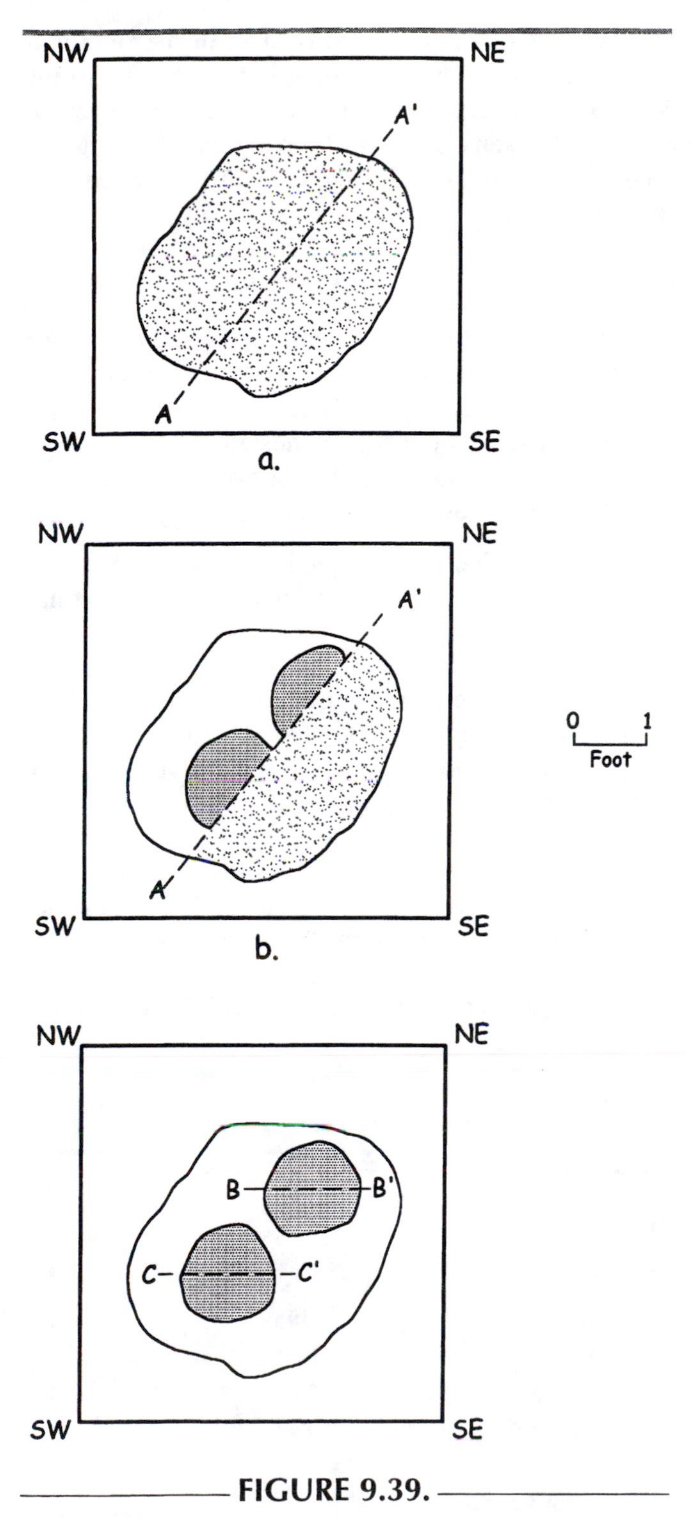

FIGURE 9.39.

A large sediment/soil anomaly (*a*) masks the top of two distinctive oval-shaped features (*b*) that are revealed as the cross section is removed. With the complete excavation of the larger anomaly, the two smaller features are fully defined and can be independently sectioned (*c*).

The unraveling of features that consist of artifact clusters proceeds in concert with the excavation of level floors, even though each is removed and recorded separately (Figure 9.40). Point provenience excavation techniques are employed. The provenience of the unit level is preserved in the provenience of the feature, e.g., Stratum 2, Level 2, Feature 1, Level 1. At the conclusion of a level, exposed portions of the feature are mapped, photographed and removed. A new level is dug across the unit, uncovering any remaining portions of the feature, e.g., Stratum 2, Level 3, Feature 1, Level 2.

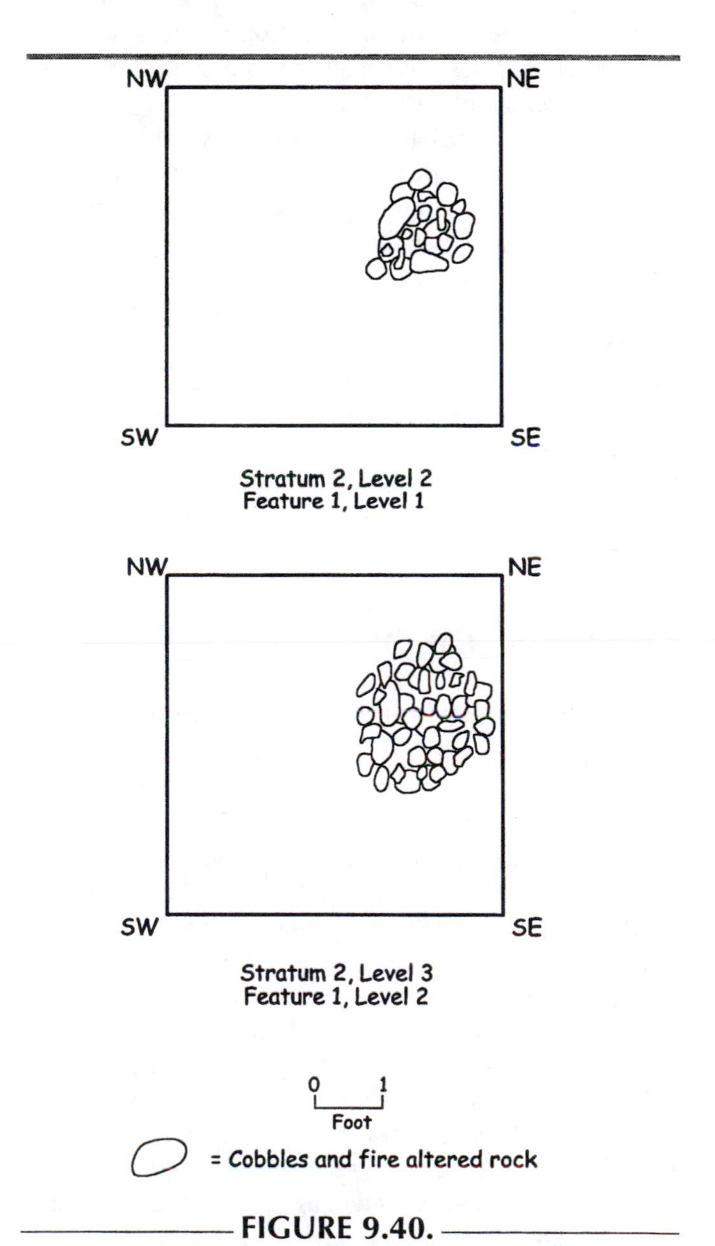

FIGURE 9.40.

The exposure of artifact clusters defined as features proceeds in unison with the removal of levels across the floor of a unit. A feature and its contents are recorded and screened separately from the general excavation level.

COMPLETING A UNIT

At the conclusion of the excavation of a unit, the profile of at least two adjoining walls should be drawn. Recording the profile of all four walls ensures that thorough documentation is at hand should any problems later arise with the interpretation of stratigraphy or stratigraphic correlations. Where the stratigraphy is extremely redundant this may not be necessary. All four walls should automatically be profiled when stratigraphy varies across a unit. A coring device, auger, or posthole digger can be used in the floor of an excavation to extend the record of stratification if necessary. If a bucket auger or posthole digger is employed, the matrix can be screened to search for artifacts.

Backfilling completed excavations by hand can be a real treat and will force you to re-evaluate your earlier decisions about positioning screening operations and backdirt. Recognizing that excavations may be intentionally or unintentionally reopened in the future, many archaeologists leave a modern token, like a penny bearing the date of the current year, at the base of their excavations to alert the diggers of the future. When the reopening of an excavation is likely as part of future research, its base can be covered with a sheet of plastic before backfilling. In this way no one will have to wonder whether or not they have reached the base of the former excavation.

FIELD RECORDS

The efficiency and standardization of data collection during excavations relies on the use of preprinted data sheets or forms. Forms may be adapted or customized for a particular region, site, or deposit. The examples here reflect general trends and basic types of data that are collected. You have probably heard this before but I will say it again—neatness counts! The forms that you complete must be readable by anyone. Many forms are double-sided to accommodate the range of observations to be made. Many are also supplemented by maps and plan views on separate sheets. It is important that all sides of any form, map, or plan view bear provenience data so that they can always be associated with a context. Once out of the field, records will be photocopied to insure that if one set is damaged or lost, documentation of fieldwork will still exist. That's why even double-sided forms require the entry of redundant information on provenience. Electronic notebooks employ digital versions of record forms, greatly enhancing the duplication and manipulation of data in the field and beyond. Even the most thorough standardized forms don't preclude the need for keeping certain types of notes in a personal field book or journal. The suggestions for note taking presented in chapter 8 are equally relevant here.

PROVENIENCE TAGS OR CARDS

The simplest of standardized forms are provenience tags or cards that must accompany any artifacts or samples collected (Figure 9.41). Most of the data categories used are self-evident. They are designed to function in a variety of field situations so not every data category will always be relevant and receive an entry.

- ☑ **Cat No.** The field specimen catalog number assigned to a provenience after its excavation has been completed.
- ☑ **Bag.** When the tags are used to label the outside of bags containing artifacts or samples from the same provenience, the numbers of bags involved are listed, e.g., bag 1 of 3, bag 2 of 3.
- ☑ **Unit.** Any type of excavation can be substituted here (e.g., trench, shovel test, auger boring) including collection units used for surface surveys.
- ☑ **Strat.** Surface may be substituted here where appropriate.

Site **Cat No** **Bag**
Unit **Quad**
Strat **Level**
Feature **Level**
Artifact No **Type**
Coord: x **y** **z**
Datum Coord
Excavator **Date**

FIGURE 9.41.

A preprinted artifact or sample provenience tag. It is designed to function in a variety of field situations.

☑ **Artifact No, Coord, Datum Coord.** These categories are used when the point provenience of an artifact has been recorded. Every mapped artifact has to be assigned a number (artifact number) that links it with a position on a map or record form and distinguishes it from other mapped artifacts. The artifact number is not the same as the catalog number.

Some archaeologists transfer the structure of the provenience card onto a stamp that can be used on the outside of paper bags used to store artifacts. Others, myself included, have converted them into peel-off, adhesive labels. I have them produced in bright colors so that anything labeled is not easily missed or lost in the field.

EXCAVATION LEVEL FORMS

Excavation level forms (Figure 9.42), like provenience cards, are designed to function in a variety of situations. Not all data categories will receive a notation or comment every time the form is used.

EXCAVATION LEVEL FORM

Project______ **Site or Deposit Designation**______
Unit Designation ______ **Horizontal Dimensions**______ **Other Provenience** (area, locus, trench, block)______
Stratum______ **Level**______ **Thickness of Arbitrary Level**______ **Quadrant or Other Subdivision**______
GPS Location______

Dates Open______ **Date Closed**______ **Field Catalog Number for Context**______
Excavators/Recorders______

Elevation Datum Location______ **Elevation** ______ **Converted?** Yes___No___
Opening Elevations of Level: NW______ **NE**______ **SW**______ **SE**______ **CENTER**______
Closing Elevations of Level: NW______ **NE**______ **SW**______ **SE**______ **CENTER**______

Excavation Method______
Screened? Yes____ No____All?____Sample? (note size)______ Dry screened?___Wet screened?___ Screen mesh size(s)____

Matrix Color (Munsell values and description)______
Matrix Texture______ **Matrix Structure**______ **Gravel Content**______

Samples Collected and Location (soil, C14, T-L, etc.)______

Features Encountered or Present (feature number, point of origin, brief description, location)______

______

Plan view (on separate sheet or attached) Yes____ No____ **Profile** (on separate sheet or attached) Yes____ No____
Photos: B&W (Roll designation)____ Color Prints (roll number) ____Color Slides (roll designation)____Digital (memory card designation)____

ARTIFACTS RECOVERED

Artifact Type/Class	Qty	Artifact Type/Class	Qty	Artifact Type/Class	Qty

Artifacts Discarded or Sampled______

COMMENTS______

______

______

______

______

______

______

______

______

(over)

FIGURE 9.42.

An example of an excavation level form used to make data collection more efficient and standardized.

Site or Deposit Designation______________**Field Catalog Number for Context**__________**Unit Designation**________________
Other Provenience______________**Stratum**__________ **Level**__________**Quadrant or Other Subdivision**________________

COMMENTS (continued)____

____

____

____

____

____

____

____

____

____

____

____

SPOT COORDINATES AND ELEVATIONS

Artifact #	*X*	*Y*	*Z*	Artifact #	*X*	*Y*	*Z*

FIGURE 9.42a.

☑ **Dates Open.** An excavation may be open for a number of days. A listing of relevant dates provides a link to entries that may be in the field books of the unit excavators or the field supervisor.

☑ **Excavators/Recorders.** These names aren't just for posterity's sake. Should a data entry need to be qualified or a discrepancy resolved, these are the people to talk to.

☑ **Elevation, Converted?** It is understood that many of the datums used for recording elevations or depths may be unit specific. These datums may or may not have been tied into a common elevation datum for a site prior to their use. It is essential to know if they have been converted to the common datum prior to use. When elevations are recorded within an excavation using a surveying instrument, it is assumed that readings will be converted and the "yes" category should be checked. More details can be provided in the "Comments" section.

☑ **Excavation Method.** Note whether shovel/trowel skimmed, chunked, spot mapping and elevations, exact or point provenience.

☑ **Screened, Sample?** If all of the excavated matrix from a context is not screened it is important to note the size of the sample that is screened. An explanation for only screening a sample of the matrix should be included in the "Comments" section.

☑ **Samples Collected and Location.** Any samples that are collected should be afforded the same attention to provenience given to any significant artifact, i.e., point provenience if possible. How a sample is handled in the field can have a bearing on laboratory analysis and interpretation. Sample handling can be described in the "Comments" section of the form.

☑ **Features Encountered or Present.** Separate forms are used to document features. This category on the excavation level form serves as a notice of, and cross reference for these other forms. It also provides an idea of how a general excavation level may have been impacted by feature-related activities.

☑ **Photos.** A separate and more detailed photo log is kept (see Chapter 8). The entry here serves as a notice that photos were taken and is a cross reference to the photo log.

☑ **Artifacts Recovered.** This section is not meant to be a detailed catalog but a summary of the artifacts found, organized by general type/class/category and material type where appropriate.

☑ **Artifacts Discarded or Sampled.** There will be times when not all artifacts found in an excavation will be collected. This is often the case with modern materials or certain types of historic debris that appear in plowzones and near-surface contexts. It may be useful to know that shot gun shells and fragments of coal were found in an excavation level, but retaining them for laboratory processing and curation is not necessary.

☑ **Comments.** Qualifications of any of the preceding data entries can be made in this section of the form along with any other insights or observations that you believe are worth noting. Don't assume that because you passed on a pithy comment to the field supervisor that it will have been entered in his/her field book or remembered months later when analysis and report preparation are underway. Some of the things that you should be thinking about and commenting upon include: where are artifacts concentrated, horizontally or vertically, in the level; is the matrix (soil/sediment) the same color and texture across the level floor; has the nature of artifacts (type, raw material represented, activities represented) and their number changed relative to what you saw in the last level; how does what you've seen in this level compare with similar levels in excavations that you have already completed; is your unit typical or atypical. You are the one who has had your nose in the level from start to finish. Don't let your thoughts and impressions go unrecorded.

EXCAVATION SUMMARY FORM

A synthesis of what has been observed and collected during the excavation of a single unit is embodied in the ***excavation summary form*** (Figure 9.43). The completion of the form requires input from the excavators and a field supervisor, fostering communication and interpretation while things are fresh in everyone's mind. Having to review level forms while still in the field is also a great way to identify problems that may exist with them. A summary of the changing nature and number of artifacts and the location of features by stratum and level is the basis for initial attempts at defining cultural stratigraphy and comparing it with the natural stratification that has been documented.

EXCAVATION SUMMARY FORM

Project__________ **Site or Deposit Designation**__________
Unit Designation __________ **Horizontal Dimensions**__________ **Other Provenience** (area, locus, trench, block)__________
GPS Location__________
Elevation Datum Location__________ **Elevation** __________ **Benchmark Tie-in?** Yes___No___
Depth of Excavation_____ **Number of Strata Excavated**_____ **Number of Levels Excavated**_____ **Number of Plan Views**_____
Number of Profiles_____ **Feature Types and Numbers Assigned**__________
____
Dates Open__________ **Date Closed**__________ **Field Catalog Numbers Used**__________
Excavators/Recorders__________
Field Supervisor__________ **Date of Review with Excavators/Recorders**__________

Stratum	Level	Artifact Frequency exclude FCR	FCR or Large Artifact Frequency	Diagnostic Artifacts Type and Age	Feature Origin (list number)	Feature Base (list number)

Samples for Chronometric Dating (type and provenience)__________
____
____
____
____

(over)

FIGURE 9.43.

An excavation summary form (double-sided) provides a structure for synthesizing and interpreting data from all of the levels of a single excavation.

Project____________________________________ **Site or Deposit Designation**____________________________
Unit Designation ____________ **Horizontal Dimensions**_______ **Other Provenience** (area, locus, trench, block)___________

PROPOSED CULTURAL STRATIGRAPHY____

CORRELATION WITH NATURAL STRATIGRAPHY____

COMMENTS (e.g., purpose of excavation; reason for halting excavation; notable disturbances; alterations in general excavation methodology; interpretations of artifacts and features; recommendations for adjusting excavation levels in future excavations; recommendations for expanding excavation)____

FIGURE 9.43a.

FEATURE LEVEL AND SUMMARY FORMS

Feature level forms (Figure 9.44a) track the excavation of features by level using the same structure as excavation level forms. And like an excavation summary, the ***feature summary form*** (Figure 9.44b) organizes and begins the interpretation of the data collected.

Photograph all features before, during, and after excavation, and any excavation floors of interest. Closeup shots should include: a sign board noting the provenience and subject of the photograph; some type of scale, whether it is a photographic bar scale or simply an outstretched tape; and a north arrow or a trowel with its point oriented to the north. Remember that many shots will need to be done both with color slide and black-and-white film to cover the need for images in basic documentation, public presentations, and publications. Recommendations for general field photography are presented in Chapter 8 and are applicable to excavation projects as well as surface surveys.

ARTIFACT COLLECTION

In most cases all artifacts found in excavations are documented and kept, although there are exceptions. The occurrence of modern artifacts in an excavated context, like a plowzone, should be recorded in field notes but the materials can be discarded in the field. Only a sample of artifacts, generally those that are diagnostic of a particular time period, may be saved from the removal of landfill (modern or historic) in an excavation. Some prehistoric archaeologists count and/or weigh fire-altered rock or hearth rocks in the field and discard them. This hardly takes full advantage of the many types of analyses that can be performed with this type of artifact. Collecting, transporting, and processing the large amount of fire-altered rock that an excavation may produce can be challenging. Ultimately, they may be culled from a collection, but not before their information potential has been fully explored. In general, if an artifact collection is to be culled, I believe that it is best done in the laboratory and at a time when investigators understand a deposit and can make well-informed judgements about what constitutes extraneous materials.

The attention given an artifact once it is discovered must extend beyond assigning it an appropriate provenience and linking it with field records. Artifacts are either accustomed to, or in the process of adjusting to the physical environment in which they are buried. Exposing and removing them from a buried context establishes new conditions to which they will react and that can result in their chemical, biological, or mechanical degradation. For some types of artifacts, like stone tools or glass, the effects may be negligible. For others, like those based on organic materials, it can be catastrophic. How an artifact is handled in the field can also have an impact on the types of laboratory analyses to which it can be subjected.

CONSERVATION OF MATERIALS

Do no additional harm, exercise patience and restraint, the best treatment is the least treatment, do nothing that can't be reversed—all of these phrases characterize the philosophy of conservators toward the treatment of materials. Deciding whether to conserve and how to conserve materials in the field can be a complex process. It requires an understanding of the chemistry and environment of a deposit, the nature of all types of artifacts, the types of analysis that may be performed on artifacts in the laboratory, and the potential effect of conservation measures on these analyses. Advances in conservation have resulted in some reversals of opinions about what constitutes reasonable treatments, and debates about some procedures continue (cf. Dowman 1970; Hester 1997; Joukowsky 1980:244–275; Sease 1994). Issues of personal safety must also be considered in the transport, storage, and use of chemicals and solvents that might be used in field conservation.

Numerous references exist on the subject of conservation and I draw on some of the more basic here. Cronyn (1990), Dowman (1970), and Sease (1994) provide important overviews for field archaeologists. Hamilton (1996), Robinson (1998), and Singley (1988) focus upon materials derived from different types of waterlogged and underwater environments. Artifacts made from plant materials are addressed by Florian (1990). Relevant article-length studies of interest can be found in the *Journal of the American Institute for Conservation, Journal of Field Archaeology,* and the *Journal of Archaeological Science.* A little knowledge can be a dangerous thing when it comes to field conservation, so I will review some general guidelines here rather than describe specific recipes for treatment.

- ☑ Be prepared and informed. The types of artifacts that may be encountered in the field and require special attention can be anticipated on the basis of previous experience and reference to the published literature of a region. Bone, wood, or any organic artifacts readily spring to mind.

FEATURE LEVEL FORM

Project________________________________**Site or Deposit Designation**______________________**Feature Number**________
Feature Exposed In: Unit (s) __________________**Other Provenience(s)** (area, locus, trench, block)______________________
Type of Feature____
Percent Feature Exposed____________________**Top of Feature Encountered in: Stratum**__________**Level**__________
Feature Shape___________________________________**Feature Dimensions: Length**_____**Width**_____**Diameter**_____
Elevation Datum Location___________________________**Elevation Top of Feature** ________**Converted?** Yes___ No____

Feature Level Designation__________________**Type of Level Excavated:** entire feature_____ arbitrary______ natural______
Elevations Opening Level __________________**Elevations Closing Level**__________________**Converted?** Yes___ No____
Cross Section? Yes___No___ **Axis?** N-S___E-W___Other____**Describe**____
Bisection? Yes___No___**Axis?** N-S___E-W___Other____**Quarter section?** Yes___No___**Orientation of Axes**__________
Dates Open_________________________**Date Closed**_______________**Field Catalog Number for Context**____________
Excavators/Recorders____

Screened? Yes____ No____All?___Sample? (note size)________ Dry screened?____Wet screened?____ Screen mesh size(s)_____
Matrix Color (Munsell values and description)____
Matrix Texture____________________________**Matrix Structure**____________________**Gravel Content**______________
Samples Collected and Location (soil, flotation, C14, T-L, etc.)____
____
Plan view (on separate sheet or attached) Yes____ No____ **Profile** (on separate sheet or attached) Yes____ No____
Photos: B&W (Roll designation)____ Color Prints (roll number) ____Color Slides (roll designation)____Digital (memory card designation)____

ARTIFACTS RECOVERED

Artifact Type/Class	Qty	Artifact Type/Class	Qty	Artifact Type/Class	Qty

COMMENTS____
____
____
____
____
____
____
____
____
____
____

(over)

FIGURE 9.44a

Feature level forms mimic the structure and types of data collected for general excavation levels.

FEATURE LEVEL FORM

Project________________________**Site or Deposit Designation**____________________**Feature Number**________
Feature Exposed In: Unit (s) __________________**Other Provenience(s)** (area, locus, trench, block)__________________
Feature Level Designation__________________**Type of Level Excavated: entire feature**______**arbitrary**______ **natural**______
Field Catalog Number for Context____________

COMMENTS CONTINUED____

SPOT COORDINATES AND ELEVATIONS

Artifact #	*X*	*Y*	*Z*	Artifact #	*X*	*Y*	*Z*

FIGURE 9.44a.

Continued.

FEATURE SUMMARY FORM

Project________________________________**Site or Deposit Designation**________________________**Feature Number**________
Feature Exposed In: Unit (s) ____________________**Other Provenience(s)** (area, locus, trench, block)______________________
Type of Feature____
Percent Feature Exposed________________________**Top of Feature Encountered in: Stratum**__________**Level**____________
Base of Feature in: Stratum________**Level**________**Number of Levels Excavated**_____**Type of Levels** arbitrary____natural____
Strata, Levels, and Other Features Cut by Feature____
Elevation Datum Location____
Elevation Top of Feature _________**Converted?** Yes___No___**Elevation Base of Feature**______**Converted?** Yes___No____
Field Catalog Numbers Used____
Dates Open____**Date Closed**______________
Excavators/Recorders____
Field Supervisor_______________________________________**Date of Review with Excavators/Recorders**________________

Feature Shape-Horizontal__________________________________**Feature Shape-Vertical**________________________
Feature Dimensions: Length________**Width**________**Diameter**________**Depth**________**Feature Profile Drawn?** Yes___No___
Plan Views Drawn (number and associated stratum/level)____

Description of Feature (matrix - color, texture, and variations; contents; disturbances; evidence for stratification, etc.)____________

____

____

____

____

____

____

____

____

____

____

Photographic Records (type, roll designation, memory card designation)____

____

____

Samples for Chronometric Dating (type and level provenience)____

____

____

Other Samples Collected (type and level provenience)____

____

____

(over)

FIGURE 9.44b.

FEATURE SUMMARY FORM

Project______________________________**Site or Deposit Designation**________________________**Feature Number**________
Feature Exposed In: Unit (s) ____________________**Other Provenience(s)** (area, locus, trench, block)____________________
Field Catalog Numbers Used____

Diagnostic Artifacts Recovered (type and level provenience)____

____

____

____

INTERPRETATION/COMMENTS____

____

____

____

____

____

____

____

____

____

____

____

____

____

____

____

____

____

____

____

____

____

FIGURE 9.44b.

Continued.

☑ Be cognizant of the factors that affect the preservation of materials (see discussion in chapter 2). Pit features often preserve artifacts that are otherwise deteriorated or absent in the general deposits of a site. Preservation of materials also may be enhanced in caves, rockshelters, and shell middens.

☑ Think about the types of analyses that are typically performed on specific classes of artifacts. The effectiveness of use wear and edge damage studies, residue detection and analysis, and dating techniques can be impacted by how objects are handled and treated in the field.

☑ Familiarize yourself with a basic text on field conservation. If nothing else, this will convince you to proceed cautiously when considering the treatment and handling of unstable and fragile artifacts. Make contact with a conservator prior to fieldwork, get their advice about basic procedures, and make arrangements for in-field consulting if needed. The ideal is to have a conservator on the field team, but this may not be feasible for survey and testing projects. Find out about conservators working in your area through your state museum, SHPO, or the American Institute for Conservation of Historic and Artistic Works (AIC). The AIC can assist individuals in selecting and finding a conservator depending on the type of service required and geographic area (contact information is provided at the end of this chapter). As Rye (1981:9–10) notes, four general rules should be kept in mind when considering conservation of artifacts in the field: 1. Conservation should stabilize rather than change an artifact; 2. All treatments should be reversible; 3. All treatments should be described in detail; and 4. Unstable materials should be kept in an environment similar to the one form which they were removed.

☑ As an article is exhumed and exposed to the air, is its physical state unstable or fragile? Is it crumbling, flaking, or exfoliating? As the artifact dries, are there dramatic changes in surface color or texture; do lines or cracks appear? You usually won't think twice about this instability when dealing with artifacts that are presumed to be relatively durable and stable, like those fashioned from stone. Pay close attention to any artifact that is damp or wet when exposed. The loss of moisture in some materials, especially organic ones, can lead to shrinkage and their mechanical breakdown or collapse. Things that are wet when exposed should be kept wet. Don't rewet artifacts that have already dried. The cycle of shrinking and swelling associated with this practice may also contribute to the mechanical breakdown of an artifact. Materials that are impregnated with soluble salts are subject to a similar type of degradation related to the growth of salt crystals. If already wet, they should be kept wet. If dry, care must be taken to package and store them in such a way as to prevent the accumulation of moisture, which can promote crystal growth.

Some artifacts may survive exposure but be too fragile to be moved and transported without treatment, like the charred remnants of fabric or basketry. Artifacts recognized as unstable or fragile should be kept covered and shielded from sunlight while strategies for their handling are devised. Any treatment or special handling of objects should be well documented, including before and after *in situ* photographs and drawings. Describe the rationale for a treatment or handling procedure, the chemicals or materials involved, and the procedure itself.

One strategy for dealing with a recognizably unstable or fragile artifact is to bring a conservator into the field and assist them with any recommended treatments. Deciding to stabilize an object, with or without a conservator, involves treatment with consolidants or adhesives. Stabilizing an artifact means treating it in such a way as to maintain its physical integrity and minimize deterioration. A ***consolidant*** is a resin in a liquid solution applied to an object to strengthen it. Polyvinyl acetate (PVA) and Acryloid B72 are widely available, reversible, all-purpose consolidants. PVA is soluble in acetone and alcohol, Acryloid B72 in acetone and toluene. Consolidants dissolved in a solvent are for artifacts that are dry, but emulsions of consolidants are used on artifacts that are damp or wet. An ***emulsion*** is a suspension of a resin-solvent solution in water. Cyclododecan is a volatile, reversible sealing wax currently experiencing wider use in this country as a consolidant. It has a low melting point (58–61 degrees Celsius) and can be applied without the use of a solvent. It eventually removes itself from the material to which it has been applied by evaporating from its solid state (Bruckle et al 1999; Gary McGowan, Cultural Preservation and Restoration, Inc. personal communication 2001; http://www.kremerpigmente.de/englisch/87100e.htm). An ***adhesive*** is something applied to pieces of an artifact to rejoin or reconsolidate them. Consolidants or adhesives should be used with caution. A basic rule in treating any type of artifact is do nothing that can't be reversed. This means that any consolidants or adhesives used on artifacts could be removed from an artifact after application without damage. Cyclododecan is excellent in this regard since it basically removes itself from a treated object, unlike most other consolidants or adhesives. If consolidants or adhesives are to be applied, then the object should be cleaned as gently and as thoroughly as possible prior to application. The analysis of residues on artifacts and radiocarbon dating can be impaired by the application of consolidants or adhesives.

Removal and transportation to the laboratory for treatment is another option for dealing with fragile or unstable artifacts. The artifact is left *in situ* and a block of matrix encompassing it is isolated, undercut and removed en masse, and finally packaged for transport. Artifacts may also be jacketed with Plaster of Paris after being well covered with gauze so that the plaster does not directly adhere to the artifact. Polyurethane foam may be substituted for plaster if a person is experienced with its use.

☑ The judicious handling of artifacts in general can prevent damage and support subsequent laboratory analyses. Don't excessively clean any type of artifact. The matrix inside of any type of container should be collected along with the artifact. There should be an established policy before entering the field about how to handle and/or sample lithic, ceramic, or any other artifacts that might possess organic or other types of residues. In general, do not clean these artifacts, handle them as little as possible, and bag them separately from other materials. Objects that will be examined for adhering pollen, phytoliths, or starch grains should be placed in clean/sterile bags or containers to prevent contamination. The wearing of latex gloves while handling these objects can also prevent any accidental contamination.

☑ Take a common-sense approach to the ways in which artifacts are bagged and packed for transport to the laboratory. Don't include large and heavy artifacts with smaller and more fragile materials in the same bag or container. As you might imagine, bagging hammerstones and pottery together can have some disastrous consequences.

☑ Use stiff containers for objects that are fragile and easily broken.

☑ Separate and individually bag artifacts that you know will be subjected to microscopic examination in the laboratory, like stone tools, or artifacts that have been targeted for any type of residue analysis.

☑ Artifacts that are damp when collected, especially bone and other organics, should be placed in breathable containers so that they will dry out slowly and mold or other fungi won't develop. If plastic bags are used, they should be punctured in places.

☑ Don't place metals in any type of container that will sweat moisture, like plastic bags, since this will promote corrosion.

☑ Specially tag artifacts whose continued stability is in doubt so that they can receive attention quickly when they reach the laboratory. Bubble wrap, acid-free tissue, and aluminum foil are helpful for packaging a variety of materials in the field and insuring their stability during transport.

SAMPLE COLLECTION

There are many types of useful samples that can be collected from an excavation (Table 9.1). What gets collected, and in what quantity, varies depending on the stage of fieldwork and the type of deposit. Samples for dating purposes are the most consistently collected during any type of fieldwork. They should be seen, recorded, and collected *in situ* so that their context is not in doubt. Samples for radiocarbon dating should be collected using a clean trowel and wrapped in clean aluminum foil or placed inside of clean plastic or glass containers. Floral or faunal remains that might be submitted for radiocarbon dating should first be identified as to species. Circumstances may arise where it becomes necessary to use charcoal or other organics from screened or flotation matrix for radiocarbon dating. Some types of samples, like those for optically stimulated luminescence or paleomagnetism, are best collected by the specialists or under their guidance. It is essential to collect pollen and phytolith samples using clean implements and store them in containers that are dry and sterile. Any sediment/soil, bulk, or flotation samples collected that may be stored for some period of time should be opened in a sheltered area and allowed to dry. This will prevent or slow down the growth of fungi and microorganisms in the samples.

HEALTH, SAFETY, AND GENERAL LOGISTICAL CONCERNS

The many things that you can do to insure your personal health and safety in the field are discussed in chapter 5. Remember that any project needs a health and safety plan that has been shared with the field crew. Be wary of entering any excavation that is over your head. There are OSHA safety standards detailing how excavations over five feet deep should be shored depending on their size and depth. Anything four feet or deeper needs to have a ladder or means of egress within it and handy to the excavator.

TABLE 9.1

Some Useful Samples to Collect from Excavations

Type	What to Collect	Uses
sediment/soil	sediment/soil from representative strata, levels, or features	chemistry, particle size, trace element analysis; definition of natural stratigraphy; recognition of decomposed materials and human activities
column samples	incremental and constant volume samples of matrix and everything within it down the wall of an excavation	as above; defining cultural stratigraphy, examining the degree of postdepositional movement of artifacts
flotation	sample or constant volume of matrix from excavation levels; sample or all of feature matrix	recovery of micro artifacts, especially floral and faunal remains; reconstructing the local environment and subsistence
bulk sample	constant volume of matrix from an excavation level or strata or feature	archival sample of matrix, or held in reserve for any variety of analyses needed in the future
pollen/phytolith	sediment from newly exposed deposits on level floors or in features, interior sections of side walls of excavation, or beneath select artifacts	reconstructing the local environment, identifying decomposed botanical remains
oxidizable carbon ratio	organic sediments from buried soil or feature	determining age of relevant strata or feature
optically stimulated luminescence	sediment from unexposed portions of level floors or side walls of excavation	determining age of relevant strata or feature
thermoluminescence	burned or heated lithic artifacts or clay, ceramics, and sample of associated sediment	determining age of relevant strata or feature
paleomagnetism or archaeomagnetism	burned sediments, clay, or rock collected *in situ* with their current compass/magnetic orientation preserved	determining age of relevant strata or feature
radiocarbon, carbon 14	wood, bone, shell, or any organic material	determining age of relevant strata or feature
dendrochronology	wood, wood charcoal	determining age of relevant strata or feature

Thorough background research should alert a field crew to the possibility that hazardous materials might be found within a project area. Specialists must identify and develop a program to deal with hazardous materials following OSHA guidelines. Archaeological fieldwork does occur on sites where hazardous materials have been identified (e.g., Poirier and Feder 2000). Field personnel should have OSHA training regarding hazardous materials and may be required to wear different types of protective gear depending on the type and level of hazard (e.g., Azizi 1995).

Consider seasonal changes in weather and their impact on digging conditions when scheduling any type of subsurface testing. With prolonged dry weather, sediments that have a high clay content can become extremely hard and require the use of picks or mattocks for excavation. Maintaining excavation controls and observing artifacts *in situ* are difficult under these conditions. Moderate to extreme force will be needed to crush excavated chunks of matrix for screening, which could harm any artifacts present. The same constraints obtain when dealing with frozen soils. When excavations are concentrated in well-defined areas, such as during intensive testing or mitigation, shelters can be erected and furnished with heaters powered by portable generators (Figures 9.45 and 9.46). Prefabricated shelters in a variety of sizes are available for purchase or can be crafted on an ad hoc basis using readily available materials like PVC pipe, duct tape, and rope. Blanketing areas to be excavated with hay (especially salt hay) or insulation batting, and keeping them covered with a tarp, are other ways that archaeologists have attempted to prevent ground from freezing. Rainy seasons in some areas of the Americas also make it impractical to carry out sustained excavation efforts. Keeping excavations dry and in a state where sediments, soils, and stratigraphy can be adequately read, and promoting suitable conditions for note taking, record keeping, and mapping become impractical. Again, investment in portable shelters and other equipment can overcome these problems, but their cost in time and dollars must be weighed against the return in information and the progress of an investigation.

Archaeological surveys require all necessary tools and equipment to be carried to some degree. You will be responsible for transporting not only the tools that you will use personally, but some share of the equipment needed by the crew at large. Planning is the key to both minimizing and spreading out the load of materials to be carried. Equipment may be stored in the field for convenience sake on long-term excavations. Security then becomes an issue and may involve the use of chains and padlocks, lockable storage boxes (Figure 9.47), on-site trailors, and even nighttime security personnel.

FIGURE 9.45.

Prefabricated shelters equipped with portable heaters make it possible to work under a variety of extreme weather conditions in some portions of the country. Pictured is a Hansen Arctic Weatherport type of shelter in use at the Howarth Nelson Site (36FA40) located in southwestern Pennsylvania. It enabled excavations to continue throughout the snowy winter season.

FIGURE 9.46.

Archaeologists have been quite innovative in building field shelters using PVC pipe, duct tape, and rope, as can be seen in this example from excavations at the Raritan Landing Site in New Jersey.

FIGURE 9.47.

This type of storage box is common on many construction jobs and is used by archaeologists for the on-site storage of smaller types of equipment. The box is hard to damage, heavy and not very portable, waterproof, and can be locked.

DISCOVERING BURIED ARCHAEOLOGICAL DEPOSITS

Find deposits or sites as efficiently as possible, and in a way that results in the least amount of impact on them. This should be the mantra guiding archaeological survey and testing. The ideal goal of a subsurface testing program is to locate all archaeological deposits that exist within a project area and generally define their horizontal and vertical boundaries. In many cases this will require the use of excavations systematically placed on a grid that blankets an area. In others, combinations of systematic and subjective strategies may be appropriate. We've already discussed the impact that sampling strategies and an understanding of geomorphology can have on fieldwork and searching for archaeological deposits. Factors affecting the discovery of archaeological deposits have also been considered (see chapter 8, Table 8.1). Still other factors impinge on the reliability of subsurface testing techniques for discovering deposits, and have been the focus of much discussion in the literature (cf. Ives 1982; Kintigh 1988; Krakker et al. 1983; Lightfoot 1986, 1989; McManamon 1984; Nance 1983; Nance and Ball 1986, 1989; Shott 1985, 1989; Stone 1981; Wobst 1983). The more basic of these factors are summarized as follows:

- ☑ horizontal size and shape of the deposit/site
- ☑ density of artifacts within the boundaries of a deposit/site
- ☑ distribution of artifacts within the boundaries of a deposit/site
- ☑ size of artifacts
- ☑ size and shape of the area being surveyed
- ☑ degree to which a deposit/site falls within the area being surveyed
- ☑ horizontal size of individual tests
- ☑ distribution of tests
- ☑ probability that a test falls within the boundary of a deposit/site
- ☑ probability that a test encompasses the location of an artifact
- ☑ probability that an artifact will be recognized or observed during excavation

Kintigh's (1988) synthesis of these issues and his simulation studies are insightful and I draw on them here. A logical assumption is that large sites or deposits are more easily detected than smaller ones. But even within larger deposits, artifacts are not evenly distributed, and do not occur in the same densities throughout (Figure 9.48). So having a subsurface test fall within the boundary of any site doesn't guarantee that an artifact or feature will be encountered in the excavation, thereby signaling the discovery of the deposit. As artifact density increases and density distributions become more uniform, the chance of finding a site increases. Even when a test excavation encompasses the location of an artifact, the size of the artifact, the excavation technique, and the type of screen mesh employed can determine whether or not a discovery is made. For example, a stone flake only an eighth of an inch long may not be seen during excavation and may easily fall through the quarter-inch mesh of the typical screen.

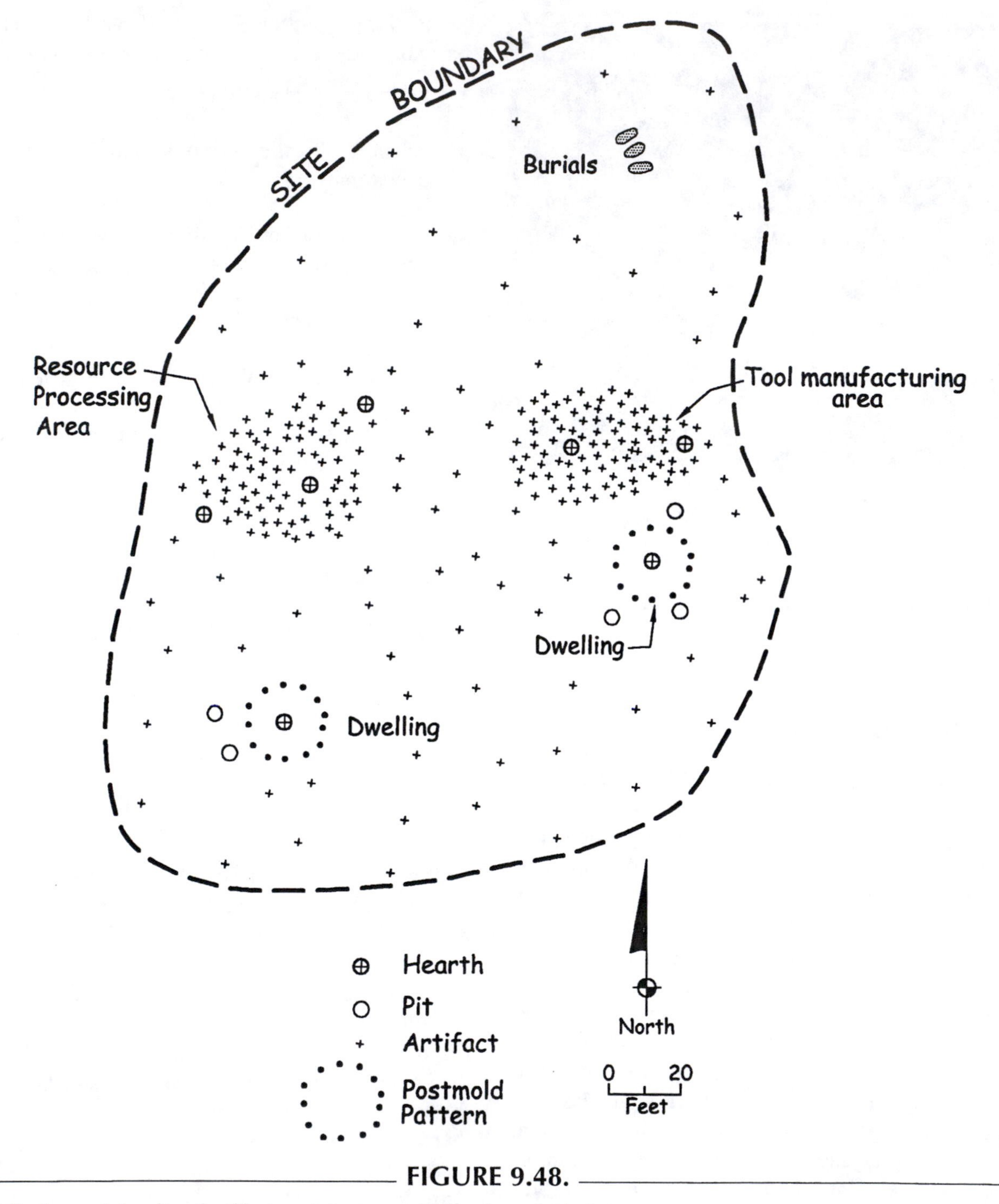

FIGURE 9.48.

The distribution and density of artifacts and features within a hypothetical site.

The size of individual tests and the way in which they are distributed are important considerations, but are intrinsically linked to the size and artifact densities of the sites being sought. Units are much better at detecting low-density sites of a variety of sizes. In contrast, deposits with high artifact densities and uniform density distributions can consistently be detected with small excavations like bucket auger borings or shovel tests. Gridded layouts of excavations, in particular hexagonal ones, are the most effective for intersecting some portion of a site. The spatial interval employed in the layouts in turn influences the size of sites that will consistently be found. Imagine that only the northwestern margin of the hypothetical site shown in Figure 9.50 was located within the area being surveyed. Given all of the considerations that have been noted, the probability of detecting it are much less than if the entire site fell within the project area.

In order to devise the most appropriate subsurface testing strategy, the archaeologists needs to make assumptions about the size and artifact densities of sites anticipated to occur within a project area and make the difficult decision about whether it is feasible to locate all of them. Kintigh (1988:708) echoes the conclusion of many when he states that "executing anything approaching an adequate subsurface testing program for a substantial-sized area generally seems out of practical reach" (p. 708). Kintigh's (1988) example of the level of effort needed to locate small sites is sobering:

> *If one wishes to find all sites on the order of 5 m in diameter, hexagonally placed test units about 4.3 m apart would be needed. Test units 1 × 1 m in size would be required to have an 87 percent chance of finding at least one artifact where the average density is two artifacts/m², to say nothing of the lower densities that appear common. (p. 708)*

The time and cost of implementing a strategy are part of the practical reality that must be faced in making decisions. The investment implied in Kintigh's example would be substantial were the project area large (cf. Carnes et al. 1986; McManamon 1984:262–268). All of the above discussions and simulations of subsurface strategies are based on the subsurface testing of relatively shallow deposits. The practicality of similar levels of subsurface testing flies out the window when the potential for archaeological deposits buried five feet or more below the surface exists (cf. Brown 1975). At these depths the prescribed use of shoring and other safety measures and the time required to conduct any type of controlled excavation dramatically increase the duration and cost of fieldwork. Similar logistical problems arise when examining areas that are currently poorly drained but may not have been so in the past. In the end, strategies for finding archaeological deposits must be flexible and take into account much more than the factors important in assessing the theoretical ability of a technique to locate them.

Decisions regarding the size of excavations, the depth of excavations, and the number and distribution of excavations are based on an evaluation of a variety of information, as summarized in Table 9.2. The types of sites or cultural features being sought influences the size of subsurface tests used in the field. For example, trenching may be more appropriate for locating architectural features, rather than the use of shovel tests or excavation units. If anomalies detected through remote sensing or other means are to be examined, the size of an excavation will have to be adjusted to accommodate the size of the anomaly, or the portion/section of the anomaly selected for scrutiny (Figure 9.49).

a

b

FIGURE 9.49.

Adjusting the size of an excavation unit to examine a subsurface anomaly or feature. a) The feature is detected in a shovel test. A split tube sampler is used to determine its horizontal extent, indicated by the white boundary line. b) The completed unit and excavated feature. The unit was oriented to fully encompass the feature and include adjacent areas of undisturbed subsoil for contrast.

TABLE 9.2

Some Factors Affecting the Use of Excavations for Deposit/Site Discovery

Attribute	Factors
Excavation Size	anticipated range of the size of deposits, associated artifact densities, and density distributions; the type of site or cultural feature being sought; size of anomalies detected through remote sensing, probing, coring, or other means; anticipated depth to which excavations will need to be taken; OSHA standards regarding excavations 5 feet deep and deeper
Excavation Depth	evaluation of geomorphology and degree of soil formation expressed in stratigraphy; depth of water table, bedrock, or other obstructions; state guidelines for archaeological survey
Number of Excavations	sampling strategy; size of survey area; use of systematic testing; geomorphic variability within project area; topography of project area; natural obstructions; size and density of sites/deposits; types of deposits/sites being sought; size and depth of excavations to be completed; surface indications of cultural deposits; number of anomalies detected through remote sensing, probing, coring, or other means; state guidelines for archaeological survey
Distribution of Excavations	sampling strategy; size of survey area; use of systematic testing; geomorphic variability within project area; topography of project area; natural obstructions; size and density of sites/deposits; types of deposits/sites being sought; size and depth of excavations to be completed; surface indications of cultural deposits; number of anomalies detected through remote sensing, probing, coring, or other means; state guidelines for archaeological survey

For specific details on the correlation of systematic types of testing, excavation size, survey area size and shape, site size, artifact densities and density distributions with the probability of site discovery see Kintigh (1988).

Excavation units smaller than 5′ × 5′ cannot be dug to any great depth because of the lack of room to maneuver within them, especially once shoring and ladders have been installed. According to OSHA standards, once an excavation has reached four feet in depth there must be a ladder or other means of egress within it. Excavations five feet deep and deeper must be shored in an appropriate manner. In-field evaluations of geomorphology, stratigraphy, and the estimated age of deposits will determine the depth to which excavations are taken. The goal is to extend an excavation through any sedimentary context that has the potential to contain evidence of human activity. In North America this generally means digging into deposits that are at least 12,000 years old. If evidence of pre-Clevis occupations of the Americas is taken seriously, then contexts 15,000 to 20,000 years old should be sought (cf. Adovasio et al. 1977; Dillehay 1989, 1996; McAvoy 1997). Unless already prepared with appropriate equipment, digging below the water table in a deposit or initiating excavations into waterlogged deposits will not be possible during survey-level investigations.

The number and distribution of excavations employed during a survey reflect the same range of factors. The degree to which systematic subsurface testing is used is a major factor and dependent on a host of other related concerns. Surface finds may largely define the horizontal boundaries of a deposit where geomorphic and stratigraphic information suggest that there is little chance for cultural material to be buried. Excavations are still needed to test for a subsurface component of the surface manifestation and to gather additional geomorphic and stratigraphic data. Accomplishing this with subsurface tests systematically arranged in a grid blanketing the area is excessive in the context of a survey-level investigation. Finds made in tree falls, animal burrows, or bank cuts during surface walkovers can provide clues as to both the horizontal and vertical limits of an archaeological deposit. Again, it may not be necessary to blanket such areas with systematically placed excavations in order to determine the boundaries of the deposit. Some combination of systematic and subjective placement of subsurface tests may be both more appropriate and more efficient. The need for deep testing often precludes the use of systematic strategies as rigorous as those used in seeking shallow deposits, as has been noted. Geomorphic variability within a project area may support systematic subsurface testing in some areas and preclude it in others. Subjectively placed excavations may be needed to adequately document stratigraphic variability across a landscape. Geomorphic and stratigraphic data may indicate that some landforms within a project area are too young to contain any archaeological deposits and don't require systematic testing. Steep topography precludes the need for excavations in general.

The number and location of excavations used to explore subsurface aspects of any type of surface find or feature will depend on the results of surface surveys. The number and location of features or anomalies located through remote sensing, probing, coring, or other means likewise will influence the subjective use of excavations.

A number of states have developed guidelines for the general conduct of archaeological surveys as part of cultural resource management studies. Some contain recommendations about the minimum depth to which an excavation must be dug and the way that environmental settings should be examined, especially floodplains or those considered to have a high probability for the occurrence of archaeological deposits. Guidelines are viewed as a type of quality control for fieldwork, ensuring that a minimal level of research is performed. They are not meant to constrain research designs developed to meet the special conditions of a given project.

More on Types of Excavations and Techniques

Many types of tests are useful during a survey, including probes, auger borings and cores, postholes, bank cuts or vertical exposures, shovel tests, units, and trenches. Trenches excavated with heavy machinery are important in areas where deposits that need to be examined are deeply buried. On landscapes where sediments are dominated by silts and sands and typically devoid of gravel, probes may be used to locate large artifacts and features incorporating rock or any type of solid or resistant material. Probing is best done systematically along transects. Pin the "0" end of a tape into the ground and pull a 100' or 200' tape in the direction desired. Probe at regular intervals along the tape marking locations where "hits" were made with the probe. Use shovel tests or small units to investigate promising locations. The usefulness of probes, auger borings, or cores is heightened once there is an initial indication of an archaeological site. Under the conditions already described, probes can provide some indication of the internal makeup of a deposit viz potential locations of features and artifact concentrations. Systematic coring with a split tube sampler can locate soil anomalies that may be features.

Bucket Augers or Postholers

Bucket augers or postholers may be used as a primary means of searching for archaeological deposits. Postholers are only effective to depths of three feet at most, while bucket augers can be used to examine much deeper deposits. The nature of the devices and their operation means that the provenience of artifacts will be limited to stratum and/or gross depth below surface. Level lines or datum stakes are not used in recording depth. Readings are taken from the existing surface. All excavated matrix is screened. Profiles are not drawn of completed tests because of the limited exposure provided. The stratigraphy is constructed using notes and measurements taken during excavation. Because of the small window into deposits that a bucket auger or postholer provides, even their systematic use on small grid intervals can miss small sites with low artifact densities. It can also be difficult to recognize features unless they are associated with a dramatic artifact concentration or soil anomaly. In contrast, the speed with which a bucket auger or posthole test can be completed relative to a shovel test or larger excavation means that more of them can be excavated, and that smaller intervals can be used in their systematic distribution. They can therefore be an efficient way to locate and define the boundaries of moderate- to large-sized sites, or deposits with moderate to high artifact densities. Bucket augers also furnish a look at deep deposits that may not be possible or practical to examine through the excavation of units during a survey.

Cut Banks

Cutting down banks and examining any available vertical exposure provides information on the stratigraphy associated with a landscape and an opportunity to recover artifacts from subsurface contexts. Cut banks may be one of the few ways to safely and efficiently search for artifacts and features five feet and more below the surface in survey situations where materials for shoring traditional excavation units are not available. When choosing where to cut down a bank or exposure, select a spot where the slope is as steep or near vertical as possible, and the horizontal distance between the top of the slope and the toe of the slope is as small as possible. This will lessen the amount of slumped sediments that you will have to remove in order to create a truly vertical exposure revealing undisturbed stratification. Avoid areas near large trees since their roots will impair your work and muddle the view of the stratigraphy. Begin your cut at the top of the slope (Figures 9.50–9.52). You won't be screening the sediments that you remove in making the cut, but watch for artifacts or features as you go and try to relate any encountered to a specific stratum. The cut should be at least shoulder width (about 2.5 feet) so that you can work comfortably within it when it is completed. Creating a single vertical face is suitable for short banks five feet or less in height with a relatively steep slope. For longer banks, or those with moderate or slight slopes, the cut can be stepped with a series of vertical faces. The side walls of the cut reveal the interface between slumped sediments and intact deposits and can be used as a guide to determine where to create steps and how long to make them. The steps make it easier to work with the different sections of the exposure. They also provide a measure of safety, allowing easy movement in and out of the cut. Stepped cuts can be fashioned so that no matter on which step a person stands, his/her head is always above the surface of the slope, and at the deepest part of the cut, a person can simply walk out of it should trouble arise.

Clean and record the profile of the vertical face(s) exposed (see chapter 7). Excavation into the face of the cut can be organized in the same way as that of a typical unit, i.e., by natural strata, arbitrary levels within natural strata, or strictly arbitrary levels depending on the nature of the stratigraphy revealed in the cut. Decide how far into the face of the cut you want to go (e.g., 0.5', 1.0'). All excavated matrix is dry screened. Begin at the top of the exposure. With a trowel, thinly slice the face of the first stratum or level that you have defined. You will continue to vertically slice the exposed face farther back into the bank. At the outset you will need to use a dust pan positioned below each slice made to collect the matrix that is removed. As your progress into the face continues, matrix will accumulate on the small "level floor" or shelf that results from the excavation, making it easier to handle. Subsequent levels or strata can be removed through the slicing of both the vertical and horizontal faces of the exposure (Figure 9.53). Since a relatively small horizontal area is being examined in this process, the horizontal provenience of artifacts is not usually recorded. This excavation technique makes it very easy to see the vertical position of artifacts within a stratum or level. Depths or relative elevations can be recorded off the same level line used to record the profile of the cut bank.

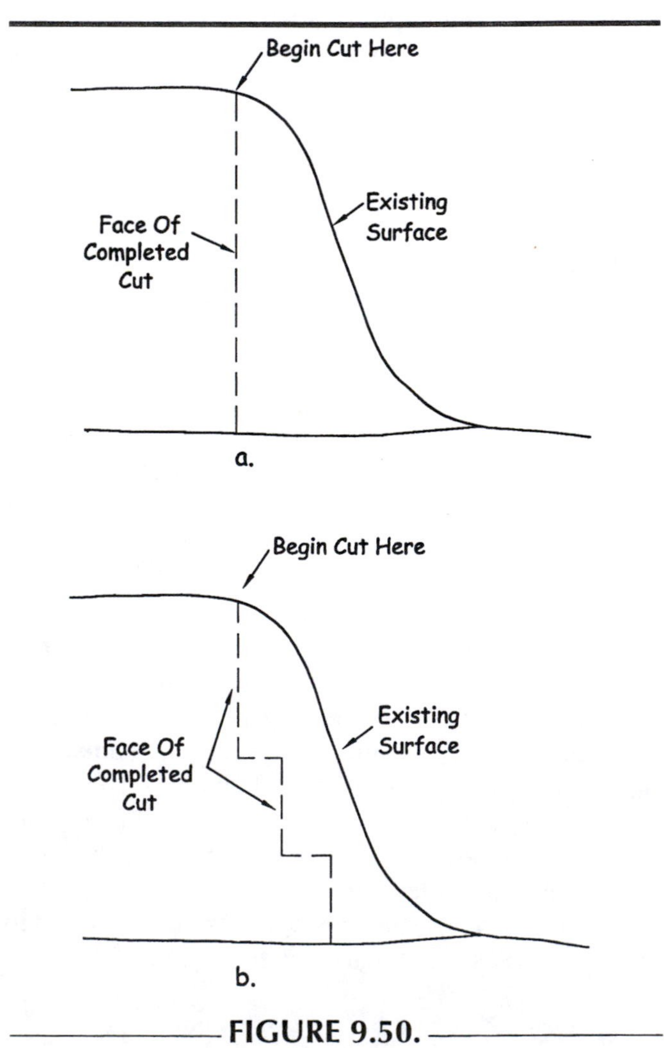

FIGURE 9.50.

Different approaches to cutting down banks for examination. Creating a single vertical face (a) is suitable for short banks five feet or less in height with a relatively steep slope. For longer banks, or those with moderate or slight slopes, the cut can be stepped with a series of vertical faces (b).

FIGURE 9.51.

A vertical face cut through a relatively short bank. The cleaned profile provides valuable information on stratigraphy and the geomorphology of the landscape. Once the profile has been recorded, the face of individual layers can be peeled with a trowel in search of artifacts and features.

FIGURE 9.52.

Stepping a cut through a long bank. The side walls of the cut reveal the interface between slumped sediments and intact deposits and can be used as a guide to determine where to create steps and how long to make them.

Shovel Tests

Shovel tests (Figure 9.54) are probably the most frequently used form of subsurface testing during an archaeological survey. Although relatively uniform in their small size, their shapes vary from round to square depending on the practices of a given organization. My preference is for square shovel tests because I think they make it easier to read the profile being exposed, a process already made difficult by the excavation's small size. Whether round, oval, or square, it is important to be consistent about the size of the test excavated if you're going to use generated data (e.g., artifact densities, density distributions) to make assumptions about the larger deposit and the patterning of remains within it.

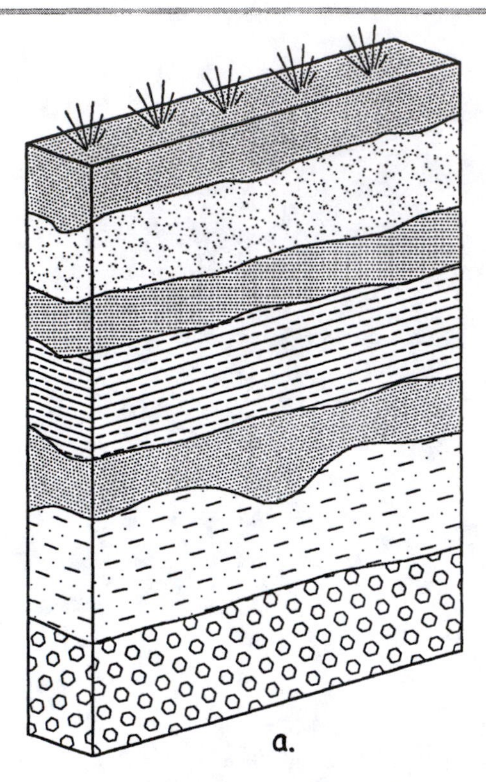

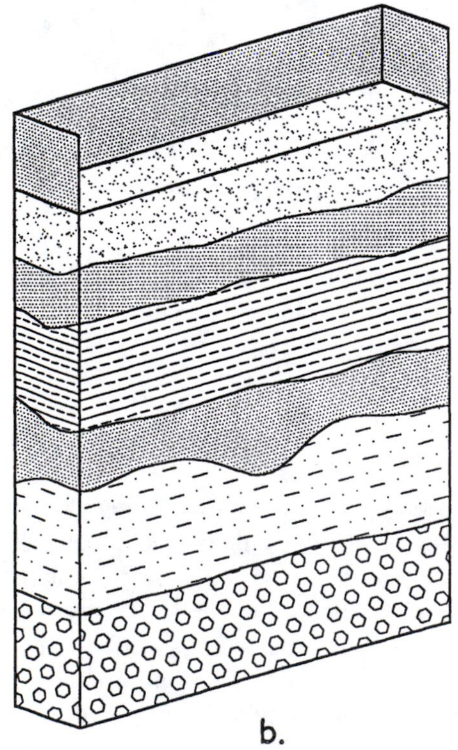

FIGURE 9.53.

Excavating the face of a cut bank or vertical exposure: a) the cleaned and profiled face; b) the face after the first level or stratum has been excavated by vertical peeling or slicing, creating a "level floor" or shelf. Subsequent levels or strata can be removed through the slicing of both the vertical and horizontal faces of the exposure.

FIGURE 9.54.

A typical shovel test under excavation. Because of their small horizontal dimensions, shovel tests can rarely be excavated more than three feet below the surface.

The excavation of a shovel test is a compromise of sorts reflecting the need of archaeologists for a relatively quick and yet reliable means of learning about subsurface deposits. Their shape and dimensions are established using a tape and the estimations of the excavator. There is no formal procedure for laying out a shovel test the way there is for an excavation unit or trench. Excavation typically proceeds by natural strata and all matrix removed is dry screened. Gross arbitrary levels (ones thicker than what might otherwise be employed in the excavation of a unit) are used where preliminary coring indicates that some strata are very thick or obvious stratification is lacking. Attempts to remove thin and tightly controlled arbitrary levels are eschewed in such a small excavation. Excavation relies most heavily on the shovel and not the trowel. There is generally no attempt to map any artifacts encountered by point provenience; stratum or level provenience is deemed to be sufficient. Depths are recorded with a tape from the existing surface. A level line or elevation datum stake is not established. Attention must still be directed toward the side walls of the test as it would in the excavation of a unit. Unfortunately, this can be forgotten given the perception of the shovel test as an expedient type of excavation. I've seen too many cases where excavators have blasted through a feature because of inattention to the side walls and their zeal to complete the excavation. Place bucket auger borings in the base of shovel tests where there is a need to examine deposits beyond the reach of the shovel test.

When a feature is recognized in a shovel test, excavations should be halted, and a decision made about whether to expand the test into a unit in order to better define the feature. If the feature is a sediment/soil anomaly, a split-tube sampler or small auger can be used in radial patterns around the shovel test to determine its horizontal size and shape. Features that contain solid materials can be traced with a probe. If the excavation is not to be expanded, complete the paperwork for the test and close it. It is foolish to attempt to section or study a feature solely in the context of a shovel test. Placing a piece of plastic at the top of the feature prior to backfilling will make it easier to relocate during future fieldwork.

Shovel tests may be impractical for exploring areas where sediments are extremely hard, compact, or gravelly. It is difficult to use the tools necessary to deal with these matrices, like picks and mattocks, when confined to such a small space. Using shovel tests in urban settings or areas where placed fill may be encountered is also problematic.

Small Units

Small units (3′ × 3′ or meter squares) may be used as the primary excavation when searching for sites during a survey, or used in concert with shovel tests. Their horizontal dimension increases the probability of encountering artifacts associated with low-density deposits. Their size also allows them to be excavated to greater depths (5 feet below surface) than shovel tests without the need for shoring side walls. Multiple bucket auger borings or excavations with a postholer can be placed in the basal level of units to extend searches into much deeper contexts. All excavated matrix is typically dry screened. Point provenience excavation techniques are rarely employed in general excavations given the exploratory nature of an archaeological survey. Recording spot elevations and coordinates for select artifacts is a compromise between expediency and the need for more detailed information in the interpretation of cultural stratigraphy and site formation processes.

Large Units

Units with larger dimensions, used in conjunction with shoring for side walls, permit controlled level excavations to advance to considerable depths (Figure 9.55; also see Figure 5.12). I have had the unique pleasure of participating in the manual excavation of a 5′ × 7′ unit to a depth of 23 feet below the surface (Stewart et al. 1991). Prior to installing any type of shoring, the profiles of side walls must be recorded and any relevant samples collected. With shoring installed, the excavations are taken deeper until additional shoring is required. Again, side walls are profiled and needed samples collected prior to installation. When the excavation is completed the shoring is removed from the bottom upwards.

Variably sized units and trenches may be used to search for subsurface features or anomalies indicated by background research, surface survey, remote sensing, or the results of shovel testing, coring, or augering. Situate units in areas where shovel tests reveal high artifact frequencies and layered deposits in order to get a better perspective on the natural and cultural stratigraphy. Strategically place units to better document the range of stratigraphic variability indicated by shovel tests, coring, or augering.

FIGURE 9.55.

The top of a deep test unit. The unit is surrounded with a small fence to keep people and equipment at a safe distance. One of a series of ladders to be lowered into the excavation is laying inside of the fence. Visible in the upper portions of the unit is plank and timber shoring bracing the side walls.

Mechanical Excavations

Heavy machinery, typically backhoes, plays an important role in the search for deeply buried deposits during surveys. The opening of deep (10′–20′) trenches whose side walls are then inspected and excavated to some degree is the customary way of using a backhoe. The use of heavy machinery is a compromise. While a backhoe trench may lead to the discovery of deeply buried archaeological deposits, its excavation disturbs that deposit. Machine-cut trenches should be kept to a minimum. Selecting the locations to place machine-cut trenches must be done with care, taking into consideration what is already known about the geomorphology and stratigraphy of the landscape, what other techniques will be used to search for deeply buried deposits, and the size of the area where deeply buried deposits may exist.

A backhoe trench can be opened in a way to salvage artifacts and features, and gather information relevant to cultural stratigraphy. Determine the location of a trench and position the backhoe. With a split-tube sampler or auger get a look at the natural stratigraphy in the upper 2.5 to 3.0 feet of the deposit. If the natural strata are not too thin, a skillful machine operator, under your direction, can strip the trench layer by layer. Review your plan for dissecting the trench with the machine operator. Arrange to dump the sediments stripped from individual layers in discrete piles. In this way you or other crew members can trowel through them or screen samples to look for artifacts. Periodically halt the stripping and check the floor of the trench for any signs of features. If any are found the machine stripping of the trench should be halted, leaving the exposed feature to be handled in the standard, controlled fashion. Reposition the backhoe at a new trench location and begin the process again. Otherwise, use a split tube sampler or auger to recheck the stratigraphic changes to come and continue with the stripping. As the machine-cut trench approaches five feet in depth there are some procedural options to consider. One approach is to temporarily halt the excavation to clean and profile one of the long wall side walls of the trench. At this depth protective systems for the trench are not required by OSHA (if sediments in the deposit are deemed to be stable) and the work can proceed unencumbered. One or more sections of the side wall face may be excavated at this time using the vertical slicing or peeling technique employed for cut banks. With these tasks completed the stripping can continue to the desired base of the trench, protective shoring installed, and the lower sections of the side walls cleaned, profiled, and sampled with excavations into the vertical face. An alternative is to first complete the stripping of the trench to its desired base, put protective measures in place, then follow up with the wholesale documentation of the profile and selective excavations. This second approach may make more efficient use of the backhoe and operator. Once the trench is completely cut, the archaeologist, operator, and machine can move on to the next trench while other members of the field team work in the open trench.

There are a variety of protective systems or shoring that can be used in a backhoe trench. Shoring can be constructed on-site using wood, or combinations of wood and trench jacks, following OSHA standards. Shoring systems also can be purchased or rented. Shoring can be installed to leave sufficient portions of side walls exposed for study (see Figures 5.12 and 5.15). Trench boxes (Figure 9.56) are a more expensive protective system more frequently used when a deep excavation will be open for an extended period of time. They are constructed entirely of metal and are lowered into an excavation by machine when the excavation has reached an appropriate depth. Two sides of the trench box are open permitting views of excavation side walls. Any excavations deeper than 20 feet, whether done with a machine or by hand, must be designed by a professional engineer according to OSHA regulations.

During survey stages of fieldwork, safety standards for deep excavations are followed haphazardly by archaeologists. You can still see workers blithely diving into 15-foot-deep backhoe trenches that are unshored and only three feet wide. If you are intent on entering without shoring, do so with extreme caution. Wait several minutes after the trench has been cut to allow for settling and to see if the side walls are going to slump or collapse as soil moisture is lost, or if loose, deeply buried strata slough into the open excavation. Make sure that groundwater is not flowing into the excavation. Hard hats should be worn as protection from falling debris and one or more ladders should be handy. Never enter a deep excavation, shored or unshored, without someone standing by.

EXPANDING YOUR PERSPECTIVE ABOUT . . . FIELD RECORDS AND DATA COLLECTION

Here I will only mention elaborations on the guidelines and suggestions presented in earlier sections of this book. Chapter 8 reviews the various observations that should be made during any field operation and the discovery of a new archaeological deposit or site.

a

b

FIGURE 9.56.

Trench boxes, another system used for safely working within deep excavations: a) an above-ground view; b) a trench box positioned within an excavation.

Figure 9.57 is an example of a double-sided form used for recording data from shovel tests and miscellaneous excavations like cut banks, cores, auger borings, postholes, or cuts made with heavy machinery. The template for recording stratigraphy is helpful since the profiles of most of these types of excavations will not be formally drawn in the way prescribed for units and other extensive vertical exposures. Profiles for some cut banks and trenches opened with heavy machinery will be recorded more formally and on separate sheets. A comparison of this data collection sheet with that used for units (see Figures 9.42 and 9.43) is further illustration of the expedient nature of shovel test excavations.

It is advantageous to construct and maintain maps of the distribution of bucket auger borings, and posthole and shovel tests while in the field. Use the maps to plot the distribution and frequency of finds, data that will figure into decisions about how to deploy unit and trench excavations during the survey. Multiple maps will be needed for the analysis of stratified archaeological deposits.

Collect any and all samples from an excavation that could be used for chronometric dating, emphasizing those whose context is not in doubt. Column samples from at least one representative unit should be collected, as should samples for sediment/soil analysis. Flotation samples should be taken from any feature excavated and should be supplemented with samples from contexts immediately above and below it. Column samples, soil samples, and samples for wet screening or flotation are not systematically collected from all excavations during a survey.

If periodic trips to the laboratory are made during the course of a survey, completed data forms should be sent along with any artifacts and samples collected. The forms should be copied with one set sent back into the field for continued reference. Samples for flotation or wet screening are rarely processed in the field and so will usually be carted back to the laboratory. Remember to open sample bags for airing and gradual drying if they won't be processed in the near future.

SHOVEL TEST AND MISCELLANEOUS EXCAVATION RECORD

Project____________________________________**Site or Deposit Designation**____________________________________
Shovel Test____**Core**____**Auger Boring**___**Probe**___**Strata Cut**___**Machine Cut**___**Dimensions**__________________________
Location____**Recorder**______________________________**Date**_____________
Excavation Method____
Screened? Yes____ No____All?____Sample? (note size)________Dry screened?____Wet screened?____ Screen mesh size(s)_____

Use the template below to record stratigraphy. Describe the thickness and depth of each stratum, its color and texture. Note artifact finds, their type and number. Provide additional comments as necessary.

Scale: 1" = 1.0 foot

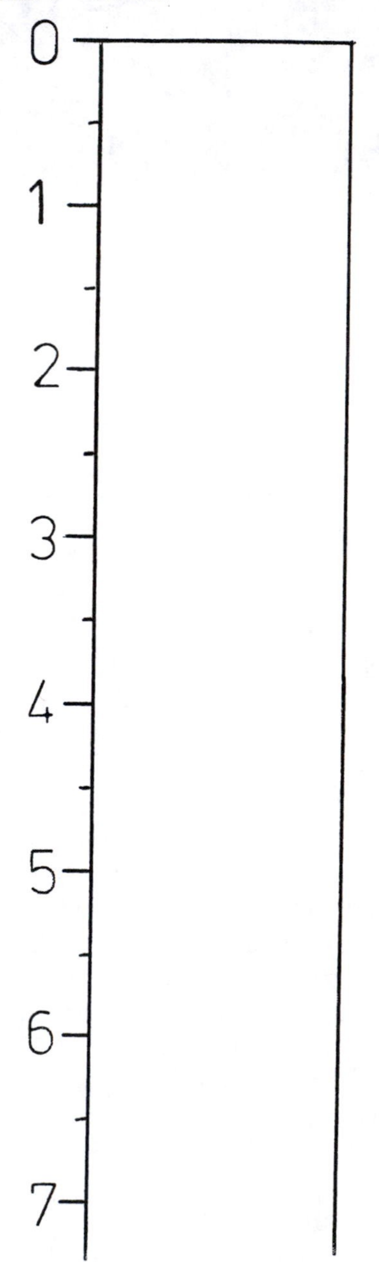

FIGURE 9.57.

An example of a pre-printed, double-sided data collection sheet used in conjunction with shovel tests and miscellaneous types of excavations.

SHOVEL TEST AND MISCELLANEOUS EXCAVATION RECORD

Project________________________________**Site or Deposit Designation**____________________________
Shovel Test___**Core**___**Auger Boring**___**Probe**___**Strata Cut**___**Machine Cut**___**Dimensions**____________________
Location___________________________________**Recorder**_______________________**Date**__________
Excavation Method____
Screened? Yes____ No____All?____Sample? (note size)_______Dry screened?____Wet screened?____ Screen mesh size(s)_____

Use the template below to record stratigraphy. Describe the thickness and depth of each stratum, its color and texture. Note artifact finds, their type and number. Provide additional comments as necessary.

Scale: 1" = 1.0 foot

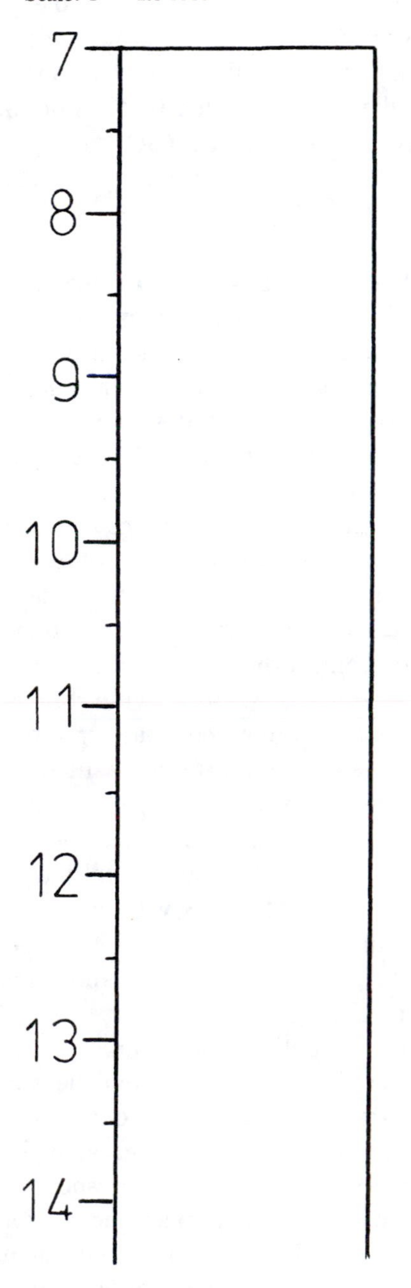

FIGURE 9.57

continued

EXPANDING YOUR PERSPECTIVE ABOUT . . . HEALTH, SAFETY, AND GENERAL LOGISTICAL CONCERNS

Since mobility is key during an archaeological survey remember to carefully plan for your personal needs. Be conservative with what you pack and carry into the field. Advance planning is critical if you will use heavy machinery, excavation shoring, or other equipment to deal with special conditions (e.g., water pumps and the generators to run them). Further, the areas where you want to use these items will have to be relatively accessible by vehicle. Review OSHA guidelines for shoring excavations and trenches before going into the field. Contact an OSHA representative when in doubt about the applicability of rules to archaeological excavations. There are illustrated published summaries of OSHA standards regarding excavations (e.g., Keller and Associates 1997) that are often easier to use than wading through the actual government regulations. However, since OSHA standards are subject to revisions, it is always good to refer to actual regulations (which can be read online) when in doubt about a procedure.

It will be necessary to leave some excavations open overnight. These should be covered both to protect the unexcavated contexts from the weather and to prevent animals and the unwary from stumbling into them. Rolls of sheet plastic are typically used for this. It may not be possible to fence these excavations but tall stakes can be placed around them and connected with strips of brightly colored flagging.

FURTHER INVESTIGATIONS OF BURIED DEPOSITS

While flexibility remains the byword for field strategies, there are both subtle and not so subtle changes in how excavations are used as deposits become the focus of increasingly more intensive study. The discovery phase of fieldwork may be combined with, or followed by investigations designed to learn more about the characteristics of a deposit and how research issues of importance might be addressed with data from future excavations. This corresponds with the determination of eligibility stage of a CRM project. The most intensive studies are reserved for sites whose research value is exceptional. Sites threatened with destruction may also be examined in detail, regardless of their perceived research value, in order to compile a documentary record for the future. Excavations on sites that are both significant and threatened constitutes the mitigation phase of fieldwork in the context of CRM (see chapter 3 for a discussion of the organization of fieldwork).

The ways in which excavations are used builds upon the methodology and results of the discovery phase of fieldwork (Table 9.3). For testing programs the goal remains one of designing research that collects needed data while disturbing as little of the deposit as possible. When intensive studies are motivated by the imminent destruction of a site, or as part of a CRM mitigation, disturbance ceases to be an issue. Intensive excavations designed solely for research purposes need to take the preservation ethic to heart.

Locating and examining a sample of features in detail is one goal of any testing program. These contexts often produce data important for determining the age and nature of occupations. Features may also be the one context where perishable types of artifacts are found. Excavation size will therefore need to be adjusted to the size and configuration of features. The type of features known or anticipated to occur have implications for the internal structure or patterning of evidence within a deposit. For example, activities may be focused around hearth areas. Architectural remains imply a relatively long-term occupation that will undoubtedly be associated with a wide range of domestic activities and their material fingerprint. Archaeological site reports for a region will provide examples of the range of known site types and associated patterning. In part, sampling strategies for testing and intensive studies are developed to reveal the internal patterning of a deposit. The discovery of a particular type of feature may result in a modification of the sampling strategy because of what the feature implies about the nature of a deposit.

The number and distribution of excavations used to test a deposit reflect the degree to which the survey effort defined its horizontal and vertical boundaries and provided information about its age, internal makeup, and research potential. When testing is a response to a threat to a site or part of a CRM project, the portion of the site that is in danger, falls within the project area, or will be most severely impacted by proposed construction will affect where excavations are placed. Sampling strategies must also take into account the geomorphic variability that exists within the site.

TABLE 9.3

Some Factors Affecting the Use of Excavations for Testing and Intensive Study

Attribute	Factors	
	Testing	**Intensive Study**
Excavation Size	known depth of deposits; OSHA standards for excavations 5 feet deep and deeper; size of known or anticipated features or activity areas	known depth of deposits; OSHA standards regarding excavations 5 feet deep and deeper; motivation for investigation (e.g., pure research, salvage, CRM); type of site; research issues to be addressed; nature of activity areas to be investigated and related; known and anticipated size of features to be exposed and related; sampling strategy for strata to be stripped
Excavation Depth	determined from previous work	determined from previous work; depth of proposed impacts (salvage of CRM)
Number of Excavations	size of deposit/site; type of deposit and internal variability anticipated (e.g., artifact densities and distributions, feature types and numbers); size of impact area (salvage or CRM); number, size, distribution and results of previous excavations; sampling strategy	size of deposit/site; type of deposit and internal variability; motivation for investigation (pure research, salvage, CRM); type of site; research issues to be addressed; nature of activity areas to be investigated and related; known and anticipated size of features to be exposed and related; number, size, distribution, and results of previous excavations; sampling strategy; size of units, trenches, or blocks to be employed; sampling strategy for strata to be stripped
Distribution of Excavations	size of deposit/site; type of deposit and internal variability anticipated (e.g., artifact densities and distributions, feature types and numbers); size of impact area (salvage or CRM); number, size, distribution and results of previous excavations; sampling strategy	size of deposit/site; type of deposit and internal variability; motivation for investigation (pure research, salvage, CRM); size of impact area (salvage, CRM); type of site; research issues to be addressed; nature of activity areas to be investigated and related; known and anticipated size of features to be exposed and related; number, size, distribution, and results of previous excavations; sampling strategy; size of units, trenches, or blocks to be employed; sampling strategy for strata to be stripped

With the outline of the physical characteristics of a deposit worked out through the process of survey and testing, research issues have a bigger impact on sampling strategies and the use of excavations during an intensive study. For example, intrasite areas that may relate to pottery production may be emphasized over those tied to the manufacturing of stone tools because the former is poorly understood in the region. An archaeologist seeking to examine economic and social differences between households in a village will want to concentrate more on dwellings and associated deposits than other areas of the village. These interests will affect strategies of how many excavations to use and where to put them. The size of related features also will dictate the size of the excavations needed to study them. In general, there is a greater focus on exposing the entirety of activity areas and features, documenting spatial relationships and contemporaneity over broad horizontal areas. Block and large-area excavations are common for these reasons (Figures 9.58a–9.61). Large-area excavations and large samples of sites are appropriate in many salvage situations and CRM mitigations given that the sites will ultimately be destroyed. Larger sample sizes, in turn, influence the number and distribution of excavations to be used.

When artifact-bearing strata are mechanically stripped to gain access to deeper deposits of greater interest the strata to be stripped must be sampled. As with an archaeological survey, the size of the units, their number, and distribution employed in the sample should reflect known or anticipated artifact densities and density distributions. It is difficult to rationalize the stripping of artifact-bearing deposits unless a site is threatened. In CRM, such stripping may be a part of the overall plan to mitigate the adverse impacts of a proposed construction project on a site through data collection. The extent of proposed impacts to a site in a salvage or CRM situation may also limit the depth to which excavations are taken. A deeply stratified archaeological site to be impacted by the construction of a parking lot may only be examined to the depth of the impact. This approach is easily rationalized in salvage situations where funding and time for fieldwork is limited and compliance with historic preservation laws is not an issue. It becomes a trickier issue in compliance situations.

FIGURE 9.58.

The process of exposing and tracking a Native American trench feature at the Playwicki Farm Site, Pennsylvania. The feature was first exposed in the unit in the foreground. Additional excavations were positioned to follow the feature.

FIGURE 9.59.

Sequential unit excavations at Playwicki ultimately revealed that the trench feature is nearly circular, 40 feet in diameter, and probably represents the outer wall of a large structure.

FIGURE 9.60.

Subsequent excavations expose the interior of the Playwicki structure for detailed study.

FIGURE 9.61.

An aerial view of large block excavations at the Shermans Creek Site, Pennsylvania. Cross sectioned postmolds defining a portion of an oval-shaped structure can be seen at the edge of the block on the right. Just above it are two large pit features. Horizontally extensive excavations are typical of intensive investigations of sites, especially those done in compliance with historic preservation laws.

As ever, available funding and personnel exert a variable influence on technical and theoretical issues for all levels of effort. State-based guidelines don't dictate specific practices for testing and intensive study as they do for archaeological surveys. Rather, they adhere to the intent of federal regulations regarding the need to determine the significance of a deposit and develop a research design that reflects its potential.

EXPANDING YOUR PERSPECTIVE ABOUT TYPES OF EXCAVATIONS, EXCAVATION TECHNIQUES, AND PROVENIENCE RECORDING SYSTEMS

During testing, shovel tests or other small excavations may be useful for "gently" detailing artifact patterning within a site. Systematic probing or the extraction of small diameter cores may be able to quickly locate features. Units or trenches are then used to examine select portions of the patterning revealed.

Interpretations of natural and cultural stratigraphy made possible by excavations completed during the discovery phase of fieldwork will determine how strata and levels are removed during testing. For example, strata initially removed separately because of differences in the particle size of sediments may prove to be part of the fining upwards sequence of a single alluvial deposit. They may be combined as a single level in subsequent excavations. Testing may lead to additional refinements in the natural and cultural stratigraphy that will be reflected in excavation strategies employed during intensive excavations. Reopening select older excavations is a practice used during testing and intensive studies to provide stratigraphic controls and references for ongoing fieldwork. Additional units or block excavations may expand off older excavations to follow up on the finds made there.

Spatial controls must be maintained regardless of the variably sized units, trenches, and blocks opened during the testing and intensive excavation of a site. Large units or trenches are subdivided into smaller sections that become themselves the focus of excavation and provenience control (Figures 9.62 and 9.63). Provenience controls (horizontal and vertical) and recording systems should reflect the integrity of a context or deposit. A buried surface with the remains of a single occupation is deserving of tighter controls than one containing multiple occupations spanning a thousand years. It makes sense then that intensive studies of significant sites customarily employ the highest degree of provenience control; integrity of context is an attribute of significance. For some testing and intensive excavations, techniques for mapping artifact locations and elevations may shift from hand measurements with tapes, unit-specific datums, and line levels to surveying instruments like total stations. The use of this equipment is more feasible during prolonged fieldwork at single locations where there will be a high volume of mapping to be done than it is during site discovery projects. Where datum stakes are relied upon to document elevations within excavations care must be taken in their location. Since excavations may be clustered or organized in blocks a single datum may serve multiple units.

Balks that preserve the stratigraphy between contiguous or clustered excavations are important in situations where deposit stratigraphy is complex and can vary dramatically over short distances (Figure 9.64). They provide a guide and reference for ongoing excavations and improve the ease and reliability of making correlations. The use of balks is relatively common in the excavation of mounds, earthworks, rockshelters, sites with numerous architectural remains and episodes of construction.

Heavy machinery may be used to remove fill or culturally sterile strata that cap archaeological deposits to be tested or intensively studied. The choice of machine depends on the amount of material to be removed, the depth to which stripping will take place, and the characteristics of the sediments being stripped. Where sediments are sandy, loose, or relatively wet, the movement of heavy machinery back and forth over the landscape during stripping can compress and distort the archaeological deposits. Draglines, gradalls, or long-armed backhoes can strip a deposit without having to ride over it. Draglines are particularly effective when large volumes of materials must be removed (Figure 9.65). Where sediments and soils are stiffer, bulldozers or backhoes with front-end loaders can be used to strip deposits. Tracked, rather than wheeled vehicles are the most effective in this type of stripping.

FIGURE 9.62.

A small block measuring 10 feet on a side has been subdivided into 16 small squares. Each of these squares is excavated and documented individually to enhance spatial control on the artifacts found in the deposit. One of these has already been excavated and a second is in progress.

FIGURE 9.63.

A large block has been gridded into meter squares for excavation and provenience control.

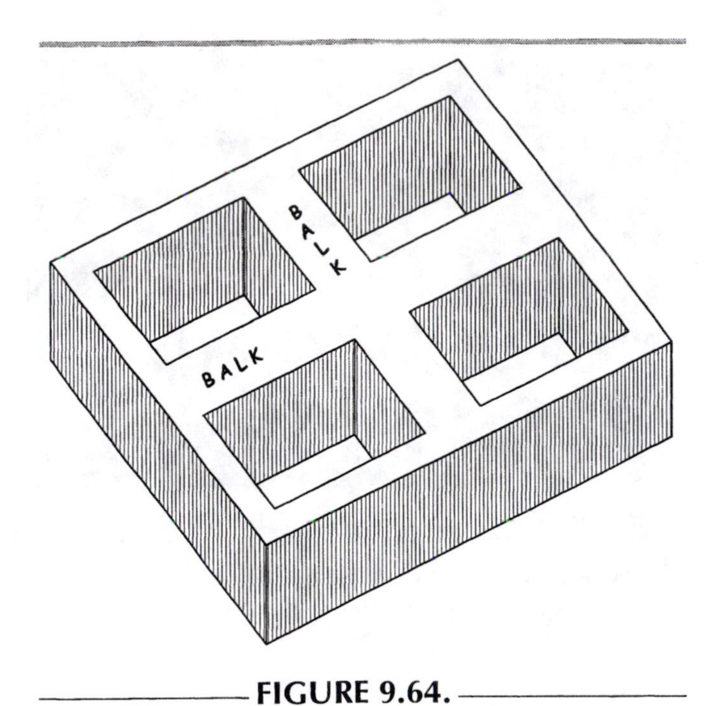

FIGURE 9.64.

Balks (unexcavated sections) between excavations preserve stratigraphy for ongoing reference and promote correlations in complex stratigraphic situations.

FIGURE 9.65.

A dragline in the early stages of stripping 10 feet of culturally sterile sediments off the top of an archaeological deposit. A dragline can strip a site without having to drive over it, averting damage to intact archaeological deposits.

Deposits may be stripped over broad areas even when they do contain artifacts. This happens most frequently in salvage situations and CRM mitigations and always requires sampling the stripped deposit. A classic example is the large site where the majority of artifacts are in a plowzone and the only thing below the plowzone are the unplowed remnants of features. A controlled surface collection and systematically placed excavations derive a sample of artifacts from the plowzone. The plowzone is then stripped off by machine, the resultant surface scraped clean with flat shovels and trowels, and examined for evidence of features. The distribution of features is mapped and units are positioned to excavate them. This procedure allows very large areas to be examined, areas large enough that systematic coring or remote sensing are not practical alternatives for locating features. A dramatic example is the Carey Farm and Island Farm sites in coastal Delaware (Custer et al. 1996). The plowzone was stripped from 25 acres of these contiguous localities revealing 835 cultural features, many of which were the remnants of pit houses. The process of sampling and then stripping artifact-bearing strata is also used to get at more deeply buried archaeological deposits. Mechanical stripping of a site may occur at the conclusion of an intensive excavation in a CRM or salvage situation as a means of squeezing the last bits of information from site deposits. The incremental stripping of unexcavated areas can provide a grab sample of artifacts and reveal additional features that can be mapped, if not excavated.

Opening large areas or block excavations into deeply buried archaeological deposits depends upon the use of heavy machinery and can be problematic because of safety concerns and OSHA standards. Protective systems that brace excavation side walls are cumbersome, expensive to use, and require some modifications of traditional excavation practices (cf., Bergman and Doershuk 1995). Sloping, stepping, or benching a large block excavation down to a smaller-sized one are alternative approaches. For example, one might strip four feet of deposits from an area 100 feet on a side in order to locate an excavation block 50 feet on a side in its center and dig archaeological deposits to depths of eight feet below the original surface.

Intensive excavations provide more opportunities and need for wet screening, the use of smaller mesh in screens, and the flotation of excavated matrix, although adjustments are made depending upon the type of site and the intensity of the excavations. The use of 1/4″ mesh in screens is simply not adequate to recover a reliable sample of the floral and faunal remains that might be preserved in a deposit. The use of 1/8″ mesh improves the recovery of animal bone, especially that representing smaller animals (see Sutton and Arkush 1998:245 for summary and additional references). Wet screening speeds up the process of using a smaller-sized mesh.

Flotation improves the recovery of all manner of micro artifacts, floral and faunal remains not the least. There is a variety of flotation systems in use by archaeologists, ranging from a simple hand-held washtub with its bottom removed and replaced with fine mesh screen, to mechanized self-contained devices that recycle the water or fluid used in the process (cf. Bodner and Rowlett 1980; Gumerman and Umemoto 1987; Pearsall 1989; Ramenofsky et al. 1986; Struever 1968; Toll 1988; Wagner 1982). Each makes use of some basic principles (Figures 9.66–9.69). A sample of excavated matrix is placed in water or a combination or water and chemicals that enhances the separation of materials. The solution is agitated in some way, separating sediments from other materials in the matrix and allowing lightweight materials (especially floral remains) to float to the surface. This ***light fraction*** is skimmed off the surface or otherwise collected. Once the sediments in a sample have been dispersed, the ***heavy fraction*** is collected from screens at the base of the apparatus. Since numerous flotation samples are generated during intensive excavations, processing in the field precludes the need for storage, transport, and additional handling in the laboratory.

FIGURE 9.66.

Flotation barrels in use in the field. This particular system requires a water hose to be attached to the side of each apparatus.

SPECIAL FEATURES AND DEPOSITS

How to deal with the special features, conditions, and deposits that can be encountered in the field is impossible to teach in an introductory textbook. It is best learned from reading site reports where archaeologists describe their methods and rationale for using them, and from supervised field experience. I'll touch on some basic themes here and provide references to direct further reading.

HUMAN REMAINS

The excavation of human remains is not to be taken lightly (see chapter 1). There is no question in the minds of many archaeologists that important information can be gained from their excavation and study (cf., Beck 1995; Chapman et al. 1981; Katzenberg and Saunders 2000; Larsen 1997; Schwartz 1998), but it is not the central, nor the only issue that must be considered. When human remains are expected to be encountered, negotiations with members of descendant communities and any relevant legal authorities can take place before fieldwork begins. If the excavation of human remains is to be allowed then protocols for the continued interaction of interested parties, excavation, removal, study, and the

FIGURE 9.67.

Interior view of a flotation barrel prior to filling. A removable fine mesh screen will be laid over the quarter-inch screen attached to the barrel. The flotation sample is then poured into the tub. Jets of water from the PVC piping below the screen fill the tub with water and agitate the flotation sample.

FIGURE 9.68.

As sediments are separated from other materials in the flotation sample, the light fraction floats to the surface and along with water begins to flow down the drain channel at the top of the barrel.

FIGURE 9.69.

The light fraction from the flotation sample is caught in a fine-mesh bag attached to the drain channel. The heavy fraction will be collected from the screens inside of the barrel once the water is turned off and the barrel partially drained.

dissemination of results can be developed. Excavations may go forward under special circumstances, e.g., with on-site monitors, attendant ceremonies, in-field analysis of remains, and immediate reburial. Excavations aside, ethical dilemmas surround the care and treatment of human remains and associated mortuary objects (e.g., McGowan and Laroche 1996). Unplanned-for discoveries of human remains require that similar consultations and negotiations eventually take place.

A basic familiarity with the human skeleton (Figure 9.70) and the ways in which human remains can be treated after death will allow you to recognize remains and related features. Depending on how a body was treated before burial and the nature of the burial environment, even soft tissue may be preserved. There are a number of identification manuals that are handy to have in the field (e.g., Bass 1995; Bennett 1993; Metress 1989; Schwartz 1995). Other texts provide guidelines for identifying skeletal elements and making critical observations of bone and related material, and include discussions of types of internments and excavation techniques (e.g., Brothwell 1981; Brown 1971; Mays 1998; Ortner and Putschar 1985; O'Shea 1984; Reichs and Bass 1998; Rogers and Waldron 1995; Sprague 1968; Stewart 1979; Ubelaker 1999). Ethnographic and historical accounts are also invaluable for understanding and anticipating the archaeological expression of mortuary practices.

There are two general categories of internments or burials. ***Primary burials*** are those where remains are found in the initial context in which they were placed after death. ***Secondary burials*** are those where all or portions of the human remains from a primary internment are collected and reinterred in a different context. Secondary internments contain nonarticulated collections of bones, i.e., skeletal elements are not found in the orientation or position that they would have had an entire body (or bodies) been buried in the flesh and not dismembered. A classic example of secondary internments is related to the "feast of the dead" practiced by some Native Americans in the Eastern Woodlands (e.g., Curry 1999). Periodically the remains of previously interred individuals would be collected, cleaned, and reburied in a communal pit as part of an elaborate ceremony.

Basic ways to describe or refer to internments are

☑ ***extended burial*** a positioning of the body where the legs are extended and in general alignment with the trunk of the body.

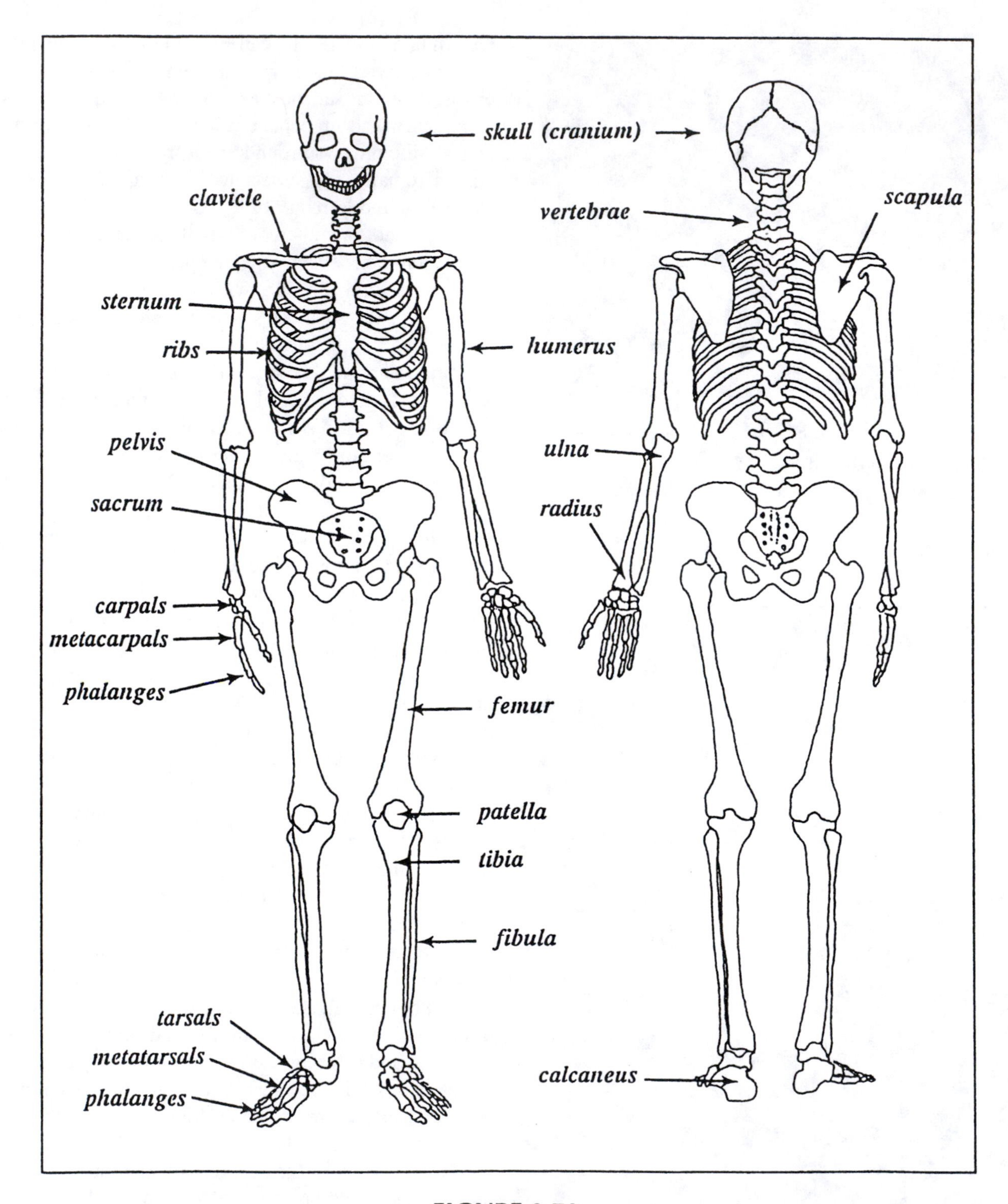

FIGURE 9.70.

Views of the human skeleton identifying some major bones (from Sutton and Arkush 1998: Figure 102).

☑ ***flexed burial*** a positioning of the body where the upper legs (femurs) are at an angle of less than 90 degrees with the trunk of the body, i.e., the body is bent or in the "fetal" position.

☑ ***semiflexed burial*** a positioning of the body where the upper legs are bent at an angle of between 90 and 180 degrees.

☑ ***bundle burial*** disarticulated collections of bones are arranged in discrete piles within a burial feature. Multiple bundles can occur within a single burial feature (mass burial/ossuary).

☑ ***cremations*** internments of burned human bone. Depending on the degree to which the body has been consumed, it may be difficult to visually recognize surviving fragments of bone as human.

☑ ***mass burials/ossuaries*** the internment of the remains of multiple individuals in a common grave pit or feature. Mass burials can involve all of the types of burials listed above.

There is no formal way to refer to miscellaneous skeletal elements that are found in contexts (feature or non-feature) that archaeologists do not readily perceive as intentional burials or disposal. Many site reports contain references to the odd human tooth or portion of long bone found in what otherwise appear to be pits used for storage or trash disposal, or in general excavation levels mixed with the artifacts of daily life. We may be overstepping ourselves in interpreting these occurrences as unintentional or accidental. Granted, archaeological deposits and burials may be disturbed and redistributed by a variety of processes. But what insights might we gain if we take the perspective that any undisturbed context containing human remains is a special context (Custer and Clark 2000)?

Once human remains or a burial feature are recognized in the field, their excavation should be halted and appropriate parties contacted for consultation. Begin with representatives of the State or Tribal Historic Preservation Office. They may refer you to descendant communities that should be contacted or may take the lead in organizing further consultations about what is to be done (or not done) with the human remains. It is also customary to contact the coroner for the county or region, although they rarely become involved in any further aspects of a project.

Back to the field. In general, the excavation of human remains should employ point provenience and pedestaling of all remains and associated artifacts, and the documentary procedures afforded any feature. Take photographs at all stages of the excavation. Do not use any metal implements in the excavation since these can easily mar bone. Be very aware of variations in the texture and color of the sediments surrounding remains and in the larger feature since they might be the remnants of organic and other materials that have decomposed. Plan on saving all excavated matrix from around the remains for flotation and other types of analysis (pollen, phytoliths, chemistry), and remaining feature fill for flotation or wet screening through 1/8″ or smaller mesh. It is best to have a specialist in human remains on-site to supervise excavations and to make observations and measurements before anything is removed from the ground. If the remains appear to be unstable as they are exposed, a conservator should be consulted and a specialist should be on-site to perform various analyses *in situ*. If it is possible to easily and safely isolate the remains in a block of matrix, they can be removed, crated, and transported to the laboratory for treatment and full excavation.

Knowing which bone of the body you have exposed in the beginning of the excavation should alert you to the probable orientation of other bones in the feature and perhaps the type of internment. Carefully expose and pedestal one bone before moving on to any others. Don't undercut any of the bones until you are ready to remove them. It is helpful to be able to see as much of the burial *in situ* as possible before making scaled drawings and removing anything. If the bone is in questionable shape, this allows specialists to make observations and take measurements that might not be possible once the remains are moved. It also provides a better sense of the nature of the internment and its relationship with other materials in the feature. However, it may be necessary to expose, map, and remove remains in several stages.

Creating a scaled drawing of a burial is a time-consuming process. A mapping frame or improvised small-scale grid will definitely make the job easier. Because of the potential intricacy of the drawing, use a scale for the drawing that will enable you to adequately reveal small details. The scale that you are employing to map excavation level floors or other types of features will not suffice. Map numbers should be assigned to individual bones in the drawing to insure that each can always be associated with a specific position. You can't always rely on the detail of your drawings to guarantee good associations, especially when the burial may be exposed, mapped, and removed in stages. A camera mounted on a bipod that can be positioned vertically over a feature and leveled can produce scaled photographs that are an excellent complement to scaled drawings. Elevations should be taken for major bones and for points representing high, mid-, and low points in the distribution of the remains. Continue to monitor the condition of the bones throughout the mapping and documentation process.

The data categories and forms used to document features also are useful for recording many of the aspects of burials. Other basic observations to be made are listed in Table 9.4. There are a variety of other observations that could be made prior to the removal of remains from an excavation, but require a specialist to do so. Burial record sheets are customized to suit the project, the types of internments and conditions typically encountered in a region, and the individual specialist. Figure 9.71 is an example of a data sheet used in the field to document historic period burials. Assign sequential numbers to burial features, beginning with 1. This sequence is separate from the one used to designate other types of features.

Make sure that a bone is completely exposed before undercutting its pedestal and removing it from the excavation. In other words, you never want to pull a bone from the matrix, you want to be able to simply lift it. Like any organic material from a context that is not waterlogged, human bone should not be packaged in airtight plastic bags or containers, or anything that might sweat moisture from the bone. Remember to always treat remains respectfully. These are your fellow human beings.

There seems to be little danger for infection from diseases that might be encountered during the excavation of ancient prehistoric remains. This may not always be the case when digging internments made during more recent historic times. If there is some doubt about potential risk follow these common sense rules (Crist 2000):

- ☑ keep any open wounds that you may have covered;
- ☑ wear gloves;
- ☑ wash hands and face with soap and water prior to eating, drinking, or smoking;
- ☑ keep the work area well ventilated;
- ☑ wear a surgical mask during excavation.

TABLE 9.4

Basic Observations for Burial Features for the Nonspecialist

Burial Type	individual, bundle, mass grave/ossuary, cremation
Burial Position	extended, flexed, semiflexed; left arm - extended at side, crossed over chest, crossed over pelvis, raised towards head; right arm - extended at side, crossed over chest, crossed over pelvis, raised towards head
Deposition	lying on back, face, right side, left side; sitting, standing, kneeling
Orientation	compass alignment of body; compass direction faced by skull
Completeness	estimate percent of skull, postcranial remains
Preservation	good (e.g., bone is fairly hard with surfaces intact, can be handled); fair (e.g., bone varies from slightly soft to slightly hard, few surfaces cracked and flaked, can be handled with care); poor (e.g., bone is very soft, crumbling, may not be removed as coherent mass)
Associations	objects or materials on or in direct contact with remains; other objects and materials in proximity or within feature

Note: These observations are in addition to those commonly made for any type of feature

Burial Recording Form

Provenience

Cemetery Number______ Burial Number_______ Name_____________________________

Burials

Orientation of Burial____________________ Orientation of Face__________________________

Number of Individuals in Burial_______ Age & Sex Estimates____________________________

Skeleton

Preservation	*Completeness*	Skull	PC	*Measurements (metric)*
Good______	75-100% Complete	___	___	Humerus______ Radius______
Fair_______	50-75% Complete	___	___	Femur _______ Tibia_______
Poor______	25-50% Complete	___	___	Maximum Length,
	Less than 25%	___	___	crown to heel_____________

Anatomical Peculiarities____

____

____

Note: Sketch the skeleton in situ on the reverse side of this form

Position of Skeleton

Extended	*Flexed*	*Disturbed*
On back ____	Left_______	Yes______
On face_____	Right______	No______

Position of Limbs

Left arm	*Right arm*	*Legs*
At side_____	At side______	Straight_____
Crossed_____	Crossed_____	Flexed______
		Other____________

Coffin

Shape Hexagonal______ Rectangular_______ Other____________ Not present______

*Construction Materials*____

Measurements Maximum length______________ Maximum width ______________

Width at headboard Maximum ______ Left board thickness_____ Right board thickness_______

Height of headboard____________ Height of footboard___________

Elevation at top ______________ Elevation at bottom___________

Note: Sketch the coffin and headboard on the reverse side of this form

Associated artifacts/clothing____

____

____

Associated hair/soft tissue____

____

Taphonomic features:____

Photograph Numbers:____

Soil sample Numbers:Show on plan on reverse____________________________

FIGURE 9.71.

An example of a data sheet used to document historic period burials.

The best course of action is to prepare a health and safety plan in consultation with a public health or infectious disease specialist prior to beginning fieldwork.

MOUNDS AND OTHER EARTHWORKS

Mounds and other earthworks present challenges that archaeologists have met with different approaches throughout the history of the discipline (cf. Dragoo 1963; Fowler et al. 1999; Greber 1984; Johnston 1968; Moore 1903; Pauketat 1993; Perino 1968; Rolingson 1998; Wilford et al. 1969; Willey and Sabloff 1993:143–147). Any excavation should begin with the construction of a detailed topographic map of the feature and the establishment of surveying datums for the subsequent establishment of grids and other mapping.

The construction of mounds/earthworks with basket loads of sediment and other materials can result in a good deal of variation in the color, texture, and composition of the fill. Add to this the effects of soil-forming processes over the millennia, intrusive features, the backfilled pits of treasure seekers and looters, and even plowing in some cases. Using variations in the matrix as a guide for excavations is difficult without some knowledge of stratification over large portions of the feature, especially when the goal is to expose horizontal expanses of common surfaces or construction stages. Such horizontal exposures offer the best opportunities for interpretation and correlation of internal features and artifact deposits.

Needed stratigraphic information can be gained by systematic coring. Older excavations or looters holes can be cleaned out and their side walls examined. Excavations positioned to include the visible margin of the feature and that extend away from it will reveal the actual base of the mound/earthwork, any associated ditches, and the natural stratigraphy of the landscape. An understanding of local deposits will be useful for interpreting the variation in sediments found within the mound/earthwork.

Trenches are frequently used in the initial exploration of mounds or earthworks. The orientation of trenches (e.g., transection, bisection, quarter section) depends on the size and shape of the feature. Trenches provide the needed stratigraphic information to plan the excavation of units and the exposure of horizontal surfaces across the face of the mound/earthwork. There are drawbacks to this approach, however. Any features exposed can only be examined in the context of the trench; the expansion of a trench to fully expose a feature at this stage of investigation defeats the purpose of using a trench in the first place. A trench can prove to be an inadequate context in which to examine large and intricate features or complexes of features. Imagine a trench that cuts through the edge of a burial chamber and the feet of an extended internment. Encountering large features in trenches is less of a problem when the mound/earthwork is low in height and more extensive excavations will follow on the heels of the trenching. Problematic features can be left unexcavated and effectively pedestaled as excavations proceed in adjacent sections of the trench.

In order to create and preserve a stratigraphic reference for ongoing excavations and to assist correlations, linear balks are incorporated into the grid system of units distributed across the face of a mound or earthwork during an intensive excavation. The preservation of these linear balks also makes it easier to re-establish grid controls as the surrounding surface of the mound or earthwork is lowered through excavation. If the mound or earthwork is of a substantial height, excavations may reach depths well below the top of the balks. Preserving the vertical faces of the balks from drying and collapse, and ensuring that they don't become a safety problem may be difficult. Periodically documenting the vertical faces of the balks and excavating them to levels more comparable to the elevation of ongoing excavations is a way to circumvent these difficulties. Small, low-to-the-ground mounds or earthworks may not require the use of balks.

Other built aspects of the landscape, like terraced gardens and fields, or filled or graded topography can be investigated using standard archaeological techniques. Such studies are a part of what is sometimes referred to as landscape archaeology (e.g., Aston and Rowley 1974; Cunliffe 1978; Kelso and Most 1990; Yamin and Metheny 1996).

ARCHITECTURAL REMAINS

Summaries of approaches for dealing with sites with substantial above- and below-ground architectural or structural remains include Barker (1996), Noel-Hume (1983), and Palmer and Neaverson (1998). As with other special types of sites, reading detailed reports of excavations at specific localities provides the greatest insights. Preserved fragments of posts, postmolds, prepared floors, pits, cellar holes, wall trenches, and foundations and walls of rock, shell, or adobe are some of the ways in which structures may be represented archaeologically. Knowing something about the variety of construction techniques and building forms used by ethnographically known or historic cultures provides a definite advantage in recognizing evidence of them in the field and designing excavation strategies for their study (for useful summaries and illustrations dealing with Native American built forms see Morgan 1965; Nabokov and Easton 1989).

Trenching is a technique useful for locating or intersecting buried architectural remains when their approximate location has already been determined through other means (e.g., documentary research, remote sensing, surface indications, probing). It is also effective for determining the configuration of structures, especially locating corners or angles points, once a portion of a structure has been located. The configuration of a structure, or its internal divisions define the boundaries within which grid systems and excavation units are organized and conformed, i.e., the structure or room becomes the context or "excavation block" that is subdivided into smaller units. This is only logical. A structure and any internal divisions bound the activities that would have been performed within them and any resultant archaeological evidence. Site-formation processes, and therefore stratigraphy, need not be the same inside of a structure as outside of it. Because of the complexities arising from destruction, rebuilding, and the burial of architectural features, balks may be used to preserve stratigraphy between excavations within and adjacent to structures.

SHELL DEPOSITS

Shell-bearing sites are a common feature of the archaeological record of coastal areas around the world. Shell deposits can be thick and range up to acres in size. Their chemistry can promote the preservation of organic materials that would not otherwise be preserved in the sediments of an area. The problems and prospects of the archaeology of shell-bearing sites are addressed in a number of publications (e.g., Claassen 1991, 1998; Stein 1992; Waselkov 1982, 1987; also see Kerber 1991:213–276 for a compilation of references). The generic term often used for such sites, shell middens, can be misleading and obscures the variety of ways and reasons that people use shell (Claassen 1991).

Ethnographic data, ethnoarchaeology, and the accumulated insights from excavations make it clear that the stratigraphy and structure of a shell deposit can be highly variable. There is no reason to assume that shell deposits build up evenly over a landscape through time, thus no reason to assume that identifiable strata will be continuous across a site. Identifying stratification within a shell deposit draws upon a number of observations including: sediments associated with shell, the ratio of shell to sediment, the orientation of shell, the degree of shell weathering, and the condition of shell (e.g., relatively whole, highly fragmented). Column samples, auger borings, and trenches distributed at close intervals to approximate continuous linear profiles across a deposit are tactics for documenting stratigraphic variation, sampling the deposit, and setting the stage for unit or block excavations. The analysis of shell has the potential to support reconstructions of habitat, gathering and consumer behavior, and nutrition. Obtaining an adequate sample of shell from a deposit and excavated contexts is crucial for reliable reconstructions. Screening should employ small mesh sizes and as much excavated matrix as possible should be processed by flotation. This reflects a shell deposit's ability to preserve floral and faunal remains.

WET SITES

Wet sites, those that are continually or periodically saturated, present a myriad of logistical and conservation problems that make traditional approaches to excavation impractical. In contrast, their potential to contain artifacts and materials not preserved under other circumstances can be great. Many sites are wet as a result of environmental change and its dramatic impact on ancient landscapes. As such, wet sites represent the human use of habitats and resources not always well represented by other terrestrial sites. For good examples of fieldwork on wet sites see Coles and Goodburn (1990), Croes (1976, 1995), and Purdy (1988).

An example from my personal experience illustrates in a small way how complicated work on a wet site can be. Site 28ME1-D is a deeply stratified locality situated on a narrow parcel of fast ground flanked by freshwater tidal marsh near Trenton, New Jersey (Wall et al. 1996). Over 6,000 years of archaeological deposits are distributed through 15 feet of alluvial sediments. More than half of this sequence lies below the water table and is saturated year-round. Excavation of the dry portion of the deposit proceeded in standard fashion. Heavy machinery was then used to strip a large area to within two feet of the existing water table. A network of large aluminum pipe was created around the margins of the excavation area (Figure 9.72). Adjacent to this framework, and at intervals of approximately six feet, well points 12 feet long were put into the ground and coupled to the network of pipe (Figure 9.73). The system was powered by a large diesel pump that ran 24 hours a day. Water flowing into the individual well points was drawn upwards into the larger network of pipes at the surface and pumped into the nearby marsh. The system effectively drained the subsurface deposits within the excavation area and prevented the infiltration of additional water. Units were laid out and excavations continued as if it were a dry terrestrial site (Figure 9.74).

FIGURE 9.72.

Laying out aluminum drainpipe and a series of well points around the edge of the area slated for excavations below the water table at Site 28ME1-D.

FIGURE 9.73.

A well point 12 feet long is attached to a high-pressure water hose. Using water pressure, the well point will be blown into the ground adjacent to the drain pipe in the foreground and eventually coupled to it.

FIGURE 9.74.

The excavation area is bordered by dozens of well points and the large piping to which they are coupled. The system is continuously powered by a large diesel pump (left background). Water flows into the individual well points, is drawn upwards into the larger network of pipes at the surface, and is pumped into the nearby marsh. The system effectively drains the subsurface deposits within the excavation area and prevents the infiltration of additional water.

CAVES AND ROCKSHELTERS

Caves and rockshelters are well-defined and enclosed (partially or fully) spaces that can foster the preservation of perishable materials better than open sites. For an appreciation of excavation strategies check out reports for specific sites (e.g., Flannery et al. 1986; Jennings 1980; Lynch 1980; Thomas 1983; Watson 1974). An investigator needs to have an understanding of the external and internal natural processes that affect rockshelters and caves (cf. Butzer 1971:204–214; Donahue and Adovasio 1990; Farrand 1985; Laville 1976; Schiffer 1987; Straus 1990; Waters 1992: 240–247). The sediments in them can derive from external processes like sheet wash and flooding, and internal processes like the weathering and spalling of fragments off the roof and walls. They are essentially sediment traps. Within caves you may have to deal with chemical precipitates that can seal archaeological deposits below levels of "rock."

From a stratigraphic perspective, every rockshelter and cave is unique owing to its lithology, weathering processes, hydrologic conditions, and the nature of the environment outside of the shelter (Waters 1992:243). Stratigraphy can vary dramatically over small distances and be difficult to relate to strata beyond the rockshelter or cave mouth. In addition, entrances to caves can be horizontal or vertical, with different implications for the natural and cultural deposition of sediments and artifacts. The shape and size of the entrance to a cave or rockshelter can change over time, altering the area available for habitation and the deposition of sediments (Figure 9.75). The interior of the cave environment is different from that at the mouth of the cave, which is comparable to that of a rockshelter. The periodic reuse of caves and rockshelters, because of the constraints they place upon habitable space, can result in culturally disturbed deposits, the activities of one set of inhabitants impacting the archaeological evidence of the activities of previous inhabitants.

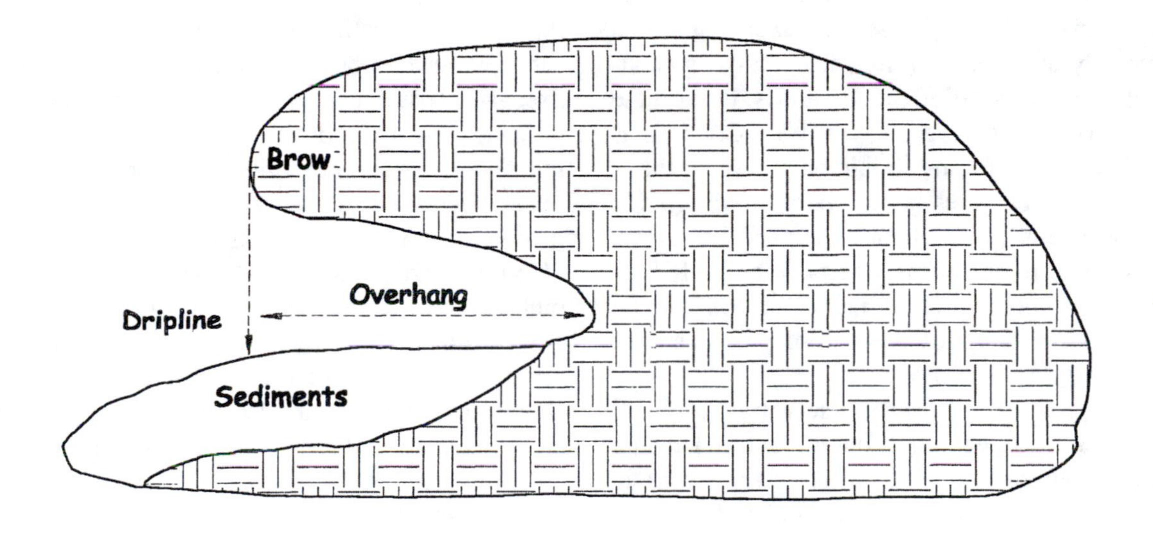

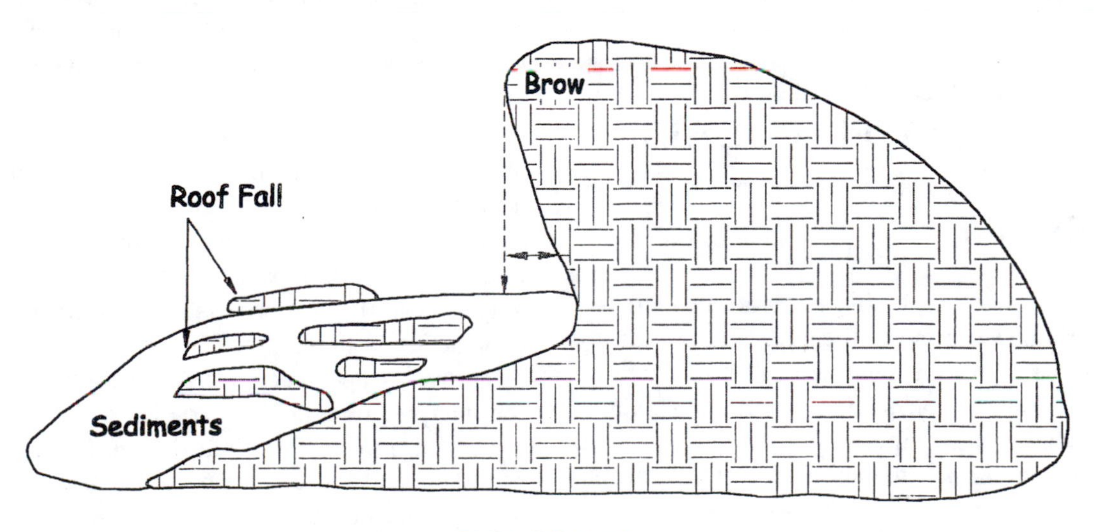

FIGURE 9.75.

Schematic of a rockshelter or the opening of a cave. Through time the shape of the shelter or opening can change as the rock face weathers and spalls.

The basic challenge for the excavator is to cope with what can be a highly variable stratigraphy and relatively confined work space. Excavations typically begin with small units positioned on the outer side of the brow and dripline. It is tough to open large horizontal areas and take level floors down in concert because of the variability in stratigraphy that may occur over short distances. Units can be expanded off the initial tests moving toward the brow of the shelter/cave entrance and under the overhang. The vertical face of the completed unit provides the control for the removal of levels in the new excavation. In this way a trench is gradually excavated into the shelter/cave. Subsequent excavations can expand laterally off the initial trench, again using existing vertical faces for stratigraphic controls. Record the profile of every wall of every excavation. Continuous linear profiles are a must and should be preserved for as long as possible. Features when discovered provide good control points from which to expand horizontal excavations since they indicate the position of former living surfaces. The progress of excavations can be frustrated by large slabs of rock from previous roof falls that are too heavy to move.

Because of the potential for preservation of floral and faunal remains, screening through fine mesh and an emphasis on flotation should be the rule. For work in the interior of caves, artificial lighting and perhaps ventilation will be necessary. Dust and fungus, and bacteria adhering to dust, can present health hazards and may require the wearing of masks (see Appendix 4).

EXPANDING YOUR PERSPECTIVE ABOUT . . . MAPPING, DATA COLLECTION, AND PROCESSING

The extent of mapping within excavations increases as field efforts become more intensive, more units are opened, or more features encountered. Heavier reliance on mapping using a total station rather than manual techniques can make operations more efficient. Photographic mapping with bipods or other rigs that permit scaled overhead shots to be taken is an alternative mapping technique. Figure 9.76 is a photo mosaic map of excavation floors composed in this way.

The collection of column samples, samples for screening through fine mesh and flotation, and samples for pollen and phytolith analysis should be an integral part of site testing. Constant volume flotation samples are regularly taken from all excavated contexts in every unit as part of an intensive study. Ideally, dry or wet screening, employing mesh smaller than 1/4″, should be undertaken for large samples, if not all excavated matrix. The size of constant-volume samples should reflect the nature of the deposit and a project's research goals. Strategies for collecting pollen and phytolith samples from excavations need to reflect the multiple ways that these data may be used. They provide a view of the site environment, resource use, and season during which occupations or activities took place. Control samples need to be collected from recent contexts in excavations, not just those of archaeological interest, in order to assess the degree of post-depositional movement of pollen and phytoliths, and for comparative purposes. This is best done in column samples. Where the goal is environmental reconstruction, off-site sampling of pollen and phytoliths is necessary for an accurate picture, since site environs arguably reflect the activities of humans and their impact on the environment. Off-site sampling localities should be chosen with care (Jacobson and Bradshaw 1981). The comparison of on-site and off-site data sets will bring the nature of human activities into clearer focus.

There should be a well-defined procedure for selecting or sampling artifacts for the analysis of blood and other residues during an intensive field project. Since large volumes of artifacts may be generated, and in-field or laboratory processing of materials may be keeping pace with excavations, a predetermined strategy insures an efficient operation.

In-field analysis and collation of data should occur on an ongoing basis so that excavation strategies can be adapted to make the most of a deposit. Minimally this should include the tracking and plotting of the horizontal frequency distributions of artifacts (overall and by category or class) by appropriate stratigraphic context. In conjunction with maintaining a master plan of features, this provides a perspective on site structure that is a basis for interpretations and comparisons with the expectations of a project's research design. Multiple copies of site plans and the completed data sheets from excavations are needed to accomplish these tasks. Electronic versions of these data, of course, make the work so much easier. Running interpretations of site deposits should be a daily component of field book entries.

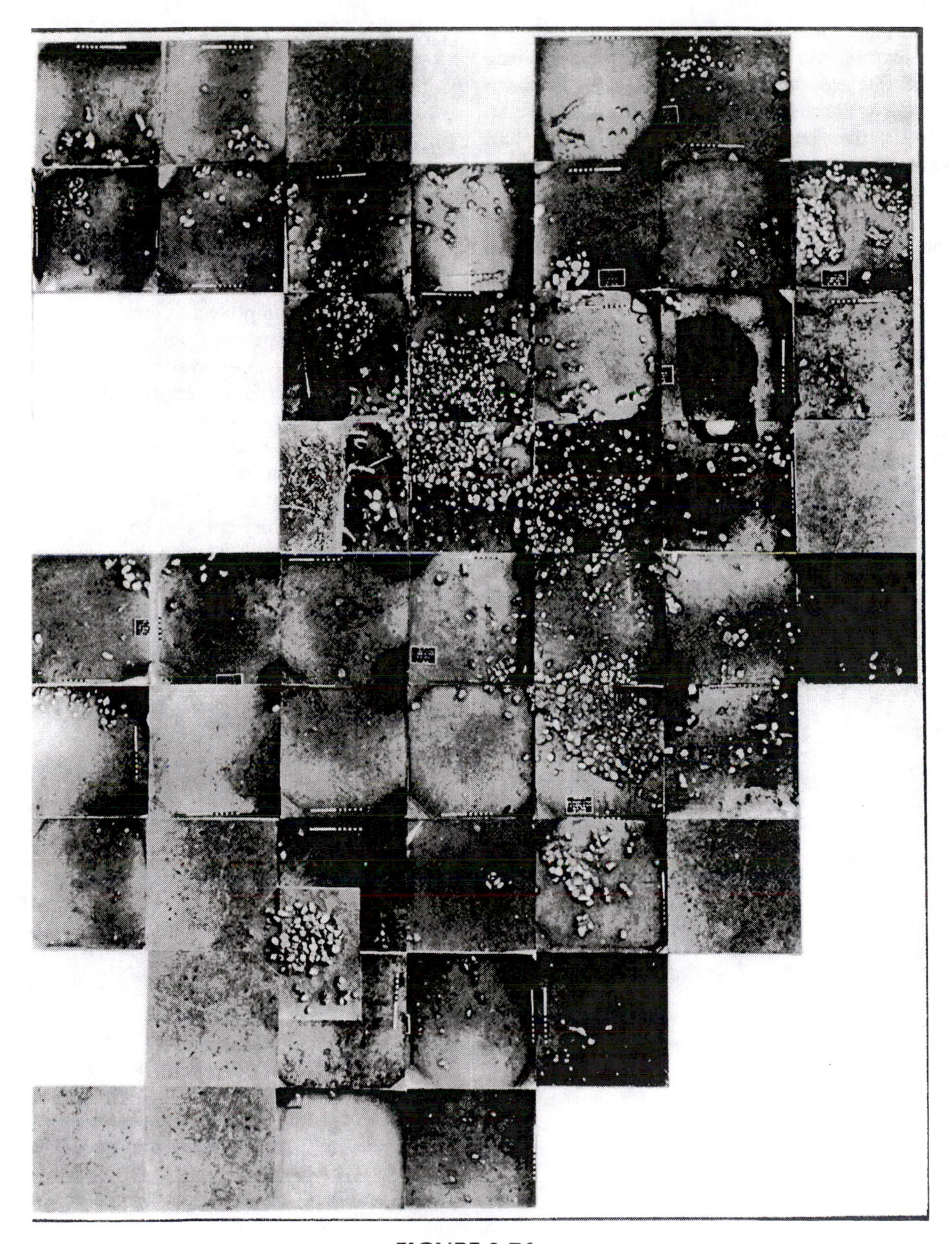

FIGURE 9.76.

A photo mosaic map of a series of excavations, all with the same level exposed and artifacts *in situ.* Each unit measures five feet on a side (from Cavallo 1987).

Establishing a field laboratory may only be possible in the context of well-funded, intensive investigations, owing to the assortment of equipment and facilities needed to complete even the most basic levels of artifact and data processing (cf. Joukowsky 1980:227–228, 246–254; Neuman et al. 1998; Nichols and Evans 1979). Given the benefits of on-site artifact and data processing, it can be argued that a field laboratory should be a mandatory part of any funded project (Nichols and Evans 1979). A field laboratory is especially critical when dealing with sites that will produce large numbers of artifacts that will need special handling and immediate conservation. Large projects generally maintain a close working relationship with an off-site laboratory facility regardless of whether or not a field laboratory is established. Again, the importance of analysis to ongoing excavations is a factor. For CRM projects, the tight time schedules often allotted to the completion of fieldwork and the preparation of reports demand that data processing and analysis keep pace with fieldwork.

EXPANDING YOUR PERSPECTIVE ABOUT ON-SITE LOGISTICS

Intensive excavations can mean lots of people working in the same block or area for prolonged periods. The constant traffic between excavations, screening areas, equipment caches, artifact and record storage can muck up clean excavation floors and cause other inadvertent damage to excavations. Working clean is very important. Boardwalks or well-defined pathways for travel are one way to protect excavations from the constant movement of people (Figures 9.77 and 9.78). Do not step down onto an excavation floor with muddy or soiled boots. Avoid letting excavated matrix accumulate on a level floor in progress.

The sediments and soils that are exposed in the walls and floor of any unit will dry out during the course of any prolonged excavation. Clayey sediments, although they take longer to dry out, will become hard and may crack. Digging them in this condition will be difficult.

FIGURE 9.77.

Sheets of plywood are used to create walkways through work areas to ease the movement of equipment and minimize damage to excavations.

FIGURE 9.78.

An example of a clean, protected, and well-organized excavation. The large excavation block is protected from the weather by a shelter. A boardwalk runs the length of the block. Equipment storage boxes are convenient to the boardwalk and excavation area. Excavation grid points are easily identified and well maintained. Screening operations and backdirt are removed from the excavation area.

Drying may also wash out the colors of sediments, making it hard to recognize some types of features. When excavations are in coarser sediments like sands and sandy loams, the drying of the matrix can result in the collapse of excavation walls or the sloughing of some strata. Of course you want to protect excavations from the ravages of the weather in general. Keep excavations covered with tarps or large sheets of plastic when shelters are not employed, and support coverings in such a way as to direct water away from excavations.

Crafting excavation strategies for deeply buried deposits that are safe, cost effective, and adhere to standards for data collection and provenience control is demanding to say the least. Review OSHA standards in conjunction with developing the project's research design. Develop alternative protective systems if possible. It may be useful to consult with an OSHA representative since regulations were developed with the construction industry in mind, not the field methods of archaeologists.

Be cognizant of the effect that intensive field investigations may have on the surrounding landscape. Wet screening operations should be situated so that runoff from them does not lead to erosion of surrounding areas. This may require surrounding the screening area with silt fence and hay bales. It may also be necessary to surround areas of stripped fill or stockpiled backdirt to prevent the erosion of sediments into nearby drainages or wetlands.

The tight schedules for fieldwork to which some salvage situations and CRM mitigation projects must adhere present logistical and organizational challenges. "We need it done yesterday!" Mobilizing and maintaining large crews in the field requires greater numbers of supervisors and higher levels of overall interaction and management. It makes it more difficult to assimilate the information produced by excavations and use it to evaluate and perhaps modify field strategies. The role of the field laboratory or in-field processing of data is heightened. For the field technician the result may be a feeling of being out of touch with the big picture of a project.

Finally, intensive investigations inevitably put archaeologists in full view of the public. Regular tours of excavations may be scheduled and there will undoubtedly be interest on the part of various news media. The efforts of all individuals involved in a project can make a big difference. Dress and act like the professional field technician that you are, make sure that your work area reflects the pride that you take in your work, and be enthusiastic about the interest that others have in your activities.

APPLYING YOUR KNOWLEDGE

1. Explore the archaeological literature for your region and compile information on the size of sites and the depth to which they are buried. What types of field methods would you have to employ in order to consistently find such sites? Do certain kinds of sites seem to be underrepresented? How might you explain this?

2. To better understand how subsurface testing methods correlate with the goals of a field project, visit your SHPO or THPO and examine technical reports resulting from discovery, testing, and intensive types of investigations.

3. Contact your state's SHPO or a THPO to find out about guidelines for performing different levels or types of fieldwork. Critique these guidelines from the perspective of site-formation processes, the types of sites known for the region, geomorphology, sampling strategies, and logistical concerns.

4. Explore the archaeological literature for your region and create a list of the types of artifacts found that might require conservation in the field. Investigate strategies that would be appropriate for their handling, conservation, or in-field analysis.

DIG DEEPER

- ❑ The *Journal of Field Archaeology* specializes in articles addressing aspects of all levels of field investigations and analysis.

- ❑ The following incorporate CD-ROMs that allow the reader to learn about and experience the step-by-step excavation, analysis, and interpretation of an archaeological site.
  - Davis, R.P. Stephen, Jr., Patrick C. Livingood, H. Trawick Ward, and Vincas Steponaitis. 1998. *Excavating Occaneechi Town: Archaeology of an Eighteenth-Century Indian Village in North Carolina.* Chapel Hill: University of North Carolina Press.
  - Dibble, Harold L., Shannon P. McPherron, and Barbara J. Roth. 1999. *Virtual Dig: A Simulated Archaeological Excavation of a Middle Paleolithic Site in France.* Mountain View, CA: Mayfield Publishing.
  - Sept, Jeanne M. 1997. *Investigating Olduvai Gorge: Archaeology of Human Origins CD-ROM.* Bloomington: Indiana University Press.

- ❑ Information on artifact and data conservation and management.
  - The American Institute for Conservation of Historic and Artistic Works, 1717 K Street NW, Suite 301, Washington, D.C. 20006 (202-452-9545). This national organization provides basic information to the public about conservation and help in selecting and finding a conservator to meet individual needs. Their website is at **http://aic.stanford.edu/**
  - *Managing Archaeological Collections,* the National Park Service, Archaeology and Ethnology program's website at **http://www.cr.nps.gov/aad/collections/index.htm**> provides a discussion of all aspects of collections, data conservation, and curation with links to other useful websites.
  - The National Center for Preservation Technology and Training focuses on the technical aspects of preservation. Their website is at **http://www.neptt.nps.gov/**

CHAPTER 10

TRAINING AND PROFESSIONAL PRACTICE

Training is everything. The peach was once a bitter almond; cauliflower is nothing but cabbage with a college education.

Mark Twain, The Tragedy of Pudd'nhead Wilson

There are many different ways to pursue a career in archaeology and equally different degrees of academic training that go along with them. Field schools and methodological training are only part of the bigger picture that you need to see in order to prepare yourself. Some useful essays on careers in archaeology are listed at the end of the chapter. There has been an effort in the profession to evaluate professional standards, the status of academic training, and how it relates to the skills needed by archaeologists working in cultural resource management and other nonacademic spheres (e.g., Athens 1993; Davis 1989; Gray et al. 1999; Krass 1998; Lipe and Steponaitis 1998; Lynott 1997; Lynott and Wylie 1995; Schuldenrein 1995). Regarding academic training and professional practice, significant results are summarized in *Teaching Archaeology in the Twenty-First Century* (Bender and Smith 2000). Examples of job descriptions and the expectations that employers have of entry-level archaeologists are in Appendix 5. They provide an interesting foil to the Bender and Smith text in terms of revamping educational programs to provide a better fit with the job market. In planning your education, field training, and career you need to be aware of the realities of both the academic world and professional practice.

FIELD SCHOOLS AND VOLUNTEER SITUATIONS

You may be reading this book out of a general interest in archaeology, or it may be a text for a class on archaeological field methods in which you are enrolled. There's no substitute for actually doing fieldwork to bring home the lessons of any book or course work. Many people gain their first field experience as students in a field school or as volunteers on a sponsored project. For those of you who have already participated in a field school, you know that there is more to be learned and experience to be gained. Enrolling in a specialized field school or field program is an option for further training. Here are some suggestions for how to locate field schools or volunteer situations, and pick the one that is right for you.

Numerous colleges, universities, and organizations offer seasonal programs that teach archaeological field methods in a field setting. On an annual basis the American Institute of Archaeology (AIA) based in Boston, Massachusetts, publishes the *Archaeological Fieldwork Opportunities Bulletin.* It contains listings and detailed information for field schools and volunteer opportunities, including the availability of paid staff positions. Anyone willing to provide the AIA with appropriate information on their program can be listed in the *Bulletin.* Their website is at http://www.archaeological.org and the publication can be ordered over the phone from Kendall/Hunt Publishing Company (1-800-228-0810). Domestic and international fieldwork opportunities are also listed on the website of the Costen Institute of Archaeology at the University of California, Los Angeles (http://www.sscnet.ucla.edu/ioa/afs/testpit.html). Although many colleges and universities that offer field schools will be listed in one or more of the sources I have mentioned, some will not. If you are interested in a particular institution that isn't listed, try contacting the relevant anthropology department directly. You can use a search engine to locate web pages for anthropology departments on the Internet. Or you can consult the *AAA Guide to Departments of Anthropology* published annually by the American Anthropological Association, Washington D.C. (http://www.aaanet.org/ pubs/guide.htm).

FIELD SCHOOLS

There are many things to consider in choosing a field school, both academic and practical. Much of the information needed to make an informed decision will be in printed literature or on websites. A phone conversation with the field school director or supervisor is a must. There will always be things you need to know that won't be found in even the most thorough promotional material (see below). Talk to other people who have already taken the field school or have classroom experience with those running the program. You might even ask the field school director for names and phone numbers of previous students so that you can get a better perspective on what to expect.

How comprehensive is the program in terms of the skills and field operations that it presents to students? Does it provide experience in both archaeological survey and excavations? Will you learn mapping techniques and the use of surveying instruments? The Society of Professional Archaeologists (SOPA), now the Register of Professional Archaeologists (ROPA), formulated guidelines and standards for archaeological field schools (http://www.smu.edu/~anthrop/sopaold.html#FIELD_SCHOOLS). While these guidelines need to be updated, they provide an overview of what you might reasonably expect from a field school. Of course, this book certainly conveys the range of basic concepts and skills needed by the field archaeologist.

Will historic or prehistoric sites be investigated and to what degree will background on relevant cultures be presented? What types of research issues are to be addressed by the field investigations? Learning the mechanics of field techniques is one thing, but seeing the connection between methods, the data generated, and the solution of research problems makes them more understandable. Every field project should have an explicit research design. However, the focus of some projects can be so specific that your exposure to a wide range of skills and situations is not guaranteed. Some field schools are designed intentionally with an overriding topical or methodological focus in mind, like the intensive excavation of a special site, geoarchaeology, faunal studies, or remote sensing applications. These will appeal to the person who already has field experience or has been through a basic field school. Programs that deal with historic sites, prehistoric sites, particular cultures, or geographic areas can be designed to be comprehensive in the presentation of field methods. In any case, there is enough variety that you should be able to find a field school in a geographic area where you would like to work and with a research focus that interests you.

Most field schools offer college credits. The number of credits varies from as few as 3 to 9 or more, as does the length of the field school. The two variables are related; the fewer the credits the shorter and less intense the program. A 3-credit field school that runs for 3–4 weeks cannot theoretically approach the scope of one that extends for 6–9 weeks and offers 6 or more credits. Of course if all you are doing for 6 weeks is surface survey then the program is hardly comprehensive. Be sure to check with your departmental advisor and home institution about how credits will transfer and count toward your degree requirements before going off into the field.

Typically field schools run all day and 5 days a week, which can present problems for students who need to balance their training with gainful employment. It is a good idea to find out how much time is actually spent in the field and how much in the classroom, at lectures, or in the laboratory. There's no escaping the need to do some things indoors, but the emphasis should be on outdoor activities. It is possible to gain focused laboratory experience through other course work and opportunities. In contrast, those seeking out a specialized field school, like geoarchaeology, will want to be sure that a healthy laboratory component is part of the program.

Determine what the advertised fee for a field school actually covers and what additional costs you will incur. The basic cost will cover college credits or a fee for the instruction and experience that you are receiving. Some programs offer scholarships and will make this clear on their promotional materials. What types of personal field equipment are you expected to provide? There will undoubtedly be costs involved in getting yourself to wherever the school is being held and living expenses. The degree to which accommodations, food, and transportation are included in the price of enrollment, or provided by the sponsoring institution, vary dramatically. Will you be commuting to the field on a daily basis from home, living in a dormitory, or camping? If you will be camping, what types of equipment will be provided for you, and what will you need to bring? And don't forget about insurance coverage. Find out if you will have to sign an injury or liability waiver, whether you are covered under the sponsor's policy, or whether you must rely on your own health and safety insurance.

VOLUNTEERING

Opportunities to gain experience as a volunteer are numerous. Situations can range from ones where you are simply a body and a pair of hands helping to get needed work done, to others that are deliberately designed to be educational. Few if any volunteer situations will provide the comprehensive experiences of a field school. Still, participating in a project as a volunteer will give you a taste of archaeological fieldwork and may help you to decide whether to pursue the discipline further. For those who already have completed a field school, volunteering can be an excellent way of increasing your experience and pursuing your interest in a particular geographic area or research topic. Volunteer situations are never free in the strict sense of the word. Transportation and living expenses must always be considered. As with a field school, you need to determine the nature of insurance coverage for injury and liability.

Most opportunities to work as a volunteer occur on a regular or seasonal basis. Some field schools accept limited numbers of volunteers. The AIA bulletin identifies field schools that provide this option. Passport in Time (PIT) is a volunteer program of the Forest Service of the United States Department of Agriculture (http://www.volunteeramerica.net/usfs/PIT_Home.htm). PIT provides opportunities for individuals and families to work with professional archaeologists and historians on historic preservation projects. There are no registration fees. The list of opportunities is updated twice a year, in March and September. Earthwatch is a nonprofit organization that places interested people with scientists working on projects throughout the world. There are fees involved that may or may not include living expenses (http://www.earthwatch.org/subject/archaeology.html). A number of state archaeological societies or State Historic Preservation Offices (SHPO) sponsor seasonal fieldwork, sometimes in conjunction with a statewide archaeology month, but most frequently during the traditional summer field season. On a more sporadic basis, large-scale cultural resource management (CRM) investigations, typically Phase III or mitigation projects, may make provisions for the participation of volunteers. Contact your SHPO to learn about these opportunities or how to contact the local chapter of the state's archaeological society.

WORKING AS A FIELD TECHNICIAN

What is it like to work as a field technician out in the "real world," drawing a paycheck and being able to say that you are actually employed in a job that relates to your undergraduate degree? There are few accounts of this aspect of professional practice and none are truly comprehensive in their coverage (Hirst 1998; Kintz 1997; McGuire and Walker 1999; Wilson 2000). The short-lived newsletter of the 1990s, *The Underground,* edited by Theresa Kintz and Ed Cooper, was prepared by and for field technicians and related a sense of the field's pluses and minuses. An essay on archaeological technicians, highlighting the problems that exist between labor and management, can be found on the website of the United Archaeological Field Technicians (UAFT) at http://www.members.aol.com/UAFT/home.htm>. My comments draw from these references and my own personal experiences, including more than 15 years as a field technician, crew chief, principal investigator, and manager in CRM.

The overwhelming majority of field technicians working today are employed in CRM, and literally all of the literature and discussion of the work experience deal with these workers. This is not too surprising considering that probably 90 percent or more of funded archaeological research in the United States is related in some way to CRM (Thomas 1998:557), and that annual expenditures in some years have been $300 million and more (Patterson 1995:109; Thomas 1998:565). However, there are no accurate numbers for either the number of people working in CRM or what percentage of this total is made up of field technicians (cf. ACRA 2001a; Patterson 1995:108; Thomas 1998:565). The total number is at least 6,000 people and probably considerably more. The UAFT estimates that there are 2,000 field technicians working in the CRM industry (http://members.aol.com/UAFT/handbook.htm#UAFTHistory). The fact that so many people enter and leave the field on an annual basis makes it difficult to provide accurate estimates.

In essence, a field technician or archaeological technician performs the physical labor and standardized data collection that are a part of any field project—tasks that have been described in this book. In reality, the tasks performed by a technician, and the level of responsibility that they are afforded, vary depending on an individual's experience and the structure of the company or organization for which he/she works (see Appendix 5). There are extremes in the overall employee experience that also derive from the structure of a company and the attitudes and management styles of senior staff. A brief look at the nature and variation in company structure sets the stage for a better understanding of the field technician experience.

CRM is archaeology and a lot more (see Chapter 3). A given project can require the services of archaeologists, historians, architectural historians, anthropologists/sociologists, laboratory specialists, specialists from the natural sciences (geology, biology, soils/pedology, geomorphology), logistical specialists, managers, accountants, report production staff, and the personnel needed to keep any kind of office running. Regardless of the size of a company, each must cope with the same realities: generating business; interacting with clients before, during, and after the completion of a job; completing background research, fieldwork, artifact and data processing, analysis, and artifact and data curation; report production; and the day-to-day maintenance of an office or facility.

The larger the company, the more potentially diversified the workforce and its structure. What I would characterize as mid- to large-sized companies are those with 20 or more full-time employees. Some are linked to, or divisions of national corporations that do things like engineering, environmental studies, and planning. Some may have branch offices. In larger companies, there are as many people or staff positions as there are specialized tasks. It is difficult to say that one type of infrastructure fits all larger companies or CRM firms in general. Corporate philosophy can have a big impact. For example, the difference between wanting to constantly grow as a company versus achieving and maintaining a certain size and level of business influences the geographic and internal organization of a firm, the number of people that it strives to employ on a full-time basis, and how (or if) field staff are rotated into laboratory positions when fieldwork is not in progress. The divisions of labor summarized below reflect what can be encountered in a larger company, even though the work that needs to get done is common to all companies.

Upper management (president, vice president, principal) is people at the top of the corporate pile concerned with making the big decisions and keeping the structure of the company intact and functioning. They market the company to potential clients and agencies, watch for requests for proposals, and decide which jobs to pursue. They assign duties to staff, track the overall financial and logistical progress of jobs, watch over accountants and solve financial problems, solve personnel problems, and make final decisions about who to hire and fire and who gets raises. Upper management represents the company and clients at public meetings or meetings with regulatory personnel. They are the persons who liaise between the cultural resources component of a national or regional firm and the upper management of its other divisions. Because of their duties they travel frequently, which can impair interaction with other staff. Field crew will rarely have the opportunity to interact with upper management. Upper management positions can be filled by strictly business-type persons and/or archaeologists with business training or proclivities. In-depth knowledge of business practices, CRM, and how archaeology works makes a big difference in the performance of upper management and reflects on the success of a company, the quality of its work, and the attitudes of its staff. In large companies upper management rarely gets involved in the nitty-gritty of preparing research designs, or organizing or overseeing fieldwork, analysis, and report preparation.

Middle management (e.g., project managers, principal investigators) is people who have one foot in general management and the other in the actual practice of archaeology or other aspects of CRM. They write proposals, formulate research designs, and oversee or supervise fieldwork, data collection, and analysis. They are often the primary authors of technical reports. They help to generate new work, track the progress of individual jobs, and have moderate to intense interaction with clients and the interested public. They appear at public meetings, make presentations, give talks at local amateur societies and professional meetings. Middle management positions are very flexible and can change dramatically from firm to firm. Their involvement in the day-to-day fieldwork of a project varies depending on the management style of the company and often the inclination of the individual filling the position. Because of their other duties and cost concerns, it is rare to see a principal investigator with a large company on-site on a daily basis. Archaeologists, historians, and architectural historians with graduate degrees fill these positions. Middle management staff travel frequently, but have more residential stability than field crew. Field crew will (or should) have many opportunities to interact with middle management staff.

Supervisory field staff/lower management (e.g., field supervisor, crew chief) handle the day-to-day organization and completion of all aspects of fieldwork. They are the closest link between the field crew and middle management, and work closely with the principal investigator on a project. Along with the field crew, they interact with the interested public more frequently than any other staff members. In-field interaction with clients may be part of their responsibilities. Depending on the company, they may be the lead individual in organizing and overseeing the laboratory processing of artifacts and field data, and the preparation of rough drafts of maps and profiles for reports. They may have report-writing duties. They are rarely involved in generating business for the company, writing proposals, or formulating research designs. Archaeologists with M.A. and B.A. degrees fill these positions. They travel frequently and live on the road for extended periods.

Field crew, field technicians, or archaeological technicians carry out fieldwork and data collection under the supervision of a crew chief, field supervisor, or principal investigator. Individual responsibilities will vary depending on experience. For example, not all field crew members will use surveying equipment to map sites. People with B.A. degrees in anthropology/archaeology or a related field fill these positions, as well as people who hold no degrees but have experience and may have completed some appropriate course work. Because of the hierarchy encountered in larger companies, even when lines of communication are open, field staff may feel like an unrecognized cog in a very big wheel. Field crew travel frequently and live on the road for extended periods.

Laboratory staff handle the general processing of artifacts and field data and perform specialized types of analyses. In some companies, the size and makeup of the laboratory staff fluctuates depending on the number and scale of projects underway. Some field staff may become part-time laboratory staff depending on how field schedules are being juggled.

Office staff consists of people who keep the place of business running and perform administrative and support functions. Office staff may include: secretaries, administrative assistants, accountants, computer operators, systems managers, graphic artists, photographers, draftspersons, word processors, and editorial staff. With the possible exception of laboratory personnel, they are the most residentially stable of the whole group. Individuals filling these positions may or may not have college degrees. These are also positions that have equivalents in other types of businesses and are not unique to CRM. Therefore there tends to be a broader consensus about what constitutes adequate skills, pay rates, and benefits for these positions in comparison with positions more unique to CRM.

Because of the volume of business done by larger companies and the fact that they are often embedded in larger corporations, it can be easier for them to offer a full range of benefits to employees. The formulation of career paths and support of continuing education for employees is also something that larger companies are in a better position to offer than smaller ones. Of course the desire to do so has to exist, and this only seems to happen when a firm is as concerned about the bottom line ($$$) as it is about their employees and the quality of their work.

In smaller companies, fewer people are responsible for the same range of tasks necessary to generate business, get a project completed, and maintain an office or facility. Their internal hierarchies are not as rigid as in larger companies. An individual may take on several specialized tasks, while others may be farmed out to subconsultants. The president of the company may also be the principal investigator for a project and run field operations on a daily basis. Field crew may double as laboratory technicians, computer operators, and draftspersons. Interaction and communication between all staff levels is much greater than in larger companies. In particular, a field technician's perception of his/her relative importance may be greater. Working for a smaller company can be a great way to get a lot of useful experience. However, the small number of full-time staff may make it difficult for a smaller company to compete for large jobs on tight time schedules without hiring temporary help—help that will be laid off at the conclusion of the project and not offered the same range of benefits as permanent staff. Regardless of the intentions of the company, a two-tiered labor force like this can have a negative impact on the attitude of employees and the quality of the work that is performed, the same negative impact that larger companies may experience because of their complex hierarchies.

Any company must string projects together in such a way as to maintain cash flow, pay the bills, and maintain employees. This means that marketing and proposal writing will at times take precedence over other tasks that staff may be performing, resulting in other shifts in responsibilities. The principal investigator who is trapped in the office responding to requests for proposals that pop up at the last minute must rely more heavily on field supervisors or crew chiefs to keep fieldwork running smoothly. Production and editorial staff may be diverted from the completion of a technical report. Meanwhile, the field crew may wonder why the principal investigator isn't on-site more frequently making use of the expertise that got them their position in the first place.

CRM is a competitive business, so any company must balance its continuing need for work with staffing capabilities, pay rates, and the fact that contracts are awarded based on an evaluation of the quality and price of proposed work. In too many cases the emphasis seems to be on the price of the proposed work. It makes sense that the more skilled and experienced an employee (at any level), the greater the rate of their pay. The higher you go in a company's hierarchy, the greater your compensation. This wage scale is taken into account when competitive budgets for projects are formulated, and has led to some behaviors that have negative impacts on professional practice, or at least are perceived negatively. One of the ways to make a budget more competitive (i.e., smaller, cheaper) is to limit the amount of time that a principal investigator spends in the field, relegating more of the daily supervision of work to field supervisors or crew chiefs who are not paid at a comparably high rate. Theoretically, principal investigators with graduate degrees should have more of the skills necessary to make fieldwork run smoothly than the field supervisors and crew chiefs they oversee. Shouldn't the quality of work be the issue here and not the expense of keeping a principal investigator in the field for the duration of a project? Issues of inequities in pay versus job responsibilities arise out of this situation. Likewise, the most labor-intensive part of many projects is the field component, so the wages of field crew have a big impact on the size and competitiveness of a budget. For companies thinking only about the bottom line, keeping wages low, offering few or no benefits, and relying more heavily on entry-level and temporary personnel are ways to keep field budgets down. As I have already mentioned, the scale and scheduling of a project can dictate the use of temporary employees, even for larger companies, and brings with it potential problems in labor relations and quality of work.

Back to the field technician experience. Everyone will need some degree of on-the-job training since each company may have a slightly different structure and way of doing things. But the more skills you bring with you to the job the better off you will be, and the greater your chances for advancement. Not everyone on a field crew will be afforded the chance to perform certain tasks, like using surveying equipment to make maps, or excavating some types of archaeological features. Field crews are male dominated although gender equity in archaeology in general is improving (Zeder 1997). People of Hispanic, African American, Native American, and Asian ancestry make up a very small percentage of the profession as a whole (Zeder 1997).

The scheduling of projects and the need of companies to keep field crew employed means that a technician works in all sorts of weather and in different geographic areas, regardless of where home base is located. Since projects can range from site surveys of small parcels of land to intensive excavations of large and deeply stratified sites, the amount of time you spend in an area will vary dramatically. Mobility is a fact of life for a field technician, making it tough to maintain a family life. You will live on the road in motels (even sometimes in tents), eat in restaurants (or figure out some way to cook for yourself), and do your laundry in laundromats. It will be difficult to conduct personal business after hours when far from home. Living expenses while on a project are covered by the company, but the way in which they are covered varies. Some firms will reimburse you for your expenditures on lodging, food, and mileage. Others offer a fixed per diem rate (so much money per day for lodging, so much per day for food) and will not provide monies beyond that. Still others will arrange for lodging, directly shoulder its cost, and provide a set amount of cash up front for meals. Daily transportation from your lodging to the project site may be your responsibility or may be arranged by the company, e.g., everyone may pile into a company van every morning and again at the end of the work day. Travel from your home base to a project location is generally your responsibility although a mileage rate is often paid.

The length of work days and the work week can vary, depending on the type of project and its location. Elaborations of the typical 8-hour work day and 5-day week are either negotiated with crew or are made known prior to signing on for a project. A crew may agree to work 4 long days in exchange for 3-day weekends. This tends to happen a lot around the holidays, or when jobs are located in interesting areas offering opportunities for sight-seeing and other activities. Schedules where crews work for 10 days and have 4 days off are not uncommon. Some excavations on tight schedules have actually employed crews in shifts (Adovasio et al. 1990)! The workday may also be arranged around extremes in weather that can characterize some areas, like the far north and the southwest.

Michele Wilson, a former field technician herself, has attempted to systematically collect data about individuals working as field technicians in CRM, their relationships with managers and other staff, and the problems they face. The general profile of the field technician provided below is based on 39 responses to 122 questionnaires that were distributed as part of her study (Wilson 2000:73):

- *the average age of field technicians is 26 to 39 (72%) with an age range of 18 to over 40.*
- *Most have been working in CRM for 1 to 3 years (53%, n=19) with service ranging from less than 6 months (8%, n=3) to more than 20 years (3%, n=1).*
- *Forty-five percent (n=16) have a college degree from a four year program while 42 percent (n=15) obtained a graduate degree or took classes at the graduate level.*
- *Field technician hourly wages ranged from $8.00 per hour (8%, n=3) to over $10.00 per hour (45%, n=16).*
- *The majority of respondents are employed in the Northeastern, Midwestern, and Mid-Atlantic regions of the United States but many rotate employment between these regions and the Pacific Northwest, California, the Great Basin, the Southwest, and Polynesia. No respondents indicated working in Alaska, the Plains, or the Southeastern United States.*
- *The average field technician works for four companies per year, with some working for as few as one and some working for as many as seven companies in one year.*
- *The average distance traveled, one-way, for a field project is over 200 miles from their home base (45%, n=16).*
- *The average yearly non-reimbursed job-related expenses incurred for all respondents is $817.00.*

Two previous surveys by *The Underground*, one in 1993 (27 responses) and the other in 1995 (34 responses) reached somewhat similar conclusions. A survey of wages can also be viewed on the website of the American Cultural Resources Association (ACRA 2001b). It's obvious from these data that living on field technician wages is tough, especially since periodic unemployment is a real possibility. On the other hand, there are companies who strive to make all of their employees full time, with full benefits and a retirement plan. And there are extremes in pay rates. Any federal CRM project that falls under the Services Contract Act, and most of them do, have to abide by the job descriptions and relevant pay rates established by the Department of Labor (DOL) for archaeological technicians. Pay rates change depending on where a project takes place. Appropriates rates are specified in individual contracts. As I write this, field technicians working on a federal project in downtown Philadelphia are making in excess of $18 per hour.

Field crew perceptions of the job environment and the working conditions promoted by companies varied in some interesting ways in the surveys. In the 1993 *Underground* survey, many of the same companies that appeared on the list of best companies to work for also appeared on the list of the worst companies to work for. What this suggests is that an individual's perception changed depending on the specific project and the people who were running it. It is my impression that both the quality of the field experience and the work that gets done can rise or fall based upon the inclinations of the principal investigator, crew chief, or field supervisor. In the future, more extensive surveys, and questionnaires designed to investigate a person's experience on a company-by-company basis, are needed for a more thorough understanding of the field technician experience. There are firms who are making changes now to solve the problems of the CRM workplace and enhance the career opportunities that the industry provides. For CRM to live up to its greatest promise the willingness for change must become more pervasive.

FINDING A JOB

PREPARING RESUMES

A basic step in securing employment as a field technician is to prepare a resume that along with a cover letter will introduce you to prospective employers. A ***resume*** is a concise summary of your activities and skills relevant to the job that you are seeking. It is distinguished from a ***curriculum vitae*** (CV), which is a more formal and detailed outline of the course of your professional life. Resumes are from 1–3 pages long while CVs are typically longer. As you begin your career your resume should be able to include everything that you have done relevant to archaeology. The purpose of a resume is to secure a job interview by presenting yourself in the best way possible. A resume should be tailored to fit the type of job for which you are applying. The discussion here focuses on field technician positions and is based on general rules for resume construction, and my experiences with interviewing and hiring people during my years in CRM. Common components of a resume are listed in Table 10.1.

Always give a prospective employer the information necessary to contact you immediately. Those who move frequently will therefore want to list a current address as well as a more stable address that can be used as a mail drop or check-in point, like the home of a relative or friend. Because of the often jerky rhythm with which CRM projects are awarded and initiated, the need to hire additional field staff can happen quickly and the search and hiring process can be rapid. On the other hand, your resume may be on file with a company for quite some time before it is pulled and contact initiated. The *Objective* should be a very short statement (2–3 sentences at most) about what you are seeking and how it ties into your skill/experience level. If you are a student looking for a summer job be up-front about it, e.g., "summer employment as a field technician to enhance my existing and ongoing training in archaeology." If you are trying to get hired for a specific project note your interest in the particular geographic area, time period, or research issue. If you are responding to an advertisement paraphrasing portions of it in your statement could be useful. If you are targeting a specific company because of their reputation make this known.

Under the *Education* heading list your high school diploma if you are not in college or have not completed other degrees. For those in college but without a completed degree, the way to express your status is something like this:

> *EDUCATION: Anthropology major, Temple University, Philadelphia, Pennsylvania. B.A. degree expected May 2001.*

Some people list the year when their enrollment began in situations like this (e.g., 1997 to present, Anthropology major . . .). That's not a bad idea unless you have been in school for an extended period because of going part-time or holding down a job. These aren't things of which to be ashamed, but you don't want the person reading your resume looking at the dates and wondering why you have been in school so long when you are not there to provide clarification. This is something that can come out in an interview when you have a chance to deal with questions and impressions as they arise. Making a good impression with a resume means casting few doubts about your abilities that a reader might consider in a negative way. Continuing education classes taken can be noted here, for example:

> *1999–2000, Temple University, continuing education classes in geology, archaeology, and business management.*

Specific courses could also be listed under *Relevant Course Work.* Certificates of specialized training that you might have earned should be listed under the *Skills* heading.

Significant Work Experience refers to things directly related to archaeology and the job for which you are applying. List your participation in a field school or internship here, or volunteer work on a survey, excavation, or in the laboratory, as well as any paid positions. There is no need to state whether or not you were paid in any circumstance. Start with the most recent experience and move backward in time. Each listing should include the date(s) of the specific experience, what your

TABLE 10.1

Common Components of a Resume

Common Components of a Resume
Name
Current Address (and phone number)
Permanent Address
Objective
Education
Significant Work Experience
Other Work Experience
Skills
Relevant Course Work
Interests
Publications
Papers Presented
Awards
Relevant Memberships
Availability
References

position was, the name of the project(s) on which you worked, and its sponsor. Include a brief statement that describes the nature of the project. For example:

> *1998 (May–July):*
>
> *Crew member (as field school participant) for an archaeological survey of Hendrick's Island, Pennsylvania, Department of Anthropology, Temple University, Philadelphia, PA. Involved surface survey and test excavations of Late Archaic through Late Woodland sites.*
>
> *1997–1998 (seasonal):*
>
> *Field technician, Ancient Lives Investigations, Tomstown, Oregon. Participated in four survey and excavation projects in the greater Northwest involving Archaic Native American sites and 19th-century colonial sites. This work included the extensive Williamette pipeline project and excavations at the prehistoric Burnt Tree Site.*

You have undoubtedly held jobs in your life that have nothing directly to do with archaeology. However, these positions can be used to highlight other abilities or experiences that will be of interest to an employer, things like: responsibility for handling cash, managing the work of others, training others in various tasks, bookkeeping, writing, and interacting with the public. Other jobs may have even more relevance, albeit still indirect, to archaeology (think of all of the things that go into the logistics of field and laboratory work). These types of jobs are worth listing under the category of *Other Work Experience.* The key is to highlight the aspect(s) of a job that show that you can handle responsibility, are a problem solver or quick learner, have desirable communication and interpersonal skills, or possess a skill relevant to the logistical, health, or safety aspects of archaeological projects. For example:

> *1998–1999:*
>
> *Knowledge of basic carpentry and construction site machinery (e.g., pumps, generators, portable heaters) developed as an employee of Building Contractors, Inc., Bethel, Maine.*
>
> *2000–present:*
>
> *As night manager of a 7-11 store in Bismarck, North Dakota am responsible for the training and supervision of 3 other staff, maintaining inventories, and interacting with regional managers.*

The *Skills* category can be a distillation of useful abilities and hands-on skills embedded in your work experiences, course work, or personal life. Again, think of things that are relevant to the practice of archaeology in the field or the laboratory. The list of skills may include:

- ☑ familiarity with computers, the types of word processing, data bases, and other software you can use, and GIS experience
- ☑ experience with 35mm or larger format photography, digital photography, darkroom experience
- ☑ orienteering, experience with surveying equipment, drafting
- ☑ hands-on experience with artifact analysis, soil/sediment description and classification
- ☑ foreign languages read or spoken
- ☑ first aid, lifesaving or EMT training.

Of course, list any certificates or official acknowledgments of training that you might have relevant to any of your listed skills.

Relevant Course Work and *Interests* are headings that not everyone may include on a resume. Some people might use a relevant course work heading but not an interests heading. If you've had few relevant courses it could send up a red flag for the resume reader. On the other hand, if you have little to list in the work experience category, listing any relevant courses will put a better spin on your background. Any interests that you list should be things that are directly relevant to archaeology and not replicate something that is self-evident in your listing of work experience, skills, or related course work. It can be useful to include a short phrase with a listing that gives the reader an indication of your active pursuit of this interest, e.g., Paleo-Indian archaeology (pursued through course work, personal reading, examination of museum displays and collections). Don't just mention things to create a list. The judicious listing of a few key interests could be helpful if you are applying for a job on a specific project that dovetails with these pursuits.

Publications, Papers Presented, Awards, and *Memberships* are headings that should be used only if you have something relevant to include in them.

Remember, the purpose of the resume is to honestly highlight your abilities and achievements to get a job interview, not to focus on what you haven't accomplished. Publishing an article or presenting a paper at an archaeological meeting are impressive achievements for undergraduates or people still in training. Under *Papers Presented* you can include any lectures that you were invited to give to local archaeological or historical societies, or presentations to school groups. Note any memberships that you have in state or national organizations dealing with archaeology, anthropology, history, or related fields. Since one benefit of membership in such organizations is often a publication or newsletter, this lets an employer know that you are actively engaged in pursuing your interests and keeping up with the latest developments.

Every resume needs an *Availability* heading. Plainly state when you are available to begin work and for what period of time. If you are looking for part-time work, a summer job, or sporadic opportunities while in school, say so. State the extent to which you are willing to travel. You may apply for a job with a firm based in Michigan but be sent out on projects in Florida or Texas. If you are ready to see the world, simply note that you are willing to travel. It is also useful to mention whether or not you have your own transportation since part of the terms of your employment may require transporting yourself to the area where a project will take place.

There are different opinions about including references on resumes. Some advise simply concluding a resume with the statement, "References available upon request," but making sure that you have contacted 2–3 people about their willingness to serve as references for you if needed. My advice is to always list references on your resume, and in archaeology there are a number of good reasons to do so. One reason relates to the rapidity with which field technicians can be hired when projects have quick start-up dates and existing staff levels are insufficient. Having to contact you to learn about your references is an extra step (read extra time) for a potential employer. Simply seeing who you have listed as references, and how they relate to the education and experiences on your resume, may be the final thing an employer needs to make a decision about whether to contact you. The employer may be familiar with one or more of the people you list and have an opinion about their reputation. "Why, if Dr. X is willing to give this person a recommendation, he can't be too bad."

Select the people you use as references wisely. If you are applying for a job as a field technician, have a reference who is familiar with your abilities in the field or has been in the field with you. List the instructor with whom you have taken many classes rather than one with whom you have had less contact. Avoid using references who are not archaeologists or employed in related professions. Always ask permission to use someone as a reference. Better yet, ask the person if they would be willing to provide you with a *favorable* reference if contacted. This can help you to avoid unpleasant surprises in the future. Provide the names, titles, addresses, and phone numbers of your references on the resume. Make sure that the phone numbers are accurate. In all of the years that I have served as a reference for people, only rarely have I been contacted by mail. A phone call is quick and to the point. Likewise, an employer knows that in conversation he/she may get a more thorough evaluation of a potential employee than when asking for a written evaluation. Things can be asked and said off the official record in a conversation that wouldn't normally appear in a written statement.

As your resume grows with your experience you will eventually need to summarize some things and delete others. A concise summary should never exceed three pages in length. You eventually will find it necessary to summarize different work experiences with the same company under a single entry rather than listing them separately. Presentations to school groups and local societies inevitably get deleted in favor of papers presented at state, regional, and national meetings. As your field experience grows there is less need to list relevant course work.

Never send out a resume without a cover letter. Cover letters are brief, never more than a few paragraphs or a single page. Look at them as a way of preparing someone to read your resume so that they see and emphasize the things that you want them to notice and focus upon. Even with a killer resume, you can't assume that your worth will be self-evident to every reader. Consider restating your objective and summarize the range of skills that you possess and how they relate to the job for which you are applying. Mention your availability and provide a phone number and the best times to call.

You will rarely find an advertisement for archaeological field technicians in the newspaper. There are several ways to organize your job hunt and keep abreast of what opportunities are available. First, target a group of firms for whom you would like to work. These may be companies that you already know something about or that consistently work in a geographic area of interest. Lists of CRM contractors that work in a given state can be obtained from the SHPO. Send a cover letter and resume to each firm that you select, regardless of whether or not they are actively looking for people. Follow up with a phone call. This is a way of making a personal contact, letting someone on the inside connect the real you with a name on a paper. It's also a way of finding out if there are any prospects for employment looming in the future that have not yet been advertised. Job openings can come and go so quickly in CRM that it is impractical to advertise them in traditional ways. When positions open, employers typically go to resumes on file to begin their search, regardless of what other tactics they may employ. If someone remembers their contact with you during this process, so much the better.

You can increase your contacts and information network by attending meetings of archaeological societies and professional organizations, seeing who is active in a certain area and doing the kinds of things of interest to you. Don't hesitate to introduce yourself, even if it is just to say that you enjoyed an individual's presentation. Talk to other people in your position and see what plans and ideas they have. Keep in touch with the people with whom you have worked and developed a relationship.

Because of its "real time" quality, the Internet is increasingly used by employers to advertise positions. A number of websites list job opportunities. The website for Archaeological Field Opportunities hosted by the Costen Institute of Archaeology I have already mentioned. It's listings include notices about paid positions. Sporadic notices also appear on the ACRA listserv (http://www.acra-crm.org) and About Archaeology with K. Kris Hirst (http://archaeology.about.com/science/archaeology). The Shovelbums listserv, moderated by Joe Brandon (http://www.shovelbums.org), is devoted to listing and disseminating notices for field crew and other positions around the country. The website of Southwestern Archaeology (http://www.swanet.org/jobs.html) provides current listings of paid positions available in the southwestern United States. Employment opportunities for archaeologists at all levels are periodically updated on the website of the Society for Historical Archaeology (http://www.sha.org/nl-emp.htm). Job notices also appear on the website of the Society for American Archaeology, but these are typically for positions requiring a graduate degree.

Job interviews for field technicians may take place in-person or over the phone. The phone is often used because of the immediate need for crew that CRM firms can experience. Obviously an in-person interview gives you more of a chance to make an impression. The interview is not only an opportunity for a prospective employer to learn about you, but for you learn about the company. Be yourself and be honest about your qualifications. The interviewer will have read your resume, and may want clarification or more explanation of some items. You may start with a brief, in-your-own-words summary of your objective and background. The experienced interviewer will diplomatically discover areas where on-the-job training is necessary. There is no bluffing here. Once in the field you either know what you are doing or you don't. Be prepared with questions to ask about the firm itself. What are their expectations of you? What will be your responsibilities? Ask about pay scales, benefits, and career advancement. Ask specifically about how pay scales for field technicians are associated with levels of expertise and responsibilities so that you can see where you fit. The discussion of the field technician experience in this chapter should give you plenty to think about in this respect.

Archaeological fieldwork remains an exciting and challenging experience, warts and all. Hopefully the skills that you have gained in using this book will make your entry and passage into the field more rewarding. The best of luck!

APPLYING YOUR KNOWLEDGE

1. Evaluate the degree to which your college program and course work equips you with the skills and experiences necessary to meet the requirements of the job descriptions in Appendix 5.

2. Prepare your resume and a cover letter. Exchange materials with classmates and critique them. What is the impression you get of a person from reading the resume and is it the same one that the writer hoped to convey? What can be done to improve the resume? What does the writer see as his/her greatest strengths or outstanding qualities? Is the cover letter written in such a way as to reflect this or to guide the reader to appropriate sections of the resume?

3. Many CRM companies are anxious to work more closely with colleges and universities to improve communications and upgrade training and professional practice. Arrange to speak with a senior staff person at a CRM company about what they look for in an entry-level employee, the variety of positions held by persons with undergraduate degrees, pay rates, benefits, and career development. Lists of companies working in your state can be obtained from your SHPO (see Chapter 3; Appendix 3). Or have your instructor invite a senior staff person to address your class. A panel discussion focused on working as a field technician involving both senior staff and field crew from different companies can be very lively and informative.

DIG DEEPER

❑ There are a number of websites that provide essays on careers in archaeology.

- Archaeology & You, Chapter 3: Archaeology as a Career or Avocation is sponsored by the Society for American Archaeology at **http://www.saa.org/Whatis/arch&you/chap3.html**
- Careers in Historical Archaeology is sponsored by the Society for Historical Archaeology at **http://www.sha.org/sha_cbro.htm**> This essay also deals with underwater archaeology.
- Frequently Asked Questions About Archaeology (FAQ) prepared by David Carlson, Texas A&M University, is available at **http://www.museum.state.il.us/ismdepts/anthro/dlcfaq.html**

❑ Descriptions of aspects of the field technician's job and life:

- Hirst, K. Kris. 1998. Have trowel will travel. Serial essay appearing on the website About Archaeology with K. Kris Hirst, **http://archaeology.about.com/science/archaeology/blfieldtech.htm**
- Kintz, Theresa. 1997. View from the trenches. *Common Ground* 2(1):48–53.
- McGuire, Randall H. and Mark Walker. 1999. Class confrontations in archaeology. *Historical Archaeology* 33(1):159–183.
- Wilson, Michele. 2000. *Tales From the Trenches: The People, Policies, and Procedures of Cultural Resource Management.* Masters thesis, Oregon State University, Corvallis. This thesis focuses on the problems that confront field technicians working in CRM.
- Zeder, Melinda.1997. *The American Archaeologist: A Profile.* Walnut Creek, CA: AltaMira Press. This study looks at the reality of training and practice as reflected by the jobs currently held by archaeologists. The primary focus is on individuals with graduate degrees and related positions.

Continued

- *United Archaeological Field Technicians:* **http://www.members.aol.com/UAFT/home.htm**> Home page for an active union representing field technicians. The essay "What Is an Archaeological Technician?" focuses heavily on labor relations and the problems associated with working in CRM. *The United Archaeological Field Technicians Handbook,* at the same website, details workers' rights in terms of working conditions, wages, health, and safety.

❑ To get a sense of what it is like to have a long-term career in archaeology read a biography or autobiography of some of the notable figures in the profession, like those noted below.

- Cressman, Luther. 1988. *A Golden Journey: Memoirs of an Archaeologist.* Salt Lake City: University of Utah Press.

- Jennings, Jesse D. 1994. *Accidental Archaeologist: Memoirs of Jesse D. Jennings.* Salt Lake City: University of Utah Press.

- White, Nancy Marie, Lynne P. Sullivan, and Rochelle A. Marrinan. 1999. *Grit-Tempered: Early Women Archaeologists in the Southeastern United States.* Gainesville: University of Florida Press.

- Willey, Gordon R. 1988. *Portraits in American Archaeology: Remembrances of Some Distinguished Americanists.* Albuquerque: University of New Mexico Press.

APPENDIX 1

Ethical Principles and Codes of Conduct for Archaeologists

SOCIETY for AMERICAN ARCHAEOLOGY (reproduced from the society's webpage at http://www.saa.org/Society/Ethics/prethic.html>). Reprinted with permission of the Society for American Archaeology.

At its April 10, 1996 meeting, the SAA Executive Board adopted the Principles of Archaeological Ethics, reproduced below, as proposed by the SAA Ethics in Archaeology Committee. The adoption of these principles represents the culmination of an effort begun in 1991 with the formation of the ad-hoc Ethics in Archaeology Committee. The committee was charged with considering the need for revising the society's existing statements on ethics. A 1993 workshop on ethics, held in Reno, resulted in draft principles that were presented at a public forum at the 1994 annual meeting in Anaheim. SAA published the draft principles with position papers from the forum and historical commentaries in a special report distributed to all members, Ethics and Archaeology: Challenges for the 1990s, edited by Mark. J. Lynott and Alison Wylie (1995). Member comments were solicited in this special report, through a notice in the SAA Bulletin, and at two sessions held at the SAA booth during the 1995 annual meeting in Minneapolis. The final principles, presented here, are revised from the original draft based on comments from members and the Executive Board.

The Executive Board strongly endorses these principles and urges their use by all archaeologists "in negotiating the complex responsibilities they have to archaeological resources, and to all who have an interest in these resources or are otherwise affected by archaeological practice (Lynott and Wylie 1995:8)." The board is grateful to those who have contributed to the development of these principles, especially the members of the Ethics in Archaeology Committee, chaired by Mark. J. Lynott and Alison Wylie, for their skillful completion of this challenging and important task. The bylaws change just voted by the members has established a new standing committee, the Committee on Ethics, that will carry on with these crucial efforts.

PRINCIPLE NO. 1: STEWARDSHIP

The archaeological record, that is, in situ archaeological material and sites, archaeological collections, records and reports, is irreplaceable. It is the responsibility of all archaeologists to work for the long-term conservation and protection of the archaeological record by practicing and promoting stewardship of the archaeological record. Stewards are both caretakers of and advocates for the archaeological record for the benefit of all people; as they investigate and interpret the record, they should use the specialized knowledge they gain to promote public understanding and support for its long-term preservation.

PRINCIPLE NO. 2: ACCOUNTABILITY

Responsible archaeological research, including all levels of professional activity, requires an acknowledgment of public accountability and a commitment to make every reasonable effort, in good faith, to consult actively with affected group(s), with the goal of establishing a working relationship that can be beneficial to all parties involved.

PRINCIPLE NO. 3: COMMERCIALIZATION

The Society for American Archaeology has long recognized that the buying and selling of objects out of archaeological context is contributing to the destruction of the archaeological record on the American continents and around the world. The commercialization of archaeological objects—their use as commodities to be exploited for personal enjoyment or profit—results in the destruction of archaeological sites and of contextual information that is essential to understanding the archaeological record. Archaeologists should therefore carefully weigh the benefits to scholarship of a project against the costs of potentially enhancing the commercial value of archaeological objects. Whenever possible they should discourage, and should themselves avoid, activities that enhance the commercial value of archaeological objects, especially objects that are not curated in public institutions, or readily available for scientific study, public interpretation, and display.

PRINCIPLE NO. 4: PUBLIC EDUCATION AND OUTREACH

Archaeologists should reach out to, and participate in cooperative efforts with others interested in the archaeological record with the aim of improving the preservation, protection, and interpretation of the record. In particular, archaeologists should undertake to: 1) enlist public support for the stewardship of the archaeological record; 2) explain and promote the use of archaeological methods and techniques in understanding human behavior and culture; and 3) communicate archaeological interpretations of the past. Many publics exist for archaeology including students and teachers; Native Americans and other ethnic, religious, and cultural groups who find in the archaeological record important aspects of their cultural heritage; lawmakers and government officials; reporters, journalists, and others involved in the media; and the general public. Archaeologists who are unable to undertake public education and outreach directly should encourage and support the efforts of others in these activities.

PRINCIPLE NO. 5: INTELLECTUAL PROPERTY

Intellectual property, as contained in the knowledge and documents created through the study of archaeological resources, is part of the archaeological record. As such it should be treated in accord with the principles of stewardship rather than as a matter of personal possession. If there is a compelling reason, and no legal restrictions or strong countervailing interests, a researcher may have primary access to original materials and documents for a limited and reasonable time, after which these materials and documents must be made available to others.

PRINCIPLE NO. 6: PUBLIC REPORTING AND PUBLICATION

Within a reasonable time, the knowledge archaeologists gain from investigation of the archaeological record must be presented in accessible form (through publication or other means) to as wide a range of interested publics as possible. The documents and materials on which publication and other forms of public reporting are based should be deposited in a suitable place for permanent safekeeping. An interest in preserving and protecting in situ archaeological sites must be taken into account when publishing and distributing information about their nature and location.

PRINCIPLE NO. 7: RECORDS AND PRESERVATION

Archaeologists should work actively for the preservation of, and long term access to, archaeological collections, records, and reports. To this end, they should encourage colleagues, students, and others to make responsible use of collections, records, and reports in their research as one means of preserving the in situ archaeological record, and of increasing the care and attention given to that portion of the archaeological record which has been removed and incorporated into archaeological collections, records, and reports.

PRINCIPLE NO. 8: TRAINING AND RESOURCES

Given the destructive nature of most archaeological investigations, archaeologists must ensure that they have adequate training, experience, facilities, and other support necessary to conduct any program of research they initiate in a manner consistent with the foregoing principles and contemporary standards of professional practice.

THE REGISTER OF PROFESSIONAL ARCHAEOLOGISTS (reproduced from their web page at http://www.rpanet.org/conduct.htm). Reprinted with permission of the Register of Professional Archaeologists.

CODE OF CONDUCT

Archaeology is a profession, and the privilege of professional practice requires professional morality and professional responsibility, as well as professional competence, on the part of each practitioner.

I. The Archaeologist's Responsibility to the Public

1.1 An archaeologist shall:

a. Recognize a commitment to represent Archaeology and its research results to the public in a responsible manner;

b. Actively support conservation of the archaeological resource base;

c. Be sensitive to, and respect the legitimate concerns of, groups whose culture histories are the subjects of archaeological investigations;

d. Avoid and discourage exaggerated, misleading, or unwarranted statements about archaeological matters that might induce others to engage in unethical or illegal activity;

e. Support and comply with the terms of the UNESCO Convention on the means of prohibiting and preventing the illicit import, export, and transfer of ownership of cultural property, as adopted by the General Conference, 14 November 1970, Paris.

1.2 An archaeologist shall not:

a. Engage in any illegal or unethical conduct involving archaeological matters or knowingly permit the use of his/her name in support of any illegal or unethical activity involving archaeological matters;

b. Give a professional opinion, make a public report, or give legal testimony involving archaeological matters without being as thoroughly informed as might reasonably be expected;

c. Engage in conduct involving dishonesty, fraud, deceit or misrepresentation about archaeological matters;

d. Undertake any research that affects the archaeological resource base for which she/he is not qualified.

II. The Archaeologist's Responsibility to Colleagues, Employees, and Students

2.1 An archaeologist shall:

a. Give appropriate credit for work done by others;

b. Stay informed and knowledgeable about developments in her/his field or fields of specialization;

c. Accurately, and without undue delay, prepare and properly disseminate a description of research done and its results;

d. Communicate and cooperate with colleagues having common professional interests;

e. Give due respect to colleagues' interests in, and rights to, information about sites, areas, collections, or data where there is a mutual active or potentially active research concern;

f. Know and comply with all federal, state, and local laws, ordinances, and regulations applicable to her/his archaeological research and activities;

g. Report knowledge of violations of this Code to proper authorities;

h. Honor and comply with the spirit and letter of the Register of Professional Archaeologist's Disciplinary Procedures.

2.2 An archaeologist shall not:

a. Falsely or maliciously attempt to injure the reputation of another archaeologist;

b. Commit plagiarism in oral or written communication;

c. Undertake research that affects the archaeological resource base unless reasonably prompt, appropriate analysis and reporting can be expected;

d. Refuse a reasonable request from a qualified colleague for research data;

e. Submit a false or misleading application for registration by the Register of Professional Archaeologists.

III. The Archaeologist's Responsibility to Employers and Clients

3.1 An archaeologist shall:

a. Respect the interests of her/his employer or client, so far as is consistent with the public welfare and this Code and Standards;

b. Refuse to comply with any request or demand of an employer or client which conflicts with the Code and Standards;

c. Recommend to employers or clients the employment of other archaeologists or other expert consultants upon encountering archaeological problems beyond her/his own competence;

d. Exercise reasonable care to prevent her/his employees, colleagues, associates and others whose services are utilized by her/him from revealing or using confidential information. Confidential information means information of a non-archaeological nature gained in the course of employment which the employer or client has requested be held inviolate, or the disclosure of which would be embarrassing or would be likely to be detrimental to the employer or client. Information ceases to be confidential when the employer or client so indicates or when such information becomes publicly known.

3.2 An archaeologist shall not:

a. Reveal confidential information, unless required by law;

b. Use confidential information to the disadvantage of the client or employer;

c. use confidential information for the advantage of herself/himself or a third person, unless the client consents after full disclosure;

d. Accept compensation or anything of value for recommending the employment of another archaeologist or other person, unless such compensation or thing of value is fully disclosed to the potential employer or client;

e. Recommend or participate in any research which does not comply with the requirements of the Standards of Research Performance.

STANDARDS OF RESEARCH PERFORMANCE

The research archaeologist has a responsibility to attempt to design and conduct projects that will add to our understanding of past cultures and/or that will develop better theories, methods, or techniques for interpreting the archaeological record, while causing minimal attrition of the archaeological resource base. In the conduct of a research project, the following minimum standards should be followed:

I. The archaeologist has a responsibility to prepare adequately for any research project, whether or not in the field. The archaeologist must:

1.1 Assess the adequacy of her/his qualifications for the demands of the project, and minimize inadequacies by acquiring additional expertise, by bringing in associates with the needed qualifications, or by modifying the scope of the project;

1.2 Inform herself/himself of relevant previous research;

1.3 Develop a scientific plan of research which specifies the objectives of the project, takes into account previous relevant research, employs a suitable methodology, and provides for economical use of the resource base (whether such base consists of an excavation site or of specimens) consistent with the objectives of the project;

1.4 Ensure the availability of adequate and competent staff and support facilities to carry the project to completion, and of adequate curatorial facilities for specimens and records;

1.5 Comply with all legal requirements, including, without limitation, obtaining all necessary governmental permits and necessary permission from landowners or other persons;

1.6 Determine whether the project is likely to interfere with the program or projects of other scholars and, if there is such a likelihood, initiate negotiations to minimize such interference.

II. In conducting research, the archaeologist must follow her/his scientific plan of research, except to the extent that unforeseen circumstances warrant its modification.

III. Procedures for field survey or excavation must meet the following minimal standards:

3.1 If specimens are collected, a system for identifying and recording their proveniences must be maintained.

3.2 Uncollected entities such as environmental or cultural features, depositional strata, and the like, must be fully and accurately recorded by appropriate means, and their location recorded.

3.3 The methods employed in data collection must be fully and accurately described. Significant stratigraphic and/or associational relationships among artifacts, other specimens, and cultural and environmental features must also be fully and accurately recorded.

3.4 All records should be intelligible to other archaeologists. If terms lacking commonly held referents are used, they should be clearly defined.

3.5 Insofar as possible, the interests of other researchers should be considered. For example, upper levels of a site should be scientifically excavated and recorded whenever feasible, even if the focus of the project is on underlying levels.

IV. During accessioning, analysis, and storage of specimens and records in the laboratory, the archaeologist must take precautions to ensure that correlations between the specimens and the field records are maintained, so that provenience contextual relationships and the like are not confused or obscured.

V. Specimens and research records resulting from a project must be deposited at an institution with permanent curatorial facilities, unless otherwise required by law.

VI. The archaeologist has responsibility for appropriate dissemination of the results of her/his research to the appropriate constituencies with reasonable dispatch.

6.1 Results reviewed as significant contributions to substantive knowledge of the past or to advancements in theory, method or technique should be disseminated to colleagues and other interested persons by appropriate means such as publications, reports at professional meetings, or letters to colleagues.

6.2 Requests from qualified colleagues for information on research results directly should be honored, if consistent with the researcher's prior rights to publication and with her/his other professional responsibilities.

6.3 Failure to complete a full scholarly report within 10 years after completion of a field project shall be construed as a waiver of an archaeologist's right of primacy with respect to analysis and publication of the data. Upon expiration of such 10-year period, or at such earlier time as the archaeologist shall determine not to publish the results, such data should be made fully accessible to other archaeologists for analysis and publication.

6.4 While contractual obligations in reporting must be respected, archaeologists should not enter into a contract which prohibits the archaeologist from including her or his own interpretations or conclusions in the contractual reports, or from a continuing right to use the data after completion of the project.

6.5 Archaeologists have an obligation to accede to reasonable requests for information from the news media.

AMERICAN CULTURAL RESOURCES ASSOCIATION (reproduced from the association's Web page at http://www.acra-crm.org/ Ethics.html>). Reprinted with permission of the American Cultural Resources Association.

CODE OF ETHICS AND PROFESSIONAL CONDUCT

PREAMBLE

This Code of Ethics and Professional Conduct is a guide to the ethical conduct of members of the American Cultural Resources Association (ACRA). The Code also aims at informing the public of the principles to which ACRA members subscribe. The Code further signifies that ACRA members shall abide by proper and legal business practices, and perform under a standard of professional behavior that adheres to high principles of ethical conduct on behalf of the public, clients, employees, and professional colleagues.

THE ACRA MEMBER'S RESPONSIBILITIES TO THE PUBLIC

A primary obligation of an ACRA member is to serve the public interest. While the definition of the public interest changes through ongoing debate, an ACRA member owes allegiance to a responsibly derived concept of the public interest. An ACRA member shall:

1) Have concern for the long-range consequences of that member's professional actions.

2) Be cognizant of the relevance to the public of that member's professional decisions.

3) Strive to present the results of significant research to the public in a responsible manner.

4) Strive to actively support conservation of the cultural resource base.

5) Strive to respect the concerns of people whose histories and/or resources are the subject of cultural resources investigation.

6) Not make exaggerated, misleading, or unwarranted statements about the nature of that member's work.

THE ACRA MEMBER'S RESPONSIBILITIES TO CLIENTS

An ACRA member is obligated to provide diligent, creative, honest, and competent services and professional advice to its clients. Such performance must be consistent with the ACRA member's responsibilities to the public interest. An ACRA member shall:

1) Exercise independent professional judgment on behalf of clients.

2) Accept the decisions of a client concerning the objectives and nature of the professional services provided unless the decisions involve conduct that is illegal or inconsistent with the ACRA member's obligations to the public interest.

3) Fulfill the spirit, as well as the letter, of contractual agreements.

4) Not provide professional services if there is an actual, apparent, or perceived conflict of interest, or an appearance of impropriety, without full written disclosure and agreement by all concerned parties.

5) Not disclose information gained from the provision of professional services for private benefit without prior client approval.

6) Not solicit prospective clients through the use of false or misleading claims.

7) Not sell or offer to sell services by stating or implying an ability to influence decisions by improper means.

8) Not solicit or provide services beyond the level or breadth of the professional competence of its staff or project team.

9) Solicit or provide services only if they can responsibly be performed with the timeliness required by its clients.

10) Not solicit or accept improper compensation for the provision of judgments or recommendations favorable to its clients.

11) Not offer or provide improper compensation as a material consideration in obtaining or sustaining client or prospective client favor.

12) Disclose information identified as confidential by its client only if required by law, required to prevent violation of the law, or required to prevent injury to the public interest.

THE ACRA MEMBER'S RESPONSIBILITIES TO EMPLOYEES

As an employer, an ACRA member firm has certain responsibilities to its employees, and shall strive to:

1) Comply with all applicable employment/labor laws and regulations.

2) Provide a safe work environment in compliance with all applicable laws and regulations.

3) Appropriately acknowledge work performed by employees.

4) Provide opportunities for the professional growth and development of employees.

5) Develop clear lines of communication between employer and employee, and provide employees with a clear understanding of their responsibilities.

6) Consistently maintain fair, equitable, and professional conduct toward its employees.

THE ACRA MEMBER'S RESPONSIBILITIES TO PROFESSIONAL COLLEAGUES

An ACRA member shall strive to contribute to the development of the profession by improving methods and techniques, and contributing knowledge. An ACRA member shall also fairly treat the views and contributions of professional colleagues and members of other professions. Accordingly, an ACRA member shall:

1) Act to protect and enhance the integrity of the cultural resources profession.

2) Accurately and fairly represent the qualifications, views, and findings of colleagues.

3) Review the work of other professionals in a fair, professional, and equitable manner.

4) Strive to communicate, cooperate, and share knowledge with colleagues having common professional interests.

5) Not knowingly attempt to injure the professional reputation of a colleague.

APPENDIX 2

Summaries of Significant Federal Historic Preservation Legislation

Antiquities Act, 1906: designed to bring a measure of protection to sites of national interest. It gave the president the authority to designate as national monuments, landmarks, prehistoric, and historic structures located on government property. Penalties are prescribed for those found damaging or appropriating any object from any monuments. A permitting process is identified for any professional investigations taking place within a national monument. The Act also grants the authority for private property to be taken by the government if it contained landmarks, prehistoric, or historic structures of outstanding national interest (a rarely used provision).

Historic Sites Act, 1935: established the National Historic Landmarks Program and recognized the importance of preserving important sites, buildings, and objects as a matter of national policy. It allows the secretary of the interior, working through the National Park Service, to conduct site surveys, collect data, acquire historic properties and archaeological sites, and rehabilitate and maintain them.

Federal Highway Acts, 1956, 1958: federally sponsored highway construction must avoid using lands with historic sites unless no other alternatives are available. These acts represent the first attempts to protect historic properties from the effects of federal undertakings.

Reservoir Salvage Act, 1960: provided for the conduct and funding of archaeological and historical surveys prior to the construction of dams and related facilities by the government, or by others requiring a federal permit. Research is to be limited to sites that are of exceptional significance and artifacts are to be deposited with interested federal and state agencies, educational, scientific, and private organizations and institutions.

Department of Transportation Act, 1966: in some respects, a reiteration of the intent of the Federal Highways Act with the difference being that transportation projects should not be initiated if they require using land with sites of local, state, or national significance (unless there are no other alternatives). It incorporates the sense of resource significance found in the National Historic Preservation Act, but allows officials at local, state, or federal levels to determine significance. The effect of a project on a historic resource must be considered regardless of whether or not it is listed or eligible for inclusion in the National Register.

National Historic Preservation Act, 1966 (amended 1980, 1986, 1999): This act has been called the cornerstone of historic preservation law in the United States. It builds upon and clarifies previous laws defining the government's responsibilities regarding cultural resources related to sponsored projects. It establishes the National Register of Historic Places, a listing of objects, structures, sites, districts (composed of individual properties), archaeological resources, Indian and Hawaiian sacred sites, and landscapes that have cultural or design values. In order to be listed, a resource must be determined to be significant at the local, state, or national level. Procedures to evaluate and nominate sites to the Register are defined and the National Park Service is charged with the maintenance of the Register. As well, it outlines the procedure for the evaluation and registering of National Historic Landmarks.

The Act created State Historic Preservation Officers (SHPO). Today there are also Tribal Historic Preservation Officers (THPO) with responsibility for Indian lands. Each SHPO is appointed by the governor of a state and oversees historic preservation or cultural resource management activities in that state. The powers of the Advisory Council on Historic Preservation (ACHP) were expanded regarding the formulation of regulations for Section 106 reviews (see below) and consulting throughout this process. The ACHP is an independent federal agency that advises the president and Congress about historic preservation matters.

The Act provides for the granting of funds to states for conducting comprehensive surveys to identify historic properties and to establish matching grants-in-aid for historic preservation. Federal agencies may waive the 1 percent cap on funding projects established by the Archaeological and Historic Preservation Act, 1974.

Section 106 of the National Historic Preservation Act requires federal agencies to consider the effects of their actions on "historic properties," and if there will be an adverse effect, determine how it might be avoided, minimized, or mitigated. A historic property is defined as a property that is included in or eligible for the National Register of Historic Places. For the purposes of the National Register, an "archaeological site" is defined as the place or places where the remnants of a past culture survive in a physical context that allows for the interpretation of these remains. Archaeological sites are frequently determined to be eligible for listing in the National Register because they have yielded, or may be likely to yield, information important to prehistory or history.

National Environmental Protection Act, 1969: Section 101 (b) of the Act states "it is the continuing responsibility of the Federal Government to use all practicable means, consistent with other essential considerations of national policy" to avoid environmental degradation, preserve historic, cultural, and natural resources, and "promote the widest range of beneficial uses of the environment without undesirable and unintentional consequences." The Act makes environmental protection a part of the mandate of every federal agency and department. It requires analysis and a detailed statement of the environmental impact of any proposed federal action that significantly affects the quality of the human environment.

Executive Order 11593, 1971: federal agencies with landholdings are required to prepare inventories of cultural resources on their property and where appropriate, nominate sites to the National Register of Historic Places. This executive order is later incorporated into amended versions of the National Historic Preservation Act in 1980.

Archaeological and Historic Preservation Act, 1974: recognizes that federally supported construction activities, or activities requiring a federal license or assistance might impact historic and archaeological sites. The agency involved in the activity can authorize the recovery or protection of data and sites, using agency funds, or request help from the secretary of the interior. This Act is basically an update of the Reservoir Salvage Act, broadening the umbrella of federal activities that must consider their impact on cultural resources. It also broke new ground in stating that up to 1 percent of the funds available for a construction project may be used for historic preservation. This Act is later encompassed by the 1980 amendments to the National Historic Preservation Act.

Federal Land Policy Management Act, 1976: establishes policies for land management and its natural and cultural components. Natural and cultural resources on federal lands are to be inventoried. In terms of cultural resources, agencies responsible for federal lands must protect and preserve the scientific, historic, cultural, or archaeological value of important resources.

American Indian Religious Freedom Act, 1978: designed to protect and preserve for American Indians, Eskimos, Aleuts, and Native Hawaiians their inherent right of freedom to believe, express, and exercise their traditional religions. It allows access to sites, use and possession of sacred objects, and freedom to worship through ceremonial and traditional practices. The federal government, its agencies and departments need to consult with traditional religious leaders to see how government practices affect Native American cultural and religious practices. A presidential executive order (13007) issued in 1996 deals more specifically with Indian sacred sites.

Archaeological Resources Protection Act, 1979: recognizes that existing federal laws are not adequate to protect or prevent the destruction of archaeological sites on public and Indian lands from uncontrolled excavation and looting. Important information may be contained in private collections made from sites on public and Indian lands prior to the passage of the Act, and this information should be communicated and shared between governmental authorities, the professional archaeological community, and individuals holding collections. Professional fieldwork on public and Indian lands requires application and receipt of a permit. Procedures are established to determine the disposition of artifacts and data resulting from such work. Prohibits the removal of artifacts, the transport, exchange, or sale of archaeological resources, any type of destruction or defacement of archaeological resources, and outlines civil and criminal penalties for prohibited activities. Sites need not be of national interest or significance, but should be at least 100 years old and capable of providing scientific or humanistic understanding of past human behavior, cultural adaptation, and related topics through the application of scientific or scholarly techniques. It involves Indian groups in management decisions and activities.

Abandoned Shipwreck Act, 1987: the U.S. government asserted title to shipwrecks located within 3 miles of the coastline of the United States. Title to wrecks within state submerged lands or coralline formations protected by the state is transferred to the state with the mandate to develop management plans for these resources that take into account the needs of all interested parties, such as sport divers, archaeologists, fisherman and salvors. This legislation removes shipwrecks from the jurisdiction of the law of salvage and the law of finds.

Native American Graves Protection and Repatriation Act, 1990: requires that federal agencies inventory human remains and associated grave goods in their holdings. Anyone who has been involved in a federal undertaking, used federal money, or required a federal permit that resulted in the collection of human remains, or grave goods must do the same. The same must be done with any human remains or funerary objects discovered, or originating on federal property. Culturally affiliated tribes must be presented with inventories of sacred objects, human remains and grave goods. The return (repatriation) of sacred objects, human remains, and grave goods occurs at the request of the culturally affiliated tribe, which must be federally recognized. The discovery of human remains, grave goods, or sacred objects during an investigation (on federal or tribal lands, or involving a federally supported action) requires consultation between affiliated or potentially affiliated Native Americans and the other parties involved about the treatment and disposition of the remains or objects involved. The sale or purchase of Native American human remains is illegal, regardless of whether or not they were found on federal or Indian lands. Work is currently underway to develop rules for dealing with culturally unaffiliated remains of Native Americans.

Executive Order 13007, 1996: deals with the protection and accommodation of access to Indian sacred sites on federal lands. It requires those that are responsible for the management of federal lands to accommodate access to, and ceremonial use of Indian sacred sites by Indian religious practitioners, and avoid adversely affecting the physical integrity of such sacred sites. It stresses that attempts should be made to maintain the confidentiality of sacred sites.

The summaries above were prepared using documents created by the Advisory Council on Historic Preservation and the National Park Service.

For information about state-based efforts to protect and manage archaeological resources see:

Carnett, Carol L. 1995. *A Survey of State Statutes Protecting Archaeological Resources.* Preservation Law Reporter Special Report, Archaeological Assistance Study Number 3. Washington, D.C.: National Trust for Historic Preservation.

State Historic Preservation Offices (SHPO) Data Base, hosted by the National Conference of State Legislatures. http://www.ncsl.org/programs/arts/statehist_intro.htm

APPENDIX 3

Listing of State Historic Preservation Offices and Important Federal Agencies for Historic Preservation

The following list was current as of March, 2001. Specific personnel, phone numbers, and email addresses may change, but office addresses should remain relatively stable. This information is provided with the permission of the National Conference of State Historic Preservation Officers. Periodically updated listings may be found at http://www.sso.org/ncshpo>

ALABAMA
Dr. Lee Warner, SHPO
Alabama Historical Commission
468 South Perry Street
Montgomery, AL 36130-0900
334-242-3184 FAX: 334-240-3477
E-Mail: lwarner@mail.preserveala.org
Deputy: Ms. Elizabeth Ann Brown
E-Mail: ebrown@mail.preserveala.org
www.preserveala.org/

ALASKA
Ms. Judith Bittner, SHPO
Alaska Department of Natural Resources
Office of History and Archaeology
550 West 7th Avenue, Suite 1310
Anchorage, AK 99501-3565
907-269-8721 FAX: 907-269-8908
E-Mail: judyb@dnr.state.ak.us
Deputy: Joan Antonson
www.dnr.state.ak.us/parks/oha_web

AMERICAN SAMOA
Mr. John Enright, HPO
Executive Offices of the Governor
American Samoa Historic Preservation Office
American Samoa Government
Pago Pago, American Samoa 96799
011-684-633-2384 FAX: 011-684-633-2367
E-Mail: enright@samoatelco.com
Deputy: Mr. David J. Herdrich
E-Mail: herdrich@samoatelco.com

ARIZONA
Mr. James W. Garrison, SHPO
Arizona State Parks
1300 West Washington
Phoenix, AZ 85007
602-542-4174 FAX: 602-542-4180
E-Mail: jgarrison@pr.state.az.us
Deputy: Ms. Carol Griffith
E-Mail: cgriffith@pr.state.az.us
Deputy: Dr. William Collins
E-mail: wcollins@pr.state.az.us
www.pr.state.az.us

ARKANSAS
Ms. Cathie Matthews, SHPO
Department of Arkansas Heritage
323 Center Street, Suite 1500
Little Rock, AR 72201
501-324-9150 FAX: 501-324-9154
E-Mail: cathiem@arkansasheritage.org
Deputy: Mr. Ken Grunewald, 501-324-9357
E-Mail: keng@arkansasheritage.org
www.arkansasperservation.org/

CALIFORNIA
Dr. Knox Mellon, SHPO
Office of Historical Preservation,
Dept. Parks & Recreation
P.O. Box 942896
Sacramento, CA 94296-0001
916-653-6624 FAX: 916-653-9824
Deputy: Mr. Daniel Abeyta
E-Mail: dabey@ohp.parks.ca.gov
http://ohp.cal-parks.ca.gov

COLORADO
Ms. Georgianna Contiguglia, SHPO
Colorado Historical Society
1300 Broadway
Denver, CO 80203
303-866-3395 FAX: 303-866-4464
Deputy: Mr. Mark Wolfe, 303-866-2776,
FAX: 303-866-2041
E-Mail: mark.wolfe@chs.state.co.us
Deputy: Dr. Susan M. Collins, 303-866-2736
E-Mail: susan.collins@chs.state.co.us
Tech Ser: Ms. Kaaren Hardy, 303-866-3398
E-Mail: kaaren.hardy@chs.state.co.us
www.coloradohistory-oahp.org

CONNECTICUT
Mr. John W. Shannahan, SHPO
Connecticut Historical Commission
59 So. Prospect Street
Hartford, CT 06106
860-566-3005 FAX: 860-566-5078
E-Mail: cthist@neca.com
Deputy: Dr. Dawn Maddox, President Programs Sup
www.lib.uconn.edu:80/ArchNet/Topical/CRM/
Conn/ctshpo.html

DELAWARE
Mr. Daniel Griffith, SHPO
Division of Historical and Cultural Affairs
P.O. Box 1401
Dover, DE 19903
302-739-5313 FAX: 302-739-6711
E-Mail: dgriffith@state.de.us
Deputy: Ms. Joan Larrivee
Delaware State Historical Preservation Office
15 The Green
Dover, DE 19901
302-739-5685 FAX: 302-739-5660
E-Mail: jlarrivee@state.de.us
www.state.de.us/shpo/index.htm

DISTRICT OF COLUMBIA
Mr. Gregory McCarthy, SHPO
c/o Historic Preservation Division
801 N. Capitol Street, NE, 3rd floor
Washington, DC 20002
202-442-4570 FAX: 202-442-4860
www.dcra.org
Deputy: Mr. Stephen J. Raiche

FLORIDA
Dr. Janet Snyder Matthews, SHPO, Director
Div. of Historical Resources, Dept. of State
R. A. Gray Building, Room 305
500 S. Bronough St.
Tallahassee, FL 32399-0250
850-488-1480 FAX 850-488-3353
E-Mail: jmatthews@mail.dos.state.fl.us
800-847-7278
www.dhr.dos.state.fl.us/

GEORGIA
Mr. Lonice C. Barrett, SHPO
Historic Preservation Division/DNR
156 Trinity Avenue, SW, Suite 101
Atlanta, GA 30303-3600
404-656-2840 FAX 404-651-8739
Deputy: Dr. W. Ray Luce, Director
E-Mail: ray_luce@mail.dnr.state.ga.us
Deputy: Ms. Carole Griffith
E-Mail: carole_griffith@mail.dnr.state.ga.us
Deputy: Mr. Richard Cloues
E-Mail: richard_cloues@mail.dnr.state.ga.us
www.dnr.state.ga.us/dnr/histpres/

GUAM
Lynda B. Aguon, SHPO
Guam Historic Preservation Office
Department of Parks & Recreation
PO Box 2950 Building 13-8 Tiyan
Hagatna, Guam 96932
1-671-475-6290 FAX: 1-671-477-2822
E-Mail: laguon@mail.gov.gu
http://www.ns.gov.gu/dpr/hrdhome.html

HAWAII
Mr. Timothy Johns, SHPO
Department of Land & Natural Resources
P.O. Box 621
Honolulu, HI 96809
808-587-0401
Deputy: Ms. Janet Kawelo

Deputy: Dr. Don Hibbard
State Historic Preservation Division
Kakuhihawa Building, Suite 555
601 Komokila Boulevard
Kapolei, HI 96707
808-692-8015 FAX: 808-692-8020
E-Mail: dlnr@pixi.com
www.hawaii.gov/dlnr/hpdhpgreeting.htm

IDAHO
Steve Guerber, SHPO
Idaho State Historical Society
1109 Main Street, Suite 250
Boise, ID 83702-5642
208-334-2682
Deputy: Suzi Neitzel
208-334-3847 FAX: 208-334-2775
E-Mail: sneitzel@ishs.state.id.us
Deputy: Ken Reid
208-334-3861
www2.state.id.us/ishs/shpo.html

ILLINOIS
Mr. William L. Wheeler, SHPO
Associate Director
Illinois Historic Preservation Agency
1 Old State Capitol Plaza
Springfield, IL 62701-1512
217-785-1153 FAX: 217-524-7525
Deputy: Mr. Theodore Hild, Chief of Staff
E-Mail: thild@hpa084r1.state.il.us
Deputy: Ms. Anne Haaker
www.state.il.us/HPA/

INDIANA
Mr. Larry D. Macklin, SHPO
Director, Department of Natural Resources
402 West Washington Street
Indiana Govt. Center South, Room W256
Indianapolis, IN 46204
E-Mail: dhpa@dnr.state.in.us
Deputy: Jon C. Smith
317-232-1646 FAX: 317-232-0693
E-mail: jsmith
E-Mail: dhpa@dnr.state.in.us
Deputy: Jon C. Smith
317-232-1646 FAX: 317-232-0693
E-mail: jsmith@dnr.state.in.us
www.state.in.us/dnr/historic/index.htm

IOWA
Mr. Tom Morain, SHPO
State Historical Society of Iowa
Capitol Complex
East 6th and Locust St.
Des Moines, IA 50319
515-281-5419 FAX: 515-242-6498
E-Mail: shpo--iowa@nps.gov
Ms. Patricia Ohlerking, DSHPO
515-281-8824 FAX: 515-282-0502
pohlerk@max.state.ia.us
www.iowahistory.org/preservation.index.html

KANSAS
Dr. Ramon S. Powers, SHPO, Executive Director
Kansas State Historical Society
6425 Southwest 6th Avenue
Topeka, KS 66615-1099
785-272-8681 x205 FAX: 785-272-8682
E-Mail: rpowers@hspo.wpo.state.ks.us
Deputy: Mr. Richard D. Pankratz, Director
Historic Preservation Department 785-272-8681 x217
Deputy: Dr. Cathy Ambler
785-272-8681 x215
E-Mail: cambler@kshs.org
www.kshs.org/resource/hispres.htm

KENTUCKY
Mr. David L. Morgan, SHPO, Executive Director
Kentucky Heritage Council
300 Washington Street
Frankfort, KY 40601
502-564-7005 FAX: 502-564-5820
E-Mail: dmorgan@mail.state.ky.us
www.state.ky.us/agencies/khc/khchome.htm

LOUISIANA
Ms. Gerri Hobdy, SHPO
Dept. of Culture, Recreation & Tourism
P.O. Box 44247
Baton Rouge, LA 70804
225-342-8200 FAX 225-342-8173
Deputy: Mr. Robert Collins 225-342-8200
E-Mail: rcollins@crt.state.la.us
Deputy: Mr. Jonathan Fricker 225-342-8160
E-Mail: jfricker@crt.state.la.us
www.crt.state.la.us

MAINE
Mr. Earle G. Shettleworth, Jr., SHPO
Maine Historic Preservation Commission
55 Capitol Street, Station 65
Augusta, ME 04333
207-287-2132 FAX: 207-287-2335
E-Mail: earle.shettleworth@state.me.us
Deputy: Dr. Robert L. Bradley
www.state.me.us/mhpc/homepag1.htm

MARSHALL ISLANDS, REPUBLIC OF THE
Mr. Fred deBrum, HPO
Secretary of Interior and Outer Islands Affairs
P.O. Box 1454, Majuro Atoll
Republic of the Marshall Islands 96960
011-692-625-4642 FAX: 011-692-625-5353
Deputy: Clary Makroro
E-Mail: rmihpo@ntamar.com

MARYLAND
Mr. J. Rodney Little, SHPO
Maryland Historical Trust
100 Community Place, Third Floor
Crownsville, MD 21032-2023
410-514-7600 FAX: 410-514-7678
E-Mail: mdshpo@ari.net
Deputy: Mr. William J. Pencek, Jr.
www.MarylandHistoricalTrust.net

MASSACHUSETTS
Massachusetts Historical Commission
220 Morrissey Boulevard
Boston, MA 02125
617-727-8470 FAX: 617-727-5128
TTD: 1-800-392-6090
Deputy: Ms. Brona Simon, Dir. Technical Serv
E-Mail: Brona.Simon@sec.state.ma.us
www.state.ma.us/sec/mhc

MICHIGAN
Brian D. Conway, SHPO
State Historic Preservation Office
Michigan Historical Center
717 West Allegan Street
Lansing, MI 48918
517-373-1630 FAX: 517-335-0348
E-Mail: conwaybd@sosmail.state.mi.us
http://www.sos.state.mi.us/history/preserve/preserve.html

MICRONESIA, FEDERATED STATES OF
Mr. Rufino Mauricio, FSM HPO
Office of Administrative Services
Div. of Archives and Historic Preservation
FSM National Government
P.O. Box PS 35
Palikir, Pohnpei, FM 96941
011-691-320-2343 FAX: 011-691-320-5634
E-Mail: fsmhpo@mail.fm
FSM includes four States, whose HPOs are listed below:
Mr. John Tharngan, HPO
Yap Historic Preservation Office
Office of the Governor
P.O. Box 714
Colonia, Yap, FM 96943
011-691-350-4226 FAX: 011-691-350-3898

HPO
Div Land Mgmt & Natural Resources
Department of Commerce & Industry
P.O. Box 280
Moen, Chuuk (Truk), FM 96942
011-691-330-2552/2761 FAX: 011-691-330-4906

Mr. David W. Panuelo, HPO
Dir. Dept. of Land, Pohnpei State Government
P.O. Box 1149
Kolonia, Pohnpei, FM 96941
011-691-320-2611 FAX: 011-691-320-5599
E-Mail: nahnsehleng@mail.fm

Mr. Berlin Sigrah, Kosrae HPO
Div. of Land Management & Preservation
Dept. of Agriculture & Lands
P.O. Box 82
Kosrae, FM, 96944
011-691-370-3078 FAX: 011-691-370-3767
E-Mail: dalu@mail.fm

MINNESOTA
Dr. Nina Archabal, SHPO
Minnesota Historical Society
345 Kellogg Boulevard West
St. Paul, MN 55102-1906
651-296-2747 FAX: 651-296-1004
Deputy: Dr. Ian Stewart, 651-297-5513
Deputy: Ms. Britta L. Bloomberg
651-296-5434 FAX: 651-282-2374
E-Mail: britta.bloomberg@mnhs.org
www.mnhs.org

MISSISSIPPI
Mr. Elbert Hilliard, SHPO
Mississippi Dept. of Archives & History
P.O. Box 571
Jackson, MS 39205-0571
601-359-6850
Deputy: Mr. Kenneth H. P'Pool

Division of Historic Preservation
601-359-6940 FAX: 601-359-6955
kppool@mdah.state.ms.us
www.mdah.state.ms.us/hpres/hprestxt.html

MISSOURI
Mr. Stephen Mahfood, SHPO
State Department of Natural Resources
205 Jefferson, P.O. Box 176
Jefferson City, MO 65102
573-751-4422 FAX: 573-751-7627
Deputy: Ms. Claire F. Blackwell
Historic Preservation Prog, Div. of State Parks
100 E. High Street
Jefferson City, MO 65101
573-751-7858 FAX: 573-526-2852
E-Mail: nrblacc@mail.dnr.state.us
Deputy: Dr. Douglas K. Eiken
www.mostateparks.com

MONTANA
Dr. Mark F. Baumler, SHPO
State Historic Preservation Office
1410 8th Avenue
P.O. Box 201202
Helena, MT 59620-1202
406-444-7717 FAX: 406-444-6575
E-Mail: mbaumler@.state.mt.us
Deputy: Mr. Herbert E. Dawson
www.hist.state.mt.us

NEBRASKA
Mr. Lawrence Sommer, SHPO
Nebraska State Historical Society
P.O. Box 82554
1500 R Street
Lincoln, NE 68501
402-471-4745 FAX: 402-471-3100
E-Mail: nshs@nebraskahistory.org
Deputy: Mr. L. Robert Puschendorf
402-471-4769 FAX: 402-471-3316
www.nebraskahistory.org/histpres/index.htm

NEVADA
Mr. Ronald James, SHPO
Historic Preservation Office
100 N Stewart Street
Capitol Complex
Carson City, NV 89701-4285
775-684-3440 FAX: 775-684-3442
Deputy: Ms. Alice Baldrica
775-684-3444
E-Mail: ambaldri@clan.lib.nv.us
www.state.nv.us

NEW HAMPSHIRE
Ms. Nancy C. Dutton, Director/SHPO
NH Division of Historical Resources
P.O. Box 2043
Concord, NH 03302-2043
603-271-6435 FAX: 603-271-3433
TDD: 800-735-2964
E-Mail: ndutton@nhdhr.state.nh.us
Deputy: Ms. Linda Ray Wilson
603-271-6434 or 603-271-3558
E-Mail: lwilson@nhdhr.state.nh.us
www.state.nh.us/nhdhr

NEW JERSEY
Mr. Robert C. Shinn, SHPO
Dept. of Environ. Protection
401 East State Street
P.O. Box 402
Trenton, NJ 08625
609-292-2885 FAX: 609-292-7695
Deputy: Ms. Cari Wild
Natural and Historic Resources
501 East State Street, 3rd Fl
P.O. Box 404
Trenton, NJ 08625
609-292-3541 FAX: 609-984-0836
Deputy: Ms. Dorothy Guzzo
Natural and Historic Resources
Historic Preservation Office, 4th Fl
609-984-0176 FAX: 609-984-0578
E-Mail: dguzzo@dep.state.nj.us
www.state.nj.us/dep/hpo

NEW MEXICO
Elmo Baca, SHPO
Historic Preservation Div., Off of Cultural Affairs
228 East Palace Avenue
Santa Fe, NM 87503
505-827-6320 FAX: 505-827-6338
Deputy: Dorothy Victor
E-Mail: dvictor@lvr.state.nm.us
Deputy: Jan Biella
E-Mail: jbiella@lvr.state.nm.us
www.museums.state.nm.us/hpd

NEW YORK
Ms. Bernadette Castro, SHPO
Parks, Recreation & Historic Preservation
Agency Building #1, Empire State Plaza
Albany, NY 12238
518-474-0443
Deputy: Mr. J. Winthrop Aldrich
518-474-9113 FAX: 518-474-4492

Historic Preservation Staff:
Ms. Ruth L. Pierpont, Director
Bureau of Field Services
NY State Parks, Recreation & Historic Preservation
Peebles Island P.O. 189
Waterford, NY 12188-0189
518-237-8643 x3269 FAX: 518-233-9049
E-Mail: rpierpont@oprhp.state.ny.us
www.nysparks.com/field/

NORTH CAROLINA
Dr. Jeffrey J. Crow, SHPO
Division of Archives & History
4610 Mail Service Center
Raleigh, NC 27699-4610
919-733-7305 FAX: 919-733-8807
E-Mail: jcrow@ncsl.dcr.state.nc.us
Deputy: Mr. David Brook
Historic Preservation Office
4617 Mail Service Center
Raleigh, NC 27699-4617
919-733-4763 FAX: 919-733-8653
E-Mail: dbrook@ncsl.dcr.state.nc.us
http://www.hpo.dcr.state.nc.us

NORTH DAKOTA
Mr. Merl Paaverud, Acting SHPO
State Historical Society of North Dakota
612 E. Boulevard Ave.
Bismarck, ND 58505
701-328-2666 FAX: 701-328-3710
swegner@state.nd.us
www.state.nd.us/hist
Deputy:
701-328-2672
mpaaverud@state.nd.us

NORTHERN MARIANA ISLANDS, COMMONWEALTH OF THE
Mr. Joseph P. DeLeon Guerrero, HPO
Dept. of Community & Cultural Affairs
Division of Historic Preservation
Airport Road
Northern Mariana Islands
Saipan, MP 96950
011-670-664-2125 FAX: 011-670-664-2139
E-Mail: cnmihpo@itecnmi.com
Deputy: Mr. Scott Russell 011-670-664-2121

OHIO
Mr. Amos J. Loveday, SHPO
Ohio Historic Preservation Office
567 E Hudson Street
Columbus, OH 43211-1030
614-298-2000 FAX: 614-298-2037
E-Mail: ajloveday@aol.com
Deputy: Mr. Franco Ruffini
614-298-2002 FAX: 614-298-2037
E-Mail: fruffini@ohiohistory.org
www.ohiohistory.org/resource/histpres

OKLAHOMA
Dr. Bob L. Blackburn, SHPO
Oklahoma Historical Society
2100 N. Lincoln Blvd.
Oklahoma City, OK 73105
405-521-2491 FAX: 405-521-2492
www.ok-history.mus.ok.us
Deputy: Ms. Melvena Thurman Heisch
State Historic Preservation Office
2704 Villa Prom, Shepherd Mall
Oklahoma City, OK 73107
405-522-4484 FAX: 405-947-2918
E-Mail: mheisch@ok-history.mus.ok.us

OREGON
Mr. Michael Carrier, SHPO
State Parks & Recreation Department
1115 Commercial Street, NE
Salem, OR 97301-1012
503-378-5019 FAX: 503-378-8936
Deputy: Mr. James Hamrick
503-378-4168 x231 FAX: 503-378-6447
E-Mail: james.m.hamrick@state.or.us
www.prd.state.or.us/about_shpo/html

PALAU, REPUBLIC OF
Ms. Victoria N. Kanai, HPO
Ministry of Community & Cultural Affairs
P.O. Box 100
Koror, Republic of Palau 96940
011-680-488-2489 FAX: 011-680-488-2657

PENNSYLVANIA
Dr. Brent D. Glass, SHPO
Pennsylvania Historical & Museum Commission
300 North Street
Harrisburg, PA 17120
717-787-2891
Acting Deputy: Mr. Dan G. Deibler, Bureau for Historic Preservation
Commonwealth Keystone Building, 2nd Floor
400 North Street
Harrisburg, PA 17120-0093
717-787-4363 FAX: 717-772-0920
E-Mail: dandeibler@state.pa.us
www.phmc.state.pa.us/

PUERTO RICO, COMMONWEALTH OF
Ms. Enid Torregrosa, SHPO
Office of Historic Preservation
Box 82, La Fortaleza
Old San Juan, Puerto Rico 00901
787-721-2676 or 3737 FAX: 787-723-0957
E-Mail: etorregrosa@prshpo.prstar.net
Deputy:

RHODE ISLAND
Mr. Frederick C. Williamson, SHPO
Rhode Island Historic Preservation & Heritage Comm
Old State House, 150 Benefit St.
Providence, RI 02903
401-222-2678 FAX: 401-222-2968
Deputy: Mr. Edward F. Sanderson
E-Mail: esanderson@rihphc.state.ri.us

SOUTH CAROLINA
Dr. Rodger E. Stroup, SHPO
Department of Archives & History
8301 Parklane Road
Columbia, SC 29223-4905
803-896-6100 FAX: 803-896-6167
Deputy: Ms. Mary W. Edmonds 803-896-6168
E-Mail: edmonds@scdah.state.sc.us
http://www.state.sc.us/scdah/

SOUTH DAKOTA
Mr. Jay D. Vogt, SHPO
State Historic Preservation Office
Cultural Heritage Center
900 Governor's Drive
Pierre, SD 57501
605-773-3458 FAX: 605-773-6041
E-Mail: jay.vogt@state.sd.us
http://www.state.sd.us/state/executive/deca/cultural/histpres.htm

TENNESSEE
Mr. Milton Hamilton, SHPO
Dept. of Environment and Conservation
401 Church Street, L & C Tower 21st Floor
Nashville, TN 37243-0435
615-532-0109 FAX: 615-532-0120
Deputy: Mr. Herbert L. Harper
Tennessee Historical Commission
2941 Lebanon Road
Nashville, TN 37243-0442
615-532-1550 FAX: 615-532-1549
www.state.tn.us/environment/hist/hist.htm

TEXAS
Mr. F. Lawerence Oaks, SHPO
Texas Historical Commission
P.O. Box 12276
Austin, TX 78711-2276
512-463-6100 FAX: 512-475-4872
E-Mail: l.oaks@thc.state.tx.us
Deputy: Mr. Terry Colley, 512-463-6100
E-Mail: terry.colley@thc.state.tx.us
Deputy: Mr. Stanley O. Graves, Dir. Architecture Div.,
512-463-6094 FAX: 512-463-6095
E-Mail: stan.graves@thc.state.tx.us
Deputy: Dr. James E. Bruseth, Dir. Antiquities Protection
Prot 512-463-6096
FAX: 512-463-8927
E-Mail: jim.bruseth@thc.state.tx.us
www.thc.state.tx.us

UTAH
Mr. Max Evans, SHPO
Utah State Historical Society
300 Rio Grande
Salt Lake City, UT 84101
801-533-3500 FAX: 801-533-3503
Deputy: Mr. Wilson Martin
E-Mail: wmartin@history.state.ut.us
http://history.utah.org

VERMONT
Ms. Emily Wadhams, SHPO
Vermont Division for Historic Preservation
National Life Building, Drawer 20
Montpelier, VT 05620-0501
802-828-3211
E-Mail: ewadhams@dca.state.vt.us
Deputy: Mr. Eric Gilbertson, Director
802-828-3043 FAX: 802-828-3206
E-Mail: ergilbertson@dca.state.vt.us
http://www.uvm.edu/~vhnet/hpres/org/vdhp/vdhp1.html
State Historic Sites http://www.historicvermont.org/

VIRGIN ISLANDS
Mr. Dean C. Plaskett, Esq., SHPO
Department of Planning & Natural Resources
Cyril E. King Airport
Terminal Building–Second Floor
St. Thomas, VI 00802
340-774-3320 FAX: 340-775-5706
Deputy: Ms. Claudette C. Lewis
340-776-8605 FAX: 340-776-7236

VIRGINIA
Kathleen Kilpatrick, Acting SHPO
Department of Historic Resources
2801 Kensington Avenue
Richmond, VA 23221
804-367-2323 FAX: 804-367-2391
E-Mail: kkilpatrick@dhr.state.va.us
Deputy:
www.dhr.state.va.us

WASHINGTON
Dr. Allyson Brooks, SHPO
Office of Archeology & Historic Preservation
P.O. Box 48343
420 Golf Club Road, SE, Suite 201, Lacey
Olympia, WA 98504-8343
360-407-0753 FAX: 360-407-6217
allysonb@acted.wa.gov
Deputy: Mr. Greg Griffith 360-407-0753
E-Mail: gregg@cted.wa.gov
www.ocd.wa.gov/info/lgd/oahp/

WEST VIRGINIA
Ms. Nancy Herholdt, SHPO
West Virginia Division of Culture & History
Historic Preservation Office
1900 Kanawha Boulevard East
Charleston, WV 25305-0300
304-558-0220 FAX: 304-558-2779
Deputy: Ms. Susan Pierce
E-Mail: susan.pierce@wvculture.org
www.wvcullture.org/shpo/index.html

WISCONSIN
Mr. George L. Vogt, SHPO
State Historical Society of Wisconsin
816 State Street
Madison WI 53706
608-264-6500 FAX: 608-264-6404
E-Mail: glvogt@mail.shsw.wisc.edu
Deputy: Ms. Alicia L. Goehring
E-Mail: algoehring@mail.shsw.wisc.edu
www.shsw.wisc.edu/about/index.html

WYOMING
Ms. Wendy Bredehoft, SHPO
Wyoming State Historic Preservation Office
2301 Central Avenue, 4th Floor
Cheyenne, WY 82002
307-777-7013 FAX: 307-777-3543
E-Mail: wbrede@missc.state.wy.us
Deputy: Judy K. Wolf 307-777-6311
E-Mail: jwolf@missc.state.wy.us
Sheila Bricher-Wade, Reg Ser 307-777-6179
E-Mail: sbrich@missc.state.wy.us
Mary M. Hopkins, Cultural Records 307-766-5324
http://commerce.state.wy.us/cr/shpo

ASSOCIATE MEMBERS:

NAVAJO NATION
Dr. Alan Downer, HPO
P.O. Box 4950
Window Rock, AZ 86515
520-871-6437 FAX: 520-871-7886
E-Mail: hpd_adowner@dine.navajo.org

LAC DU FLAMBEAU Of *Lake Superior* BAND CHIPPEWA INDIANS
Ms. Patricia A. Hrabik Sebby, THPO
P.O. Box 67
Lac Du Flambeau, WI 54538
715-588-3303

LEECH LAKE BAND OF CHIPPEWA INDIANS
Ms. Rose A. Kluth, THPO
Leech Lake Reservation
RR3, Box 100
Cass Lake, MN 56633
218-335-8200 FAX: 218-335-8309
E-Mail: rkluth@aol.com

TURTLE MOUNTAIN BAND OF CHIPPEWA INDIANS
Mr. Kade M. Ferris, THPO
Turtle Mountain Band of Chippewa Indians
P.O. Box 900
Belcourt, ND 58316
E-Mail: kferris@utma.com

NATIONAL PARK SERVICE—National Center
http://www.nps.gov/
Associate Director, Cultural Resources, Kate Stevenson 202-208-7625
Assistant Director & Manager, Cultural Resources 202-343-9596
Archeology and Ethnography, Frank McManamon, Program Manager 202-343-4101
HABS/HAER Division, E. Blaine Cliver, Chief 202-343-9618
Heritage Preservation Services Program, Pat Tiller, Chief 202-343-9569
Preservation Initiatives Branch, Bryan Mitchell, Chief 202-343-9558
Technical Preservation Services Branch, Sharon Park, Chief 202-343-9584

State, Tribal & Local Programs Branch, Joe Wallis, Chief 202-343-9564
Museum Management Program, Ann Hitchcock, Chief Curator 202-343-9569
National Register, History & Education, Dwight Picaithley, Chief Historian 202-343-9536
Keeper of the National Register of Historic Places, Carol Shull 202-343-9536
Park Historic Structures/Cultural Landscape Program, Randall Biallas, Chief Historical Architect 202-343-9588

NATIONAL PARK SERVICE—Systems Support Offices
Anchorage 907-257-2690
Philadelphia 215-597-0652
Denver 303-969-2875
Atlanta 404-562-3157
San Francisco 415-427-1300

ADVISORY COUNCIL ON HISTORIC PRESERVATION—http://www.achp.gov
John Fowler, Executive Director 202-606-8503
Ron Anzalone, Assistant to Executive Director 202-606-8505
Don Klima, Director, Office of Planning & Review, Eastern and Western Regions 202-606-8505

APPENDIX 4

Possible Field Ailments: Symptoms and Treatments

What follows is intended to heighten the awareness of the person going into the field about some of the difficulties that might be encountered. These summaries are intended as a general guide and should not be construed as the last word on the symptoms or treatment of any particular condition, or a substitute for first aid or emergency medical training. Cuts, injuries resulting from falls (including sprains), and muscle strain probably plague archaeologists most frequently. The information presented here is abstracted from a variety of sources including: Auerbach 1999; Centers for Disease Control (http://www.cdc.gov/); Lyme Disease Foundation (http://www.lyme.org/index2.html); Flanagan 1995; Valley Fever Center for Excellence, Arizona Research Laboratories (http://www.arl.arizona.edu/vfce/); Southwestern Archaeology Health Issues (http://www.swanet.org/health.html).

AIRBORNE HAZARDS

A small number of conditions are related to areas of the country, primarily sections of California and the Southwest, and environments (deserts, caves, and rockshelters) where fungi and bacteria adhering to dust may be inhaled by the archaeologist working in the field.

Valley Fever *is primarily a disease of the lungs caused by the inhalation of the fungus Coccidioides immitis. It is common in the southwestern United States and northwestern Mexico. Fungal spores become airborne when the soil is disturbed. The fungal spores are often found in abundance in the soil around rodent burrows and Indian ruins. Valley Fever symptoms generally occur within three weeks of exposure and most commonly involve fatigue, cough, chest pain, fever, rash, headache, and joint aches. Some people develop painful red bumps on their shins or elsewhere that gradually turn brown. Valley fever is not a contagious disease. Most cases are very mild and second infections are rare. Most patients with Valley fever recover with no treatment. Antifungal drug therapy is used in severe cases.*

Hantavirus Pulmonary (Lung) Syndrome (HPS) *results from the inhalation of viral particles. The hantaviruses that cause HPS are carried by rodents, especially the deer mouse, and are excreted through the saliva, urine, and droppings. You can become infected by exposure to their droppings in much the same way as valley fever, and the flu-like first signs of sickness (especially fever and muscle aches) appear one to six weeks later, followed by shortness of breath and coughing. HPS is not contagious. The Centers for Disease Control recommends wearing either a half mask air purifying (or negative pressure) respirator or a powered air purifying respirator (PAPR) with N-100 filters. There is no specific treatment—see a doctor quickly.*

ANAPHYLAXIS

This encompasses severe allergic reactions that can be life threatening. Severe allergic reactions can be associated with venomous insect and animal bites. Bee stings are a frequent culprit, but remember, not everyone will have a severe reaction to stings or bites. Symptoms will appear quickly and can include

- ☑ difficulty breathing;
- ☑ low blood pressure;
- ☑ swelling of the lips, tongue, and throat;
- ☑ itching;
- ☑ hives.

Respiratory distress is the most common life-threatening problem associated with this reaction and must be dealt with quickly. Subcutaneous injections of epinephrine (adrenaline) are used as treatment and can be administered in the field using the EpiPen. If you know that you have severe reactions to stings you should carry an EpiPen with you whenever you are in the field. If epinephrine is not available in the field, the victim should be taken immediately to an emergency room.

BITES/STINGS

If bitten by an *animal,* flush the wound with at least two quarts of water, clean the wound with soap and water, or Betadine if available, and flush the wound again. Pat the wound dry and cover with a sterile dressing. See a doctor. Rabies and tetanus (see "Cuts") can be contracted through animal bites.

> ***Rabies*** *virus is contracted when bitten by infected animals. Don't wait for symptoms to appear if you think that you have been bitten by an infected animal. Animals acting in abnormal ways may have rabies (e.g., a racoon wandering about in the open during the middle of the day). The rabies virus is present in the saliva of an infected animal, so if you are caring for someone who has been bitten by a rabid animal, don't let any fluid come in contact with any sores or open wounds that you may have. Wash wounds thoroughly to prevent further contamination and seek medical help immediately. Pre-exposure vaccines can be obtained for people working in high-risk situations.*

Scorpions are found in the southwestern United States in dry and warm climates. They are most active at night. Their sting is painful but rarely fatal. Nausea, double vision, difficulty breathing or swallowing, sweating, and increased salivation may result. Wash the wound and apply a cold pack to the site of the sting. Seek medical attention immediately. Antivenin may be used for treatment.

Snake bites are rarely fatal. The degree of concern and treatment depend on whether a venomous or nonvenomous snake is involved. Bites from nonvenomous snakes should be washed thoroughly with soap and water and the victim treated with dicloxacillin, erythromycin, or cephalexin. Complications can arise from any snakebite when the victim has an allergic reaction, or too much time elapses between the bite and getting attention at a medical facility. If a bite results in the breaking of the skin and the release of venom, there may be burning pain and swelling at the site of the bite (within 5–10 minutes); numbness and tingling of the affected limb, or of the lips, face, fingers, toes, and scalp (30–60 minutes), sweating, nausea, vomiting, and fainting (within 1–2 hours).

The wound should be washed and the affected portion of the body should be kept at an elevation lower than the heart. Apply the extractor suction device if available (removes venom without the need for an incision in the skin). If an extractor is not available, incising the skin should **only** be done by an experienced person and when the victim cannot receive medical attention within 1–2 hours. Immobilize the affected body part with a splint if necessary. Prevent or minimize physical activity on the part of the victim and carry them if any appreciable distances must be walked. If the snake can be captured and contained without additional injury to anyone, do so, since identifying it will be helpful. Get the victim to the nearest hospital as quickly as possible (within one hour). Other measures will be necessary if the victim cannot receive medical attention within 1–2 hours of being bitten.

Spider bites may be painful and cause localized inflammation. There are a few species that are cause for serious concern. The first is the black widow spider, which is black with a reddish marking on the underbelly that resembles an hourglass. The brown recluse spider is light brown with a darker brown violin-shaped marking on its back. The victim may experience muscle cramps, muscle twitching and some numbness, nausea and vomiting, difficulty breathing or swallowing, sweating, and increased salivation, severe pain in the affected area, and swelling. Wash the wound and apply a cold pack to the site of the bite. Seek medical attention immediately. Antivenin is needed for black widow spider bites.

Ticks are frequently encountered in the fields and woods of many areas of the country. There are a variety of ticks and the potential health hazards they represent also vary. In general, a tick bite can cause a localized reaction (inflammation, itchy nodule or ulcer). Simply finding a tick on your skin does not automatically mean that you have been bitten. Typically, the mouthparts of a tick will remain embedded in your skin until the tick becomes bloated from its blood meal. When removing a tick, it is necessary that you get its mouthparts as well as the body. Tick removal is best accomplished with tweezers or fingernails, grasping the tick as close as you can to the mouthparts. If the tick's head is buried in the skin and your efforts to remove it fail, see a doctor. You can also apply permethrin (Permanone insect repellent), which will relax the tick, allowing you to eventually pull it free. Wash the area with soap and water. You may apply an antiseptic or antibiotic ointment to help prevent infection. If the bite becomes infected, seek medical treatment.

Wearing long pants and socks and treating your clothing with an insect repellent that contains Permethrin is helpful. Pants may be tucked into the socks for additional protection. If you wear light-colored clothing, it will be easier to see ticks. Carefully inspect yourself, including scalp and ears, at the end of the day prior to and after bathing.

***Lyme Disease** is a bacterial infection caused by the spirochete* Borrelia burgdoferi *(Bb). Transmitters of the bacteria in North America include: the Western black-legged tick* (Ixodes pacificus) *in the West, and the black-legged tick* (Ixodes scapularis) *in the rest of the country. The black-legged tick was temporarily known as the "deer" tick* (Ixodes "dammini"). *An infective tick must be attached to you for a day or more before transmission of Bb occurs. Symptoms often start with flu-like feelings of headache, stiff neck, fever, muscle aches, and fatigue. About 60 percent of light-skinned victims notice a unique enlarging rash days to weeks after the bite. On dark-skinned people, this rash resembles a bruise. The rash may appear within a day of the bite or as much as a month later. This rash may start as a small, reddish bump about one-half inch in diameter. It may be slightly raised or flat. It soon expands outward, often leaving a clearing (normal flesh color) in the center. It can enlarge to the size of a thumb-print or cover a person's back. A small inflamed skin bump or discoloration that develops within hours of a bite and over the next day or two is not likely to be due to infection, but rather a local reaction to the disruption of the skin. Treatment varies and depends on how early a diagnosis is made. Oral antibiotics may be sufficient for early stages of non-disseminated infection. A preventive vaccine has been developed that requires three treatments over a period of six months. See your doctor for details.*

***Rocky Mountain Spotted Fever,** despite its name, is most frequently encountered in eastern sections of the country. It is caused by a tick-borne parasite carried typically by the dog tick* (Dermacentor variabilis) *or western wood tick* (Dermacentor andersoni). *Symptoms include a fever beginning abruptly 2–14 days after being bitten, headache, chills, joint and muscle aches, and a red-spotted rash may develop on the hands and feet, spreading to other parts of the body. See a doctor. Tetracycline or doxycycline are often prescribed as treatment.*

***Colorado Tick Fever,** also transmitted by the western tick* (Dermacentor andersoni), *results in the onset of fever, severe headache, muscle aches and fatigue from 3 to 6 days after being bitten. See a doctor.*

Bee, wasp, and *hornet* stings are painful and result immediately in some form of redness or swelling and itching at the site of the sting. Remove the stinger if it is still present. An ice pack may be applied to the sting. Benadryl may be taken orally if a mild allergic reaction occurs. If a severe allergic reaction is suspected, epinephrine must be administered (see Anaphylaxis).

BURNS

Along with contact with flame, radiated heat, or hot objects, burns can also be caused by contact with chemicals, electricity, or frozen surfaces. Superficial or first-degree burns involve the reddening of the skin. A second-degree burn, one that affects deeper layers of skin is accompanied by blisters. Third-degree burns are deep and affect the entire thickness of the skin. They involve the destruction of tissue and nerves. The destruction of nerves make this type of burn feel less-painful than first- and second-degree ones. The skin will be blackened and there may be white or waxy-looking areas. Clotted blood vessels may be visible below the surface.

Cool/irrigate the area of a heat burn with water (never apply ice). Chemical burns require irrigation with gallons of water to insure that the chemical has been flushed from the skin. Cold compresses can be used to reduce pain, as can an anti-inflammatory drug like aspirin or ibuprofen. Don't cool the affected area to the degree that the victim begins to shiver. Cover the burn with a clean, non-adherent dressing. Don't attempt to remove any material adhering to a burn that doesn't wash off during the irrigation process. Don't prick any blisters that form. Burns to the face may have an effect on the victim's breathing. Any victim who has inhaled smoke, fumes, or superheated air, been burnt on the face, has second-degree burns covering 5% or more of the body, or has third-degree burns should seek immediate medical attention.

CUTS/EXTERNAL BLEEDING

The advise given here is for severe cuts and external bleeding. After determining where the bleeding is coming from, apply direct pressure with the cleanest folded bandage/cloth/gauze available. Add additional cloths as the ones in use become soaked with blood, but do not remove them. Pressure should be maintained for 10 minutes. Elevate the injury, if possible, above the level of the heart. When the bleeding subsides apply a pressure dressing. See a doctor.

If necessary, pressure points or a tourniquet may be used if the bleeding cannot be stopped. Use a pressure point nearest the wound. A tourniquet should only be used in life-threatening situations since it may result in the eventual loss of the limb to which the tourniquet has been applied. Never loosen a tourniquet once applied. If massive wounds expose internal organs or bone, keep them covered with moistened bandages and in place with gentle pressure. Seek immediate medical help. When wounds are associated with objects that become imbedded in the body (e.g., arrow, tree limb, metal object) do not remove them. Stabilize the object as best as possible and dress the wound around the object. Seek immediate medical help.

Tetanus *is a bacterial infection of wounds involving cuts and breaks in the skin which can result in painful abdominal muscle contractions, spasms, and painful muscle contractions in the neck and body. Prevention with a tetanus shot is the best strategy to deal with this condition. Immunization should occur every 10 years, although a 5-year cycle of shots is recommended for people working in high-risk situations. Animal bites that break the skin are also of concern. Almost all dogs carry* Clostridium teteni *(the agent of tetanus) on their teeth, so bites should be taken seriously.*

MUSCLE-RELATED HAZARDS

This category includes torn muscles, strains, and sprains. Signs of a torn ("pulled") muscle are sudden pain associated with vigorous activity. This may be associated with bruising, swelling, or loss of mobility. Immobilize the affected area and apply cold packs. After a few days, heat may be applied and gentle movement of the affected area may be initiated. Avoid pulled muscles by stretching and warming up prior to engaging in intense physical activity.

Strains and sprains are injuries to ligaments and tendons (attachments for bone and muscle) and are accompanied by severe shooting pain, bruising, and swelling. The application of ice or cold packs, compression of the affected area with firmly applied wrappings, rest and elevation of the affected area are recommended. Heat may be applied a day or so after the incident, but not before any swelling has subsided. Hot and cold compresses may be alternately applied to promote healing. Most sprains take at least 6–8 weeks to heal. Ankle sprains are probably the most common and can be prevented by wearing well-fitting, ankle-height boots. Back strains can result from improper lifting of objects. Think about what you are doing; bend your knees and lift with your legs.

Bursitis/Tendinitis: *Bursitis is the irritation and inflammation of the lubricating sacs associated with muscles and joints. Tendinitis is inflammation or irritation of a tendon. Both can be initiated as a result of repeated, redundant motions/overuse, and falls. For bursitis, areas typically affected are the knees, elbows, shoulder and outside of the hip. Both conditions may involve redness and swelling, pain associated with movement, or weakness of the affected area. Anti-inflammatory drugs like aspirin or ibuprofen may be taken. Rest and elevate the affected area.*

POISONOUS PLANTS

The discussion here is limited to poison ivy, oak, and sumac, plants that can cause skin rashes when contacted. There are a variety of plants that are poisonous if ingested, or that cause nonallergic irritation to the skin when contacted. Check field guides to plants and local first-aid manuals for the area in which you are working for additional information.

The itchy rashes associated with poison ivy, oak, and sumac result from the contact of the skin with the resin or oil produced by these plants. The plant extrudes the oils when plant parts are injured or bruised, and smoke from burning plants can also transport it. Individual reactions vary considerably. A rash can appear within 8 hours or even days after exposure. If the resin/oil is removed from the skin shortly after exposure, a rash may not develop. Washing exposed areas should occur within 30 minutes of exposure to be effective. If a rash develops, there are a variety of topical treatments that help to reduce itching and dry the rash. Don't forget that the resin/oil can adhere to clothing, gloves, backpacks, and tools. Take care how you handle and clean these items in the field and at home so as not to inadvertently expose yourself. There are topical treatments that, when applied to the skin, provide a barrier between it and any resin/oil that you might come in contact with. Learning to recognize the plants and avoid them is the best method of prevention.

WEATHER-RELATED HAZARDS

Extremes in weather and your exposure to them can result in a number of conditions. These include hypothermia, frostbite, heat exhaustion or heatstroke.

Hypothermia is a condition associated with the lowering of the body's core temperature (moderate hypothermia involves core body temperatures between 95 and 90 degrees; severe hypothermia, 90 degrees and lower). It is recognizable as vigorous, uncontrollable shivering and eventually dizziness, lightheadedness, muscular stiffness, and difficulty in moving. If treatment is not forthcoming the victim will exhibit slurred speech, slow pulse, memory loss, and unconsciousness. It can be fatal. The body temperature must be raised slowly, which can be done by moving the victim to warm shelter, removing wet clothing and replacing it with dry clothing, wrapping in blankets, and drinking warm fluids. Avoid caffeine and seek immediate medical attention.

Frostbite is the actual freezing of tissues as a result of exposure to extreme temperatures. Poor circulation can predispose some people to frostbite. Bodily extremities are usually the first to be affected. At low temperatures, blood vessels constrict and the skin becomes numb, occasionally preceded by itching and prickly pain. Frostbitten portions of the body will appear to be white and waxy. Thawed frostbite can be ranked in much the same way as burns: numbness, redness and swelling with no tissue loss = first-degree frostbite; superficial blisters with milky fluid, and surrounded by redness and swelling = second-degree frostbite; deep blistering with purple, blood-containing fluid = third-degree frostbite; fourth-degree frostbite = the deepest and most severe involvement, which can also affect bone.

Treatment of frostbite requires the rapid rewarming of affected areas. Slow or partial rewarming, and especially refreezing following warming can cause more damage. Don't rewarm a person until they have been moved to a situation where they can stay warm. If toes are frostbitten, do not let the victim walk until the toes are warm. Remove and replace wet clothing. Immerse frostbitten parts of the body in hot water that normal skin could stand for prolonged periods without discomfort. If water is not available, wrap the body in sheets and blankets. The skin needs to be warmed to the point where it regains a pink color. Never rub frostbitten skin or use heat from a stove, fire, or vehicle exhaust to warm the skin. Given the degree of frostbite, blisters will appear 6 to 24 hours after warming the skin and should be left intact. Blisters can be covered with aloe vera lotion or cream and covered with fluffy sterile bandages. Aspirin or ibuprofen can be taken, along with warm liquids. Avoid alcohol. Seek medical help immediately.

Dressing properly, keeping clothing dry, drinking sufficient fluids, and eating regularly are ways to prevent frostbite. Underwear that is designed to wick moisture away from the skin is useful in regulating skin temperature. Noncotton clothing is generally better for dealing with moisture under cold conditions than cotton-based clothing.

Heat exhaustion and heat stroke are different levels of conditions resulting from the elevation of the body's core temperature. Heat exhaustion (some body temperatures up to 105°F) results in confusion, a rapid pulse, dizziness, headache, nausea, and diarrhea. The victim may or may not be sweating and the skin may feel cool to the touch. Heatstroke (core body temperature exceeding 105°F) is recognizable as extreme confusion, unconsciousness, low blood pressure or shock, shortness of breath, darkened urine, vomiting, and diarrhea. The victim may or may not be sweating and the skin can feel cool.

The treatment for both conditions is to cool the victim as quickly as possible by removing them to shelter, removing clothing, placing them in cool water, sponging them with water or rubbing alcohol, or wrapping them with wet towels or clothing, and fanning them. If available, ice packs may be placed in the armpits, behind the neck and in the groin. If the victim is alert, offer fluids to begin rehydration of the body. Begin to taper off the cooling effort as the victim's temperature drops below 101 degrees, but be aware that their temperature could start to rise again. Monitor the victim's temperature for 3 to 4 hours following cooling. If heatstroke is involved, get the victim to a hospital as soon as possible. Avoid heat exhaustion and heatstroke by dressing properly and drinking plenty of fluids before and during fieldwork.

APPENDIX 5

Examples of Job Descriptions for Archaeological Technicians, Entry/BA Degree Level Staff

UNITED STATES DEPARTMENT OF LABOR

EMPLOYMENT STANDARDS ADMINISTRATION

WAGE HOUR DIVISION

SERVICE CONTRACT ACT - DIRECTORY OF OCCUPATIONS

29000 TECHNICAL OCCUPATIONS

This category includes occupations concerned with providing technical assistance to engineers and scientists in both laboratory and production activities as well as occupations concerned with independently operating and servicing technical equipment and systems. Characteristic of occupations in this category are the requirements for a knowledge of scientific, engineering, and mathematical theories, principles and techniques that are less than full professional knowledge but which nevertheless enables the technician to understand how and why a specific device or system operates.

The technician solves practical problems encountered in fields of specialization, such as those concerned with development of electrical and electronic circuits, and establishment of testing methods for electrical, electronic, electro-mechanical, and hydro-mechanical devices and mechanisms; application of engineering principles in solving design, development, and modification problems of parts or assemblies for products or systems; and application of natural and physical science principles to basic or applied research problems in fields, such as metallurgy, chemistry, and physics. May perform technical procedures and related activities independently.

Workers with the title of Technician who are concerned primarily with maintenance and repair are classified with Mechanics and Maintenance and Repair Occupations.

29023 ARCHAEOLOGICAL TECHNICIAN I

Under the direct supervision of archaeological crew chiefs and under the general supervision of field director/project archaeologist performs unskilled and semi-skilled tasks at archaeological field sites. Assists crew chief in activities associated with the excavation of project areas and found features. Walks over project searching for archaeological materials such as historic and prehistoric remains. Excavates, screens, back-fills excavated areas. Assists in preparation of sketch maps and forms, and field photography. Conducts simple surveys using compass, topographical map and aerial photographs. Determines the exact locations of sites and marks them on maps and/or aerial photographs.

Records information on archaeological site survey form and prepares simple reports. Cleans, packages, and labels artifacts recovered from inventories and excavations and assists in the flotation of soil samples.

29024 ARCHAEOLOGICAL TECHNICIAN II

Under the general supervision of field director/project archaeologist, performs skilled tasks. Conducts hand excavations, completes plan and profile maps of excavated units, completes standard feature and level forms, screens soils to recover artifacts. Performs flotation of soil samples, walk overs, and shovel testing. Catalogs, packages/labels archaeological artifacts. Maintains field equipment and supplies. Conducts inventories of cultural resources in areas of proposed projects. Researches reference materials such as state and National Register files, historic documents, archaeological reports, maps, and aerial photos, and interviews source individuals concerning project areas. Performs on-the-ground area searches for surface and subsurface evidence of historic and prehistoric archaeological remains. Identifies and records historic and prehistoric cultural resource sites. Prepares Archaeological Reconnaissance Reports and maps. Insures that archaeology work assignments are carried out in safe, timely manner according to established standards and procedures. Maintains the Archaeological Reconnaissance schedule by estimating and reporting an expected time of completion of each project and updating the project planning board. Reviews work in progress to see that standards for pre-field research, survey design, site recording, graphics and final report are being met. Advises other employees on methods of cultural resource inventory and provides written instructions, research materials and supplies to all involved in planning and operation of natural resource activities.

29025 ARCHAEOLOGICAL TECHNICIAN III

Serves as lead archaeological technician, under the general supervision of field directory/project archaeologist, and performs skilled tasks at archaeological field sites. Conducts hand excavations, completes plan and profile maps of excavated units, completes standard feature and level forms, screens soils to recover artifacts. Performs flotation of soil samples, and shovel testing. Packages/labels archaeological artifacts. Maintains field equipment and supplies. Conducts inventories of forest cultural resources in areas of proposed forest service projects. Researches reference materials such as state and National Register files, historic documents, archaeological remains. Identifies and records historic and prehistoric cultural resource sites. Prepares Archaeological Reconnaissance Reports and maps. Insures that archaeology work assignments are carried out in safe, timely manner according to established standards and procedures. Maintains the Archaeological Reconnaissance schedule by estimating and reporting an expected time of completion of each project and updating the project planning board. Reviews work in progress to see that standards for pre-field research, survey design, site recording, graphics and final report are being met. Advises other employees on methods of cultural resource inventory and provides written instructions, research materials and supplies to all involved in planning and operation of natural resource activities. Provides site recording and implements field data strategies. Provides leadership to at least three lower graded Archaeological Aids or Technicians. Leadership responsibilities are regular and recurring and occupy about 25 percent of the work time. As crew leader, assures that work assignments of employees are carried out. Assigns tasks, monitors status, and assures timely accomplishment of workload. Instructs employees in special tasks and job techniques. Checks work in progress and amends or rejects work not meeting established standards. Reports performance, progress, etc., of employees to supervisor.

The above descriptions were derived from the Department of Labor's online version of the Service Contract Act, Directory of Occupations at http://www.dol.gov/dol/esa/public/regs/compliance/whd/wage/main.htm> For a discussion relevant to the derivation of these descriptions and their integration with CRM practices visit the website of the American Cultural Resources Association (ACRA) at http://www.acra-crm.org/wagedetermination.html

NATIONAL PARK SERVICE TRAINING AND DEVELOPMENT DIVISION ESSENTIAL COMPETENCIES FOR NATIONAL PARK SERVICE EMPLOYEES

Competency: A combination of knowledge, skills, and abilities in a particular Career Field, which, when acquired, allows a person to perform a task or function at a specifically defined level of proficiency.

Entry Level: Is just starting in the Career Field; has appropriate academic preparations but little or no work experience.

ARCHAEOLOGY ESSENTIAL COMPETENCIES

Following is a list of the competencies and the knowledge, skills, and abilities (KSAs) needed to perform in this particular discipline at the Entry level. The competencies are in **boldface** print and are followed by a brief definition. The definitions are then followed by a list of the knowledge, skills, and abilities needed to be effective at each level. The competencies and KSAs of the previous level(s) are also required at the next higher level.

Archaeological Technician (GS-0102)

Archaeological technicians are specialists in archaeology-related fields such as photography; fieldwork, excavation, surveying and mapping; artifact collection, cleaning, sorting and labeling; automated data base management; field logistics; equipment management; and other assistance functions. Technicians possessing the competencies of this level have the knowledge and analytical skills equivalent to an advanced undergraduate educational level or a Bachelor's Degree in anthropology, archaeology, history, or a related field with specialized training in archaeology. Archaeological technicians perform under the direct supervision of a professional archaeologist.

Field experience as part of an archaeological field crew or field school.

Archaeologist, Entry Level (GS-0193)

Archaeologists possessing the competencies of the Entry Level have the knowledge and analytical skills equivalent to a Bachelor's Degree in anthropology, archaeology, history, or a related field with specialized training in archaeology. At this level, archaeologists serve on the professional staff at a park or other facility under the direction of a supervisory archaeologist. They deal with archaeological resource identification, documentation, protection, interpretation, and preservation; and carry out or assist in limited scope monitoring, survey, testing, and excavating archaeological sites; lab work and managing of field data; and the production of small-scale reports.

I. PROFESSIONAL DISCIPLINE

Provides general information and knowledge about archaeology.

Fundamental knowledge of archaeology equivalent to the completion of an accredited curriculum leading to a Bachelor's Degree in anthropology or archaeology-related field with basic competency acquired through field school and/or experience.

Supervisory field experience at the level of crew chief.

Familiarity with techniques for archaeological survey, testing, excavation and data retrieval, condition/integrity assessment, remote sensing of archaeological sites, archaeological sampling strategies, and other basic field and laboratory procedures.

II. PRESERVATION LAW, PHILOSOPHY, AND PRACTICE

Provides general information and knowledge on the identification, evaluation, documentation, treatment, and management of cultural resources, especially those archaeological in nature.

Possesses broad, but basic understanding and knowledge in preservation law, philosophy, and practice.

Familiarity with national cultural resource laws and regulations, policies, and National Park Service regulations, policies, and guidelines relating to archaeological and cultural resources.

III. RESEARCH AND INVENTORIES

Under the direct supervision of a professional archaeologist, conducts basic research on archaeological topics and participates in archaeological surveys and excavations and documents all work in accordance with professional standards.

A. Archaeological Investigations

Ability to conduct small-scale, limited scope archaeological investigations using a variety of techniques and preparing a wide range of archaeological documentation.

Ability to assist other professional archaeologists in conducting archaeological monitoring, surveys, and excavations; including the location of sites, recording of archaeological and environmental data, and summarization of information collected.

Knowledge of the techniques involved in maintaining field notes and preparing field descriptions, drawings, maps, surveying instruments and their appropriate uses, photographs, and video recordings related to the archaeological fieldwork.

Knowledge of basic professional procedures in organizing hard and digitally generated records such as site files, base maps, and other data.

Knowledge of basic professional procedures and operations in conducting archaeological site monitoring, archaeological surveys, archaeological investigations and testing.

Knowledge of local and regional prehistory and/or history needed to assist in analyzing and processing archaeological data and material resulting from fieldwork.

B. Laboratory Analysis/Conservation of Field Collections

Carries out limited scope laboratory procedures, including analyzing, accessioning, cataloging, and preserving artifacts, and data generated by the field activities.

Ability to assist other professional archaeologists in preparing site information for updating the Archaeological Sites Management Information System (ASMIS), the Cultural Sites Inventory (CSI); the NPS Geographic Information System (GIS); and the Automated National Catalog System (ANCS+).

Knowledge of laboratory analysis and artifact curation to accession, catalog, analyze, and preserve artifacts and data generated by the survey.

Skills in photography, darkroom techniques, and graphic recording techniques to prepare photographs and other visual displays for recording the results of archaeological surveys, and preparing acceptable reports.

IV. PRESERVATION, TREATMENT, AND MAINTENANCE

Assists in planning and implementing archaeological projects and provides technical assistance.

Hands-on experience in recording soil depositional sequences, site formation processes, agents of deterioration, and recommendations for enhanced documentation, treatment, monitoring, and protection programs.

Knowledge of applicable management documents such as area management reports and preservation plans.

V. PROGRAM AND PROJECT MANAGEMENT

Under direct supervision, may assist in the development and execution of a park, center, cluster, or office level program or project.

Ability to assist, under direct supervision, in the development and execution of a park, center, cluster, or office level program or project.

VI. WRITING AND COMMUNICATION

Communicates, interprets, and presents information pertinent to the preservation of archaeological sites and materials.

A. WRITING

Ability to prepare limited scale scientific reports that will disseminate the cultural resource data derived from projects in accordance with service and agency policies.

Knowledge of basic techniques of writing technical and professional reports on the results of archaeological surveys that meet professional and National Park Service standards.

Administrative skills such as preparing scopes of work, cooperative agreements, and contracts.

B. PUBLIC INTERPRETATION/PRESENTATION/ OUTREACH

Knowledge and understanding of the importance of public interpretation of archaeological sites and materials, outside consultations, developing partnerships, and contacts with the professional community.

Basic knowledge of techniques of conveying technical archaeological information to the lay public.

Ability to work as a team member in the design and implementation of effective public interpretation programs such as popular histories, brochures, pamphlets, videos, exhibits, posters, lesson plans, and other public interpretation devices.

Knowledge of public speaking techniques.

VII. TRAINING

Assists others in presenting training.

Basic knowledge of current policies, guidelines, standards, and technical information related to archaeology.

Ability to assist others in coordinating and conducting a training session.

VIII. SAFETY

Insures on-the-job safety and health of all employees.

Knowledge of on-the-job safety and health considerations of the work place.

Knowledge of job safety and health hazards and safety requirements for job assignments.

The above was abstracted from the National Park Service's online versions of "Essential Competencies for National Park Service Employees" at http://www.nps.gov/training/npsonly/ npsescom.htm and "Archaeology: Essential Competencies" at http://www.nps.gov/training/npsonly/RSC/archeolo.htm>

WRITTEN JOB DESCRIPTIONS HUNTER RESEARCH, INC. TRENTON, NEW JERSEY

TITLE: FIELD ASSISTANT/LAB ASSISTANT

GENERAL CHARACTERISTICS:

Working knowledge of archaeological survey and excavation procedures and equipment; basic artifact processing systems; basic field and laboratory recording systems. Familiarity with archaeological provenance principles and basic stratigraphic analysis. Basic field drawing and artifact marking skills; knowledge of basic artifact classifications and materials for sorting and cleaning; knowledge of appropriate artifact storage; familiarity with computers and basic word processing systems.

DIRECTION RECEIVED:

From President, Vice President, Data Manager, Report Manager/Historian, other principals and senior staff.

TYPICAL DUTIES AND RESPONSIBILITIES:

Performance of field survey and excavation. Initial processing of archaeological materials recovered on field projects. Occasional transport of field equipment and personnel. Responsible for maintenance and repair of field equipment, maintenance of basic field records, photo inventory sheets, archaeological provenance records, field drawings and field photography on-site. Performance of occasional drafting tasks. Bagging, sorting, appropriate cleaning of artifactual materials recovered during projects. Marking of artifacts. Word processing of tables of field data.

RESPONSIBILITY FOR DIRECTION OF OTHERS:

No.

EDUCATION:

High school diploma.

EXPERIENCE:

Prior participation in archaeological survey and excavation projects and/or completion of a course of study in Anthropology or Archaeology.

TITLE: ASSISTANT ARCHAEOLOGIST

GENERAL CHARACTERISTICS:

Technical and practical knowledge of both survey and excavation procedures and equipment. Field and laboratory supervisory experience. Management of field records and archaeological materials, field photography, field drawing, basic stratigraphic and artifact analysis. Familiarity with computers and basic word processing systems. Basic drafting skills related to preparation of graphics for project reports.

DIRECTION RECEIVED:

From President, Vice President, Data Manager, Report Manager/Historian, other principals and senior staff in the direction of project mobilization and implementation of field strategies.

TYPICAL DUTIES AND RESPONSIBILITIES:

Liaison between senior archaeological staff and laboratory staff for the organization and management of field equipment throughout project schedule. Responsible for field equipment checkout and return. Occasional transport of field personnel. Execution of project fieldwork, both survey and excavation. Assist in maintenance of field records, photo inventory sheets, archaeological provenance records, field drawings, and field photography on-site. Liaison with laboratory staff during and after fieldwork for management of archaeological materials. Consolidation and preparation of field records for report writing.

RESPONSIBILITY FOR DIRECTION OF OTHERS:

Supervision of field assistants on-site.

EDUCATION:

BA degree in Anthropology/Archaeology or related discipline.

EXPERIENCE:

Two years experience in various aspects of archaeological field projects.

TITLE: SENIOR ARCHAEOLOGIST

GENERAL CHARACTERISTICS:

Overall on-site management of field projects; strategic planning and organization for particular projects. Project liaison with both on-site and off-site staff. Supervision and management of assistant archaeologists and field assistants. Stratigraphic analysis and site interpretation, basic knowledge of archival materials, artifact identification and analysis. Report preparation and writing, basic familiarity with computers and word processing and database software systems.

DIRECTION RECEIVED:

From President, Vice President, Principal Investigators, Report Manager/Historian, and other senior staff.

TYPICAL DUTIES AND RESPONSIBILITIES:

Collaboration in the assessment of projects in connection with the preparation of responses to client requests for proposals and cost estimates. Selected background research tasks related to specific archaeological projects. Assist in the development and implementation of field strategies for specific projects. Overall site direction and day-to-day management of field projects.

RESPONSIBILITY FOR DIRECTION OF OTHERS:

Supervision of and liaison with Assistant Archaeologists concerning the compilation and curation of field records, archaeological provenance records, field drawings, and on-site field photography. Supervision of all field crew.

EDUCATION:

B.A. degree in Anthropology/Archaeology or related discipline.

EXPERIENCE:

Three years experience in all supervisory and documentation aspects of archaeological field projects.

The above used with permission of Hunter Research, Inc., Trenton, New Jersey.

WRITTEN JOB DESCRIPTIONS URS CORPORATION FLORENCE, NEW JERSEY

TITLE: TECHNICIAN I

GENERAL CHARACTERISTICS:

Field Technicians excavate and record archaeological sites. Laboratory Technicians process artifacts for analysis.

DIRECTION RECEIVED:

Follows standard work methods or explicit instructions. Techniques/procedures for non-routine work identified by supervisor.

TYPICAL DUTIES AND RESPONSIBILITIES:

In the field, excavates soils, records findings, collects artifacts. In the laboratory, washes, marks, and sorts artifacts.

RESPONSIBILITY FOR DIRECTION OF OTHERS:

None.

EDUCATION:

Bachelors Degree in Anthropology, History, or related field and attended an archaeological field school.

EXPERIENCE:

A minimum of six months of fieldwork experience on archaeological sites in the United States.

TITLE: TECHNICIAN II

GENERAL CHARACTERISTICS:

Field Technicians excavate and record archaeological sites. Laboratory Technicians process artifacts for analysis.

DIRECTION RECEIVED:

Receives initial instructions and advice from supervisor.

TYPICAL DUTIES AND RESPONSIBILITIES:

In the field, excavates soils, records findings, and collects artifacts. In the laboratory, washes, marks, and sorts artifacts. May be asked to perform specific problem solving tasks in the field and laboratory.

RESPONSIBILITY FOR DIRECTION OF OTHERS:

Coordinates the work of other lower-level technicians.

EDUCATION:

Bachelors Degree in Anthropology, History, or related field and attended an archaeological field school.

EXPERIENCE:

A minimum of 1 year of fieldwork experience on archaeological sites in the United States.

TITLE: CREW CHIEF

GENERAL CHARACTERISTICS:

Responsible for assisting the Field Supervisor by managing and conducting specific aspects of an archaeological investigation.

TYPICAL DUTIES AND RESPONSIBILITIES:

A crew chief is assigned specific and often complicated field tasks (e.g., feature excavation, soil profiles and plan drawings) beyond the normal duties of a field technician. Receives close supervision from the Field Supervisor. Supervises field technicians and collects and inspects field records. Teaches general field methods and techniques to new field crew. Reports progress to the Field Supervisor. Also helps in specific analytical laboratory tasks.

EDUCATION:

Bachelors degree in Anthropology, History, or related field, and attended an archaeological field school.

EXPERIENCE:

A minimum of two years experience as a field technician, three months of which was served as an acting Crew Chief. Should be proficient in field photography and the use of surveying equipment.

TITLE: FIELD SUPERVISOR

GENERAL CHARACTERISTICS:

Responsible for managing the day-to-day aspects of an archaeological investigation.

DIRECTION RECEIVED:

Supervisor screens assignments for unusual or difficult problems and selects techniques and procedures to be applied to non-routine work. Receives close supervision on new aspects of assignments.

TYPICAL DUTIES AND RESPONSIBILITIES:

Supervises crew chiefs and field technicians, collects and inspects field records and artifacts, reporting progress to the Principal Archaeologist. Also helps in the preparation of reports and other documents.

RESPONSIBILITY FOR DIRECTION OF OTHERS:

Supervises and coordinates the work of crew chiefs and field technicians.

EDUCATION:

Bachelors Degree in Anthropology, History, or related field, and attended an archaeological field school.

EXPERIENCE:

A minimum of two years experience in the United States, six months of which was in a supervisory role.

The above used with permission of URS Greiner Woodward Clyde Corporation, Florence, New Jersey.

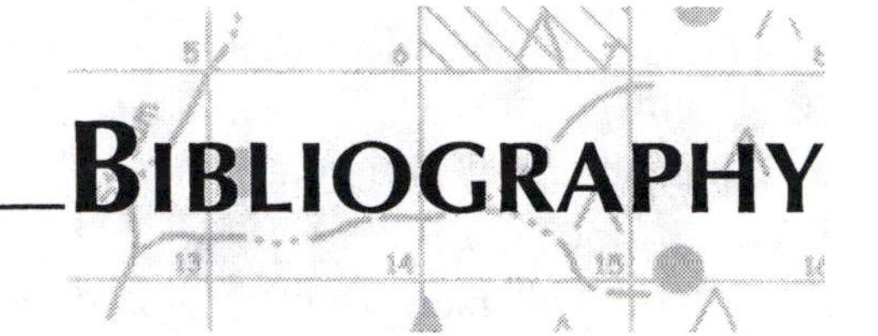

BIBLIOGRAPHY

Adkins, Lesley and Roy Adkins. 1989. *Archaeological Illustration.* New York: Cambridge University Press.

Adovasio, James M. and Ronald C. Carlisle. 1988. Some thoughts on cultural resource management archaeology in the United States. Antiquity 62(1):72–87.

Adovasio, J. M., J. D. Gunn, J. Donahue, and R. Stuckenrath. 1977. Meadowcroft Rockshelter: Retrospect 1976. *Pennsylvania Archaeologist* 47(2–3):1–93.

Adovasio, J. M., K. J. Shaunessy, W. C. Johnson, W. P. Athens, A. T. Boldurian, R. C. Carlisle, D. C. Dirkmaat, J. Donahue, D. R. Pedler, and E. J. Siemon III. 1990. Archaeology at the Howarth-Nelson Site (36FA40), Fayette County, Pennsylvania. *Pennsylvania Archaeologist* 60(1):32–67.

Akerson, Louise E. (ed.). 1995. The effects of OSHA regulations on archaeology. *Journal of Middle Atlantic Archaeology* 11:1–10 (Thematic compilation of articles).

Aldenderfer, Mark S. and Herbert D. G. Maschner (eds.). 1996. *Anthropology, Space, and Geographic Information Systems.* New York: Oxford University Press.

Allen, Kathleen, Stanton Green, and Ezra Zubrow (eds.). 1990. *Interpreting Space: GIS and Archaeology.* London: Taylor and Francis.

American Anthropological Association. 1998. *Code of Ethics of the American Anthropological Association, Approved 1998.* American Anthropological Association, Washington, D.C. http://www.aaanet.org/committees/ethics/ethcode.htm>

American Cultural Resource Association (ACRA).
2001a. Frequently Asked Questions. http://www.acra-crm.org/ faq.html#anchor189081>
2001b. CRM Salaries and Other Statistics. Based on 29 CRM firms from across the country in 1996. http://www.acra.crm.org/BusinessStats.html>

Amerman, A.J. 1985. Plow-zone experiments in Calabria, Italy. *Journal of Field Archaeology* 12:33–40.

Ames, Kenneth M. and Herbert D. G. Maschner. 1999. *Peoples of the Northwest Coast: Their Archaeology and Prehistory.* New York: Thames and Hudson.

Anderson, James and Edward Mikhail. 1998. *Surveying: Theory and Practice* (7th ed.). Boston: WCB/McGraw-Hill.

Andrefsky, William, Jr. 1998. *Lithics: Macroscopic Approaches to Analysis.* Cambridge, U.K.: Cambridge University Press.

Aston, Michael and Trevor Rowley. 1974. *Landscape Archaeology: An Introduction to Fieldwork Techniques on Post-Roman Landscapes.* Newton Abbot, England: David and Charles.

Athens, J. Stephen. 1993. Cultural resource management and academic responsibility in archaeology: A further comment. *Society for American Archaeology Bulletin* 11(2):6–7.

Auerbach, Paul S. 1999. *Medicine for the Outdoors: The Essential Guide to Emergency Medical Procedures and First Aid.* New York: Lyons Press.

Azizi, Sharla C. 1995. It's Level D! Now what do we do? In the Field and in the Lab: Dealing with a Hazmat Site. *Journal of Middle Atlantic Archaeology* 11:3–15.

Barber, Russell J. 1994. *Doing Historical Archaeology: Exercises Using Documentary, Oral, and Material Evidence.* Upper Saddle River, NJ: Prentice-Hall.

Barber, Russell J. and Frances F. Berdan. 1998. *The Emperor's Mirror: Understanding Cultures Through Primary Sources.* Tucson: University of Arizona Press.

Barker, Philip. 1996. *Techniques of Archaeological Excavation* (3rd ed.). London: B.T. Batsford Ltd.

Bass, William M. 1995. *Human Osteology: A Laboratory and Field Manual* (3rd ed.). Special Publication No. 2. Columbia, MO: Missouri Archaeological Society.

Beck, Charlotte (ed.). 1994. *Dating in Exposed and Surface Contexts.* Albuquerque: University of New Mexico Press.

Beck, Lane Anderson. 1995. *Regional Approaches to Mortuary Analysis.* New York: Plenum Press.

Bello, Charles A. and John H. Cresson. 1995. A fluted point from Island Beach State Park. *Bulletin of the Archaeological Society of New Jersey* 50:53–56.

Bender, Susan J. and George S. Smith (eds.). 2000. *Teaching Archaeology in the Twenty-First Century.* Washington, D.C.: Society for American Archaeology.

Bennett, Kenneth A. 1993. *A Field Guide for Human Skeletal Identification.* Springfield, IL: Charles C. Thomas.

Berg, Thomas M. and Christine M. Dodge. 1981. *Atlas of Preliminary Geologic Quadrangle Maps of Pennsylvania.* Harrisburg, PA: Department of Environmental Resources, Bureau of Topographic and Geologic Survey.

Bergman, C. A. and J. F. Doershuk. 1995. OSHA regulations and the excavation of the deeply stratified Sandts Eddy Site (36NM12). *Journal of Middle Atlantic Archaeology* 11:17–29.

Bettes, F. 1992. *Surveying for Archaeologists.* University of Durham, England: Department of Archaeology.

Bettis, E. Arthur, III. 1992. Soil morphologic properties and weathering zone characteristics as age indicators in Holocene alluvium in the Upper Midwest, pp. 119–144. In Vance T. Holliday (ed.), *Soils in Archaeology: Landscape Evolution and Human Occupation.* Washington, D.C.: Smithsonian Institution Press.

Bilzi, A. F. and E. J. Ciolkosz. 1977. A field morphology rating scale for evaluating pedological development. *Soil Science* 124:45–48.

Binford, Lewis R. 1968. Archaeological perspectives, pp. 5–32. In Sally R. Binford and Lewis R. Binford (eds.), *New Perspectives in Archaeology.* Chicago: Aldine Publishing Company.

Birkeland, Peter W. 1999. *Soils and Geomorphology.* New York: Oxford University Press.

Bleed, Peter. 2000. Will the Marshalltown be forever golden? *Society for American Archaeology Bulletin* 18(4):21.

Bodner, Connie Cox and Ralph M. Rowlett. 1980. Separation of bone, charcoal, and seeds by chemical flotation. *American Antiquity* 45(1):110–116.

Boismier, William A. 1997. *Modeling the Effects of Tillage Processes on Artefact Distributions in the Ploughzone: A Simulation Study of Tillage Induced Pattern Formation.* BAR British Series 259, Oxford.

Brady, Nyle C. and Raymond R. Weil. 1998. *The Nature and Property of Soils.* Upper Saddle River, NJ: Prentice-Hall.

Brothwell, Don R. 1981. *Digging Up Bones: The Excavation, Treatment and Study of Human Skeletal Remains.* Third edition. British Museum and Cornell University Press, Ithaca, New York.

Brown, James A. 1971. *Approaches to the Social Dimensions of Mortuary Practices.* Memoir 25. Washington, D.C.: Society for American Archaeology.

Brown, James A. 1975. Deep-site excavation strategy as a sampling problem, pp. 155–169. In J. Mueller (ed.), *Sampling in Archaeology.* Tucson: University of Arizona Press.

Bruckle, Irene, Jonathan Thornton, Kimberly Nichols, and Gerri Strickler. 1999. Cyclododecan: Technical Note on Some Uses in Paper and Objects Conservation. *Journal of the American Institute for Conservation* 38(2):162–175.

Brunswig, Robert H., Jr. 1999. *An Evaluation of Archaeological Applications of Mapping Grade Global Positioning Systems: Field Test in Northeastern Colorado's Plains and Mountains.* U.S. Department of the Interior, National Park Service, National Center for Preservation Technology and Training Publication no. 1999-03.

Buccellati, Giorgio. 1992. Stratigraphic sections, pp. 66–76. In B. D. Dillon (ed.), *The Student's Guide to Archaeological Illustrating.* Archaeological Research Tools, Volume 1. Los Angeles: Institute of Archaeology, University of California.

Bureau of Land Management. 1973. *Manual of Instructions for the Survey of the Public Lands of the United States.* Washington, D.C.: U.S. Government Printing Office.

Butler, William B. 1979. The no-collection strategy in archaeology. *American Antiquity* 44(4):795–799.

Butler, William B. 1987. Significance and other frustrations in the CRM process. *American Antiquity* 52(4):820–829.

Butzer, Karl W. 1971. *Environment and Archaeology: An Ecological Approach to Prehistory.* Chicago: Aldine/Atherton.

Butzer, Karl W. 1976. *Geomorphology From the Earth.* New York: Harper and Row.

Butzer, Karl W. 1980. Context in archaeology: An alternative perspective. *Journal of Field Archaeology* 7(4):417–422.

Campbell, Sarah K. 1981. *The Duwamish No.1 Site: A Lower Puget Sound Shell Midden.* Research Report 1. Seattle: University of Washington Office of Public Archaeology.

Canti, M. G. and F. M. Meddens. 1998. Mechanical coring as an aid to archaeological projects. *Journal of Field Archaeology* 25:97–105.

Carnes, Linda F., Roy S. Dickens, Jr., Linda France, and Ann Long. 1986. *Cost Analysis of Archaeological Activities at Eight Southeastern Sites.* Chapel Hill: Research Laboratories of Anthropology, University of North Carolina, Chapel Hill. Washington, D.C.: National Park Service, Archaeological Assistance Division.

Carnett, Carol L. 1995. *A Survey of State Statutes Protecting Archaeological Resources.* Preservation Law Reporter Special Report, Archaeological Assistance Study Number 3. Washington, D.C.: National Trust for Historic Preservation.

Cavallo, John A. 1987. *Area B Site (28Me1-B): Data Recovery.* Trenton Complex Archaeology, Report 8. Trenton: Federal Highway Commission and New Jersey Department of Transportation.

Chapman, Robert, Ian Kinnes, and Klavs Randsborg. 1981. *The Archaeology of Death.* Cambridge, London: Cambridge University Press.

Chatters, James C. 1997. Encounter with an ancestor. *Anthropology Newsletter,* pp. 9–10.

Childe, V. Gordon. 1944. *Progress and Archaeology.* London: Watts.

Claassen, Cheryl. 1991. Normative thinking and shell-bearing sites. *Archaeological Method and Theory* 3:249–298.

Claassen, Cheryl. 1998. *Shells.* Cambridge Manuals in Archaeology. Cambridge, London: Cambridge University Press.

Clark, Anthony. 1996. *Seeing Beneath the Soil: Prospecting Methods in Archaeology.* London: B.T. Batsford Ltd.

Clark, Grahame. 1957. *Archaeology and Society* (3rd ed.). London: Methuen.

Coles, J. M. and D. M. Goodburn. 1990. Wet site excavation and survey: Proceedings of a conference at the museum of London. WARP Occasional Paper No. 5. England: University of Exeter.

Conyers, Lawrence B. and Catherine M. Cameron. 1998. Ground penetrating radar techniques and three-dimensional computer mapping in the American Southwest. *Journal of Field Archaeology* 25:417–430.

Conyers, Lawrence B. and Dean Goodman. 1997. *Ground-Penetrating Radar: An Introduction for Archaeologists.* Walnut Creek, CA: AltaMira Press.

Cressman, Luther. 1988. *A Golden Journey: Memoirs of an Archaeologist.* Salt Lake City: University of Utah Press.

Crist, Thomas A.J. 2000. Smallpox and other scourges of the dead, pp. 79–106. In D. A. Poirier and K. L. Feder (eds.), *Dangerous Places: Health, Safety, and Archaeology.* Westport, CT: Greenwood Publishing Group.

Croes, Dale R. 1976. *The Excavation of Water-Saturated Archaeological Sites: Wet Sites on the Northwest Coast of North America.* Mercury Series No. 50. Ottawa: National Museum of Canada, Archaeological Survey of Canada.

Croes, Dale R. 1995. *The Hoko River Archaeological Complex: The Wet/Dry Site (45CA213), 3000–1,700 B.P.* Pullman: Washington State University Press.

Cronyn, J. M. 1990. *The Elements of Archaeological Conservation.* London and New York: Routledge.

Cross, Dorothy. 1941. *Archaeology of New Jersey, Volume 1.* Trenton: Archaeological Society of New Jersey and the New Jersey State Museum.

Cunliffe, B. (ed.). 1978. Landscape Archaeology. *World Archaeology,* Volume 9.

Curry, Dennis C. 1999. *Feast of the Dead: Aboriginal Ossuaries in Maryland.* Crownsville, MD: Archaeological Society of Maryland and Maryland Historical Trust.

Custer, Jay F. 1979. *An Evaluation of Sampling Techniques for Cultural Resources Reconnaissance in the Middle Atlantic Area of the United States.* Ph.D. dissertation, The Catholic University of America. Washington, D.C.

Custer, Jay F. 1988. Lithic scatter sites of the Piedmont Zone of Pennsylvania, Maryland, and Delaware. *Pennsylvania Archaeologist* 58(1):30–42.

Custer, Jay F. and Charles Clark. 2000. I've got biases I don't even know about: Rethinking Middle Atlantic archaeology. Paper presented at the Annual Meeting of the Middle Atlantic Archaeological Conference, Ocean City, MD.

Custer, Jay F., Scott C. Watson, and Barbara H. Silber. 1996. *Final Archaeological Investigations at the Carey Farm (7K-D-3) and Island Farm (7K- C-13) Sites, State Route 1 Corridor, Kent County, Delaware.* Delaware Department of Transportation Archaeology Series No. 146. Dover, DE.

Dancey, William S. 1981. *Archaeological Field Methods: An Introduction.* Minneapolis, MN: Burgess Publishing Company.

Daniels, Raymond B. and Richard D. Hammer. 1992. *Soil Geomorphology.* New York: John Wiley and Sons.

Darsie, Richard F. and James D. Keyser (eds.). 1985. *Archaeological Inventory and Predictive Modeling in the Pacific Northwest.* Studies in Cultural Resource Management No. 6, United States Department of Agriculture, Forest Service, Pacific Northwest Region.

Darvill, Timothy. 1995. Value systems in archaeology, pp. 40–50. In Malcolm A. Cooper, Anthony Firth, John Carman and David Wheatley (eds.), *Managing Archaeology.* London: Routledge Press.

Daugherty, Richard D. 1988. Problems and responsibilities in the excavation of wet sites, pp. 15–30. In B. Purdy (ed.), *Wet Site Archaeology.* Caldwell, NJ: Telford Press.

Davidson, Donald A. 1985. Geomorphology and archaeology, pp. 25–26. In G. Rapp and J. Gifford (eds.), *Archaeological Geology.* New Haven, CT: Yale University Press.

Davis, Hester. 1989. Learning by doing: This is no way to treat archaeological resources, pp. 275–279. In Henry F. Cleere (ed.), *Archaeological Heritage Management in the Modern World.* London: Unwin Hyman.

Davis, R. P. Stephen, Jr., Patrick C. Livingood, H. Trawick Ward, and Vincas Steponaitis. 1998. *Excavating Occaneechi Town: Archaeology of an Eighteenth-Century Indian Village in North Carolina.* Chapel Hill: University of North Carolina Press.

Department of the Interior, Office of the Secretary. 1993. 43 CFR Part 10. Native American Graves Protection and Repatriation Act Regulations; Proposed Rule. *Federal Register, Volume 58, No. 102, Friday, May 28, 1993.*

Dibble, Harold L., Shannon P. McPherron, and Barbara J. Roth. 1999. *Virtual Dig: A Simulated Archaeological Excavation of a Middle Paleolithic Site in France.* Mountain View, CA: Mayfield Publishing.

Dillehay, Tom D. (ed.). 1989. *Monte Verde: A Late Pleistocene Settlement in Chile, Volume 1: A Paleoenvironment and Site Context.* Smithsonian Series in Archaeological Inquiry, Washington, D.C.

Dillehay, Tom D. (ed.). 1996. *Monte Verde: A Late Pleistocene Settlement in Chile, Volume 2: The Archaeological Context and Interpretation.* Smithsonian Series in Archaeological Inquiry, Washington, D.C.

Dillon, Brian D. (ed.). 1992. *The Student's Guide to Archaeological Illustrating.* Archaeological Research Tools, Volume 1. Los Angeles: Institute of Archaeology, University of California.

Donahue, Jack and James M. Adovasio. 1990. Evolution of sandstone rockshelters in Eastern North America: A geoarchaeological perspective, pp. 231–251. In N. Lasca and J. Donahue (eds.), *Archaeological Geology of North America.* Centennial Special Volume 4. Boulder, CO: Geological Society of America.

Doran, Glen H. and David N. Dickel. 1988. Multidisciplinary investigations at the Windover Site, pp. 263–289. In Barbara A. Purdy (ed.), *Wet Site Archaeology.* Caldwell, NJ: Telford Press.

Doran, Glen H., David N. Dickel, and Lee A. Newsom. 1990. A 7,290-year-old bottle gourd from the Windover Site, Florida. *American Antiquity* 55(2):354–360.

Dorrell, Peter. 1989. *Photography in Archaeology and Conservation.* New York: Cambridge University Press.

Dowman, Elizabeth. 1970. *Conservation in Field Archaeology.* London: Methuen.

Dragoo, Don W. 1963. *Mounds for the Dead.* Annals of the Carnegie Museum, Volume 37. Pittsburgh, PA.

Drake, Avery A., Jr., Dean B. McLaughlin, and Robert E. Davis. 1967. *Geologic Map of Riegelsville Quadrangle, Pennsylvania-New Jersey.* Scale 1:24,000. Map GQ-593. Washington, D.C.: United States Geological Survey.

Drennan, Robert D. 1996. *Statistics for Archaeologists: A Commonsense Approach.* New York: Plenum Press.

Dunnell, Robert D. 1978. Style and function: A fundamental dichotomy. *American Antiquity* 43(2):192–202.

Dunnell, Robert D. 1988. Low-density archaeological records from plowed surfaces: Some preliminary considerations. *American Archaeology* 7(1):29–38.

Dunnell, Robert D. and William S. Dancey. 1983. The siteless survey: A regional scale data collection strategy. *Advances in Archaeological Method and Theory* 6:267–287.

Dunnell, Robert D. and James K. Feathers. 1994. Thermoluminescence Dating of Surficial Archaeological Material, pp. 115–137. In Charlote Beck (ed.), *Dating in Exposed and Surface Contexts.* Albuquerque: University of New Mexico Press.

Dunnell, Robert D. and Jan F. Simek. 1995. Artifact size and plowzone processes. *Journal of Field Archaeology* 22:305–319.

Easterbrook, Don J. 1999. *Surface Processes and Landforms.* Upper Saddle River, NJ: Prentice Hall.

Ebert, James I. 1988. Modeling human systems and "predicting" the archaeological record: The unavoidable relationship of theory and method. *American Archaeology* 7(1):3–8.

Ebert, James I. 1992. *Distributional Archaeology.* Albuquerque: University of New Mexico Press.

Elon, Amos. 1996. Politics and archaeology, pp. 183–188. In Linda L. Hasten (ed.), *Annual Editions: Archaeology 96/97.* Guilford, CT: Dushkin Publishing Group/Brown & Benchmark Publishers.

Farrand, William R. 1985. Rockshelter and cave sediments, pp. 21–39. In J. K. Stein and W. R. Farrand (eds.), *Archaeological Sediments in Context.* Orono: Center for the Study of Early Man, University of Maine.

Ferguson, T. J. 1996. Native Americans and the practice of archaeology. *Annual Review of Anthropology* 25:63–79.

Flanagan, Joseph. 1995. What You Don't Know Can Hurt You. *Federal Archaeology* 8(2):10–13.

Flannery, K. V., C. L. Moser, and S. Maranca. 1986. The Excavation of Guila Naquitz, pp. 65–96. In K. V. Flannery (ed.), *Guila Naquitz: Archaic Foraging and Early Agriculture in Oaxaca, Mexico.* New York: Academic Press.

Florian, Mary-Lou E. 1990. *The Conservation of Artifacts Made From Plant Materials.* Marina del Ray, CA: Getty Conservation Institute.

Folk, Robert L. 1954. The Distinction Between Grain Size and Mineral Composition in Sedimentary Rock Nomenclature. *Journal of Geology* 62:344–359.

Folk, Robert L. 1974. *Petrology of Sedimentary Rocks.* Austin, TX: Hemphill Publishing.

Folsom, Franklin and Mary Elting Folsom. 1993. *America's Ancient Treasures: A Guide to Archaeological Sites and Museums in the United States and Canada.* Albuquerque: University of New Mexico Press.

Foss, John E. and Antonio V. Segovia. 1984. Rates of soil formation, pp. 1–17. In R.G. LaFleur (ed.), *Groundwater as a Geomorphic Agent.* Binghamtom Symposia in Geomorphology, International Series No.13. Boston: Allen and Unwin.

Foss, J. E., R. J. Lewis, M. E. Timpson, M. W. Morris, and J. T. Ammons. 1992. Pedologic approaches to archaeological sites of contrasting environments and ages, pp. 19–22. In J. E. Foss, M. E. Timpson, and M. W. Morris (eds.), *Proceedings of the First International Conference on Pedo-Archaeology.* Knoxville: University of Tennessee Agricultural Experiment Station.

Foss, John E., Fred P. Miller, and Antonio V. Segovia. 1985. *Field Guide to Soil Profile Description and Mapping.* Moorhead, MN: Soil Resources International, Inc.

Foss, J. E., C. E. Robinette, R. G. Darmody, and D. P. Wagner. 1976. *A Laboratory Manual for Soil Science* (4th ed). College Park: Department of Agronomy, University of Maryland.

Foss, J. E., W. R. Wright and R. H. Coles. 1975. Testing the accuracy of field textures. *Proceedings of the Soil Science Society of America* 39:800–802.

Fowler, Don D. 1982. Cultural resources management. *Advances in Archaeological Method and Theory* 5:1–50.

Fowler, Don D. 1986. Conserving American Archaeological Resources pp. 135–162. In D. Meltzer, D. Fowler, and J. Sabloff, (eds.), *American Archaeology: Past and Future.*

Fowler, M. L., J. Rose, B. Vander Leest, and S. R. Ahler. 1999. *The Mound 72 Area: Dedicated and Sacred Space in Early Cahokia.* Report of Investigations, Illinois State Museum, No. 54. Springfield, IL.

Frink, Douglas S. 1984. Artifact behavior within the plow zone. *Journal of Field Archaeology* 11:356–363.

Frink, Douglas S. 1992. The chemical variability of carbonized organic matter through time. *Archaeology of Eastern North America* 20:67–79.

Frink, Douglas S. 1994. Application of the newly developed OCR dating procedure in pedo-archaeological studies, pp. 149–157. In Albert C. Goodyear, John E. Foss, and Kenneth E. Sassaman (eds.), *Proceedings of the Second International Conference on Pedo-Archaeology.*. Anthropological Studies 10, Occasional papers of the South Carolina Institute of Archaeology and Anthropology, University of South Carolina.

Gardner, William M. 1983. Stop me if you've heard this before: The Flint Run Paleoindian Complex revisited. *Archaeology of Eastern North America* 11:49–59.

Goldberg, Paul, David T. Nash, and Michael D. Petraglia (eds.). 1993. *Formation Processes in Archaeological Context.* Monographs in World Prehistory No. 17. Madison, WI: Prehistory Press.

Gorospe, Kathy. 1985. *American Indian Cultural Resources: A Preservation Handbook.* Salem, OR: Commission on Indian Services.

Grant, Campbell. 1983. *The Rock Art of the North American Indians.* New York: Cambridge University Press.

Gray, Marcy, Joseph Schuldenrein, Michael Stewart, and Terry Klein. 1999. Plenary Session Panel: The Changing CRM Workforce. Annual Meeting of the American Cultural Resources Association, Mt. Laurel, NJ.

Greber, N'omi (ed.). 1984. *Recent Excavations at the Edwin Harness Mound, Liberty Works, Ross County, Ohio.* MCJA Special Paper No. 5. Kent, OH: Kent State University Press.

Green, William and John F. Doershuk. 1998. Cultural resource management and American archaeology. *Journal of Archaeological Research* 6(2):121–167.

Gruber, Anna. 1990. *The Archaeology of Mr. Jefferson's Slaves.* MA thesis, University of Delaware, Newark.

Grumet, Robert S. 1992. *Historic Contact: Early Relations Between Indian People and Colonists in Northeastern North America, 1524–1783.* Cultural Resources Planning Branch, National Register Programs Division, Mid-Atlantic Region, National Park Service, Philadelphia.

Grumet, Robert S. 1995. *Historic Contact: Indian People in Today's Northeastern United States in the Sixteenth through Eighteenth Centuries.* Norman: University of Oklahoma Press.

Gumerman, George, IV and Bruce S. Umemoto. 1987. The siphon technique: An addition to the flotation process. *American Antiquity* 52(2):330–336.

Hall, George M. 1973. *Ground Water in Southeastern Pennsylvania.* Harrisburg, PA: Department of Environmental Resources, Bureau of Topographic and Geologic Survey.

Hamilton, D. L. 1996. *Basic Methods of Conserving Underwater Archaeological Material Culture.* Washington, D.C.: U.S. Department of Defense, Legacy Resource Management Project.

Harden, Jennifer W. 1982. A quantitative index of soil development from field descriptions: Examples from a chronosequence in Central California. *Geoderma* 28:1–28.

Harden, Jennifer W. 1986. *Soil Chronosequences in the Western United States.* United States Geological Survey, Denver, CO: U.S. Government Printing Office.

Harris, Edward. 1975. The stratigraphic sequence: A question of time. *World Archaeology* 7(1):109–121.

Harris, Edward. 1979. *Principles of Archaeological Stratigraphy.* New York: Academic Press.

Hassan, Fekri A. 1978. Sediments in archaeology: Methods and implications for paleoenvironmental and cultural analysis. *Journal of Field Archaeology* 5:197–212.

Hayden, Brian. 1993. *Archaeology: The Science of Once and Future Things.* New York: W.H. Freeman and Company.

Heimmer, Don H. 1992. *Near-Surface, High Resolution Geophysical Methods for Cultural Resource Management and Archaeological Investigations.* Manual prepared for Interagency Archaeological Services, Rocky Mountain Regional Office, National Park Service, Denver, CO. Golden, CO: Geo-Recovery Systems, Inc.

Heizer, Robert F. and John A. Graham. 1967. *Guide to Field Methods in Archaeology.* Palo Alto, CA: National Press.

Hester, Thomas R. 1997. The handling and conservation of artifacts in the field, pp. 143–158. In T. R. Hester, H. J. Shafer, and K. L. Feder (eds.), *Field Methods in Archaeology.* Mountain View, CA: Mayfield Publishing Company.

Hester, Thomas R., Robert F. Heizer and John A. Graham. 1975. *Field Methods in Archaeology* (6th ed). Palo Alto, CA: Mayfield Publishing Company.

Hester, Thomas R., Harry J. Shafer and Kenneth L. Feder. 1997. *Field Methods in Archaeology* (7th ed). Mountain View, CA: Mayfield Publishing Company.

Hill, G. F. 1929. Introductory, pp. 13–15. In G. F. Hill (ed.), *How To Observe In Archaeology: Suggestions for Travelers in the Near and Middle East* (2nd ed.). London: The British Museum.

Hirst, K. Kris. 1998. "Have Trowel Will Travel." About Archaeology with K. Kris Hirst Web Site (1998): . Online. Internet. http://archaeology.about.com/science/archaeology/blfieldtech.htm

Hodder, Ian. 1991. Postprocessual archaeology and the current debate, pp. 30–41. In Robert Preucel (ed.), *Processual and Postprocessual Archaeologies: Multiple Ways of Knowing the Past.* Center for Archaeological Investigations, Occasional paper No. 10, Southern Illinois University, Carbondale.

Hodges, Henry. 1971. *Artifacts: An Introduction to Early Materials and Technology* (4th ed). London: John Baker.

Holliday, Vance T. (ed.). 1992a. *Soils in Archaeology: Landscape Evolution and Human Occupation.* Washington, D.C.: Smithsonian Institution Press.

Holliday, Vance T. 1992b. Soil formation, time, and archaeology, pp. 101–118. In Vance T. Holliday (ed.), *Soils in Archaeology: Landscape Evolution and Human Occupation.* Washington, D.C.: Smithsonian Institution Press.

Howell, Carol L. and W. Blanc. 1995. *A Practical Guide to Archaeological Photography.* Archaeological Research Tools, Volume 6. Los Angeles: Institute of Archaeology, University of California.

Howell, Nancy. 1990. *Surviving Fieldwork: A Report of the Advisory Panel on Health and Safety in Fieldwork, American Anthropological Association.* Washington, D.C.: American Anthropological Association.

Hranicky, William J. 1991 *Using USGS Topographic Maps.* Special Publication Number 20. Richmond: Archaeological Society of Virginia.

Hunt, Charles B. 1972. *Geology of Soils: Their Evolution, Classification, and Uses.* San Francisco: W.H. Freeman and Company.

Ives, Edward D. 1995. *The Tape Recorded Interview: A Manual for Field Workers in Folklore and Oral History.* Knoxville: University of Tennessee Press.

Ives, J. 1982. Evaluating the effectiveness of site discovery techniques in boreal forest environments, pp. 95–114. In P. Francis and E. Poplin (eds.), *Directions in Archaeology: A Question of Goals.* Alberta: University of Calgary.

Jacobson, Cliff. 1999. *Basic Essentials of the Map and Compass.* Guilford, CT: Globe Pequot Press.

Jacobson, G. L., Jr. and R. H. W. Bradshaw. 1981. The selection of sites for paleoenvironmental studies. *Quaternary Research* 16(1):80–96.

Jennings, Jesse D. 1980. *Cowboy Cave.* Anthropological Papers No. 104. University of Utah. Salt Lake City: University of Utah Press.

Jennings, Jesse D. 1994 *Accidental Archaeologist: Memoirs of Jesse D. Jennings.* Salt Lake City: University of Utah Press.

Jensen, John R. 2000. *Remote Sensing of the Environment: An Earth Resource Perspective.* Upper Saddle River, NJ: Prentice Hall.

Johnston, Richard B. 1968. *The Archaeology of the Serpent Mounds Site.* University of Toronto, Royal Ontario Museum, Art and Archaeology Occasional Paper 10, Ontario.

Joukowsky, Martha. 1980. *A Complete Manual of Field Archaeology.* Englewood Cliffs, NJ: Prentice-Hall.

Joyce, Arthur A., William Sandy, and Sharon Horan. 1989. Dr. Charles Conrad Abbott and the question of human antiquity in the New World. *Bulletin of the Archaeological Society of New Jersey* 44:59–70.

Kalin, Jeffrey. 1981. Stem point manufacture and debitage recovery. *Archaeology of Eastern North America* 9:134–175.

Katzenberg, M. Anne and Shelley R. Saunders (eds.). 2000. *Biological Anthropology of the Human Skeleton.* New York: Wiley.

Kauffman, Judson. 1990. *Physical Geology* (8th ed.). Englewood Cliffs, NJ: Prentice-Hall.

Keller, J.J. and Associates, Inc. 1997. *OSHA Excavation Standard Handbook, 29 CFR Parts 1926.650 Through 1926.652.* Neenah, WI: Keller and Associates, Inc.

Kelso, William M. 1997. *Archaeology at Monticello.* Monticello Monograph Series. Charlottesville, VA: Thomas Jefferson Memorial Foundation.

Kelso, William and Rachel Most (eds.). 1990. *Earth Patterns: Essays in Landscape Archaeology.* Charlottesville: University of Virginia Press.

Kennedy, Michael. 1996. *The Global Positioning System and GIS: An Introduction.* Chelsea, MI: Ann Arbor Press, Inc.

Kerber, Jordan E. 1991. *Coastal and Maritime Archaeology: A Bibliography.* Metuchen, NJ: The Scarecrow Press, Inc.

King, Thomas F. 1987. Prehistory and beyond: The place of archaeology, pp. 235–264. In R. E. Stipe and A. J. Lee (eds.), *The American Mosaic: Preserving a Nation's Heritage.* Washington, D.C.: United States Committee, International Council on Monuments and Sites.

King, Thomas F. 1998. *Cultural Resource Law and Practice: An Introductory Guide.* Walnut Creek, CA: AltaMira Press.

Kingerly, W. David (ed.). 1996. *Learning From Things: Method and Theory of Material Culture Studies.* Washington, D.C.: Smithsonian Institution.

Kintigh, Keith. 1988. The effectiveness of subsurface testing: A simulation approach. *American Antiquity* 53(4):686–707.

Kintz, Theresa. 1997. View from the trenches. *Common Ground* 2(1):48–53.

Kjellstrom, Bjorn. 1994. *Be Expert with Map and Compass: The Complete Orienteering Handbook.* New York: Maxwell Macmillan International.

Kohl, Philip. 1983. Archaeology and prehistory, pp. 25–28. In Thomas B. Bottomore (ed.), *Dictionary of Marxist Thought.* Cambridge, MA: Harvard University Press.

Kollmorgan Instruments Corporation, Macbeth Division. 1994. *Munsell Soil Color Charts.* NY: New Windsor.

Kraft, Herbert C. 1993. Dr. Charles Conrad Abbott, New Jersey's Pioneer Archaeologist. *Bulletin of the Archaeological Society of New Jersey* 48:1–12.

Krakker, J. J., M. J. Shott, and P. D. Welch. 1983. Design and evaluation of shovel test sampling in regional archaeological survey. *Journal of Field Archaeology* 10:469–480.

Krass, Dorothy S. 1998. *SAA Survey of Departments Regarding CRM/Public Archaeology Teaching.* Paper prepared for the Society for American Archaeology Workshop, "Enhancing Undergraduate and Graduate Education and Training in Public Archaeology and Cultural Resource Management," Wakulla Springs, Florida.

Kutsche, Paul. 1998. *Field Ethnography: A Manual for Doing Cultural Anthropology.* Upper Saddle River, NJ: Prentice-Hall, Inc.

Kvamme, Kenneth L. 1989a. Determining empirical relationships between the natural environment and prehistoric site locations: A hunter-gatherer example, pp. 208–238. In Christopher Carr (ed.), *For Concordance in Archaeological Analysis: Bridging Data Structure, Quantitative Technique and Theory.* Prospect Heights, IL: Waveland Press, Inc.

Kvamme, Kenneth L. 1989b. Geographic information systems in regional archaeological research and data management, pp. 139–203. In Michael B. Schiffer (ed.), *Archaeological Method and Theory,* volume 1. Tucson: University of Arizona Press.

Lake, Griffing and Stevenson. 1877. *An Illustrated Atlas of Washington County, Maryland.* Philadelphia: H. J. Toudy.

Lambert, Joseph B. 1997. *Traces of the Past: Unraveling the Secrets of Archaeology Through Chemistry.* Reading, MA: Addison-Wesley.

Larsen, C. S. 1997. *Bioarchaeology: Interpreting Behavior From the Human Skeleton.* Cambridge, England: Cambridge University Press.

Laville, Henri. 1976. Deposits in calcareous rock shelters: Analytical methods and climatic interpretation, pp. 137–155. In D. A. Davidson and M. L. Shackley (eds.), *Geoarchaeology: Earth Science and the Past.* London: Duckworth.

Lee, Georgia. 1991. *Rock Art and Cultural Resource Management.* Calabasas, CA: Wormwood Press.

LeeDecker, Charles, Edward Morin, Ingrid Wuebber, Meta Janowitz, Marie-Lorraine Pipes, Nadia Shevchuk, Mallory Gordon, and Diane Dallal. 1993. *Meadows Site: Historical and Archaeological Investigations of Philadelphia's Waterfront.* Report prepared for the Federal Highway Administration and the Pennsylvania Department of Transportation, District 6-0. East Orange, NJ: Cultural Resource Group, Louis Berger and Associates, Inc.

Leick, Alfred. 1995. *GPS Satellite Surveying.* New York: John Wiley and Sons.

Leone, Mark P. and Parker B. Potter, Jr. 1992. Legitimation and the classification of archaeological sites. *American Antiquity* 57(1):137–145.

Lewarch, Dennis E. and Michael J. O'Brien.

1981a. Effect of short term tillage on aggregate provenience surface pattern, pp. 7–49. In M. J. O'Brien and D. E. Lewarch (eds.), *Plowzone Archaeology: Contributions to Theory and Technique.* Nashville, TN: Vanderbilt University Publications in Anthropology 27.

1981b. The expanding role of surface assemblages in archaeological research. *Advances in Archaeological Method and Theory* 4:297–342.

Lightfoot, Kent. 1986. Regional surveys in the Eastern United States: The strengths and weaknesses of implementing subsurface testing programs. *American Antiquity* 51(3):484–504.

Lightfoot, Kent. 1989. A defense of shovel-test sampling: A reply to Shott. *American Antiquity* 54(2):413–416.

Lipe, William D. 1984. Value and Meaning in Cultural Resources, pp. 1–11. In H. Cleere (ed.), *Approaches to the Archaeological Heritage.* London: Cambridge University Press.

Lipe, William and Vincas Steponaitis. 1998. SAA to Promote Professional Standards through ROPA Sponsorship. *Society for American Archaeology Bulletin* 16(2):1,16–17.

Lock, Gary and Zoran Stanic (eds.). 1995. *Archaeology and Geographic Information Systems.* Bristol, PA: Taylor and Francis.

Lynch, Thomas. 1980. *Guitarrero Cave: Early Man in the Andes.* New York: Academic Press.

Lynott, Mark J. 1997. Ethical principles and archaeological practice: Development of an ethics policy. *American Antiquity* 62(4):589–599.

Lynott, Mark J. and Alison Wylie. 1995. *Ethics in American Archaeology: Challenges for the 1990s.* Special Report. Washington, D.C.: Society for American Archaeology.

Lyon, Edwin A. 1996. *A New Deal for Southeastern Archaeology.* Tuscaloosa: University of Alabama Press.

MacCord, Howard A., Sr. and C. Lanier Rodgers. 1966. The Miley Site, Shenandoah County, Virginia. *Quarterly Bulletin of the Archaeological Society of Virginia* 21(1):9–20.

Madrigal, Lorena. 1998. *Statistics for Anthropology.* Cambridge, U.K.: Cambridge University Press.

Marquardt, William H., Anita Montet-White, and Sandra C. Scholtz. 1982. Resolving the Crisis in Archaeological Collections Curation. *American Antiquity* 47 (2):409–418.

Mays, Simon. 1998. *The Archaeology of Human Bones.* New York: Routledge.

McAvoy, Joseph. 1997. *Archaeological Investigations of Site 44SX202, Cactus Hill, Sussex County, Virginia.* Survey and Planning Report Series No.8. Richmond: Virginia Department of Historic Resources.

McDonald, Jerry N. and Susan L. Woodward. 1987. *Indian Mounds of the Atlantic Coast: A Guide to Sites from Maine to Florida.* Newark, OH: McDonald & Woodward Publishing Company.

McGimsey, Charles R. III. 1995. Standards, Ethics, and Archaeology: A Brief History, pp. 11–13. In M. Lynott and A. Wylie (eds.), *Ethics in American Archaeology: Challenges for the 1990s.* Special Report. Washington, D.C.: Society for American Archaeology.

McGowan, Gary S. and Cheryl J. Laroche. 1996. The Ethical Dilemma Facing Conservation: Care and Treatment of Human Skeletal Remains and Mortuary Objects. *Journal of the American Institute for Conservation* 35:109–121.

McGuire, Randall H. and Mark Walker. 1999. Class confrontations in archaeology. *Historical Archaeology* 33(1):159–183.

McManamon, Francis P. 1984. Discovering sites unseen. *Advances in Archaeological Method and Theory* 7:223–292.

McManamon, Francis P. 1991. The many publics for archaeology. *American Antiquity* 56(1):121–130.

McManamon, Francis P. 1992. Managing repatriation: Implementing the Native American Graves Protection and Repatriation Act. *Cultural Resource Management, No. 5:* 9–12.

McManamon, Francis P. and Alf Hatton. 1999. *Cultural Resource Management in Contemporary Society: Perspectives on Managing and Presenting the Past.* London: Routledge Press.

Meighan, Clement W. 1961. *The Archaeologist's Note Book.* San Francisco, Chandler Publishing Company.

Meighan, Clement W. 1996. Burying american archaeology, pp. 209–213. In Karen D. Vitelli (ed.), *Archaeological Ethics.* Walnut Creek, CA: AltaMira Press.

Meixner, R. E. and M. J. Singer. 1981. Morphology Rating System to Evaluate Soil Formation and Discontinuities. *Soil Science* 131:114–123.

Meltzer, David J. 1983. The antiquity of man and the development of american archaeology, pp. 1–51. In Michael Schiffer (ed.), *Advances in Archaeological Method and Theory,* Volume VI. New York: Academic Press.

Metress, Seamus P. 1989. *Human Osteology for the Archaeologist.* Occasional Publications in Northeastern Anthropology, No.10. Rindge, NH.

Micozzi, Marc S. 1991. *Postmortem Change in Human and Animal Remains: A Systematic Approach.* Springfield, IL: Charles C. Thomas.

Miller, George L., Olive Jones, Lester A. Ross, and Terecita Majewski (compilers). 1991. *Approaches to Material Culture Research for Historical Archaeologists.* California, PA: Society for Historical Archaeology.

Moeller, Roger W. 1980. *6LF21: A Paleo-Indian Site in Western Connecticut.* Occasional Paper Number 2. Washington, CT: American Indian Archaeological Institute.

Moore, Clarence B. 1903. Certain aboriginal mounds of the Florida Central West-Coast. *Journal of the Academy of Natural Sciences of Philadelphia,* Volume XII.

Morgan, Lewis Henry. 1965. *Houses and House-Life of the American Aborigines.* Classics in Anthropology. Chicago: University of Chicago Press. (Reprint of 1881 original published as Volume IV of *Contributions to North American Ethnology,* Government Printing Office, Washington, D.C.)

Morse, Phyllis A. and Dan F. Morse. 1998. *The Lower Mississippi Valley Expeditions of Clarence Bloomfield Moore.* Classics in Southeastern Archaeology. Tuscaloosa: University of Alabama Press.

Mueller, James W. 1974. *The use of sampling in archaeological survey.* Society for American Archaeology, Memoir 28.

Mueller, James W. 1975. *Sampling in Archaeology.* Tucson: University of Arizona Press.

Nabokov, Peter and Robert Easton. 1989. *Native American Architecture.* New York: Oxford University Press.

Nance, Jack D. 1983. Regional sampling in archaeological survey: The statistical perspective, pp. 289–356. In Michael B. Schiffer (ed.), *Advances in Archaeological Method and Theory* (Vol. 6). New York: Academic Press.

Nance, Jack D. and Bruce F. Ball. 1986. No surprises? The reliability and validity of test pit sampling. *American Antiquity* 51(3):457–483.

Nance, Jack D. and Bruce F. Ball. 1989. A shot in the dark: Shott's comments on Nance and Ball, and Lightfoot. *American Antiquity* 54(2):405–412.

Napton, L. Kyle and Elizabeth Anne Greathouse. 1997. Archaeological mapping, site grids, and surveying, pp. 177–234. In Thomas Hester, Harry Shafer, and Kenneth Feder (eds.), *Field Methods in Archaeology.* Mountain View, CA: Mayfield Publishing Company.

National Association of Practicing Anthropologists. 1988. Naitonal Association of Practicing Anthropologists, Ethical Guidelines for Practitioners. http://www.aaanet.org/napa/code.htm>

National Park Service. 1991. *36CFR Part 79: Curation of Federally-owned and Administered Archaeological Collections.* Washington, D.C.: Government Printing Office. (This can also be read online at http://www.cr.nps.gov/aad/36cfr79.htm)

National Park Service. 1996. *Collections and Curation into the 21st Century.* Special report on collections and curation. *Common Ground* 1(2).

Neuman, Thomas W., Brian D. Bates and Robert M. Sanford. 1998. Setting up the basic archaeology laboratory, pp. 343–358. In M. Q. Sutton and Brooke S. Arkush (eds.), *Archaeological Laboratory Methods: An Introduction.* Dubuque, IA: Kendall/Hunt Publishing Company.

Nichols, Jacqueline and June Evans. 1979. The aggressive field lab. *American Antiquity* 44(2):324–326.

Niquette, Charles M. 1997. Hard hat archaeology. *Society for American Archaeology Newsletter* 15(3):15–17.

Noel-Hume, Ivor. 1974. *A Guide to Artifacts of Colonial America.* New York: Alfred A. Knopf.

Noel-Hume, Ivor. 1983. *Historical Archaeology: A Comprehensive Guide for Both Amateurs and Professionals to the Techniques and Methods of Excavating Historical Sites.* New York: Alfred A. Knopf.

Noli, Dieter. 1985. Low altitude aerial photography from a tethered balloon. *Journal of Field Archaeology* 12:497–501.

O'Brien, Michael J. and R. Lee Lyman. 1999. *Seriation, Stratigraphy, and Index Fossils: The Backbone of Archaeological Dating.* Norwell, MA: Kluwer/Plenum Publishers.

Occupational Safety and Health Administration, U.S. Department of Labor. "OSHA Standards Interpretation and Compliance Letters, Standards Addressing Excavations." (1991). Online. Internet. http://www.osha-slc.gov/OshDoc/Interp_dataI19911202B.htm

Occupational Safety and Health Administration, U.S. Department of Labor. "OSHA Regulations (Standards-29 CFR) Part 1926 Subpart P-Excavations." (2000). Online. Internet. http://www.osha-slc.gov/OshStd_toc/OSHA_Std_toc_1926_SUBPART_P.html

Occupational Safety and Health Administration, U.S. Department of Labor. "OSHA Regulations (Standards-29 CFR). Part 1910 Subpart I-Personal Protective Equipment. (2000). http://www.osha-slc.gov/OshStd_toc/OSHA_Std_toc_1910_SUBPART_I.html>

Office of Public Correspondence, United States Department of the Treasury. *The Lincoln Cent.* Fact Sheet OPC-75. United States Department of the Treasury, Washington, D.C. 1998. http://www.ustreas.gov/opc/opc0003.html>

Orser, Charles E., Jr. and Brian M. Fagan. 1995. *Historical Archaeology.* New York: Harper Collins.

O'Shea, John M. 1984. *Mortuary Variability: An Archaeological Investigation.* Orlando, FL: Academic Press.

Otte, George E., Duane K. Setness, Walter A. Anderson, Fred J. Herbert, Jr., and Clarence A. Knezevich. 1974. *Soil Survey of the Yamhill Area, Oregon.* United States Department of Agriculture, Soil Conservation Service in cooperation with the Oregon Agricultural Experiment Station. Washington, D.C.: U.S. Government Printing Office.

Ortner, Donald J. and Walter Putschar. 1985. *Identification of Pathological Conditions in Human Skeletal Remains.* Washington, D.C.: Smithsonian Institution Press.

Owens, James P. and James P. Minard. 1975. *Geologic Map of the Surficial Deposits in the Trenton Area, New Jersey and Pennsylvania.* Scale 1:48,000. Map I-884, Miscellaneous Investigations Series. Reston, VA: United States Geological Survey.

Palmer, Marilyn and Peter Neaverson.1998. *Industrial Archaeology: Principles and Practice.* London: Routledge.

Patterson, Thomas C. 1995. *Toward a Social History of Archaeology in the United States.* Fort Worth, TX: Harcourt Brace College Publishers.

Pauketat, Timothy. 1993. *Temples for Cahokia Lords: Preston Holder's 1955–1956 Excavations of Kunneman Mound.* University of Michigan, Museum of Anthropology Memoirs 26, Ann Arbor, MI.

Pearsall, Deborah M. 1989. *Paleoethnobotany: A Handbook of Procedures.* New York: Academic Press.

Perino, Gregory H. 1968. The Pete Klunk Mound Group, Calhoun County, Illinois: The Archaic and Hopewell Occupations, pp. 9–124. In J. A. Brown (ed.), *Hopewell and Woodland Site Archaeology in Illinois.* Bulletin 6. Urbana: Illinois Archaeological Survey.

Petrie, W. M. F. 1904. *Methods and Aims in Archaeology.* New York: Macmillan.

Plog, Stephen, Fred Plog, and Walter Wait. 1978. Decision making in modern surveys. *Advances in Archaeological Method and Theory* 1:383–421.

Poirier, David A. and Kenneth L. Feder (eds.). 2000. *Dangerous Places: Health, Safety, and Archaeology.* Westport, CT: Greenwood Publishing Group.

Powers, M. C. 1953. A new roundness scale for sedimentary particles. *Journal of Sedimentary Petrology* 23:117–119.

Purdy, Barbara A. (ed.). 1988. *Wet Site Archaeology.* Caldwell, NJ: Telford Press.

Ramenofsky, Ann F., Leon C. Standifer, Ann M. Whitmer, and Marie S. Standifer. 1986. A new technique for separating flotation samples. *American Antiquity* 51(1):66–72.

Rapp, George, Jr. and Christopher L. Hill. 1998. *Geoarchaeology: The Earth-Science Approach to Archaeological Interpretation.* New Haven, CT: Yale University Press.

Redman, Charles L. 1987. Surface collection, sampling, and research design: A retrospective. *American Antiquity* 52(2):249–265.

Redman, Charles L. and Patty J. Watson. 1970. Systematic intensive surface collection. *American Antiquity* 35:279–291.

Reichs, Kathleen J. and William Bass. 1998. *Forensic Osteology: Advances in the Identification of Human Remains.* Springfield, Illinois: Charles Thomas Publishing Ltd.

Renfrew, Colin. 1976. Archaeology and the earth sciences, pp. 1–5. In D. A. Davidson and M. L. Shackley (eds.), *Geoarchaeology: Earth Science and the Past.* Boulder, CO: Westview Press.

Renfrew, Colin and Paul Bahn. 1996. *Archaeology: Theory, Methods, and Practice.* New York: Thames and Hudson.

Rice, Prudence M. 1987. *Pottery Analysis: A Sourcebook.* Chicago: University of Chicago Press.

Rick, John. 1996. Total stations in archaeology. *Society for American Archaeology Bulletin* 14(4):24–27.

Rick, John. 1999. Digital still cameras and archaeology. *Society for American Archaeology Bulletin* 17(3):37–41.

Ritter, Dale F., R. Craig Kochel, and Jerry R. Miller. 1995. *Process Geomorphology.* Dubuque, IA: W. C. Brown.

Robinson, Wendy. 1998. *First Aid For Underwater Finds.* London: Archetype Publications, Ltd. and Portsmouth, U.K.: Nautical Archaeology Society.

Rogers, Juliet and Tony Waldron. 1995. *A Field Guide to Joint Disease in Archaeology.* New York: Wiley.

Rolingson, Martha Ann (ed.). 1998. *Toltec Mounds and Plum Bayou Culture: Mound D Excavations.* Arkansas Archaeological Survey Research Series No. 54, Fayetteville.

Roper, Donna C. 1976. Lateral displacement of artifacts due to plowing. *American Antiquity* 41(3):372–375.

Russell, Caroline H. (compiler). 1990. *Secretary of the Interior's Standards and Guidelines for Architectural and Engineering Documentation.* Washington, D.C.: National Park Service.

Rye, Owen S. 1981. *Pottery Technology.* Taraxacum Manuals in Archaeology 4. Washington, D.C.: Taraxacum.

Sanger, Kay K. and Clement W. Meighan. 1990. *Discovering Prehistoric Rock Art: A Recording Manual.* Calabasas, CA: Wormwood Press.

Schaafsma, Curtis F. 1989. Significant until proven otherwise: Problems versus representative samples, pp. 38–51. In H. F. Cleere (ed.), *Archaeological Heritage Management in the Modern World.* London: Unwin Hyman.

Schaafsma, Polly. 1985. Form, content, and function: Theory and method in North American rock art studies. *Advances in Archaeological Method and Theory* 8:237–277.

Schiffer, Michael. 1972. Archaeological context and systemic context. *American Antiquity* 37(2):156–165.

Schiffer, Michael. 1987. *Formation Processes of the Archaeological Record.* Albuquerque: University of New Mexico Press.

Schiffer, Michael. 1996. *Formation Processes of the Archaeological Record.* (2nd ed.). Salt Lake City: University of Utah Press.

Schiffer, Michael B. and George J. Gumerman. 1977. *Conservation Archaeology: A Guide for Cultural Resource Management Studies.* New York: Academic Press.

Schiffer, Michael B., Alan P. Sullivan, and Timothy Klinger. 1978. The design of archaeological surveys. *World Archaeology* 10(1):1–28.

Schlereth, Thomas J. 1980. *Artifacts and the American Past.* Nashville, TN: American Association for State and Local History.

Schneiderman-Fox, Faline and A. Michael Pappalardo. 1996. A paperless approach toward field data collection: An example from the Bronx. *Society for American Archaeology Bulletin* 14(1):1, 18–20.

Schoenwetter, James. 1981. Prologue to a contextual archaeology. *Journal of Archaeological Science* 8:367–379.

Schuldenrein, Joseph. 1991. Coring and the identity of cultural-resource environments: A comment of Stein. *American Antiquity* 56(1):131–137.

Schuldenrein, Joseph. 1995. The care and feeding of archaeologists: A plea for pragmatic training in the 21st century. *Society for American Archaeology Bulletin* 13(3):22–24.

Schwartz, Jeffrey H. 1995. *Skeleton Keys: An Introduction to Human Skeletal Morphology, Development, and Analysis.* New York: Oxford University Press.

Schwartz, Jeffrey H. 1998. *What Bones Tell Us.* Tucson: University of Arizona Press.

Sease, Catherine. 1994. *A Conservation Manual for the Field Archaeologist.* Archaeological Research Tools, Volume 4. Los Angeles: Institute of Archaeology, University of California.

Sept, Jeanne M. 1997. *Investigating Olduvai Gorge: Archaeology of Human Origins.* CD-ROM. Bloomington: Indiana University Press.

Sharer, Robert and Wendy Ashmore.1993. *Archaeology: Discovering Our Past.* Mountain View, CA: Mayfield Publishing.

Shaw, Joey N. 1993. *Soils Developed in Freshwater Marl Sediments in the Hagerstown (Great) Limestone Valley.* M.S. thesis, University of Maryland, College Park.

Shott, Michael J. 1985. Shovel-test sampling as a site discovery technique: A case study from Michigan. *Journal of Field Archaeology* 12:458–469.

Shott, Michael J. 1989. Shovel test sampling in archaeological survey: Comments on Nance and Ball, and Lightfoot. *American Antiquity* 54(2):396–404.

Shott, Michael J. 1995. Reliability of archaeological records on cultivated surfaces: A Michigan case study. *Journal of Field Archaeology* 22:475–490.

Singley, Katherine. 1988. *The Conservation of Archaeological Artifacts from Freshwater Environments.* South Haven, MI: Lake Michigan Maritime Museum.

Skinner, Alanson and Max Schrabisch. 1913. *A Preliminary Report of the Archaeological Survey of the State of New Jersey.* Geological Survey of New Jersey, Bulletin 9. Trenton.

Soil Survey Staff. 1951. *Soil Survey Manual.* Agricultural Handbook No. 18. Washington, D.C.: United States Government Printing Office.

Soil Survey Staff. 1998. *Keys to Soil Taxonomy* (8th ed.). U.S. Department of Agriculture, Natural Resources Conservation Service, Washington, D.C.

Soil Survey Staff. 1999. *Soil Taxonomy: A Basic System of Soil Classification for Making and Interpreting Soil Surveys.* Agricultural Handbook No. 436. U.S. Department of Agriculture, Natural Resources Conservation Service, Washington, D.C.

Slayman, Andrew. 1997. A battle over bones. *Archaeology* 50(1):16–23.

Spier, Robert F. G. 1970. *Surveying and Mapping: A Manual.* New York: Holt, Rinehart, and Winston, Inc.

Sprague, Roderick. 1968. A suggested terminology and classification for burial description. *American Antiquity* 33(4):479–485.

Stafford, C. Russell. 1995. Geoarchaeological perspectives on paleolandscapes and regional subsurface archaeology. *Journal of Archaeological Method and Theory* 2(1):69–104.

Stanzeski, Andrew. 1974. The Three Beeches: Excavations in the House of an Archaeologist. *Bulletin of the Archaeological Society of New Jersey* 31:30–32.

Staski, Edward. 1982. Advances in Urban Archaeology. *Advances in Archaeological Method and Theory* 5:97–149.

Stein, Julie K. 1986. Coring archaeological sites. *American Antiquity* 51(3):505–527.

Stein, Julie K. 1987. Deposits for Archaeologists. *Advances in Archaeological Method and Theory* 11:337–395.

Stein, Julie K. (ed.). 1992. *Deciphering a Shell Midden.* New York: Academic Press.

Sterud, Eugene L. and Peter P. Pratt. 1976. Archaeological intra-site recording with photography. *Journal of Field Archaeology* 2:151–167.

Stewart, Hilary. 1996. *Stone, Bone, Antler and Shell: Artifacts of the Northwest Coast.* Seattle: University of Washington Press.

Stewart, R. Michael. 1980. *Prehistoric Settlement and Subsistence Patterns and the Testing of Predictive Site Location Models in the Great Valley of Maryland.* Anthropology Studies No. 48, Washington, D.C.: The Catholic University of America.

Stewart, R. Michael. 1984. Archaeologically significant characteristics of Maryland and Pennsylvania metarhyolites, pp. 1–13. In Jay Custer (ed.), *Prehistoric Lithic Exchange Systems in the Middle Atlantic Region.* Center for Archaeological Research, University of Delaware, Monograph No.3, Newark.

Stewart, R. Michael. 1986. *Lister Site (28Me1-A), Data Recovery.* Trenton Complex Archaeology, Report 6. Federal Highway Administration and the New Jersey Department of Transportation, Trenton.

Stewart, R. Michael. 1994. *Prehistoric Farmers of the Susquehanna Valley: Clemson Island Culture and the St. Anthony Site.* Occasional Publications in Northeastern Anthropology, No. 13. Bethlehem, CT: Archaeological Services.

Stewart, R. Michael. 1995. Archaic Indian burials at the Abbott Farm. *Bulletin of the Archaeological Society of New Jersey* 50:73–80.

Stewart, R. Michael. 1997. *Ancient Ponds, Marl Deposits, and Prehistory in the Ridge and Valley Province of Maryland and Southcentral Pennsylvania.* Paper presented at the Third International Conference on Soils, Geomorphology and Archaeology, Luray, VA.

Stewart, R. Michael. 1999. The Indian town of Playwicki. *Journal of Middle Atlantic Archaeology* 15:35–53.

Stewart, R. Michael, Jay F. Custer and Donald Kline. 1991. A deeply stratified archaeological and sedimentary sequence in the Delaware River Valley of the Middle Atlantic Region, United States. *Geoarchaeology* 6(2):169–182.

Stewart, T. Dale. 1979. *Essentials of Forensic Anthropology.* Springfield, IL: Charles Thomas Publishing Ltd.

Stone, G. G. 1981. On artifact density and shovel probes. *Current Anthropology* 22:182–183.

Stose, George W. and Anna I. Jonas. 1973. *Geologic Map of Southeastern Pennsylvania.* Scale 1:380,160. Harrisburg, PA: Department of Environmental Resources, Bureau of Topographic and Geologic Survey.

Strahler, Arthur N. and Alan H. Strahler.1989. *Investigating Physical Geography: An Exercise Manual.* New York: John Wiley and Sons.

Straus, Lawrence Guy. 1990. Perspectives on caves and rockshelters. *Archaeological Method and Theory* 2:255–304.

Struever, Stuart. 1968. Flotation techniques for the recovery of small-scale archaeological remains. *American Antiquity* 33(3):353–362.

Sullivan, Alan P. (ed.). 1998. *Surface Archaeology: Method, Theory, and Practice.* Albuquerque: University of New Mexico Press.

Sutton, Mark Q. and Brooke S. Arkush. 1998. *Archaeological Laboratory Methods: An Introduction* (2nd ed.). Dubuque, IA: Kendall/Hunt Publishing Company.

Swidler, Nina, Kurt E. Dongoske, and Roger Anyon (eds.). 1997. *Native Americans and Archaeologists: Stepping Stones to Common Ground.* Thousand Oaks, CA: AltaMira Press.

Taylor, R. E. and Martin J. Aitken. 1997. *Chronometric Dating in Archaeology.* New York: Plenum Press.

Thomas, David Hurst. 1975. Nonsite sampling in archaeology: Up the creek without a site?, pp. 61–81. In J. W. Mueller (ed.), *Sampling in Archaeology.* Tucson: University of Arizona Press.

Thomas, David Hurst. 1983. *The Archaeology of the Monitor Valley, Volume 2, Gatecliff Shelter.* Anthropological Papers of the American Museum of Natural History 59(1).

Thomas, David Hurst. 1986. *Refiguring Anthropology: First Principles of Probability and Statistics.* Prospect Heights, IL: Waveland Press.

Thomas, David Hurst. *1998. Archaeology* (3rd ed.). Fort Worth, TX: Harcourt Brace and Company.

Thomas, David Hurst. 1999. *Exploring Ancient Native America: An Archaeological Guide.* New York: Routledge.

Thompson, Morris M. 1988. *Maps for America: Cartographic Products of the U.S. Geological Survey and Others.* Reston, VA: U.S. Geological Survey.

Toll, Mollie S. 1988. Flotation sampling: Problems and some solutions, with examples from the American Southwest, pp. 36–52. In C. A. Hastorf and V. S. Popper (eds.), *Current Paleoethnobotany.* Chicago, IL: University of Chicago Press.

Trigger, Bruce G. 1990. *A History of Archaeological Thought.* New York: Cambridge University Press.

Turnbaugh, Sarah Peabody and William A. Turnbaugh. 1997. *Indian Baskets.* Atglen, PA: Schiffer Publishing Ltd.

Ubelaker, Douglas H. 1999. *Human Skeletal Remains: Excavation, Analysis, Interpretation.* Manuals on Archaeology 2 (3rd ed.). Washington, D.C.: Taraxacum.

Van Horn, David. 1988. *Mechanized Archaeology.* Calabasas, CA: Wormwood Press.

Vento, Frank, Harold Rollins, Michael Stewart, Paul Raber, and William Johnson. 1990. *Genetic Stratigraphy, Climate Change, and the Burial of Archaeological Sites Within the Susquehanna, Delaware, and Ohio River Drainage Basins.* Report prepared for the Bureau of Historic Preservation of the Pennsylvania Historical and Museum Commission, Harrisburg.

Vitelli, Karen D. (ed.). 1996. *Archaeological Ethics.* Walnut Creek, CA: AltaMira Press.

Wagner, Gail E. 1982. Testing flotation recovery techniques. *American Antiquity* 47(1):127–132.

Walker, James W. 1993. *Low Altitude Large Scale Reconnaissance: A Method of Obtaining High Resolution Vertical Photographs for Small Areas.* Interagency Archaeological Services, Rocky Mountain Regional Office, National Park Service, Denver, CO.

Wall, Robert D., R. Michael Stewart, John A. Cavallo, and Virginia Busby. 1996. *Area D Site (28Mel-D): Data Recovery.* Trenton Complex Archaeology, Report 9. The Federal Highway Commission and New Jersey Department of Transportation, Trenton.

Wandsnider, LuAnn and Eileen L. Camilli. 1992. The character of surface archaeological deposits and its influence on survey accuracy. *Journal of Field Archaeology* 19:169–188.

Wandsnider, LuAnn and James Ebert (eds.). 1988. Issues in archaeological surface survey: Meshing method and theory. *American Archaeology* 7(1) (A thematic issue).

Waselkov, Gregory. 1982. *Shellfish Gathering and Shell Midden Archaeology.* Ph.D. dissertation, University of North Carolina, Chapel Hill.

Waselkov, Gregory. 1987. Shellfish gathering and shell midden archaeology. *Advances in Archaeological Method and Theory* 10:93–210

Waters, Michael R. 1996. *Principles of Geoarchaeology: A North American Perspective.* Tucson: University of Arizona Press.

Watkins, Joe, K. Anne Pyburn, and Pam Cressey. 2000. Community relations: What the practicing archaeologist needs to know to work effectively with local and/or descendant communities, pp. 73–81. In Susan J. Bender and George S. Smith (eds.), *Teaching Archaeology in the Twenty-First Century.* Washington, D.C.: Society for American Archaeology.

Watson, Patty Jo (ed.). 1974. *Archaeology of the Mammoth Cave Area.* New York: Academic Press.

Webster, David L., Susan Toby Evans, and William T. Sanders. 1993. *Out of the Past: An Introduction to Archaeology.* Mountain View, CA: Mayfield Publishing Company.

Wellmann, Klaus F. and Ursula Arndt. 1979. *A Survey of North American Indian Rock Art.* Akademische Druck -u.Verlagsanstadt, Graz.

Wendrich, Willemina. 1991. *Who Is Afraid of Basketry: A Guide to Recording Basketry and Cordage for Archaeologists and Ethnographers.* Leiden, the Netherlands: Centre for Non-Western Studies, Leiden University.

Wescott, Konnie L. and R. Joe Brandon. 2000. *Practical Applications of GIS for Archaeologists: A Predictive Modeling Toolkit.* Philadelphia, PA: Taylor and Francis.

White, Nancy Marie, Lynne P. Sullivan, and Rochelle A. Marrinan. 1999. *Grit-Tempered: Early Women Archaeologists in the Southeastern United States.* Gainesville: University of Florida Press.

Whitlam, Robert G. 1998. Cyberstaking archaeological sites: Using electronic marker systems (EMS) for a site datum and monitoring station. *Society for American Archaeology Bulletin* 16(2):12–15.

Whitley, David S. 2000. *Handbook of Rock Art Research.* Walnut Creek, CA: AltaMira Press.

Whittaker, John C. 1994. *Flintknapping: Making and Understanding Stone Tools.* Austin: University of Texas Press.

Wilford, Lloyd A., Elden Johnson, and Joan Vicinus. 1969. *Burial Mounds of Central Minnesota.* Minnesota Prehistoric Archaeology Series. St. Paul: Minnesota Historical Society.

Willey, Gordon R. 1988. *Portraits in American Archaeology: Remembrances of Some Distinguished Americanists.* Albuquerque: University of New Mexico Press.

Willey, Gordon R. and Jeremy A. Sabloff. 1993. *A History of American Archaeology* (3rd ed.). New York: W. H. Freeman and Company.

Wilson, Josleen. 1980. *The Passionate Amateur's Guide to Archaeology in the United States.* New York: Collier Books, A Division of MacMillan Publishing Company, Inc.

Wilson, Michele. 2000. *Tales From the Trenches: The People, Policies, and Procedures of Cultural Resource Management.* MA thesis, Oregon State University, Corvallis.

Winters, Howard D. 1969. *The Riverton Culture: A Second Millennium Occupation in the Central Wabash Valley.* Illinois State Museum, Report of Investigations No. 13, and the Illinois Archaeological Survey, Springfield.

Wisseman, Sarah U. and Wendell S. Williams. 1993. *Ancient Technologies and Archaeological Materials.* Langhorne, PA: Gordon and Breach Science Publishers.

Wobst, H. Martin. 1983. We can't see the forest for the trees: Sampling and the shapes of archaeological distributions, pp. 37–85. In J. A. Moore and A. S. Keene (eds.), *Archaeological Hammers and Theories.* New York: Academic Press.

Wood, Raymond W. and Donald L. Johnson. 1978. A survey of disturbance processes in archaeological site formation. *Advances in Archaeological Method and Theory* 1:315–382.

Woodward, Susan L. and Jerry N. McDonald. 1986. *Indian Mounds of the Middle Ohio Valley: A Guide to Adena and Ohio Hopewell Sites.* Newark, OH: The McDonald & Woodward Publishing Company.

Wylie, Alison. 1995. Archaeology and the antiquities market: The use of "looted" data, pp. 17–21. In M. Lynott and A. Wylie (eds.), *Ethics in American Archaeology: Challenges for the 1990s.* Special Report. Washington, D.C.: Society for American Archaeology.

Yamin, Rebecca and Karen Bescherer Metheny. 1996. *Landscape Archaeology: Reading and Interpreting the American Historical Landscape.* Nashville: University of Tennessee Press.

Zeder, Melinda. 1997. *The American Archaeologist: A Profile.* Walnut Creek, CA: AltaMira Press.

Zeidler, James A. 1997. ProbeCorder: Pen-based computing for sediment profile recording. *Society for American Archaeology Bulletin* 15(3):32–37.

Zingg, T. 1935. Beitrage zur Schotter Analyse. *Schweizerische Mineralogische und Petrographische Mitteilungen* 15:39–40.

INDEX

D

I

J

K

L

M

N

O

P

Q

R

T

U